VERTEBRATE EMBRYOLOGY

Vertebrate Embryology

Editor:

Dr. Q. Tian Wang

ArclerPress
www.arclerpress.com

Vertebrate Embryology
Editor: Dr. Q. Tian Wang

© 2017 Arcler Press LLC
708 3rd Avenue, 6th Floor
New York
NY 10017
United States of America
www.arclerpress.com

ISBN: 978-1-68094-517-1

Library of Congress Control Number: 2016951377

About the Editor

Dr. Q. Tian Wang, PhD

Dr. Q. Tian Wang holds a PhD in Biochemistry, Molecular Biology and Cell Biology from Northwestern University and has received postdoctoral training at Stanford University. She has extensive biomedical research experience and has directed multiple award-winning research programs studying the molecular basis of embryonic development and disease. She is currently a Science Officer in support of Congressionally Directed Medical Research Programs.

Contents

List of Contributors

Aleksandar I. Mihajlović
Laboratory of Developmental Biology & Genetics (LDB&G), Department of Molecular Biology, Faculty of Science, University of South Bohemia, Branišovská 31, 37005 České Budějovice, CZECH REPUBLIC

Vasanth Thamodaran
Laboratory of Developmental Biology & Genetics (LDB&G), Department of Molecular Biology, Faculty of Science, University of South Bohemia, Branišovská 31, 37005 České Budějovice, CZECH REPUBLIC

Alexander W. Bruce
Laboratory of Developmental Biology & Genetics (LDB&G), Department of Molecular Biology, Faculty of Science, University of South Bohemia, Branišovská 31, 37005 České Budějovice, CZECH REPUBLIC
Institute of Entomology Biology Centre of the Czech Academy of Sciences, Branišovská 31, 37005 České Budějovice, CZECH REPUBLIC

Hiroki Nagai
Laboratory for Early Embryogenesis, RIKEN Center for Developmental Biology, Kobe, Hyogo 650-0047, Japan

Maiko Sezaki
Laboratory for Early Embryogenesis, RIKEN Center for Developmental Biology, Kobe, Hyogo 650-0047, Japan

Kisa Kakiguchi
Electron Microscopy Laboratory, RIKEN Center for Developmental Biology, Kobe, Hyogo 650-0047, Japan

Yukiko Nakaya
Laboratory for Early Embryogenesis, RIKEN Center for Developmental Biology, Kobe, Hyogo 650-0047, Japan

Hyung Chul Lee
Department of Agricultural Biotechnology, Seoul National University, Seoul 151-921, Korea

Raj Ladher
Laboratory for Sensory Development, RIKEN Center for Developmental Biology, Kobe, Hyogo 650-0047, Japan

Tomohiro Sasanami
Department of Applied Biological Chemistry, Faculty of Agriculture, Shizuoka University, 836 Ohya, Shizuoka 422-8529, Japan

Jae Yong Han
Department of Agricultural Biotechnology, Seoul National University, Seoul 151-921, Korea

Shigenobu Yonemura
Electron Microscopy Laboratory, RIKEN Center for Developmental Biology, Kobe, Hyogo 650-0047, Japan

Guojun Sheng
Laboratory for Early Embryogenesis, RIKEN Center for Developmental Biology, Kobe, Hyogo 650-0047, Japan

Andra´ s Szabo´
Deparment of Cell and Developmental Biology, University College London, London WC1E 6BT, UK

Isidoro Cobo
Deparment of Cell and Developmental Biology, University College London, London WC1E 6BT, UK

Sharif Omara
Deparment of Cell and Developmental Biology, University College London, London WC1E 6BT, UK

Sophie McLachlan
Deparment of Cell and Developmental Biology, University College London, London WC1E 6BT, UK

Ray Keller
Department of Biology, University of Virginia, Charlottesville, VA 22908, USA

Roberto Mayor
Deparment of Cell and Developmental Biology, University College London, London WC1E 6BT, UK

Raphaël Thuret
Faculty of Life Sciences, Michael Smith Building, University of Manchester, Oxford Road, Manchester M13 9PT, UK

Hélene Auger
Faculty of Life Sciences, Michael Smith Building, University of Manchester, Oxford Road, Manchester M13 9PT, UK

Nancy Papalopulu`
Faculty of Life Sciences, Michael Smith Building, University of Manchester, Oxford Road, Manchester M13 9PT, UK

Barbara Murdoch
Biology Department, Eastern Connecticut State University, Willimantic, Connecticut, United States of America

Casey DelConte
Department of Molecular, Cellular and Developmental Biology, Yale University, New Haven, Connecticut, United States of America

Martı´n I. Garcı´a-Castro
Department of Molecular, Cellular and Developmental Biology, Yale University, New Haven, Connecticut, United States of America

Sam Goldsmith,
The Francis Crick Institute, Mill Hill Laboratory, The Ridgeway, Mill Hill, London NW7 1AA, UK.

Robin Lovell-Badge
The Francis Crick Institute, Mill Hill Laboratory, The Ridgeway, Mill Hill, London NW7 1AA, UK.

Karine Rizzoti
The Francis Crick Institute, Mill Hill Laboratory, The Ridgeway, Mill Hill, London NW7 1AA, UK.

Shinichi Hayashi
Department of Genetics, Cell Biology and Development, University of Minnesota, Minneapolis, Minnesota, United States of America
[2]Stem Cell Institute, University of Minnesota, Minneapolis, Minnesota, United States of America

Ryutaro Akiyama
Department of Genetics, Cell Biology and Development, University of Minnesota, Minneapolis, Minnesota, United States of America
Stem Cell Institute, University of Minnesota, Minneapolis, Minnesota, United States of America

Julia Wong
Department of Genetics, Cell Biology and Development, University of Minnesota, Minneapolis, Minnesota, United States of America

Naoyuki Tahara
Department of Genetics, Cell Biology and Development, University of Minnesota, Minneapolis, Minnesota, United States of America
Stem Cell Institute, University of Minnesota, Minneapolis, Minnesota, United States of America

Hiroko Kawakami
Department of Genetics, Cell Biology and Development, University of Minnesota, Minneapolis, Minnesota, United States of America
Stem Cell Institute, University of Minnesota, Minneapolis, Minnesota, United States of America

Yasuhiko Kawakami
Department of Genetics, Cell Biology and Development, University of Minnesota, Minneapolis, Minnesota, United States of America
Stem Cell Institute, University of Minnesota, Minneapolis, Minnesota, United States of America

Saartje Hontelez
Department of Molecular Developmental Biology, Radboud Institute for Molecular Life Sciences, Faculty of Science, Radboud University, PO Box 9101, 6500 HB Nijmegen, The Netherlands

Ila van Kruijsbergen
Department of Molecular Developmental Biology, Radboud Institute for Molecular Life Sciences, Faculty of Science, Radboud University, PO Box 9101, 6500 HB Nijmegen, The Netherlands

Georgios Georgiou
Department of Molecular Developmental Biology, Radboud Institute for Molecular Life Sciences, Faculty of Science, Radboud University, PO Box 9101, 6500 HB Nijmegen, The Netherlands

Simon J. van Heeringen
Department of Molecular Developmental Biology, Radboud Institute for Molecular Life Sciences, Faculty of Science, Radboud University, PO Box 9101, 6500 HB Nijmegen, The Netherlands

Ozren Bogdanovic
ARC Center of Excellence in Plant Energy Biology, The University of Western Australia, Perth, Western Australia 6009, Australia

Ryan Lister
ARC Center of Excellence in Plant Energy Biology, The University of Western Australia, Perth, Western Australia 6009, Australia
The Harry Perkins Institute of Medical Research, Perth, Western Australia 6009, Australia

Gert Jan C. Veenstra
Department of Molecular Developmental Biology, Radboud Institute for Molecular Life Sciences, Faculty of Science, Radboud University, PO Box 9101, 6500 HB Nijmegen, The Netherlands

Charles C. Bell
Mater Research, Translational Research Institute, University of Queensland, Brisbane, Queensland, 4101, Australia

Paulo P.Amara
Mater Research, Translational Research Institute, University of Queensland, Brisbane, Queensland, 4101, Australia
The Institute for Molecular Bioscience, University of Queensland, Brisbane, Queensland, Australia

Anton Kalsbeek
The Institute for Molecular Bioscience, University of Queensland, Brisbane, Queensland, Australia
Garvan Institute of Medical Research, Sydney, Australia

GrahamW. Magor
Mater Research, Translational Research Institute, University of Queensland, Brisbane, Queensland, 4101, Australia

Kevin R.Gillinder
Mater Research, Translational Research Institute, University of Queensland, Brisbane, Queensland, 4101, Australia

PierreTangermann
The Institute for Molecular Bioscience, University of Queensland, Brisbane, Queensland, Australia

Lorena di Lisio
Mater Research, Translational Research Institute, University of Queensland, Brisbane, Queensland, 4101, Australia

SethW. Cheetham
Mater Research, Translational Research Institute, University of Queensland, Brisbane, Queensland, 4101, Australia
Diamantina Institute; Translational Research Institute, University of Queensland, Brisbane, Queensland, 4102, Australia

FranziskaGruhl
Mater Research, Translational Research Institute, University of Queensland, Brisbane, Queensland, 4101, Australia
Diamantina Institute; Translational Research Institute, University of Queensland, Brisbane, Queensland, 4102, Australia

Jessica Frith
The Institute for Molecular Bioscience, University of Queensland, Brisbane, Queensland, Australia
The Australian Institute for Bioengineering and Nanotechnology, University of Queensland, Brisbane, Queensland, 4102, Australia

Michael R.Tallack
Mater Research, Translational Research Institute, University of Queensland, Brisbane, Queensland, 4101, Australia
The Institute for Molecular Bioscience, University of Queensland, Brisbane, Queensland, Australia

Ke-Lin Ru
The Institute for Molecular Bioscience, University of Queensland, Brisbane, Queensland, Australia

The Australian Institute for Bioengineering and Nanotechnology, University of Queensland, Brisbane, Queensland, 4102, Australia

JoannaCrawford
The Institute for Molecular Bioscience, University of Queensland, Brisbane, Queensland, Australia

John S. Mattick
Garvan Institute of Medical Research, Sydney, Australia
St Vincents Clinical School, Faculty of Medicine, UNSW Australia, Sydney, Australia

Marcel E. Dinger
Garvan Institute of Medical Research, Sydney, Australia
Diamantina Institute; Translational Research Institute, University of Queensland, Brisbane, Queensland, 4102, Australia
St Vincents Clinical School, Faculty of Medicine, UNSW Australia, Sydney, Australia

AndrewC. Perkins
Mater Research, Translational Research Institute, University of Queensland, Brisbane, Queensland, 4101, Australia
The Institute for Molecular Bioscience, University of Queensland, Brisbane, Queensland, Australia
The Princess Alexandra Hospital, Brisbane, Queensland, 4102, Australia

Jamie Trott
Department of Genetics, University of Cambridge, Cambridge CB2 3EH, UK
Wellcome Trust Centre for Stem Cell Research, University of Cambridge, Cambridge CB2 1QR, UK

Alfonso Martinez Arias
Department of Genetics, University of Cambridge, Cambridge CB2 3EH, UK

Kaori Isono
Department of Neurology, Graduate School of Medical Science, Kumamoto University, Kumamoto, Japan
Department of Transplantation and Pediatric Surgery, Graduate School of Medical Science, Kumamoto University, Kumamoto, Japan

Hirofumi Jono
Department of Diagnostic Medicine, Graduate School of Medical Science, Kumamoto University, Kumamoto, Japan
Department of Pharmacy, Kumamoto University Hospital, Kumamoto University, Kumamoto, Japan

Yuki Ohya
Department of Transplantation and Pediatric Surgery, Graduate School of Medical Science, Kumamoto University, Kumamoto, Japan

Nobuaki Shiraki
Department of Stem Cell Biology, Institute of Molecular Embryology and Genetics, Kumamoto University, Kumamoto, Japan

Taiji Yamazoe
Department of Stem Cell Biology, Institute of Molecular Embryology and Genetics, Kumamoto University, Kumamoto, Japan

Ayaka Sugasaki
Department of Neurology, Graduate School of Medical Science, Kumamoto University, Kumamoto, Japan

Takumi Era
Department of Cell Modulation, Institute of Molecular Embryology and Genetics, Kumamoto University, Kumamoto, Japan

Noemi Fusaki
DNAVEC Corporation, Ibaraki, Japan
Japan Science and Technology Agency (JST), PRESTO, Saitama, Japan
Ophthalmology, School of Medicine, Keio University, Japan

Masayoshi Tasaki
Department of Neurology, Graduate School of Medical Science, Kumamoto University, Kumamoto, Japan
Department of Diagnostic Medicine, Graduate School of Medical Science, Kumamoto University, Kumamoto, Japan

Mitsuharu Ueda
Department of Diagnostic Medicine, Graduate School of Medical Science, Kumamoto University, Kumamoto, Japan

Satoru Shinriki
Department of Diagnostic Medicine, Graduate School of Medical Science, Kumamoto University, Kumamoto, Japan

Yukihiro Inomata
Department of Transplantation and Pediatric Surgery, Graduate School of Medical Science, Kumamoto University, Kumamoto, Japan

Shoen Kume
Department of Stem Cell Biology, Institute of Molecular Embryology and Genetics, Kumamoto University, Kumamoto, Japan

Yukio Ando
Department of Neurology, Graduate School of Medical Science, Kumamoto University, Kumamoto, Japan
Department of Diagnostic Medicine, Graduate School of Medical Science, Kumamoto University, Kumamoto, Japan

Bastian Zimmer
The Center for Stem Cell Biology
Developmental Biology Program Sloan Kettering Institute for Cancer Research, 1275 York Avenue, New York, NY 10065, USA

Jinghua Piao
The Center for Stem Cell Biology
Department of Neurosurgery, Memorial Sloan Kettering Cancer Center, 1275 York Avenue, New York, NY 10065, USA

Kiran Ramnarine
The Center for Stem Cell Biology
SKI Stem Cell Research Facility, 1275 York Avenue, New York, NY 10065, USA

Mark J. Tomishima
The Center for Stem Cell Biology
SKI Stem Cell Research Facility, 1275 York Avenue, New York, NY 10065, USA

Viviane Tabar
The Center for Stem Cell Biology
Department of Neurosurgery, Memorial Sloan Kettering Cancer Center, 1275 York Avenue, New York, NY 10065, USA

Lorenz Studer
The Center for Stem Cell Biology
Developmental Biology Program Sloan Kettering Institute for Cancer Research, 1275 York Avenue, New York, NY 10065, USA
Department of Neurosurgery, Memorial Sloan Kettering Cancer Center, 1275 York Avenue, New York, NY 10065, USA

Brian C. Gibbs
Department of Developmental Biology, University of Pittsburgh School of Medicine, Pittsburgh, PA 15201, USA

Rama Rao Damerla
Department of Developmental Biology, University of Pittsburgh School of Medicine, Pittsburgh, PA 15201, USA

Eszter K. Vladar
Department of Pathology, Stanford University School of Medicine, Stanford, CA 94305, USA

Bishwanath Chatterjee
Department of Developmental Biology, University of Pittsburgh School of Medicine, Pittsburgh, PA 15201, USA

Yong Wan
Department of Oral Biology, University of Pittsburgh School of Dental Medicine, Pittsburgh, PA 15261, USA

Xiaoqin Liu
Department of Developmental Biology, University of Pittsburgh School of Medicine, Pittsburgh, PA 15201, USA

Cheng Cui
Department of Developmental Biology, University of Pittsburgh School of Medicine, Pittsburgh, PA 15201, USA

George C. Gabriel
Department of Developmental Biology, University of Pittsburgh School of Medicine, Pittsburgh, PA 15201, USA

Maliha Zahid
Department of Developmental Biology, University of Pittsburgh School of Medicine, Pittsburgh, PA 15201, USA

Hisato Yagi
Department of Developmental Biology, University of Pittsburgh School of Medicine, Pittsburgh, PA 15201, USA

Heather L. Szabo-Rogers
Department of Oral Biology, University of Pittsburgh School of Dental Medicine, Pittsburgh, PA 15261, USA

Kaye L. Suyama
Department of Pathology, Stanford University School of Medicine, Stanford, CA 94305, USA

Jeffrey D. Axelrod
Department of Pathology, Stanford University School of Medicine, Stanford, CA 94305, USA

Cecilia W. Lo
Department of Developmental Biology, University of Pittsburgh School of Medicine, Pittsburgh, PA 15201, USA

Swapnalee Sarmah
Department of Biology, Indiana University-Purdue University Indianapolis, 723 West Michigan Street, Indianapolis, IN 46202-5130, USA

Pooja Muralidharan
Department of Biology, Indiana University-Purdue University Indianapolis, 723 West Michigan Street, Indianapolis, IN 46202-5130, USA

Courtney L. Curtis
Department of Biology, Indiana University-Purdue University Indianapolis, 723 West Michigan Street, Indianapolis, IN 46202-5130, USA

Jeanette N. McClintick
Department of Biochemistry and Molecular Biology, Indiana University School of Medicine, Indianapolis, IN 46202, USA

Bryce B. Buente
Department of Biology, Indiana University-Purdue University Indianapolis, 723 West Michigan Street, Indianapolis, IN 46202-5130, USA

David J. Holdgrafer
Department of Biology, Indiana University-Purdue University Indianapolis, 723 West Michigan Street, Indianapolis, IN 46202-5130, USA

Osato Ogbeifun
Department of Biology, Indiana University-Purdue University Indianapolis, 723 West Michigan Street, Indianapolis, IN 46202-5130, USA

Opeyemi C. Olorungbounmi
Department of Biology, Indiana University-Purdue University Indianapolis, 723 West Michigan Street, Indianapolis, IN 46202-5130, USA

Liliana Patino
Department of Biology, Indiana University-Purdue University Indianapolis, 723 West Michigan Street, Indianapolis, IN 46202-5130, USA

Ryan Lucas
Department of Biology, Indiana University-Purdue University Indianapolis, 723 West Michigan Street, Indianapolis, IN 46202-5130, USA

Sonya Gilbert
Department of Biology, Indiana University-Purdue University Indianapolis, 723 West Michigan Street, Indianapolis, IN 46202-5130, USA

Evan S. Groninger
Department of Mathematics, Wabash College, Crawfordsville, IN 47933, USA

Julia Arciero
Department of Mathematics, Indiana University-Purdue University Indianapolis, Indianapolis, IN 46202, USA

Howard J. Edenberg
Department of Biochemistry and Molecular Biology, Indiana University School of Medicine, Indianapolis, IN 46202, USA

James A. Marrs
Department of Biology, Indiana University-Purdue University Indianapolis, 723 West Michigan Street, Indianapolis, IN 46202-5130, USA

Keisuke Nakajima
Division of Embryology and Genetics, Institute for Amphibian Biology, Graduate School of Science, Hiroshima University, Higashihiroshima 739-8526, Japan

Yoshio Yaoita
Division of Embryology and Genetics, Institute for Amphibian Biology, Graduate School of Science, Hiroshima University, Higashihiroshima 739-8526, Japan

M. Crispo
Unidad de Animales Transgénicos y de Experimentación (UATE), Institut Pasteur de Montevideo, Montevideo, Uruguay

A. P. Mulet
Unidad de Animales Transgénicos y de Experimentación (UATE), Institut Pasteur de Montevideo, Montevideo, Uruguay

L. Tesson
INSERM UMR 1064, Center for Research in Transplantation and Immunology-ITUN, Nantes, France

N. Barrera
Instituto de Reproducción Animal Uruguay, Fundación IRAUy, Montevideo, Uruguay

F. Cuadro
Instituto de Reproducción Animal Uruguay, Fundación IRAUy, Montevideo, Uruguay

P. C. dos Santos-Neto
Instituto de Reproducción Animal Uruguay, Fundación IRAUy, Montevideo, Uruguay

T. H. Nguyen
INSERM UMR 1064, Center for Research in Transplantation and Immunology-ITUN, Nantes, France

A. Crénéguy
INSERM UMR 1064, Center for Research in Transplantation and Immunology-ITUN, Nantes, France

L. Brusselle
INSERM UMR 1064, Center for Research in Transplantation and Immunology-ITUN, Nantes, France

I. Anegón
INSERM UMR 1064, Center for Research in Transplantation and Immunology-ITUN, Nantes, France

A. Menchaca
Instituto de Reproducción Animal Uruguay, Fundación IRAUy, Montevideo, Uruguay

Preface

The study of vertebrate embryonic development became a separate science in the early 1800s thanks to advances in microscopy. Embryologists collected and dissected specimens from chicken, frogs, turtles, fish, mice as well as humans, observed and documented the anatomy and physiology. These largely observatory work produced many of the important concepts in modern developmental biology, such as germ layers (von Baer, 1828), the theory of natural selection (Darwin, 1859), and the Organizer (Spemann and Mangold, 1934). After World War II, especially after the demonstration of DNA as the genetic material and advancement in recombinant DNA technology, the focus of studies shifted to the cellular, molecular and genetic processes of development, the mechanisms of regulation, and the basis for developmental defects and diseases. In recent years, advancements in stem cell biology and gene editing technology ushered vertebrate embryology into a new era as scientists use the tools and paradigms of embryonic development to tackle diseases of all ages, bridging the embryo and the adult.

There is a vast body of research on vertebrate embryology, enough to fill a whole library. It is not the goal of this book to provide a comprehensive picture. Rather, I draw an extremely simplified sketch, a "tasting menu", of current research topics. The book is organized into five sections. Section 1 illustrates a number of major research questions on the three developmental stages - cleavage, gastrulation, and organogenesis. Section 2 presents the frontiers in the study of genetic and epigenetic regulation: the role of genetic interactions, chromatin-based regulation, and non-coding RNAs. Section 3 describes the current research and application of stem cells. Section 4 introduces the role of genetic and environmental factors in developmental defects. Finally, Section 5 highlights the exciting new field of gene editing, using TALENs and CRISPR/Cas9 technologies as two examples

Editor

Dr. Q. Tian Wang, PhD

SECTION I. CURRENT RESEARCH IN VERTEBRATE EMBRYOGENESIS: CELL FATE, LINEAGES, AND MOLECULAR BASIS

1

CHAPTER

THE FIRST TWO CELL-FATE DECISIONS OF PREIMPLANTATION MOUSE EMBRYO DEVELOPMENT ARE NOT FUNCTIONALLY INDEPENDENT

Aleksandar I. Mihajlovic[1], Vasanth Thamodaran[1], and Alexander W. Bruce[1,2]

[1]Laboratory of Developmental Biology & Genetics (LDB&G), Department of Molecular Biology, Faculty of Science, University of South Bohemia, Branišovská 31, 37005 České Budějovice, CZECH REPUBLIC

[2]Institute of Entomology Biology Centre of the Czech Academy of Sciences, Branišovská 31, 37005 České Budějovice, CZECH REPUBLIC

ABSTRACT

During mouse preimplantation embryo development, three distinct cell lineages are formed, represented by the differentiating trophectoderm (TE), primitive endoderm (PrE) and the pluripotent epiblast (EPI). Classically, lineage derivation has been presented as a two-step process whereby outer TE cells are first segregated from inner-cell mass (ICM), followed by ICM refinement into either the PrE or EPI. As ICM founders can be produced following the fourth or fifth cleavage divisions, their potential to equally contribute to EPI and PrE is contested. Thus, modelling the early sequestration of ICM founders from TE-differentiation after the fourth cleavage division, we examined ICM lineage contribution of varying sized cell clones unable to initiate TE-differentiation. Such TE-inhibited ICM cells do not equally contribute to EPI and PrE and are significantly biased to form EPI. This bias is not caused by enhanced expression of the EPI marker Nanog, nor correlated with reduced apical polarity but associated with reduced expression of PrE-related gene transcripts (Dab2 and Lrp2) and down-regulation of plasma membrane associated Fgfr2. Our results favour a unifying model were the three cell lineages are guided in an integrated, yet flexible, fate decision centred on relative exposure of founder cells to TE-differentiative cues.

Citation: Mihajlović, A. I. et al. The first two cell-fate decisions of preimplantation mouse embryo development are not functionally independent. Sci. Rep. 5, 15034; doi: 10.1038/srep15034 (2015).

INTRODUCTION

The preimplantation period of embryogenesis encompasses the first cell-fate decisions of mammalian development and results in the derivation of three distinct cell lineages by the time of uterine implantation at the late blastocyst stage. Two of these lineages, the trophectoderm (TE) and primitive endoderm (PrE), are differentiating and ultimately yield extraembryonic tissues. The third epiblast (EPI) lineage is pluripotent and is a progenitor pool for all subsequently derived foetal cells[1,2,3]. Classically, this process of lineage separation has been presented as occurring in two distinct independent cell-fate decisions. The first decision involves the spatial separation of outer-residing TE progenitor cells from inner-cell mass (ICM) during the fourth/8- to 16-cell and fifth/16- to 32-cell cleavages, via so-called 'asymmetric/differentiative' cell-divisions, that are distinct from 'symmetric/conservative' divisions yielding two outer-cesslls[4,5]. Additionally, this spatial separation can occur when atypical apolar outer-cells are physically forced inside the embryo by neighbouring cells exhibiting pronounced apical-basolateral polarity[6]. The second decision is classically defined as involving gene expression refinements and active cell sorting within the ICM that ultimately results in EPI cells, residing deep within the ICM, and the PrE comprising a monolayer of blastocoel-facing cells at the surface of the ICM (reviewed in[1,2,3,7]). The collective result of much research into these two cell-fate decisions has led to the identification of key transcription-factor genes (*e.g. Tead4*[8,9]and *Cdx2*[10] in the TE, *Nanog*[11] in EPI and *Gata6/4*[12,13,14] and *Sox17*[15] in the PrE) required to regulate the necessary gene expression patterns to support these emerging cell-types. Additionally, the differential activity of the hippo-signalling pathway in functionally regulating Tead4 output between outer and inner-cells[16,17,18,19], and PrE promoting extra-cellular signalling provided by Fgf4 (fibroblast growth factor 4), via the Fgfr2 (fibroblast growth factor receptor 2) receptor in the ICM[20,21], have also emerged as important mechanisms.

However, the classical view of a functionally independent 'two-step' model of cell-fate derivation has been questioned[2,22,23]. The keystone of such questioning has been the observation that in time-lapse cell lineage tracing experiments, the majority of inner-cells generated after the fourth cleavage division are biased to ultimately form EPI. Whereas, inner-cells generated during the fifth cleavage division are strongly biased towards PrE, with inner-cells derived by an infrequent third wave of asymmetric division, being invariably fated to PrE[22]. These data, although not universally accepted[24,25,26], suggest that not all ICM cells are generated with equal potential and suggests that ancestral cell history plays a role during ICM cell lineage separation. Moreover, it implies that rather than being a 'two-step' process, the emergence of the three blastocyst cell lineages is more an inter-related continuum, whereby mechanisms influencing TE and ICM separation, in the 'first cell-fate decision', have knock-on consequences in the ICM during the 'second cell-fate decision'. Expressed alternatively, cell-fate can be influenced by the timing of inner-cell derivation, ultimately resulting in three possible outcomes, TE, PrE and EPI. This 'integrated cell-fate model' postulates that the earliest possible sequestration of EPI progenitor cells to the ICM is required to maintain their pluripotent properties, whereas the extra developmental 'time' parental PrE progenitors reside on the outside of the embryo, receiving TE-differentiative and apical-basolateral polarisation cues[16,17,18,19,27,28,29], prior to their internalisation during the fifth cleavage division, predisposes their ICM progeny to form PrE; a model sometimes called the 'time-outside hypothesis'[23,30,31] but herein referred to as the 'integrated cell-fate model'.

Accordingly, to test the integrated cell-fate model, we report our experiments inhibiting TE cell-fate initiation within defined sub-populations of cells (*i.e.* clones) in the developing preimplantation mouse embryo, assaying the frequency at which such cell clones, within the ICM, are capable of contributing to PrE; therefore modelling the early removal of inner-cells derived during the fourth cleavage division from TE-promoting differentiative signals, such as that provided by inhibited hippo-signalling[16,17,18,19]. We show, from observing PrE/EPI contribution in clones of varying size, that TE-inhibited ICM cells preferentially contribute progeny to the EPI rather than PrE, in a statistically significant manner. Moreover, the biased contribution is not because of non-physiological inductions in the expression of the EPI associated gene *Nanog*, but rather correlates with reduced expression of PrE markers (*Dab2* and *Lrp2*). Additionally, we observe decreased expression of *Fgfr2* and down-regulation of Fgfr2 protein from the plasma membrane, within TE-inhibited clones. Our results indicate that the ability to initiate and respond to TE-differentiation cues/primes blastomeres to contribute future PrE progenitors and that preventing TE-differentiation favours eventual EPI formation. Consequently, the data are consistent with the integrated cell-fate model stating that the early removal of cells from TE-differentiation, by their internalisation at or shortly after the fourth cleavage, predisposes their progeny to populate EPI; whereas later internalisation resulting from the fifth cleavage, whereby the outer-residing parental cells are exposed to additional differentiative signals, such as inhibited hippo-signalling[17], biases development towards PrE. However, it is not impossible for TE-inhibited cells to yield PrE progeny, suggesting the observed relationship is not rigid and reflects the remarkable regulative capacity of the developing embryo to respond to additional concurrent, and potentially stochastic, cell-fate inputs, possibly relating to overall ICM cell number.

RESULTS

ICM founder cells are generated during or shortly after the fourth (the 8- to 16-cell transition) and fifth cleavage (the 16- to 32-cell transition) divisions[4,6]. The time between the completion of these divisions is approximately twelve hours[32], during which outer-residing 16-cell stage blastomeres remain apical-basolaterally polarised and exposed to TE-differentiative cues, such as suppressed hippo-pathway signalling, whilst apolar inner-cells are protected from TE-differentiation by active hippo-pathway signalling[16,17,18,19,27,28]. As these outer-residing blastomeres can also generate further ICM founders after the fifth cleavage, it is questionable whether ICM progenitors produced by the fourth and fifth cleavage divisions have equal potential to contribute to EPI and PrE[24,25,26]. In order to test if ICM cells are generated with equal potential, irrespective of the extent of TE induction their parental cells received, we assayed ICM lineage contribution of TE-inhibited cell clones in the embryo. We hypothesised if the extent of TE induction was unimportant for PrE differentiation in the ICM, TE-inhibited clones would not be impaired in their potential to contribute to PrE. Conversely, if being able to initiate TE-differentiation facilitates PrE differentiation, such clones would be disadvantaged in populating the PrE, therefore supporting the integrated cell-fate model.

Tead4 Down-Regulation Using Long dsRNA Phenocopies the Zygotic _Tead4_ $^{-/-}$ Knock-Out

We employed an RNAi microinjection based strategy, targeting the *Tead4* gene, to inhibit TE-differentiation within defined cell clones. We reasoned clonal *Tead4* down-regulation would mimic the naturally occurring removal of cells from Tead4 regulation that occurs during their internalisation after the fourth cleavage division. We chose to target Tead4 as it is the earliest known transcription-factor to function in TE specification and its transcriptional activating properties are known to be regulated by hippo-signalling, thereby confining its regulatory output to polarised outer-cells[8,9,16]. Accordingly, we synthesised a *Tead4* specific long double-stranded RNA (Tead4-dsRNA) for use in single cell microinjection experiments that could be used to elicit TE-inhibited cell clones in the preimplantation mouse embryo. We first confirmed the efficacy of the construct by microinjecting recovered 2-cell (E1.5) stage embryos, in both blastomeres, with RDBs (rhodamine dextran conjugated beads, lineage tracer)±the Tead4-dsRNA. As shown in Fig. 1b, we observed >95% reduced Tead4 mRNA expression, at the 16- (E3.1) and 32-cell (E3.6) stages, with accompanying undetectable levels of Tead4 protein, in Tead4-dsRNA injected embryos (*Tead4*-KD embryos) compared to microinjection controls (Fig. 1c); confirming the efficacy of the Tead4-dsRNA construct. We also observed robust reductions in the expression of two other TE-specific transcription factor genes known to function downstream of Tead4, namely *Cdx2*[10] (Fig. 1b,c) and *Gata3*[33,34] (Fig. 1b), demonstrating a successful block in TE-specification. Moreover, when we *in vitro* cultured such *Tead4*-KD embryos, they developed in-step with controls until the 32-cell (E3.6) stage but subsequently failed to initiate blastocoel formation and displayed considerable cell-death by the late blastocyst stage (E4.5); recapitulating the phenotype exhibited by genetically null *Tead4*$^{-/-}$ zygotic knockout embryos[8,9] (Fig. 1d).

a)

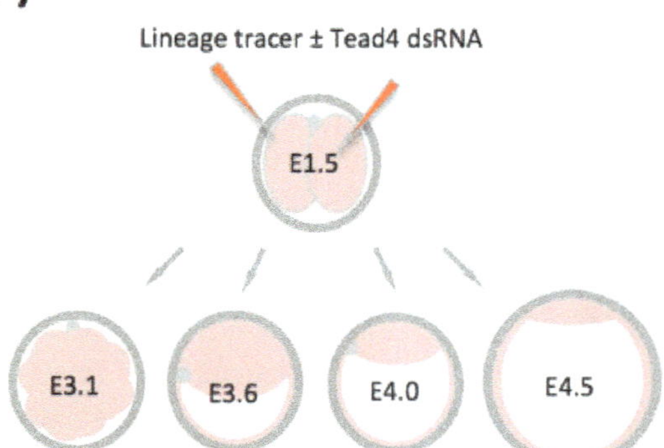

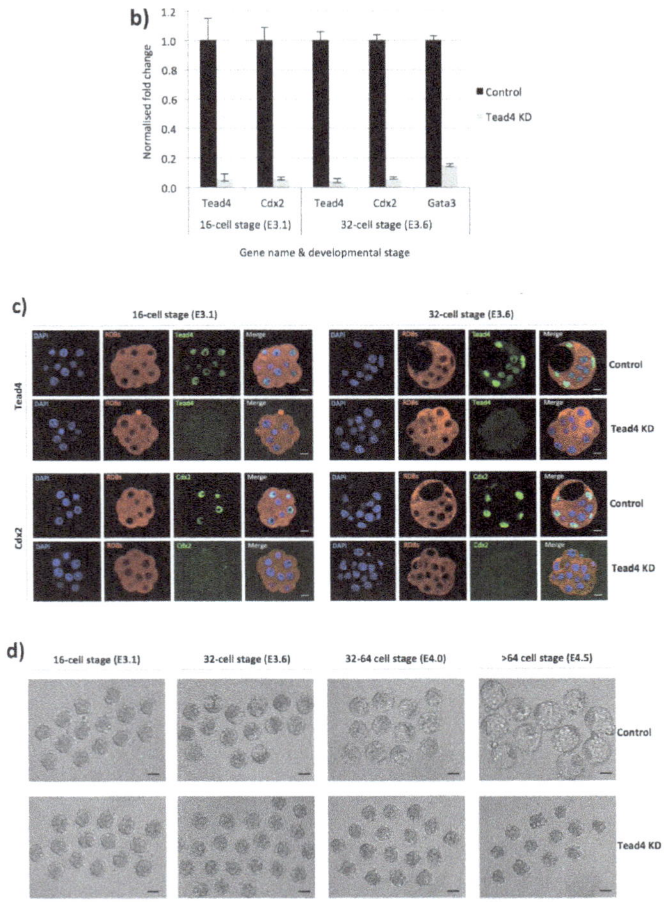

Figure 1: Long dsRNA mediated *Tead4* **down-regulation** phenocopies the zygotic *Tead4*$^{-/-}$ **null TE-deficit phenotype.**

(a) Schematic representation of experimental strategy. Embryos were microinjected with RDB injection marker (red)±Tead4-dsRNA in both cells at the 2-cell stage (E1.5) and *in vitro*cultured until the mid-16-cell (E3.1), 32-cell (E3.6), 32–64-cell (E4.0) or >64-cell (E4.5) stages, prior to Q-RTPCR/microscopic analyses. (b) Q-RTPCR data detailing normalised average fold changes in mRNA expression of *Tead4, Cdx2* and *Gata3* in embryos microinjected with Tead4-dsRNA, relative to microinjection control embryos. Individual gene mRNA levels were normalised against Rpl23 and/or H2afz within control and experimental knockdown conditions and the fold change associated with *Tead4* KD calculated. Errors are given as s.e.m. n=at least 2 for biological replicates and 3 for technical replicates. (c) Representative single confocal immuno-fluorescence microscopy sections of embryos microinjected with RDB injection marker±Tead4-dsRNA immuno-stained for Tead4 or Cdx2 protein (green) and DNA co-stained with DAPI (blue). RDB microinjection marker is visible (red). Scale bars = 10 µm. (d) Bright-field micrographs of control and Tead4-dsRNA microinjected embryos at various preimplantation stage developmental time-points in *in vitro* culture. Note that the*Tead4*-KD embryos fail to initiate blastocoel formation and starting from the E4.0 time-point exhibit cell death; a phenotype consistent with that observed in zygotic genetic *Tead4*$^{-/-}$null preimplantation embryos. Scale bars = 50 µm.

Clonal Down-Regulation of *Tead4* Expression Inhibits PrE Formation

We next asked what would be the consequence for PrE derivation in a clone of *Tead4*-KD cells in the developing embryo? First, we microinjected 2-cell (E1.5) stage embryos in one blastomere with RDBs ±Tead4-dsRNA, to elicit a fluorescently marked and TE-inhibited clone comprising half the embryo. As shown in Fig. 2, we observed complete down-regulation of Tead4 and Cdx2 protein, only within the cells derived from the Tead4-dsRNA microinjected blastomere and throughout the preimplantation developmental period; confirming our clonal TE-inhibition approach. However, in contrast to the global down-regulation of *Tead4* (Fig. 1), clonal *Tead4*-KD embryos were able to initiate appropriate blastocoel formation, illustrating the regulative capacity of the preimplantation mouse embryo.

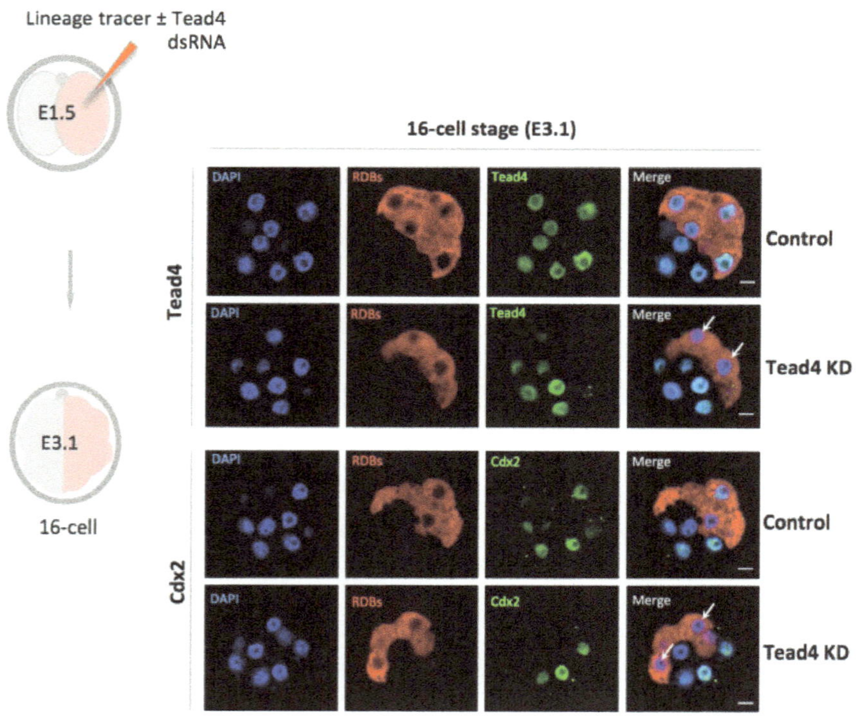

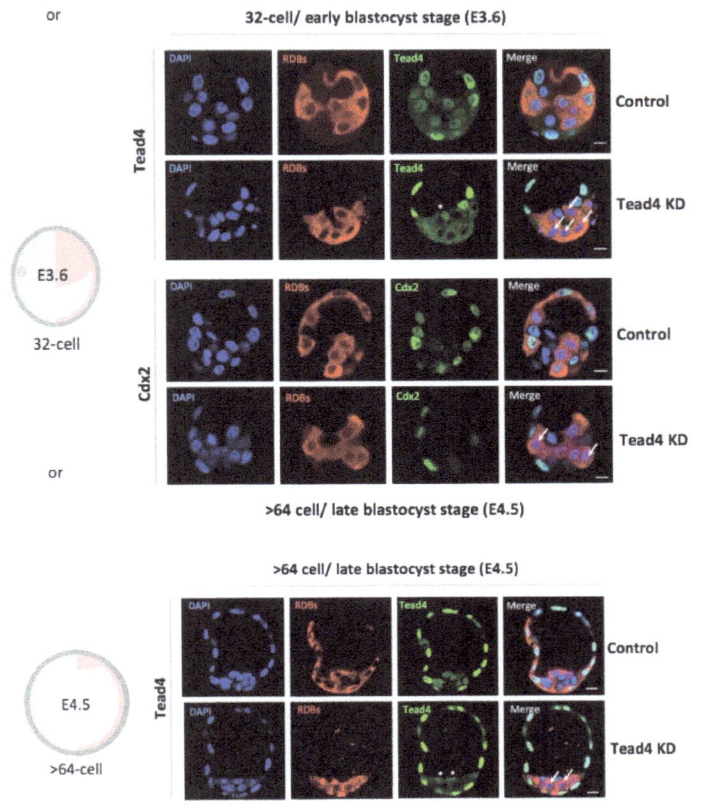

Figure 2: Clonal down-regulation of *Tead4* **expression and TE-differentiation inhibition.**

A schematic of experimental strategy to effect clonal *Tead4* knockdown (KD) and TE-inhibition in one-half of the embryo using microinjected RDBs ± Tead4-dsRNA (see materials and methods) is given on the left. Representative single z-plane confocal micrographs of control and *Tead4*-KD embryos at either the mid-16-cell (E3.1), 32-cell (E3.6) or >64-cell (E4.5) stages immuno-stained for Tead4 or Cdx2 (green) are given. Cells derived from the microinjected 2-cell stage clone are distinguishable by the co-injected RDB fluorescence (red). DNA counter-stain (blue) is also shown. In merged images the arrows denote cells exhibiting a lack of Tead4 or Cdx2 expression in the *Tead4*-KD microinjected cell clone, thus confirming the efficacy and the functional and clonal inhibition of TE-differentiation by Tead4-dsRNA until the late blastocyst stage (E4.5). ICM cells not from the microinjected clone, expressing Tead4 protein are marked with asterisks (in Tead4 alone micrographs). Note, in contrast to global *Tead4*-KD embryos (Fig. 1), such clonal *Tead4*-KD embryos initiate blastocoel formation in a manner indistinguishable from control microinjected embryos. Scale bars = 10 μm.

We next assayed the frequency that *Tead4*-KD cells contributed to each late blastocyst (E4.5) cell lineage, compared with the non-injected clone and the equivalent clone from control embryos in which RDBs alone had been microinjected (Fig. 3). It is important to note that blastocyst lineage contribution and the incidence of apoptosis in such control embryos were statistically identical to that observed in a second group of control embryos that had been similarly microinjected at the 2-cell stage with RDBs plus a GFP specific dsRNA (GFP-dsRNA; lacking an endogenous mouse mRNA target—Supplementary Fig. S1 andSupplementary Tables ST1 & ST2). The lack of any statistically significant difference between these two groups, confirmed the appropriateness of using the RDB alone control in all the subsequently described assays.

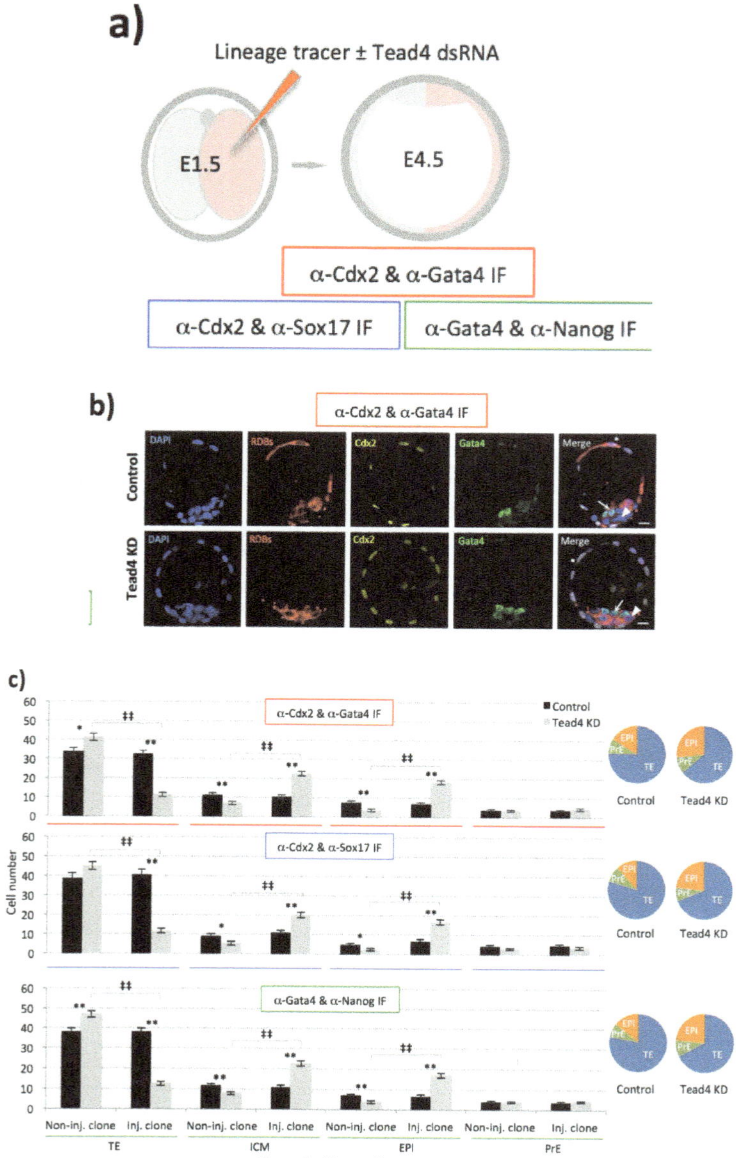

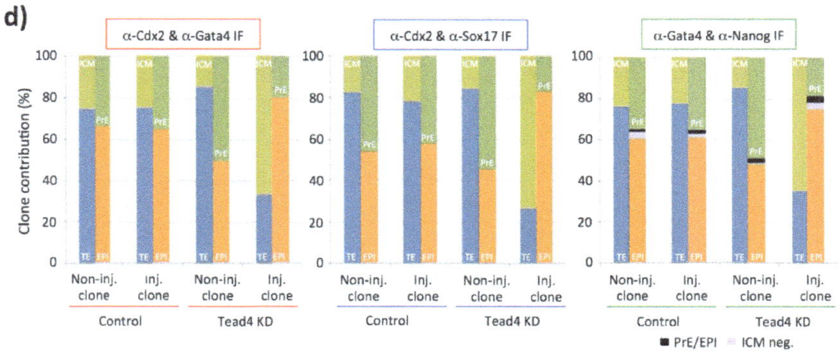

Figure 3: Clonal inhibition of TE-differentiation preferentially biases cells to EPI rather than PrE fates.

(a) Experimental strategy to effect clonal *Tead4*-KD and TE-inhibition in one-half of the embryo and assess cell lineage allocation in late blastocysts (E4.5) via immunofluorescence detection of marker gene expression, using; i) Cdx2 (TE) & Gata4 (late PrE) - red, ii) Cdx2 & Sox17 (early PrE)—blue, and iii) Gata4 & Nanog (EPI)—green, (*n.b.* inner-cells devoid of either lineage marker in i) and ii) were classified as EPI and outer cells devoid of immuno-reactivity in iii) were designated as TE). (b) Representative single z-plane confocal micrographs of microinjection control and clonal *Tead4*-KD late blastocyst (E4.5) embryos immuno-stained for Cdx2 (pseudo-coloured yellow) and Gata4 (green) protein expression. Progeny of the microinjected cell are distinguishable by co-injected RDB fluorescence (red). DNA is counterstained with DAPI (blue). Merged image asterisks represent exemplar cells classified in our analyses as TE, arrows as PrE cells and arrow-heads as EPI. Note similar exemplar micrographs for the alternative immuno-staining regimes are given inSupplementary Figs S3 and S4. Scale bars = 10 µm. (c) Average number of cells from either non-microinjected or microinjected cell clones contributing to late blastocyst (E4.5) lineages, in control and clonal *Tead4*-KD embryos, immuno-stained in each of the three regimes outlined in a). Error bars represent s.e.m; */** and ‡/‡‡ denote statistically significant differences between equivalent cell clones of control and clonal *Tead4*-KD embryos, or between cell clones within control and clonal *Tead4*-KD embryo groups, respectively ($p<0.05$ and $p<0.005$, 2-tailed student t-tests). The relative average percentage contribution of total cell number to each late blastocyst lineage in both control- and *Tead4*-KD embryos is also provided as a pie-chart. (d) Averaged percentage contribution of non-microinjected and microinjected cell clones, of control and clonal *Tead4*-KD embryos, immuno-stained according to the three regimes outlined in a), between the TE (blue) & ICM (yellow) and the PrE (green) & EPI (orange) of analysed late (E4.5) blastocysts. In the anti-Gata4/Nanog immuno-stained embryo groups, the contribution of ICM cells either positive or negative for both PrE and EPI marker gene expression are shown in black and violet, respectively. Overall, for control embryos n = 24, 13 & 25 and for clonal *Tead4*-KD embryos n = 24, 9 & 23 in each of the three immuno-staining regimes outlined in a), respectively. Note, further analysis and variant presentation of the above data is provided inSupplementary Figs S2–S4 and Supplementary Tables ST3–ST8.

Therefore, lineage contribution was assayed in RDB alone microinjected control and clonal *Tead4*-KD late blastocysts (E4.5) that had been double immuno-stained for either i) Gata4, a PrE marker35, and Cdx2, a marker of TE cells10, or ii) Sox17, an early PrE marker15,22, and Cdx2 (under each immuno-staining protocol, ICM cells devoid of either marker were classified as EPI, as ICM lineage segregation is reported to be completed by this time36), or, iii) Gata4 and Nanog, an EPI marker11, (thus directly assaying both ICM cell lineages whilst designating outer cells negative for either marker as TE). Further comprehensive analyses relating to the relative spatial location of cell clones, ICM versus outer-TE location, and cells with fragmented nuclei typical of apoptosis were also undertaken. Note that each of the three respective immuno-staining regimes described above were associated with dedicated and matched control and *Tead4*-KD embryo groups that were microinjected during the same experimental sessions. Accordingly, n=24, 13 and 25 for RDB alone control groups and n=24, 9 and 23 for *Tead4*-KD embryo groups per respective immuno-staining regime. Figure 3 summarises these data and Supplementary Figs S2, S3, and S4 provide expanded summaries from each immuno-staining method used. Supplementary Tables ST3, ST4 and ST5 provide individual embryo data for each of immuno-staining group and data relating to apoptosis can be found inSupplementary Fig. S5 and Supplementary Tables ST6, ST7 and ST8.

Our analyses showed that whilst clonal *Tead4*-KD embryos were able to develop into morphologically normal (E4.5) late blastocysts (Fig. 3b), the clonal origin of these cells, irrespective of how they were assayed in the differing immuno-staining regimes, were not equal. Specifically, *Tead4*-KD clones preferentially and significantly contributed to the ICM versus the outer TE with partial compensation coming from the non-microinjected clone (Fig. 3c); although *Tead4*-KD embryos presented with significantly less TE and more ICM than control embryos (Fig. 3cand Supplementary Figs S2–S4 and Supplementary Tables ST3–ST5). Cells derived from the Tead4-dsRNA microinjected clone that remained in an outer TE position, did not express Cdx2 protein (with very few isolated exceptions detailing very weak anti-Cdx2 immuno-fluorescence) and were almost exclusively spatially restricted to the polar TE, overlying the ICM, rather than mural TE surrounding the expanding blastocoel. Albeit marked, this phenotype is entirely consistent with the well-characterised *Tead4* role in TE-differentiation16. However, unlike the situation in control embryos, ICM cell lineage segregation was also unequal in *Tead4*-KD embryos, with the Tead4-dsRNA microinjected cell clones preferentially and significantly contributing to EPI rather than PrE (Fig. 3c). Whilst the ICMs of *Tead4*-KD embryos were significantly larger than control embryos, the number of PrE cells remained equal in both groups, with all the extra *Tead4*-KD ICM cells contributing to increased EPI cell number (Fig. 3c and Supplementary Figs S2–3 and Supplementary Tables ST3–ST5). Moreover, the smaller non-microinjected ICM clone in TE-inhibited embryos (for example comprising 7.0 ± 0.8 cells versus 22.6 ± 1.1 cells for the microinjected clone in embryos immuno-stained with Gata4/Cdx2—Supplementary Fig. S2 and Supplementary Table ST3), contributed a statistically equal number of PrE cells than the microinjected clone (equating, in the Gata4/Cdx2 immuno-stained example, but a trend repeated in the other staining groups, to 50.3% of its overall size, versus 19.0% in the microinjected *Tead4*-

KD ICM clone—Fig. 3d), emphasising the bias within the microinjected *Tead4*-KD clone against contributing to the PrE, despite forming the majority of the total cell number of the ICM. We also noted significant reductions in total cell number in *Tead4*-KD embryos (Supplementary Figs S2–S4and Supplementary Tables ST3–ST5) accountable for by increased apoptosis within outer-residing cells of the microinjected clone, presumably unable to adapt to a spatial location requiring TE-differentiation (Supplementary Fig. S5 and Supplementary Tables ST6–ST8).

Overall, we found that individual ICM cells from either the non-microinjected and microinjected TE-inhibited clones of *Tead4*-KD embryos did not segregate between the EPI and PrE in a manner that would be consistent with them being generated with equal potential. Rather TE-inhibited ICM cells preferentially allocate to the EPI, with the non-microinjected cell clone contributing a greater proportion of its cells to the PrE. Importantly, this result was consistently observed irrespective of the combination of specific antibody markers used to assay the emerging late blastocyst (E4.5) cell lineages, thus validating the functional relevance of the ICM allocation phenotype observed. It has previously been reported that the ICMs of late blastocyst (E4.5) stage embryos can present with cells expressing neither an EPI or PrE marker21·35. Consistent with such reports we also observed incidences, using the anti-Gata4/Nanog antibody combination, of late blastocyst (E4.5) stage ICM cells that were either negative or positive for both ICM markers, albeit at very low frequencies and in both control and *Tead4*-KD embryo groups (on average less than one cell per embryo in both conditions—Fig. 3d, Supplementary Fig. S4 and Supplementary Table ST5). Although there was a statistically significant increased frequency of cells exhibiting negative immuno-fluorescence for each marker in the microinjected clone of *Tead4*-KD embryos, when compared with both the non-microinjected clone or the equivalent clone of control embryos, it was in overall numbers extremely modest (0.9 ± 0.2 cells per embryo versus 0.0 and 0.3 ± 0.1, respectively). Moreover, it was insufficiently large to explain the increased contribution of cells classified as EPI in embryos that were assessed by immuno-fluorescent staining for either Gata4/Cdx2 or Sox17/Cdx2 (Fig. 3 andSupplementary Figs S2 and S3 plus Supplementary Tables ST3 and ST4) as being derived from unspecified ICM cells.

We therefore interpret the collective sum of these experimental datasets as demonstrating that the clonal inhibition of TE cell-fate biases ICM progeny towards EPI rather than PrE formation, albeit with a very small number of ICM cells remaining uncommitted to either lineage. Moreover, this correlates with TE-differentiation potentiating later PrE differentiation, as suggested by the integrated cell-fate model. However, inhibition of TE-differentiation is not an impermeable block to future PrE formation, as cells derived from the TE-inhibited clone can populate the PrE but at much reduced frequency.

TE-Inhibition within Smaller Sized Cell Clones also Inhibits PrE Formation

We next created embryos containing smaller sized TE-inhibited clones to assay if the observed bias against PrE contribution, at the late blastocyst (E4.5) stage, would persist.

This was because, the above described experiments yielded embryos with relatively larger ICMs (*e.g. Tead4*-KD; 29.6±1.5, versus controls; 21.8±1.0 cells, in embryo groups immuno-stained for Gata4/Cdx2 - Supplementary Fig. S2 andSupplementary Table ST3) and we wanted to exclude the possibility that E4.5 stage late blastocysts could only support a finite number of PrE cells and that our results were not due to excessive developmental regulation (*i.e.* that once a finite number of PrE cells had been specified, any extra ICM cells would be defaulted to form EPI). Our first approach involved the direct microinjection of single individual blastomeres of mid-4-cell stage (E2.0) embryos, causing *Tead4*-KD in one quarter of the embryo (Supplementary Fig. S6 and Supplementary Table ST9, for apoptosis data see Supplementary Fig. S5 and Supplementary Table ST10; n=19 control embryos and n=29 *Tead4*-KD embryos, immuno-stained for Gata4 and Cdx2). We observed the same bias, as for the above described 2-cell microinjections, in the *Tead4*-KD microinjected cell clone. However, the average size of such *Tead4*-KD embryo ICMs was still greater than that observed in control embryos. Therefore, we adopted another strategy to characterise late blastocyst (E4.5) cell lineage contribution of still smaller clones of TE-inhibited cells, involving the creation of chimeric embryos from non-compacted 8-cell stage (E2.5) embryos aggregated with single developmentally matched donor blastomeres in which *Tead4* expression was down-regulated. Such chimeras contained marked TE-inhibited clones (*Tead4*-KD-chimeras) equivalent to one-ninth of their initial total cell number and when *in vitro* cultured to the last blastocyst stage were morphologically indistinguishablefromcontrol-clonecontainingchimeras(referredtoascontrol-chimeras, see summarised data in Fig. 4, expanded inSupplementary Fig. S7 with Supplementary Tables ST11 and ST12documenting individual embryo allocation and apoptosis data, respectively; n=30 control-chimeras and n=17 *Tead4*-KD-chimeras, immuno-stained for Gata4/Cdx2). Importantly, such *Tead4*-KD-chimeras had a total number of ICM cells statistically equal to that observed in control–chimeras (20.9±1.0 versus 19.0±0.9 cells, respectively) and were more similar in size to the ICMs of the non-perturbed *in vitro* cultured control blastocysts (22.4±1.5 cells—Supplementary Table ST13). When we assayed late blastocyst (E4.5) cell lineage contribution, we found that the marked TE-inhibited clone of *Tead4*-KD-chimeras was not only biased to populate the ICM, it was also biased in a statistically significant manner to contribute to the EPI rather than the PrE. Therefore, we found that the inhibition of TE-differentiation within discreet small clones, within ICMs of more physiological appropriate size, also biases cell-fate against PrE formation. Thus, demonstrating that the observed anti-PrE biases observed in TE-inhibited ICM residing clones, derived from any of the three experimental strategies described, are not simply a function of generating unusually large ICMs. Moreover, the data reflect unequal cell-fate potential between such TE-inhibited clones when compared with non-manipulated, non-TE-inhibited, ICM cells of the same embryo. Expressed alternatively, clonal inhibition of TE-differentiation inhibits the progeny of that clone to differentiate towards a PrE cell-fate, when competing with otherwise unperturbed cells outwith the clone.

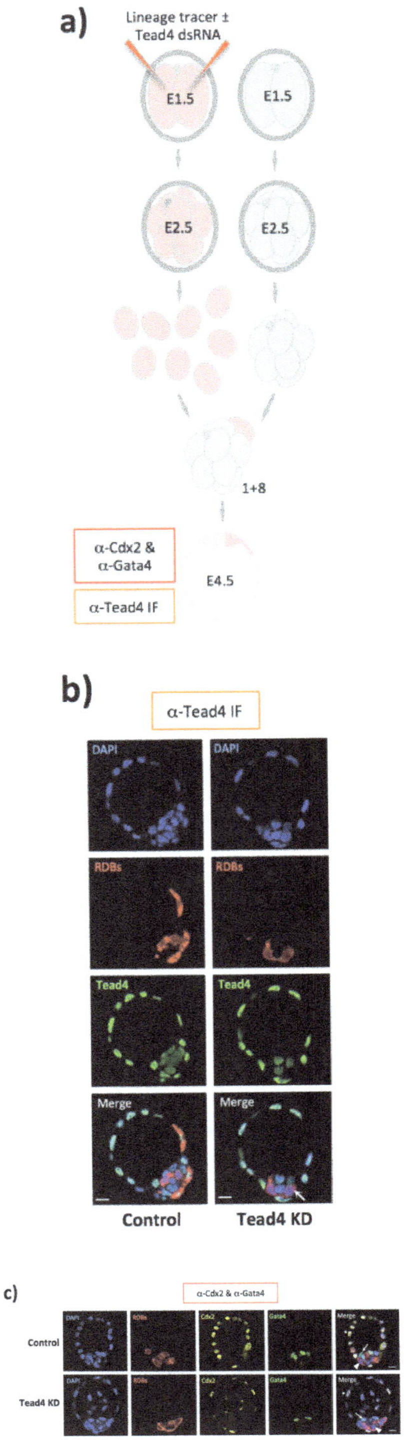
a)
Lineage tracer ±
Tead4 dsRNA
E1.5
E1.5
E2.5
E2.5
1+8
α-Cdx2 &
α-Gata4
E4.5
α-Tead4 IF
b)
α-Tead4 IF
DAPI
DAPI
RDBs
RDBs
Tead4
Tead4
Merge
Merge
Control
Tead4 KD
c)
α-Cdx2 & α-Gata4
Control
Tead4 KD

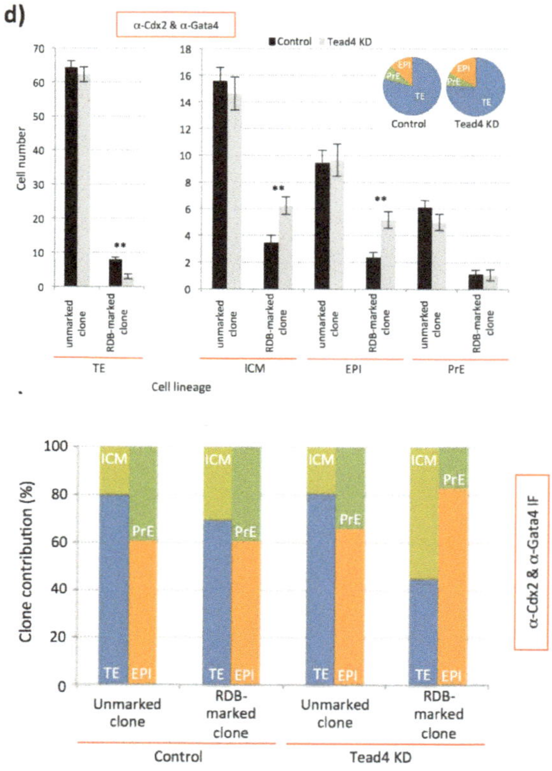

Figure 4: TE-inhibition within small chimeric ICM clones also biases against ultimate PrE cell-fate.

(a) Experimental strategy to generate embryo chimeras containing marked TE-inhibited cells equivalent to one ninth of the embryo and to asses late blastocyst (E4.5) lineage formation, via immuno-fluorescent staining for Cdx2 and Gata4 (*n.b.* inner-cells devoid of either lineage marker were classified as EPI and immuno-staining using anti-Tead4 antibody was also used to confirm *Tead4* KD in the marked chimeric clones). (b) Representative single z-plane confocal micrographs of control and *Tead4*-KD clone containing chimeras immuno-stained for Tead4 (green) expression. In the merged image, arrows denote cells not expressing detectable levels of Tead4 derived from original *Tead4*-KD donor blastomere (c) Further, representative single z-plane confocal micrographs of late blastocyst (E4.5) chimeras immuno-stained for Cdx2 (pseudo-coloured yellow) and Gata4 (green). Merged image asterisks represent exemplar cells classified in our analyses as belonging to the TE, arrows PrE and arrow-heads EPI. In (b,c) cells deriving from donor blastomeres, themselves originally derived from control or Tead4-dsRNA microinjected 2-cell (E1.5) stage embryos, within chimeras are distinguishable by co-injected RDB (red). DNA is counterstained (DAPI, blue). Scale bars = 10 μm. (d) Average number of cells from either RDB-marked or non-marked cell clones in TE lineage (left) and ICM, EPI and PrE lineages (right), in control and*Tead4*-

KD-chimeras; error bars represent s.e.m and */** denote statistically significant differences between equivalent cell clones of control- and *Tead4*-KD-chimeras ($p<0.05$ and $p<0.005$, respectively - 2-tailed student t-tests). Inset pie-charts denote the relative average contribution of cells within each embryo group to each of the three late blastocyst (E4.5) lineages. (e) Averaged percentage contribution of unmarked and RDB-marked cell clones, of control and clonal *Tead4*-KD chimeric embryos, immuno-stained for Cdx2 and Gata4, between the TE (blue) & ICM (yellow) and the PrE (green) & EPI (orange) of analysed late blastocysts (E4.5). Overall, for control chimeras $n=30$ and for clonal *Tead4*-KD chimeras $n=17$. Note, further analysis and variant presentation of the above data is provided inSupplementary Fig. S7 and Supplementary Tables ST11 and ST12.

Molecular Characterisation of Attenuated PrE Formation in TE-Inhibited Cell Clones

The extent of 8-cell stage blastomere, and subsequent outer-cell, apical-basolateral polarisation has previously been correlated with cell-division orientation and outer-cell engulfment, whereby reduced polarity is associated with inner-cell generation and active hippo-signalling6,28. We therefore assayed for apical polarisation and hippo-signalling deficits within TE-inhibited cell clones using previously described anti-phospho-ezrin/radixin/moesin (pERM) and Yap1 antibodies6,19, respectively. However, we observed no apical polarisation defects by the mid-16 cell (E3.1) stage (Fig. 5) although we sometimes documented outer mid-16-cell stage blastomeres with substantially reduced apical pERM (not correlating with any clone) at frequencies similar to recent reports6. Yap1 immuno-staining of clonal *Tead4*-KD embryos, detailed nuclear and unusually enriched cytoplasmic expression in outer-residing TE-inhibited cells of mid-16-cell stage embryos, contrasting with more typical nuclear enriched localisation in non-microinjected outer-cells (or control embryos); a similar pattern of Yap1 mis-localisation was also observed at the 32-cell (E3.6) stage, although outer cell staining was more prominently cytoplasmic by this stage. Using anti-sera previously reported16 to only recognise phosphorylated Yap1 (pYap1—Supplementary Fig. S8), we determined that the mis-localised Yap1 was non-phosphorylated, confirming the atypical cytoplasmic localisation was not because of the induction of active hippo-signalling. We speculate it may be associated with a lack of mechanism to retain Yap1 in the nucleus given its binding partner's absence (*i.e.* Tead4), and thus reflects an equilibrium of the active nuclear import and passive export of non-phosphorylated Yap1. Curiously, we also observed nuclear localised pYap1 in outer mid-16-cell stage blastomeres both within and outside the Tead4-dsRNA microinjected clone (Supplementary Fig. S8a,b). pERM immuno-staining of 32-cell (E3.6) stage embryos revealed robust apical expression within outer-residing TE-inhibited clones, associated with rounded non-TE like morphology (Fig. 5) at frequencies consistent with subsequent apoptosis (see Supplementary Fig. S5 and Supplementary Tables ST6–ST8). Overall, we did not observe any apical polarisation deficits/or intracellular mis-localisations within TE-inhibited clones that would be consistent with the generation of the extra ICM founder cells (Fig. 3), nor did we observe such deficits when assaying other apical polarity (Prkcz and Pard6b), basolateral polarity (Scrib) or cell adhesion molecular markers.

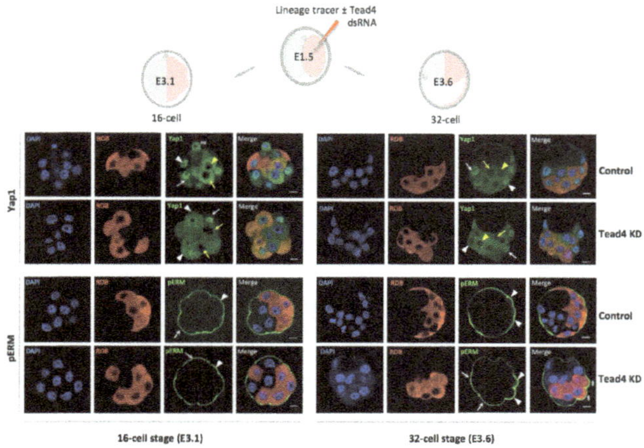

Figure 5: Clonal TE-inhibition causes outer-cell Yap1 mis-localisation and enhanced 32-cell stage apical polarity.

A schematic of experimental strategy to assay, via confocal immuno-staining, Yap1 and phospho-ezrin/radixin/moesin (pERM) expression in TE-inhibited clones (comprising half the embryos cells) at the mid-16-cell (E3.1) or 32-cell (E3.6) stages is shown (top). Lower panels; representative single confocal z-plane micrographs of control and *Tead4*-KD embryos immuno-stained for Yap1 (anti-sera does not discriminate between phosphorylated or non-phosphorylated forms) and pERM expression at the mid-16-cell and 32-cell stages (both in green). DNA DAPI counter-stain (blue) and RDB marked microinjected clones (red) are also shown. Arrows and arrow-heads highlight exemplar Yap1/pERM protein expression within non-microinjected and microinjected clones respectively. In Yap1 images the white arrows or arrow-heads indicate staining in outer-cells and yellow variants in inner-cells (second polar body = 'PB'). Double crosshairs denote outer 32-cell stage cells in the *Tead4*-KD group exhibiting atypical morphology with enhanced apically localised pERM immuno-staining (also seen in Prkcz and Pard6b immuno-staining—Supplementary Fig. S9). Scale bars = 10 μm.

We next ascertained if TE-inhibited clone segregation towards EPI cell-fate was associated with precocious and/or increased expression of the EPI marker gene *Nanog*37. We first assayed *Nanog* derived transcript levels in 16- (E3.1) and 32-cell (E3.6) stage embryos derived from embryos microinjected in both blastomeres at the 2-cell (E1.5) stage with RDBs ± Tead4-dsRNA, to effect global TE-inhibition in all cells of the embryo (Fig. 6a). We found that Nanog mRNA levels remained unchanged at both stages. We therefore assayed Nanog protein expression, but utilising the clonal down-regulation model by microinjecting one blastomere at the 2-cell (E1.5) stage to permit side-by-side comparison of Nanog protein levels in non-microinjected and TE-inhibited microinjected clones (Fig. 6b). Similarly, we did not detect any difference in Nanog protein expression levels between the microinjected and non-microinjected clones in both control and *Tead4*-KD embryos or between the embryo groups, confirming TE-inhibition by *Tead4*-KD is not associated with enhanced or precocious *Nanog*expression.

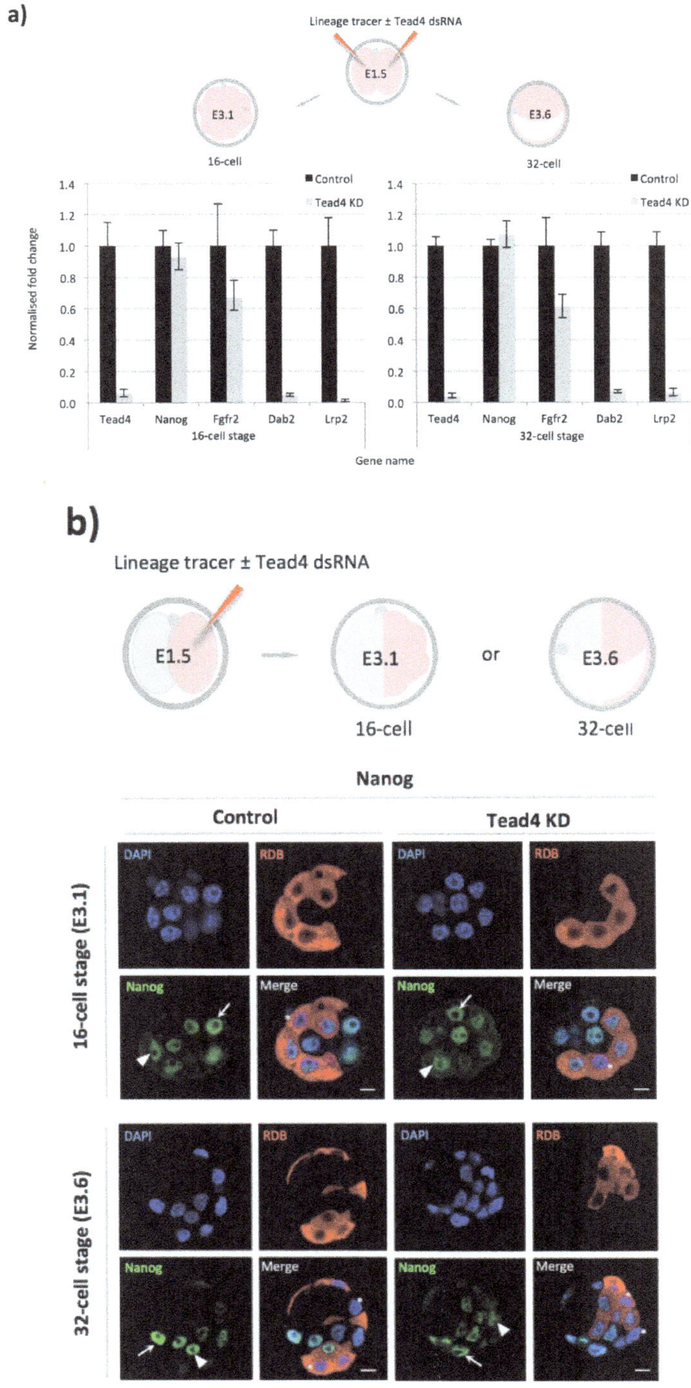
a)
Lineage tracer ± Tead4 dsRNA
E1.5
E3.1
E3.6
16-cell
32-cell
Control
Tead4 KD
Normalised fold change
Tead4
Nanog
Fgfr2
Dab2
Lrp2
16-cell stage
32-cell stage
Gene name
b)
Lineage tracer ± Tead4 dsRNA
E1.5
E3.1
or
E3.6
16-cell
32-cell
Nanog
Control
Tead4 KD
16-cell stage (E3.1)
32-cell stage (E3.6)
DAPI
RDB
Nanog
Merge

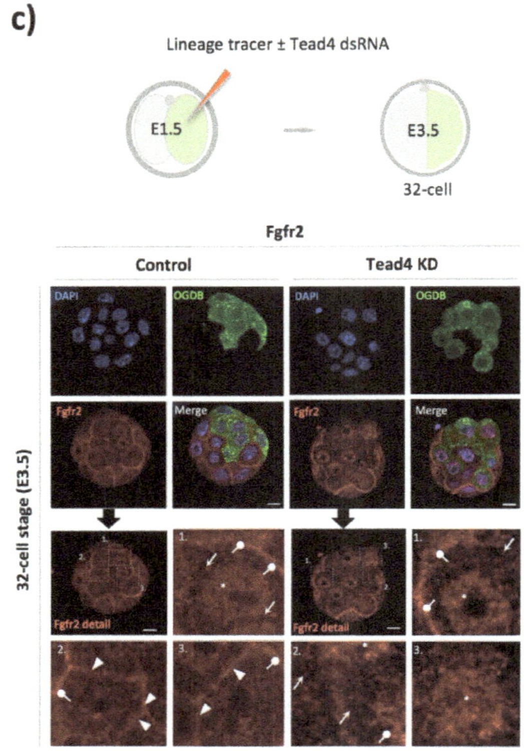

Figure 6: Global and clonal TE-inhibition; no enhanced *Nanog*expression prior to 32-cell stage but attenuated PrE-specific marker expression.

(a) The experimental strategy to down-regulate *Tead4* and inhibit TE-differentiation throughout all cells of the embryo prior to Q-RTPCR analysis (upper). Normalised expression fold changes, resulting from *Tead4*-KD, of the stated transcripts at the mid-16-cell (E3.1) or 32-cell (E3.6) stages (lower panels). Individual gene mRNA levels were normalised against Rpl23 and/or H2afz transcript levels within control and experimental knockdown conditions prior to fold change calculation. Errors = s.e.m, n = at least 2 for biological and 3 for technical replicates (*n.b. Tead4* specific data is repeated from Fig. 1 as Q-RTPCR was performed from same cDNA preparations). (b) Confocal microscopy analysis of Nanog (green) expression after clonal *Tead4*-KD at the mid-16-cell (E3.1) and 32-cell (E3.6) stages. Arrows and arrow-heads denote exemplar Nanog expression in non-microinjected and microinjected clones, respectively. Asterisks highlight TE cells without Nanog expression reflecting previously characterised inter-cell heterogeneity. (c) Confocal microscopy analysis of Fgfr2 (red) expression after clonal *Tead4*-KD at the non-cavitated 32-cell (E3.5) blastocyst stage. Representative single z-plane confocal micrographs are shown (middle panels) with the lower 4 panels detailing magnified anti-Fgfr2 immuno-stained images, according to numbered regions of interest. Arrow-heads highlight plasma membrane associated Fgfr2 between neighbouring ICM cells (control embryos) and arrows approximate equivalent regions of other neighbouring

ICM cells without anti-Fgfr2 signal (illustrating heterogeneous *Fgfr2* expression within control embryo ICMs). Similarly, arrows show ICM cell boundaries between cells of the microinjected clone devoid of Fgfr2 in *Tead4*-KD embryos. Asterisks and lollipop markers, in both control and *Tead4*-KD embryos, show nuclear Fgfr2 protein (especially in Tead4-KD embryos) or expression at the interface of TE and ICM cells, respectively. In both (b,c) progeny cells of microinjected clones are distinguishable by co-injected RDBs (red) or OGDBs (Oregon-green dextran beads—green). DNA was counterstained with DAPI (blue) and scale bars = 10 μm.

We next hypothesised that the EPI bias of TE-inhibited ICM residing clones may be associated with attenuated PrE-related gene expression, and thus indicative of a potential obstacle for such cells to initiate or maintain PrE differentiation. Accordingly, we investigated the expression levels of PrE-related gene mRNAs following TE-inhibition (Fig. 6a). We found that, in contrast to *Nanog*, the normalised transcript levels for the PrE marker genes *Lrp2*[38], and *Dab2*[39] were down-regulated >95%. Moreover, we also detected a ~40% reduction in *Fgfr2*mRNA transcripts, a gene required to mediate PrE-promoting Fgf4-based signalling in the ICM[20,21]. We also observed the same trend of attenuated PrE-related gene mRNA expression in clonally TE-inhibited/*Tead4*-KD late blastocysts (E4.5), derived by microinjection of a single blastomere at the 2-cell (E1.5) stage (Supplementary Fig. S10). Given the centrally important role of Fgf4 signalling within maturing ICMs[20,21], we decided to assay Fgfr2 protein expression after clonal TE-inhibition (Fig. 6c—note microinjected clones were marked with Oregon-green-conjugated dextran beads/OGDBs rather than RDBs). In microinjection control embryos, we observed Fgfr2 in regions demarking approximate boundaries between TE and ICM cells, characteristic of plasma membrane association (the microscopic resolution used prevented exact cellular origin to be determined). Similarly, we also detected Fgfr2 expression at regions between some fellow ICM cells but notably not all potential ICM-to-ICM cell contact regions. We interpret these data to indicate heterogeneous inter-cell, membrane associated, Fgfr2 protein expression in the ICMs of control embryos; an expression pattern consistent with a concurrent and similar heterogeneous and reciprocal inter-cell 'salt and pepper' expression pattern for EPI (Nanog) and early PrE (Gata6) markers, already long described[37]. Importantly, in *Tead4*-KD embryos we did not detect Fgfr2 protein at any ICM-to-ICM cell contact region between cells of the microinjected clone. Moreover, such inner-cells also exhibited prominent nuclear localised Fgfr2. Interestingly, nuclear localised Fgfr2 was seen in some, but not all, inner-cells of control embryos, with the suggestion that nuclear-enriched cells also lack plasma membrane associated Fgfr2, akin to TE-inhibited clones (Fig. 6c). It is possible that nuclear sequestration, be it in some ICM cells of control embryos or within the TE-inhibited ICM clones of *Tead4*-KD embryos, blunts their capacity to respond to Fgf4-based signals at the plasma membrane and thus differentiate towards PrE fate.

In summary, the reported attenuated PrE formation of TE-inhibited ICM clones is not associated with increased expression of the EPI marker *Nanog*, nor with reduced apical-basolateral polarity. Inhibition of TE cell-fate, does however correlate with reduced transcript levels for PrE marker genes and down-regulation of Fgfr2 protein from the plasma membrane of ICM cells.

DISCUSSION

We created variously sized cell clones within developing preimplantation mouse embryos that were unable to initiate TE-differentiation and assayed their ICM cell-fate potential to see if such clones would be predisposed towards EPI rather than PrE, in a similar manner to that described for ICM founders generated during the fourth cleavage22. In each of our experiments we found this to be true, as ICM cells derived from TE-inhibited clones preferentially contributed EPI versus PrE cells. These data indicate that level of TE-differentiation exposure ancestral cells of ICM progenitors are exposed to, is functionally relevant, although certainly not deterministic, in regard to guiding their ultimate fate. We conclude that our data are consistent with the integrated cell-fate model, whereby differing levels of exposure to TE-differentiative cues, provided by prolonged inhibition of hippo-signalling in outer-cells that yield ICM founders at the fifth cleavage division rather than those that do at the fourth, biases ICM cells to preferentially occupy either EPI or PrE.

A question arising from our clonal studies is: Are ICM progenitors generated as a result of the fourth or fifth cleavage, or more precisely from cells with differing exposures to TE-differentiation, molecularly equivalent? Others report, the first population of derived inner-cells are characterised by higher *Fgf4* and lower *Fgfr2* expression, whilst the second population exhibit relatively elevated *Fgfr2* expression26·31, consistent with our observations of reduced Fgfr2 mRNA levels after*Tead4*-KD (Fig. 6a); these data support the hypothesis that prolonged exposure to TE-differentiation, primes cells to yield ICM progeny biased towards PrE, by becoming more responsive to Fgf421. Moreover, still others report that the *Fgfr2* gene itself, plus the *Lrp2* and *Dab2*genes, are occupied by promoter proximal Tead4 protein in TS cells40. We observed a lack of detectable Fgfr2 protein expression between adjacent ICM cells within the *Tead4*-KD clone (plus increased nuclear localisation), suggesting that TE inhibited clones are less able to differentiate to PrE because of reduced ability to receive Fgf4-based membrane signals. Interestingly, a similarly heterogeneous anti-Fgfr2 immuno-staining pattern has previously been reported in mouse blastocysts, although the significance of this observation was not discussed41. The reason for enhanced Fgfr2 nuclear localisation is unclear, although not unprecedented42·43 and is observed heterogeneously in control embryo ICM cells populations (Fig. 6c).

A recent genetic study demonstrated that restricted inner-cell expression of the pluripotency-related transcription factor *Sox2* is also regulated by hippo-signalling in a *Tead4*-dependant manner44. This report is consistent with the integrated cell-fate model as it highlights the importance of the relatively early and late removal of cells from TE-differentiative environments. The reported data also suggest that the EPI bias of TE-inhibited clones (Figs 3 and 4 and Supplementary Figs S2–S4, S6 and S7) maybe due to increased *Sox2* expression. A previous study investigating ICM lineage contribution of varying ratios of marked inner-cells, in chimeric 16-cell stage embryos, demonstrates that the initial number of ICM founders has profound effects for cell-fate. Chimeras generated with fewer (*e.g.* 3) inner-cells exhibited EPI bias within these ICM founders, reminiscent of previous reports of non-manipulated *in vitro* cultured time-lapse observed embryos22, in addition to PrE bias of unmarked cells internalised following the fifth cleavage. However, generating chimeras with more initial inner-cells (*e.g.* 4 or 5) ablated the biases26. The authors speculated elevated extra-cellular Fgf4 levels in chimeras with more ICM founders, causes cells that may have otherwise contributed to EPI, to differentiate to PrE; a hypothesis also supported by mathematical modelling of ICM cell-fate45. In our data (Figs 3 and 4 and Supplementary Figs S2–S4, S6 and S7), we observed the greatest percentage PrE contribution of non-TE-inhibited ICM-resident clones when the TE-inhibited/*Tead4*-KD clone in the ICM was largest, and *vice-versa*. We interpret these results as indirect evidence that larger TE-inhibited clones may generate greater levels of extra-cellular Fgf4 within the ICM than smaller clones. As such, our data support the relevance of the number of generated ICM cells, and not just their developmental origin, as being important in biasing/directing lineage segregation26·45. In all our experimental strategies, we observed PrE contribution of cells derived from TE-inhibited clones, indicating that removal from TE-differentiation does not 'block' PrE differentiation. It is possible that elevated Fgf4 levels in ICMs containing TE-inhibited clones are sufficient to act upon cells within the clone itself. Alternatively, stochastic mechanisms could also be involved, whereby inherent self-reinforcing fluctuations in gene expression would predispose TE-inhibited ICM cells towards a profile favourable for PrE differentiation. Indeed, stochastic mechanisms are promoted as a means by which ICM lineage segregation normally occurs46. We would argue that our data indicate significant biases in ICM cell-fate derivation based on cell history but do not dismiss the existence, input nor importance of stochastic mechanisms. Indeed, early mammalian development is highly regulative, indicating, by definition, the existence of multiple complementary and overlapping pathways that must work together to guide, rather than restrict, the appropriate cell-fate decisions required47.

Theoretically our results are explainable by a direct and unprecedented PrE promoting role for *Tead4* within the ICM, potentially acting via inactivation of the hippo-signalling pathway. However, existing ES cell evidence argue against this hypothesis. Firstly, as *Tead4*$^{-/-}$ ES cell differentiation leads to the expression of multiple lineage marker genes, including the PrE markers *Gata6* and *Sox17*, the evidence suggests that*Tead4* is dispensable for PrE formation *per se* (agreeing with our clonal data)9. Moreover, when we performed double immuno-fluorescent staining of, control microinjected and *Tead4*-KD (microinjecting one blastomere at the 2-cell stage) or *in vitro* cultured control late blastocysts (E4.5) for Yap1 and Gata4 (Supplementary Fig. S11), we did not observe any nuclear co-localisation of the two factors that would have been indicative of activated hippo-signalling in specified PrE cells. Nor, did we observe any indication of nuclear Yap1 in any inner/ICM cells at earlier developmental points prior to the late blastocyst (E4.5) stage (Fig. 5 and Supplementary Fig. S8). We therefore consider it unlikely that Tead4 has a direct PrE promoting role within the ICM. Furthermore, given that it is even possible to derive *Tead4*$^{-/-}$ ES cells, a direct requirement of *Tead4* to maintain the pluripotent EPI lineage is also highly improbable9. However, it is important to note that ES cells may not share the same degree of functional similarity with cells of the early embryo48. Overall, the collective evidence points against any direct cell-fate role of *Tead4* within the ICM. Consequently, our clonal*Tead4*-KD data are best explained by the inability of the ancestors of ICM progenitors to initiate/respond to

TE-differentiative cues, rather than a lack of functionally relevant Tead4 protein within individual ICM cells.

Our observations of pERM and Yap1 immuno-staining (Supplementary Fig. S8) are completely consistent with an earlier report detailing the existence of apolar outer-cells with cytoplasmically localised Yap1 protein, indicative of active hippo-signalling6. However, we did observe unexpected nuclear pYap1 in outer-residing, mid-16-cell stage blastomeres (Supplementary Fig. S8a,b). This result is unintuitive, as phosphorylation of Yap1 is normally associated with its cytoplasmic sequestration in response to active hippo-signalling, and neither the reason nor its functional significance are clear. Possibly, the phospho-serine-127 epitope recognised by the antibody is not a sufficient/determining factor governing nuclear exclusion at this stage; for instance Yap1 has five known hippo-signalling directed (by Lats) phosphorylatable serine containing consensus motifs49.

We conclude our data are compatible with the integrated cell-fate model of early lineage segregation, whereby TE, EPI and PrE fates all begin their segregation at the fourth cleavage division and are guided by relative differences in the exposure of the founding cells to TE-differentiative cues and the timely activation (or liberation from suppression) of the hippo-signalling pathway. However, and in keeping with the remarkable regulative capacity of the early embryo, such mechanisms act to guide rather than dictate cell-fate and can be modulated by factors such as generated inner-cell number or the induction of self-reinforcing stochastic gene expression.

METHODS

All described experimental procedures on mice and mouse embryos were approved, ratified and licensed by the local ethics committee of the Biology Centre (in České Budějovice) of the Czech Academy of Sciences and by the responsible committee of the Czech Academy of Sciences on the national level, in accordance with Czech and European Union law. Accordingly, the described experiments were implemented in full accordance with the approved guidelines provided by the two above stated committees.

Embryo Culture

Embryos were recovered into M2 medium containing 4mg/ml BSA (M2+BSA) from oviducts of 10 week F1 hybrid (C57Bl6 × CBA/W) female mice (mated with F1 males) after superovulation with 7.5 IU PMSG (Sigma) and 7.5 IU hCG (Sigma; administered 48 hours later). 2- and 4-cell stage embryos were recovered from dissected oviducts 44 and 53 hours post-hCG. Unless microinjected immediately, recovered embryos were washed through KSOM drops (EmbryoMax, Millipore) and cultured under mineral oil in 5% CO_2 at 37 °C (15 embryos/20 µl drop).

Tead4-dsRNA Generation and Microinjection

A *Tead4* specific PCR primer pair, incorporating T_7-derived RNApol promoters (sen se;TAATACGACTCACTATAGGGTGTTGGAGTTCTCGGCTTTC, antisense;TAATACGACTCACTATAGGGTCGGTAGATGTGGTGCTGAG, T_7-promoter underlined) was used to derive *in vitro* transcription template (from HMI ES cell cDNA) for Tead4 mRNA specific long double-stranded RNA synthesis (Tead4-dsRNA; MEGAscript T_7; Ambion), according to manufacturers protocol. Tead4-dsRNA uniquely and specifically targets exons 7 to 11 of transcript ENST00000006311. GFP mRNA specific double stranded RNA (GFP-dsRNA) was similarly produced using specific T7-promoter linked primers (sense;TAATACGACTCACTATAGGGAGAGTACAAATTTTCTGTCAGTG GAGAGG, antisense;TAATACGACTCACTATAGGGAGATGTATAGTTCATCCAT GCCATGTGTA) using pRN3P:EGFP22 plasmid DNA as initial template. The integrity of both Tead4-dsRNA and GFP-dsRNA was confirmed on non-denaturing agarose gels.

Single blastomere microinjections were performed as previously described50 on 2- or 4-cell stage embryos in suspended M2+BSA media drops using IX71 inverted-microscope (Olympus), micromanipulators (Leica) and FemtoJet microinjection system (Eppendorf). Tead4-dsRNA (100 ng/μl) was co-microinjected with rhodamine-/Oregon-green-conjugated dextran beads (RDBs/OGDBs; 2 μg/μl and 1 μg/μl), with controls comprising RDBs/OGDBs microinjection alone or RDBs plus GFP-dsRNA (100 ng/μl). Non-microinjected embryos (1–3 per experiment) served as sentinels for appropriate *in vitro* development.

Embryo Chimeras

Fluorescent microinjected (both blastomeres at 2-cell/E1.5 stage) donor non-compacted 8-cell stage (E2.5), *zona*-less (acid tyrodes treatment), embryos were transferred into Ca^{2+}/Mg^{2+} free M2 media and disaggregated into single blastomeres. Per chimera, single fluorescently-labelled control or *Tead4*-KD blastomeres were placed in contact with non-compacted *zona*-less 8-cell stage (E2.5) embryos in PHA (phytohaemagglutin—300 μg/mL; Sigma) containing M2 media and incubated for 10 minutes (37 °C and 5% CO_2). Following confirmation of aggregation, chimeras were returned to conventional KSOM culture until the late blastocyst (E4.5) stage. All manipulations were performed in pre-warmed (37 °C) media under mineral oil on heated stereo-dissecting microscope stages.

Immuno-Fluorescent Staining

Embryos were fixed in 4% PFA (20 minutes, 37 °C) and prepared for confocal-based immuno-fluorescence microscopy as follows (at room temperature in 150 μl, unless stated); i) three 5 minute PBS washes, ii) 20 minutes in 0.5% (in PBS) Triton X-100 (Fgfr2 immuno-staining 0.1% Triton X-100), iii) three 10 minute PBS-Tween 20 (0.15%; PBS-T) washes, iv) one 10 minute NH_4Cl (50 mM in PBS) wash, v) one 4 hour 3% BSA (in PBS-T; BSA-PBS-T) blocking-step (4 °C), vi) overnight primary antibody (diluted in BSA-PBS-T) incubation (4 °C), vii) three 10 minute PBS-T washes, viii) 4 hour BSA-PBS-T blocking-step (4 °C), ix) 1 hour fluorescently-labelled secondary antibody incubation (diluted in BSA-PBS-T, 4 °C), x) three 10 minute PBS-T washes, xi) 30 minutes PBS wash, and xii) mounting (Vectashield plus DAPI, Vector Labs) on glass-bottomed, poly-L-lysine coated culture dishes. For phospho-Yap1 (pYap1), some embryos were pre-treated with 1000 units λ-phosphatase (sc-200312, Santa Cruz Biotech) according to manufacturers protocol. Primary and secondary antibody details and dilutions are given inSupplementary Table ST14. Note, that for embryos double immuno-stained for Nanog and Gata4, the rat monoclonal version of the anti-Nanog antibody was used. Embryos were imaged on IX81 upright or Fluoview Fv10i confocal microscopes (Olympus).

Image Analysis/Cell Counting

Individual E4.5 stage blastocyst cell contributions within and outwith the microinjected clone, or within inner or outer-cell populations, or within cell lineage marker protein positive or negative cells (after specific immuno-staining), plus the incidence of cells with fragmented/apoptotic nuclei, were determined in both experimental and control embryos by inspection of confocal z-sections using Fluoview ver.1.7.a (Olympus) and Imaris 6.2.1 (BitPlane) software. The above contribution criteria in *in vitro* culture control embryo groups (starting from 2-cell/E1.5 stage) was first established (Supplementary Fig. S12 and Supplementary Table ST13) to provide a baseline for the appropriate development of individual control and experimental embryos to be screened, removing those with generally poor development due to manipulation (particularly in chimeras). Accordingly, embryos with fewer than 64, or less than four Gata4 positive cells (minimum we observed in the reference data) were excluded from the analysis.

Statistical Analysis

The mean of cells within defined embryonic populations and the standard error of means (mean$\pm$s.e.m.) were calculated and statistical significance determined using 2-tailed Student's t-tests.

Quantitative RTPCR (Q-RTPCR)

Total RNA was prepared from ~30 cultured 16- (E3.1) or 32-cell stage (E3.6) *Tead4*-KD or control embryos previously microinjected at the 2-cell stages (E1.5, in either one or both blastomeres) as instructed (Arcturus Biosciences; 'PicoPure RNA isolation'). Eluted RNA (10 µl) was DNaseI treated (Ambion; 'DNA-*free*' kit) and used to derive cDNA (30 µl) using oligodT priming (Invitrogen; 'SuperscriptIII Reverse Transcriptase'). 0.5 µl of diluted cDNA (1:3, nuclease-free water) was used as template in 10 µl real-time PCR reactions (Qiagen: 'SYBR Green PCR kit') to assay specific transcripts (BioRad, 'CFX96 Real-Time System')—see Supplementary Table ST15 for oligonucleotide primer sequences (final reaction conc. 400 nM). Transcript levels were internally normalised against Rpl23 (60S ribosomal subunit component) and/or H2afz levels, and fold changes (plus s.e.m.) derived using the $\Delta\Delta$Ct method51. A minimum of 2 biological replicates of at least three technical replicates were employed.

ACKNOWLEDGEMENTS

Research in our laboratory is supported by a Marie Curie Career Integration Fellowship (IDNOVCELFAT2011) within the 7th European Community Framework Programme (awarded to A.W.B.) and a grant from the Czech Science Foundation (13-03295S). A.I.M. is the recipient of a one-year *Ph.D.* student individual project grant funded by the University of South Bohemia (004/2015/P). We acknowledge Marta Gajewska

(Department of Genetics, Cancer Center & Institute of Oncology, Warsaw, Poland) for providing founder CBA/W mice used. We also acknowledge Alena Krejčí and members of her group, at the Faculty of Sciences, University of South Bohemia, for useful discussions and advice and for pooling laboratory resources.

REFERENCES

1. Rossant, J. & Tam, P. P. Blastocyst lineage formation, early embryonic asymmetries and axis patterning in the mouse.*Development* 136, 701–713, 10.1242/dev.017178 (2009).
2. Zernicka-Goetz, M., Morris, S. A. & Bruce, A. W. Making a firm decision: multifaceted regulation of cell fate in the early mouse embryo. *Nat Rev Genet* 10, 467–477 (2009).
3. Schrode, N. *et al.* Anatomy of a blastocyst: Cell behaviors driving cell fate choice and morphogenesis in the early mouse embryo.*Genesis*, 10.1002/dvg.22368 (2013).
4. Johnson, M. H. & Ziomek, C. A. The foundation of two distinct cell lineages within the mouse morula. *Cell* 24, 71–80 (1981).
5. Sutherland, A. E., Speed, T. P. & Calarco, P. G. Inner cell allocation in the mouse morula: the role of oriented division during fourth cleavage. *Dev Biol* 137, 13–25 (1990).
6. Anani, S., Bhat, S., Honma-Yamanaka, N., Krawchuk, D. &Yamanaka, Y. Initiation of Hippo signaling is linked to polarity rather than to cell position in the pre-implantation mouse embryo. *Development* 141, 2813–2824, 10.1242/dev.107276 (2014).
7. Johnson, M. H. From mouse egg to mouse embryo: polarities, axes, and tissues. *Annu Rev Cell Dev Biol* 25, 483–512 (2009).
8. Yagi, R. *et al.* Transcription factor TEAD4 specifies the trophectoderm lineage at the beginning of mammalian development. *Development* 134, 3827–3836 (2007).
9. Nishioka, N. *et al.* Tead4 is required for specification of trophectoderm in pre-implantation mouse embryos. *Mech Dev*125, 270–283 (2008).
10. Strumpf, D. *et al.* Cdx2 is required for correct cell fate specification and differentiation of trophectoderm in the mouse blastocyst. *Development* 132, 2093–2102 (2005).
11. Mitsui, K. *et al.* The homeoprotein Nanog is required for maintenance of pluripotency in mouse epiblast and ES cells. *Cell*113, 631–642 (2003).
12. Morrisey, E. E. *et al.* GATA6 regulates HNF4 and is required for differentiation of visceral endoderm in the mouse embryo. *Genes Dev* 12, 3579–3590 (1998).
13. Koutsourakis, M., Langeveld, A., Patient, R., Beddington, R. & Grosveld, F. The transcription factor GATA6 is essential for early extraembryonic development. *Development* 126, 723–732 (1999).
14. Cai, K. Q., Capo-Chichi, C. D., Rula, M. E., Yang, D. H. & Xu, X. X.Dynamic GATA6 expression in primitive endoderm formation and maturation in early

mouse embryogenesis. *Dev Dyn* 237, 2820–2829, 10.1002/dvdy.21703 (2008).

15. Niakan, K. K. *et al.* Sox17 promotes differentiation in mouse embryonic stem cells by directly regulating extraembryonic gene expression and indirectly antagonizing self-renewal. *Genes Dev*24, 312–326, 10.1101/gad.1833510 (2010).
16. Nishioka, N. *et al.* The Hippo signaling pathway components Lats and Yap pattern Tead4 activity to distinguish mouse trophectoderm from inner cell mass. *Dev Cell* 16, 398–410, 10.1016/j.devcel.2009.02.003 (2009).
17. Hirate, Y. *et al.* Polarity-dependent distribution of angiomotin localizes Hippo signaling in preimplantation embryos. *Curr Biol*23, 1181–1194, 10.1016/j.cub.2013.05.014 (2013).
18. Cockburn, K., Biechele, S., Garner, J. & Rossant, J. The Hippo pathway member Nf2 is required for inner cell mass specification.*Curr Biol* 23, 1195–1201, 10.1016/j.cub.2013.05.044 (2013).
19. Leung, C. Y. & Zernicka-Goetz, M. Angiomotin prevents pluripotent lineage differentiation in mouse embryos via Hippo pathway-dependent and -independent mechanisms. *Nat Commun*4, 2251, 10.1038/ncomms3251 (2013).
20. Nichols, J., Silva, J., Roode, M. & Smith, A. Suppression of Erk signalling promotes ground state pluripotency in the mouse embryo. *Development* 136, 3215–3222 (2009).
21. Frankenberg, S. *et al.* Primitive endoderm differentiates via a three-step mechanism involving Nanog and RTK signaling. *Dev Cell* 21, 1005–1013, 10.1016/j.devcel.2011.10.019 (2011).
22. Morris, S. A. *et al.* Origin and formation of the first two distinct cell types of the inner cell mass in the mouse embryo. *Proc Natl Acad Sci USA* 107, 6364–6369, 10.1073/pnas.0915063107 (2010).
23. Morris, S. A. & Zernicka-Goetz, M. Formation of distinct cell types in the mouse blastocyst. *Results and problems in cell differentiation* 55, 203–217, 10.1007/978-3-642-30406-4_11 (2012).
24. Yamanaka, Y., Lanner, F. & Rossant, J. FGF signal-dependent segregation of primitive endoderm and epiblast in the mouse blastocyst. *Development* 137, 715–724, 10.1242/dev.043471 (2010).
25. Morris, S. A. Cell fate in the early mouse embryo: sorting out the influence of developmental history on lineage choice. *Reprod Biomed Online* 22, 521–524, 10.1016/j.rbmo.2011.02.009 (2011).
26. Krupa, M. *et al.* Allocation of inner cells to epiblast vs primitive endoderm in the mouse embryo is biased but not determined by the round of asymmetric divisions (8–>16- and 16–>32-cells). *Dev Biol* 385, 136–148, 10.1016/j.ydbio.2013.09.008 (2014).
27. Ziomek, C. A. & Johnson, M. H. Cell surface interaction induces polarization of mouse 8-cell blastomeres at compaction. *Cell* 21, 935–942 (1980).
28. Plusa, B. *et al.* Downregulation of Par3 and aPKC function directs cells towards the ICM in the preimplantation mouse embryo. *J Cell Sci* 118, 505–515 (2005).

29. Skamagki, M., Wicher, K. B., Jedrusik, A., Ganguly, S. & Zernicka-Goetz, M. Asymmetric localization of Cdx2 mRNA during the first cell-fate decision in early mouse development. *Cell Rep* 3, 442–457, 10.1016/j.celrep.2013.01.006 (2013).
30. Bruce, A. W. & Zernicka-Goetz, M. Developmental control of the early mammalian embryo: competition among heterogeneous cells that biases cell fate. *Curr Opin Genet Dev* 20, 485–491, 0.1016/j.gde.2010.05.006 (2010).
31. Morris, S. A., Graham, S. J., Jedrusik, A. & Zernicka-Goetz, M. The differential response to Fgf signalling in cells internalized at different times influences lineage segregation in preimplantation mouse embryos. *Open Biol* 3, 130104, 10.1098/rsob.130104 (2013).
32. Artus, J., Babinet, C. & Cohen-Tannoudji, M. The cell cycle of early mammalian embryos: lessons from genetic mouse models. *Cell cycle* 5, 499–502 (2006).
33. Home, P. *et al.* GATA3 is selectively expressed in the trophectoderm of peri-implantation embryo and directly regulates Cdx2 gene expression. *J Biol Chem* 284, 28729–28737 (2009).
34. Ralston, A. *et al.* Gata3 regulates trophoblast development downstream of Tead4 and in parallel to Cdx2. *Development* 137, 395–403, 10.1242/dev.038828 (2010).
35. Plusa, B., Piliszek, A., Frankenberg, S., Artus, J. & Hadjantonakis, A. K. Distinct sequential cell behaviours direct primitive endoderm formation in the mouse blastocyst. *Development* 135, 3081–3091 (2008).
36. Gardner, R. L. & Rossant, J. Investigation of the fate of 4-5 day post-coitum mouse inner cell mass cells by blastocyst injection. *J Embryol Exp Morphol* 52, 141–152 (1979).
37. Chazaud, C., Yamanaka, Y., Pawson, T. & Rossant, J. Early lineage segregation between epiblast and primitive endoderm in mouse blastocysts through the Grb2-MAPK pathway. *Dev Cell* 10, 615–624 (2006).
38. Gerbe, F., Cox, B., Rossant, J. & Chazaud, C. Dynamic expression of Lrp2 pathway members reveals progressive epithelial differentiation of primitive endoderm in mouse blastocyst. *Dev Biol* 313, 594–602, 10.1016/j.ydbio.2007.10.048 (2008).
39. Yang, D. H., Cai, K. Q., Roland, I. H., Smith, E. R. & Xu, X. X.Disabled-2 is an epithelial surface positioning gene. *J Biol Chem*282, 13114–13122, 10.1074/jbc.M611356200 (2007).
40. Home, P. *et al.* Altered subcellular localization of transcription factor TEAD4 regulates first mammalian cell lineage commitment.*Proc Natl Acad Sci USA* 109, 7362–7367, 10.1073/pnas.1201595109 (2012).
41. Salas-Vidal, E. & Lomeli, H. Imaging filopodia dynamics in the mouse blastocyst. *Dev Biol* 265, 75–89 (2004).
42. Bagheri-Fam, S. *et al.* Loss of Fgfr2 leads to partial XY sex reversal. *Dev Biol* 314, 71–83, 10.1016/j.ydbio.2007.11.010 (2008).

43. Martin, A. J. *et al.* FGFR2 protein expression in breast cancer: nuclear localisation and correlation with patient genotype. *BMC Res Notes* 4, 72, 10.1186/1756-0500-4-72 (2011).

44. Wicklow, E. *et al.* HIPPO Pathway Members Restrict SOX2 to the Inner Cell Mass Where It Promotes ICM Fates in the Mouse Blastocyst. *PLoS Genet* 10, e1004618, 10.1371/journal.pgen.1004618 (2014).

45. Bessonnard, S. *et al.* Gata6, Nanog and Erk signaling control cell fate in the inner cell mass through a tristable regulatory network.*Development* 141, 3637–3648, 10.1242/dev.109678 (2014).

46. Ohnishi, Y. *et al.* Cell-to-cell expression variability followed by signal reinforcement progressively segregates early mouse lineages. *Nat Cell Biol* 16, 27–37, 10.1038/ncb2881 (2014).

47. Bruce, A. W. Generating different genetic expression patterns in the early embryo: insights from the mouse model. *Reprod Biomed Online* 27, 586–592, 10.1016/j.rbmo.2013.03.024 (2013).

48. Boroviak, T., Loos, R., Bertone, P., Smith, A. & Nichols, J. The ability of inner-cell-mass cells to self-renew as embryonic stem cells is acquired following epiblast specification. *Nat Cell Biol* 16, 516–528, 10.1038/ncb2965 (2014).

49. Zhao, B., Li, L., Tumaneng, K., Wang, C. Y. & Guan, K. L. A coordinated phosphorylation by Lats and CK1 regulates YAP stability through SCF(beta-TRCP). *Genes Dev* 24, 72–85, 10.1101/gad.1843810 (2010).

50. Zernicka-Goetz, M. *et al.* Following cell fate in the living mouse embryo. *Development* 124, 1133–1137 (1997).

51. Livak, K. J. & Schmittgen, T. D. Analysis of relative gene expression data using real-time quantitative PCR and the 2(-Delta Delta C(T)) Method. *Methods* 25, 402–408 (2001).

2
CHAPTER

CELLULAR ANALYSIS OF CLEAVAGE-STAGE CHICK EMBRYOS REVEALS HIDDEN CONSERVATION IN VERTEBRATE EARLY DEVELOPMENT

Hiroki Nagai[1], Maiko Sezaki[1], Kisa Kakiguchi[2], Yukiko Nakaya[1], Hyung Chul Lee[3], Raj Ladher[4], Tomohiro Sasanami[5], Jae Yong Han[3], Shigenobu Yonemura[2], and Guojun Sheng[1]

[1]Laboratory for Early Embryogenesis, RIKEN Center for Developmental Biology, Kobe, Hyogo 650-0047, Japan

[2]Electron Microscopy Laboratory, RIKEN Center for Developmental Biology, Kobe, Hyogo 650-0047, Japan

[3]Department of Agricultural Biotechnology, Seoul National University, Seoul 151-921, Korea

[4]Laboratory for Sensory Development, RIKEN Center for Developmental Biology, Kobe, Hyogo 650-0047, Japan

[5]Department of Applied Biological Chemistry, Faculty of Agriculture, Shizuoka University, 836 Ohya, Shizuoka 422-8529, Japan

ABSTRACT

Birds and mammals, phylogenetically close amniotes with similar post-gastrula development, exhibit little conservation in their post-fertilization cleavage patterns. Data from the mouse suggest that cellular morphogenesis and molecular signaling at the cleavage stage play important roles in lineage specification at later (blastula and gastrula) stages. Very little is known, however, about cleavage-stage chick embryos, owing to their poor accessibility. This period of chick development takes place before egg-laying and encompasses several fundamental processes of avian embryology, including zygotic gene activation (ZGA) and blastoderm cell-layer increase. We have carried out morphological

Citation: Hiroki Nagai, Maiko Sezaki, Kisa Kakiguchi, Yukiko Nakaya, Hyung Chul Lee, Raj Ladher, Tomohiro Sasanami, Jae Yong Han, Shigenobu Yonemura, Guojun Sheng, Cellular analysis of cleavage-stage chick embryos reveals hidden conservation in vertebrate early development, doi: 10.1242/dev.118604.

and cellular analyses of cleavage-stage chick embryos covering the first half of pre-ovipositional development, from Eyal-Giladi and Kochav stage (EGK-) I to EGK-V. Scanning electron microscopy revealed remarkable subcellular details of blastomere cellularization and subgerminal cavity formation. Phosphorylated RNA polymerase II immunostaining showed that ZGA in the chick starts at early EGK-III during the 7th to 8th nuclear division cycle, comparable with the time reported for other yolk-rich vertebrates (e.g. zebrafish and Xenopus). The increase in the number of cell layers after EGK-III is not a direct consequence of oriented cell division. Finally, we present evidence that, as in the zebrafish embryo, a yolk syncytial layer is formed in the avian embryo after EGK-V. Our data suggest that several fundamental features of cleavage-stage development in birds resemble those in yolk-rich anamniote species, revealing conservation in vertebrate early development. Whether this conservation lends morphogenetic support to the anamniote-to-amniote transition in evolution or reflects developmental plasticity in convergent evolution awaits further investigation.

INTRODUCTION

A mature avian oocyte is fertilized in the infundibulum shortly after ovulation. From fertilization to egg-laying, the period of pre-ovipositional development covers ~25 h in the chick, during which a number of important early embryological events take place. These events include the meroblastic cleavage, cellularization, zygotic gene activation, blastoderm layer increase and reduction, radial symmetry breaking, and segregation of epiblast and hypoblast lineages (Sheng, 2014). Owing to poor accessibility, no comprehensive molecular or cellular study on pre-ovipositional embryos has been reported in the literature. General morphological characterization in the 1970s by Eyal-Giladi and Kochav provided us with a basic staging system, the EGK stages (Eyal-Giladi, 1991; Eyal-Giladi and Kochav, 1976; Kochav et al., 1980), and the ultrastructure of embryos from the later half of this developmental period was investigated in the 1990s with electron microscopy by Etches and colleagues (Watt et al., 1993).

According to the EGK staging system, pre-streak chick development is divided into 14 sub-stages, from EGK-I to EGK-XIV. The pre-ovipositional period includes the stages EGK-I to EGK-X. These 10 stages are further grouped into two developmental phases. In the first, from EGK-I to EGK-VI (the cleavage stages), a chick embryo grows mainly by means of cell proliferation (1→10,000) and blastoderm cell-layer increase (1→6). In the second, from EGK-VII to EGK-X (the area pellucida formation stages), the center of the blastoderm above the subgerminal cavity thins out from six to two cell layers, leading to the formation of two circular territories: the central area pellucida, which is detached from the yolk and contains the epiblast and hypoblast cell layers; and the peripheral area opaca, which varies in its thickness and maintains cellular interactions with the yolk. Radial symmetry breaking, one of the most important developmental events in vertebrate embryogenesis, also occurs during this phase.

With a paucity of molecular and cellular understanding, and with the exception of primordial germ cells (PGCs) (Tsunekawa et al., 2000), chick embryos during the first phase of pre-ovipositional development are generally regarded as being undifferentiated, containing equipotent cells whose morphological variations reflect their cellularization and proliferation heterogeneity rather than their fate diversification. These embryos are also considered to be inactive in their preparation for lineage segregation (except for the PGCs) or radial symmetry breaking because embryonic axis can be efficiently re-specified with physical perturbation during the second phase of pre-ovipositional development (Kochav and Eyal-Giladi, 1971). Data from other vertebrate models, however, suggest that molecular and cellular events taking place during cleavage phase of the early development are crucial for dynamic signaling, lineage specification and morphogenetic movements associated with the late blastula/early gastrula development.

Among those crucial events are the initiation of zygotic gene activation (ZGA), the separation of outside and inside blastomeres, and the formation of a yolk syncytial layer (YSL). In both mouse and zebrafish embryos, ZGA comes under direct control of key pluripotency regulators (Foygel et al., 2008;Lee et al., 2013b; Leichsenring et al., 2013) and is tightly associated with epigenetic reprogramming of parental genomes that, later in development, controls the specification of both somatic and germ cell lineages (Hackett and Surani, 2013). During mouse early development, 8→16 and 16→32 divisions produce two types of blastomeres: those located outside, which are biased towards the trophectoderm fate; and those located inside, which will become the inner cell mass (Rossant and Tam, 2009). This process is regulated by polarized cell division and post-divisional cell rearrangement (Johnson and Ziomek, 1981; Parfitt and Zernicka-Goetz, 2010; Sasaki, 2010; Watanabe et al., 2014). The YSL is a special adaptive feature of telolecithal development. Although eutherian eggs are devoid of yolk, both birds and reptiles, two other extant amniote groups, have telolecithal eggs. Studies in zebrafish, a genetic model for telolecithal embryonic development, suggest that the YSL is formed at the cleavage stage and that it undergoes highly coordinated cytokinetic movements with the overlying blastoderm cells (Carvalho and Heisenberg, 2010). It also regulates mesendoderm induction by secreting Ndr1 and Ndr2 signals (two nodal related molecules) under the transcriptional control of Mxtx2 (Hong et al., 2011), and produces the ventralizing signal BMP2b under the transcriptional control of MGA, MAX and Smad4 proteins (Sun et al., 2014). To investigate how these events occur in the chick and whether novel insights into vertebrate early development can be gained by taking a comparative approach, we collected cleavage-stage chick embryos and performed in-depth morphological and cell biological analyses.

RESULTS AND DISCUSSION

General Description of EGK-I to -V Embryos

Pre-ovipositional eggs were retrieved from laying hens using a non-invasive method described previously (Eyal-Giladi and Kochav, 1976; Lee et al., 2013a). Because the ultrastructure of late cleavage-stage embryos and the area pellucida of formation-stage embryos had been described previously (Watt et al., 1993), we focused our attention on the early and mid-cleavage stages, namely from EGK-I to EGK-V. This period covers about 8 hours of development. Together with time needed from the fertilization to the first cell division, it is considered here to represent the first half of pre-ovipositional development (~13 h out of a total of 25 h). After egg retrieval, shell and albumen were removed, and embryos were briefly fixed in 4% PFA with intact yolk for morphological preservation. They were then cut out together with a small amount of underlying yolk to retain embryo-yolk contacts, further fixed in 4% PFA and stored at 4°C for later analysis. Out of a total of 138 embryos collected, 73 were confirmed under a stereo-microscope to represent stages between EGK-I and EGK-V. Their general morphology is briefly described here. At EGK-I, all cells are connected to the yolk basally; when viewed from the apical side, they are not yet enclosed by lateral cell membrane, except for a few centrally located ones at late EGK-I. At EGK-II, two types of cells, the central ones with closed lateral membrane and peripheral ones which are open to the yolk laterally, can be easily distinguished. From EGK-III to EGK-V, the central, closed cells become progressively smaller as they divide, and the area they occupy expands as more closed cells are produced from the peripheral open cells after division. In a majority of embryos at all EGK stages, the central cell cluster is positioned eccentrically in the blastoderm. Compared with those described by Eyal-Giladi and Kochav (1976), our embryos had their morphology better preserved and, likely as a result of this, we did not notice any obvious shrinkage in blastoderm size at EGK-III or large vacuoles at EGK-I and EGK-II, as described previously (Eyal-Giladi and Kochav,1976).

Scanning Electron Microscopy of EGK-I to -V Embryos

To gain a better understanding of cellular morphogenesis during early cleavages, we performed scanning electron microscopy (SEM) analysis with EGK-I to EGK-V embryos (Fig. 1A; summarized in Fig. 1B). Under SEM, cleavage furrows and cell boundaries were easily visible. In an EGK-I embryo with eight open cells (Fig. 1A), the order of three separate cytokinetic events could be estimated based on the shape and extent of furrow progression. Fracture surface views indicated that transition from vertical furrow burrowing to horizontal furrow burrowing took place ~50-100 μm deep into the yolk The transition area had elaborate membrane protrusions, resembling the furrow based body (Bellairs et al., 1978; Gipson, 1974). The apical surface was rich in microvilli, with small vesicular bodies at the cortex. Yolk granules exhibited fine-graded distribution in their sizes. No obvious transition zone could be distinguished between yolk granules to be included in future blastoderm cells (<100 μm below the membrane) and those to be allocated to the yolk. In an EGK-II embryo with 16 laterally closed cells (Fig. 1A), the center of closed cells was located clearly off the geometric center of the blastoderm, possibly as a consequence of eccentric localization of the zygotic nucleus. It is unclear whether this eccentricity has any predictive value with regard to where the future posterior side of the epiblast (dorsal side of the embryo) will be. The shape and size of central cells was variable. Contrary to what had been suggested based on light microscopy analysis (Eyal-Giladi and Kochav, 1976), a few central cells at EGK-II had already finished the cellularization process with complete basal and lateral membranes. Numerous small, intracellular vesicular bodies were present in all cells, resembling those reported in other yolk-rich vertebrate embryos (Danilchik et al., 2003; Li et al., 2006). These vesicles are likely to be involved in the cellularization and cell division processes, which require a constant supply of new plasma membrane. It is also possible, however, that they are related to the supply of surface proteins or to the secretion of extracellular matrix molecules.

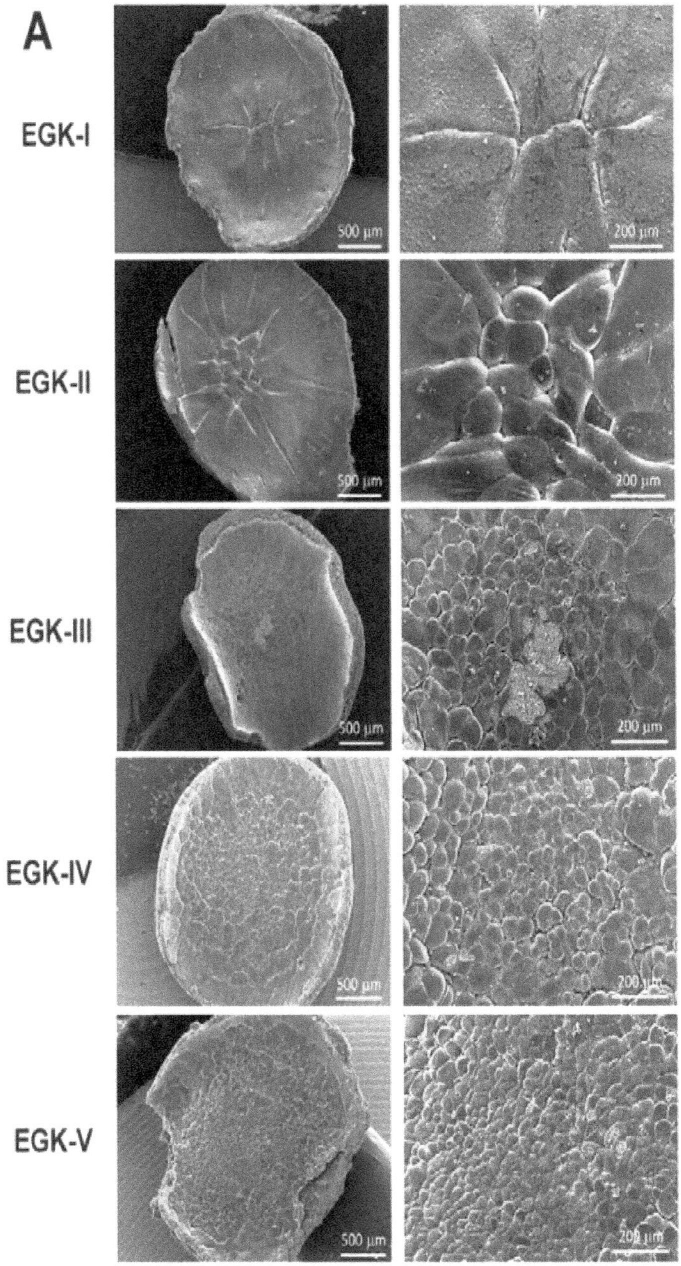

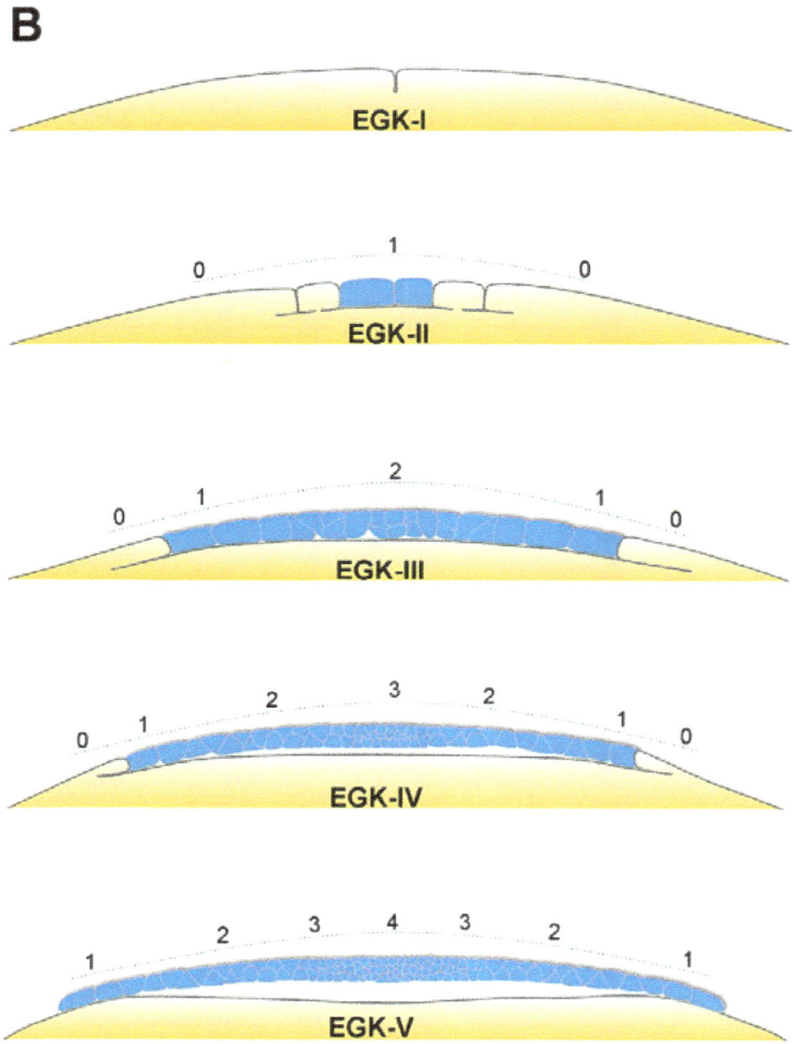

Figure 1: Scanning electron microscopy analysis of EGK-I to -V chick embryos. (A) Apical views of EGK-I to -V embryos under SEM. Scale bars: 500 μm (left); 200 μm (right). Details of the fracture surface are shown summarized in B. (B) Schematic view of cellularization and cell proliferation processes during chick development from EGK-I to -V. Completely cellularized blastomeres are shown in blue. Numbers above the blastoderm indicate the distribution of cell layer numbers.

At EGK-III (Fig. 1A), the territory of laterally closed cells from apical surface view expanded significantly as more peripheral open cells gave rise to laterally closed daughter cells after division. The apical membrane remained rich in microvilli, but with reduced abundance at cleavage furrows. A small subgerminal cavity space appeared under some central cells and two-cell thick blastoderm was observed occasionally. This trend continued at EGK-IV (Fig. 1A). At this stage, the subgerminal cavity space became obvious under most central cells and started to merge to form a bona fide cavity above the yolk cell surface. A significant proportion of the blastoderm became two cells thick, with the very central part being three cells thick. In slightly more peripheral regions, the subgerminal cavity under newly closed cells had just appeared as an isolated pocket between two closed cells and the yolk cell. At the basal surface of these cells, two closely apposed plasma membranes marked the completion of their cellularization. Interestingly, unlike the fine-graded size distribution before cellularization, yolk granules at the basal cortex of newly formed cells were much smaller than those located in the adjacent yolk cells. At EGK-V (Fig. 1A), all non-edge cells of the blastoderm had completed the cellularization process. Many edge cells had also become fully cellularized, although some of them remained open peripherally. A smooth transition in blastoderm thickness, from one cell layer at the edge to four cell layers at the center, could be observed. Cells bridging the one-cell and two-cell regions often exhibited wedge-shaped morphology. The subgerminal cavity became a continuous, expanded space below the multilayered part of the blastoderm, but the one-cell layered part remained in close association with the underlying yolk cell membrane. Cell-cell contacts among blastomeres were prominent in both superficial and deep layers.

Zygotic Gene Activation (ZGA) is Initiated at Late EGK-II/Early EGK-III

The cellular morphogenetic process from EGK-I to EGK-V is summarized in Fig. 1B. During this period of chick development, cell (nucleus) number increases from 1 to ~2000 (Park et al., 2006). ZGA in mammalian embryos (e.g. in the mouse and human) starts very early, before the 3rd cell cycle (Vassena et al., 2011; Wang and Dey, 2006; Xue et al., 2013; Yan et al., 2013). In the zebrafish, with a telolecithal egg, ZGA was reported to start at the 128-cell stage (8th cell cycle) (Aanes et al., 2011; Harvey et al., 2013; Mathavan et al., 2005). In other yolk-rich embryos, such as *Xenopus* and *Drosophila*, ZGA can be detected at about 128- to 256-cell (nucleus) stage (8th to 9th cell cycle) (Baroux et al., 2008; Tadros and Lipshitz, 2009). It has been hypothesized that instead of the absolute cell cycle number, the timing of ZGA in large, yolk-rich embryos is determined by the nucleocytoplasmic ratio and the maternal clock (Tadros and Lipshitz, 2009). We stained early EGK stage chick embryos with an anti-phosphorylated-RNA Polymerase II CTD (p-PolII) antibody to mark the zygotic gene transcription, DAPI to visualize the nuclei and E-cadherin to visualize cell boundaries. No specific p-PolII immunoreactive signals could be detected before late EGK-II stage (Fig. 2A, leftmost column panels). At late EGK-II/early EGK-III (Fig. 2A, 2nd and 3rd column panels), weak nuclear p-PolII signals were observed in a few centrally located cells. By mid-/late EGK-III stages (Fig. 2A, 4th and 5th column panels), p-PolII signals became progressively stronger and could be readily detected in many central cells. About half of p-PolII-positive cells had strong cytoplasmic signals at late EGK-III, a phenomenon possibly related to the cytoplasmic-to-nuclear translocation of RNA PolII at the onset of ZGA, as observed in the mouse embryo (Bellier et al., 1997). At EGK-IV (Fig. 2A, 6th and rightmost column panels; Fig. 2C,D), p-PolII signals became more robust and positive cells became more abundant. During the entire transition period from mid-EGK-II to late EGK-IV, p-PolII signals appeared in central, smaller cells sooner and more strongly than in peripheral, larger cells. However, it is unclear whether this is causally related to changes in the nucleocytoplasmic ratio because many central cells were weak or negative for p-PolII signals even when their similarly sized neighbors were strongly positive (Fig. 2C). Supporting our ZGA observation based on the p-PolII staining data, pre-ovipositional embryos with GFP transgene insertion under the control of a CMV promoter (Park and Han, 2012) showed no GFP signals at mid-EGK-II (Fig. 2B) and strong GFP signals at mid-EGK-IV (Fig. 2B). Taken together, our data suggest that ZGA in the chick is initiated at late EGK-II/early EGK-III (about 64-128 total cell stage; or approximately during the 7th or 8th cell cycle) and becomes readily detectable at mid- to late EGK-III (about 128-256 total cell-stage; 8th to 9th cell cycle). Chick embryos therefore initiate ZGA with a developmental timing comparable with that in the zebrafish and *Xenopus* embryos, pointing to potential conservation in molecular mechanisms controlling ZGA in yolk-rich embryos.

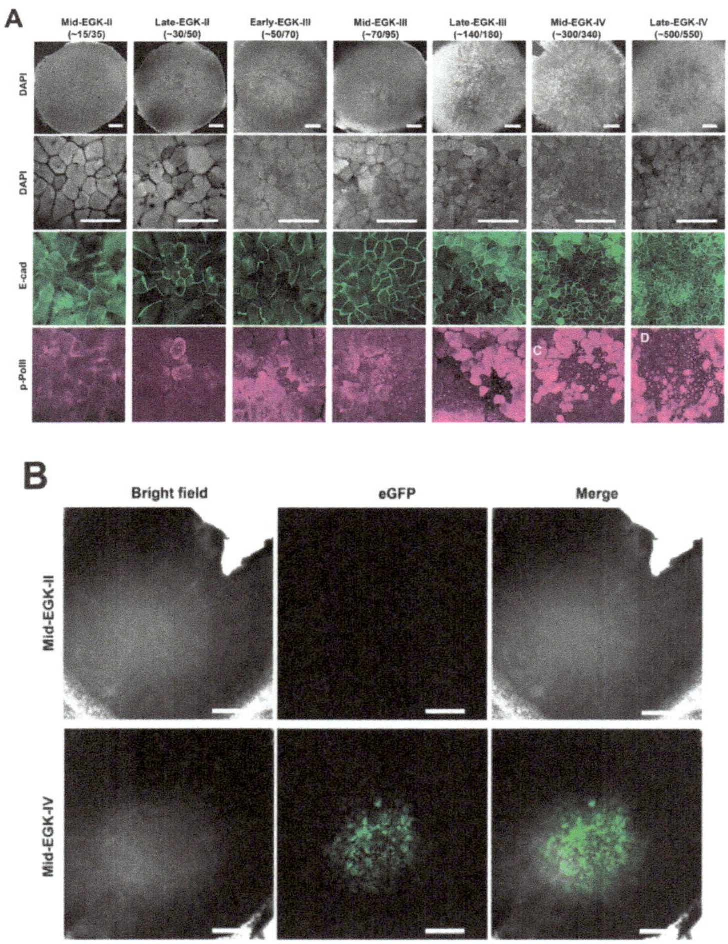

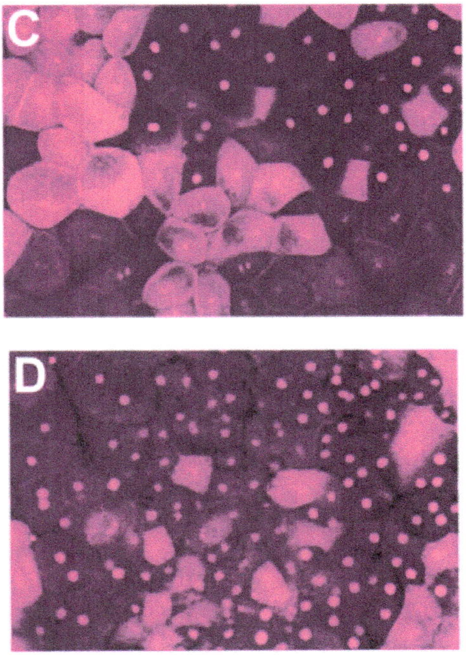

Figure 2: ZGA in EGK-II to -IV chick embryos analyzed by anti-Ser5 phosphorylation of Pol II CTD (p-PolII) antibody staining and by GFP transgenesis. (A) Wild-type embryos ranging from mid-EGK-II to late EGK-IV are stained for p-PolII signals. Embryos are co-stained for E-cadherin (E-cad; marking cell boundaries) and DAPI (marking nuclei). Numbers (x/y) under embryo stage labels indicate approximate counts of laterally closed cells (x) and total cells (y). (Top) Whole-embryo views after DAPI staining. At higher magnification: DAPI (second row), E-cad (third row) and p-PolII (bottom row) staining. Signals for p-PolII are negative at mid-EGK-II, very weak in late EGK-II, rapidly increase at EGK-III and become very strong at late EGK-IV. From late EGK-III to late EGK-IV, strong signals are detected both in the nucleus (in some central blastomeres) and in the cytoplasm (in adjacent blastomeres), although the proportion of cytoplasmic-positive cells decreases gradually (C,D). p-PolII-negative, small blastomeres are also observed in central areas next to similarly sized positive blastomeres (C). Signals in peripherally positioned blastomeres are much weaker or negative. (B) Embryos from GFP-transgenic male and wild-type female crosses show strong GFP signals at mid-EGK-IV but no GFP signals at mid-EGK-II. Scale bars: 500 μm.

Separation of Surface and Inner Cells is not caused by Oriented Cell Divisions

From our scanning EM analysis, late EGK-II/early EGK-III is also the stage when the cellularization process becomes complete in a few centrally located cells and the increase in blastoderm cell-layer number is initiated. This latter process is significant because it results in the separation of two types of blastomeres: those located on the surface and exposed to the external environment; and those located in the inside and shielded from external influences. Eyal-Giladi and colleagues hypothesized that oriented blastomere division led to this increase (Kochav et al., 1980). In eutherian embryos, this separation is directly correlated with the future fate of these cells: trophectoderm for the surface-located cells and inner cell mass for the inside ones. As mentioned in the introduction, there are two competing, but not mutually exclusive, theories to explain this process in the mouse embryo (Johnson and Ziomek, 1981; Parfitt and Zernicka-Goetz, 2010; Sasaki, 2010; Watanabe et al., 2014). To investigate whether mitotic plane orientation is correlated with the increase in cell-layer number in the chick embryo (Fig. 3A), we performed DAPI staining with EGK-III and EGK-IV embryos in which the blastoderm was increasing its thickness from 1- to 2-cells or from 2- to 3-cells (Fig. 3B). Embryos with many mitotic cells in anaphase/telophase were selected. These embryos were sectioned and mitotic plane orientation was measured as indicated inFig. 3B (embryo, n=5; mitotic nuclei pair, n=240). Data for dividing surface cells and dividing non-surface (deep) cells were plotted separately (Fig. 3C). The majority (about 3/4) of all surface cells divided with their mitotic plane oriented at less than 30° angle relative to the blastoderm surface, and about 40% of all surface cell divisions had an angle of less than 10° angle (Fig. 3C, left). Although dynamic cytokinetic behavior could not be visualized, those divisions most likely resulted in generating two surface daughter cells. Mitotic orientation of dividing deep cells was distributed more randomly (Fig. 3C, right), with 44% cells dividing at 0-30° angles and with the rest evenly represented at 30-90° angles. Collectively, these data indicate that cell layer number increase at cleavage stages in the chick embryo is not directly caused by oriented cell divisions.

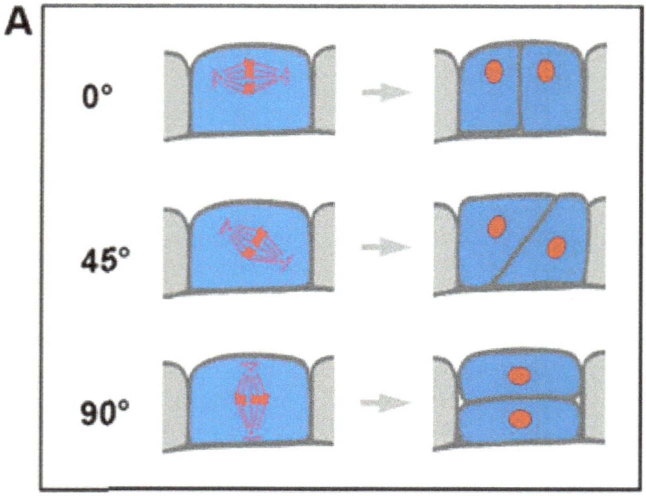

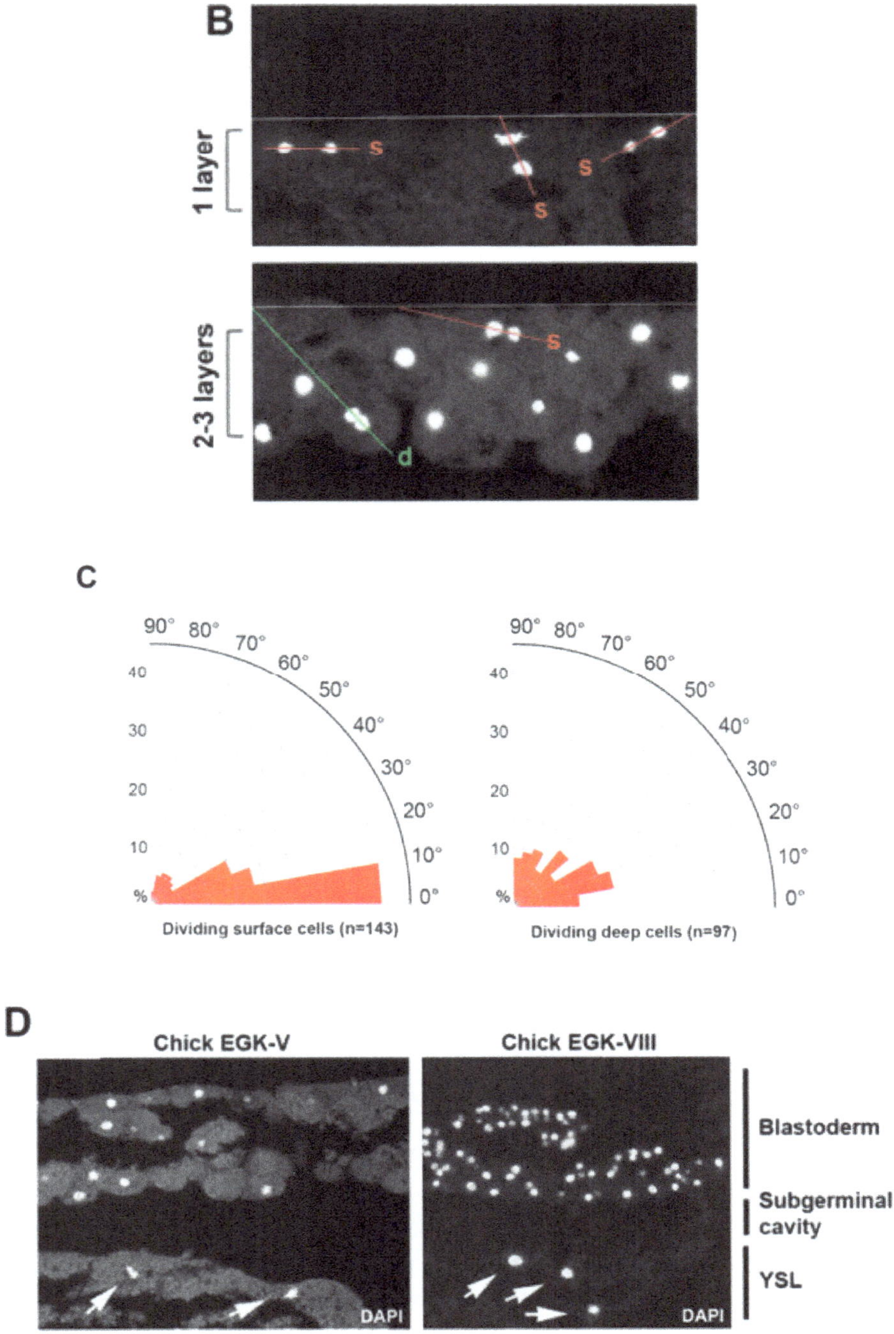
B
1 layer
2-3 layers
s
d
C
90° 80° 70° 60° 50° 40° 30° 20° 10° 0°
40
30
20
10
%
Dividing surface cells (n=143)
Dividing deep cells (n=97)
D
Chick EGK-V
Chick EGK-VIII
DAPI
Blastoderm
Subgerminal cavity
YSL

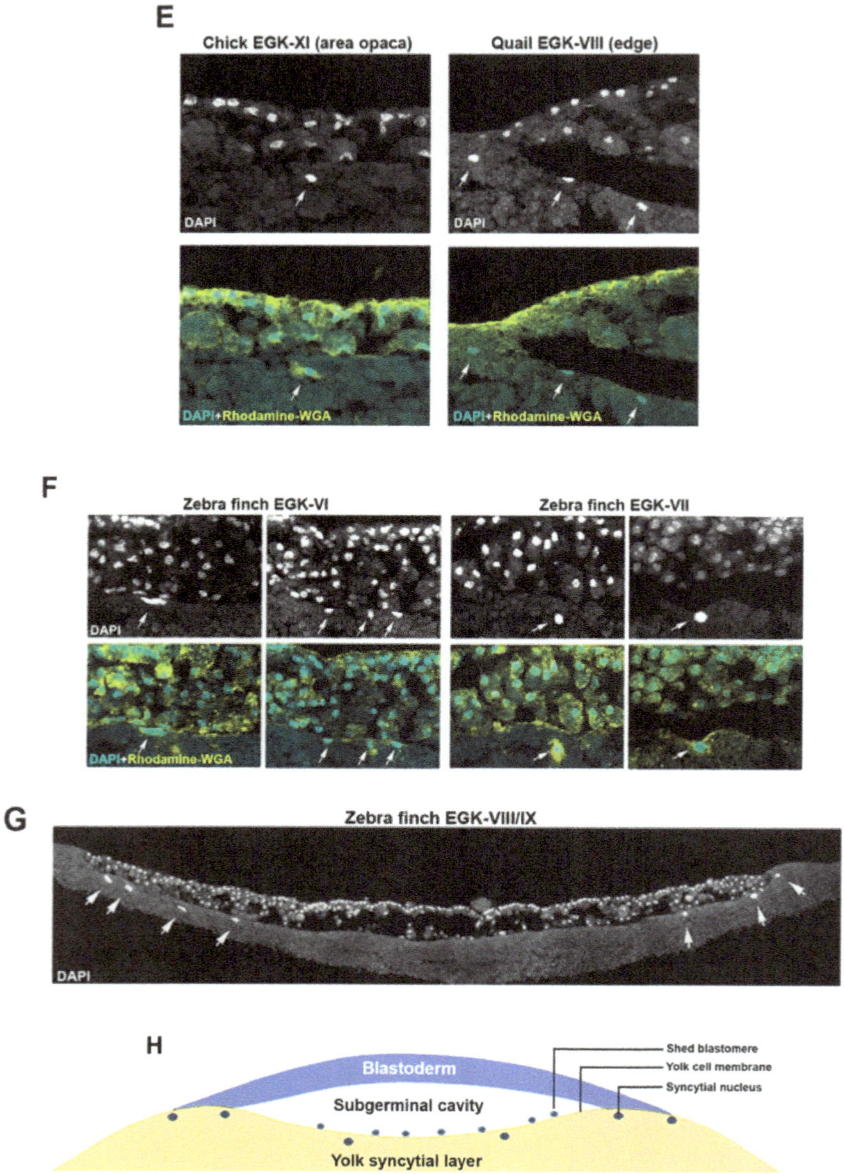

Figure 3: Mitotic division orientation and yolk syncytial nuclei. (A-C) Increase in cell layer number is not caused by oriented mitotic division. (A) Schematic view of three representative mitotic plane angles (0°, 45° and 90°). The last (90°) scenario was hypothesized as the cause for blastoderm cell-layer number increase by Eyal-Giladi and colleagues. (B) Representative section views of EGK-III to EGK-V embryos stained with DAPI to reveal mitotic cells. Top: one-cell layer region; bottom: 2- to 3-cell-layer region. Only anaphase and telophase nuclei were counted. Mitotic plane orientation was

calculated as the angle between the blastoderm surface and the line passing through two separating nuclei. Red lines: surface-located cell divisions (s). Green line: deep cell division (d). (C) Rose diagrams showing the distribution of mitotic plane orientation. Left: dividing surface cells (n=143). Right: dividing deep cells (n=97). A vast majority of surface cells divide with their mitotic planes orientated at a less than 30° angle to the surface, likely producing two surface daughter cells. Mitotic planes of non-surface cells exhibit a more randomized distribution. (D-H) Yolk syncytial nuclei are detected in three different avian species. (D) DAPI staining of EGK-V (left) and EGK-VIII (right) chick embryos. DAPI-positive structures (arrows) are detected underneath the yolk cell surface. (E) In post-ovipositional chick (EGK-XI, left panels) and pre-ovipositional quail (EGK-VIII, right panels) embryos, double-staining with DAPI (nucleus) and Rhodamine-WGA (membrane) reveals that DAPI-positive signals (arrows) are located underneath the yolk cell membrane. (F) Four examples of zebra finch embryos (two EGK-VI and two EGK-VII) showing DAPI-positive syncytial nuclei (arrows) located underneath the yolk cell membrane. (G) A composite view (assembled from four images) of a zebra finch embryo cross-section, showing an intact yolk cell membrane and many syncytial nuclei (arrows) underneath it. (H) A schematic view of the YSL in an avian embryo.

A Yolk Syncytial Layer (YSL) is formed during early Avian Development

As mentioned in the introduction, the importance of the YSL in early development has been demonstrated in the zebrafish model. However, whether a similar YSL exists in the avian embryo has not been investigated. DAPI staining and section analysis suggested that no syncytial nuclei could be found at stages EGK-I to -IV (not shown). At EGK-V, occasional syncytial nuclei were detected (Fig. 3D, left), which became brighter, larger and more abundant at later EGK stages (Fig. 3D, right, showing an EGK-VIII embryo) and persisted through post-ovipositional stages (Fig. 3E, left, showing an EGK-XI embryo). These nuclei are to be distinguished from DAPI-positive cells located above the surface of the yolk cell (e.g. as shown in Fig. 3G), which are shed from the blastoderm cell mass (as described by Eyal-Giladi and Kochav) and are frequently observed during the blastoderm thinning process from EGK-VII onwards. To confirm the existence of a YSL in the avian embryo, we performed similar analysis with quail (Fig. 3E, right) and zebra finch (Fig. 3F,G) embryos. Pre-ovipositional quail eggs were retrieved and DAPI staining of these embryos revealed the presence of syncytial nuclei, especially underneath the area opaca and close to the blastoderm margin (Fig. 3E, right). Zebra finch eggs are laid at an earlier developmental stage than chicken or quail eggs, and zebra finch embryos at stages EGK-VI to -IX can be collected easily from freshly laid eggs (S. S. Mak, C. Alev, H.N., A. Wrabel, Y. Matsuoka, A. Honda, G. Sheng and R. K. Ladher, unpublished). DAPI staining and section analysis of EGK-VI to -IX zebra finch embryos supported our chick and quail data (Fig. 3F,G). Syncytial nuclei were detected more abundantly in the zebra finch (Fig. 3G, arrows) and co-staining with a membrane marker (rhodamine-labeled WGA) revealed that these DAPI-positive signals were located underneath the yolk cell membrane (Fig. 3F). Interestingly, many peripherally located syncytial nuclei in the finch embryo could be found even in whole-mount views. There nuclei were located away from the blastoderm edge, as revealed by phalloidin co-staining and in section. Taken together, these data strongly support the hypothesis (Fig. 3H) that a YSL is present in the avian embryo. Whether this YSL plays a role in patterning the overlying blastoderm remains to be clarified.

SUMMARY

The first half of pre-ovipositional chick development, from EGK-I to EGK-V, was investigated in this study (**Table 1**). During this developmental period, a chick embryo undergoes ~11 rounds of mitotic divisions. The cell number increases from one to ~2000 and the cell-layer number increases from one to four. Meroblastic cleavages continue from EGK-I, when all cells are open, to late EGK-V, when all cells are fully cellularized. Blastomere-yolk separation starts at late EGK-II and ZGA starts at late EGK-II/early EGK-III (~7th to 8th nuclear division cycle). Separation of outside and inside blastomeres is not due to oriented cell division, as previously suggested. Syncytial nuclei located underneath the yolk cell membrane can be detected from EGK-V and become more abundant during the second half of pre-ovipositional development. The presence of yolk syncytial nuclei is also observed in quail and zebra finch embryos. Our data on blastoderm cell-layer increase support the hypothesis that this process is controlled by cellular mechanisms other than oriented mitotic division. Collectively, our data suggest that many features of cleavage-stage development in the chick resemble those in the zebrafish. Whether this conservation is a result of convergent evolution or is indicative of hardwired molecular and cellular mechanisms regulating vertebrate early development awaits future investigation.

Table 1: Summary of EGK-I to -V embryos

EGK	I	II	III	IV	V	VI	→	X
Total number of nuclei or cells (median)	1-16 (~8)	16-64 (~32)	64-256 (~128)	256-1k (~512)	1k-4k (~2k)	4k-16k (~8k)	→	~55k
Nuclear division cycle	1st-4th	4th-6th	6th-8th	8th-10th	10th-12th	12th-14th	→	15th-16th
Hours post-fertilization	5.5	7	8.5	10	13.5	15.5	→	25
Laterally closed cells (median)	0	~16	~80-90	~250-300	~All	All	→	All
Cell layer number	1	1	2	3	4	5-6	→	1 or 2
Basally closed cells	–	–/+	+	+	+	+	→	+
Zygotic gene activation	–	–/+	+	+	+	+	→	+
Yolk syncytium	–	–	–	–	–/+	+	→	+
Cellularization of edge cells	–	–	–	–	–/+	+	→	+

Avian early development from fertilization at the beginning of EGK-I to complete cellularization by the end of EGK-V is viewed using the following criteria: total number of cells or nuclei; nuclear division cycle; time post-fertilization; number of laterally closed cells; number of cell layers in central blastoderm; blastomere basal closure; zygotic gene activation; yolk syncytium and edge cell cellularization. Of practical values among them are the total number of cells/nuclei and number of laterally closed cells. However, early development is a continuous process and these criteria should be used only as a general guide.
–, negative; +, positive; –/+, weakly or partially positive.

MATERIALS AND METHODS

Embryo Collection

EGK stage chick embryos were collected from laying white leghorn hens using a non-invasive, manual retrieval (abdominal massage) method as previously described (Lee et al., 2013a). The retrieval was carried out in the afternoon or early evening and was timed to target eggs in the first half of pre-ovipositional development. Each collected egg was placed in a Petri dish. The eggshell was opened and removed, and the egg yolk was adjusted so that the embryo was positioned at the top. Excessive albumen covering the embryo was removed and a few drops of ice-cold 4% paraformaldehyde (PFA) fixative (Alev et al., 2013) were put on the vitelline membrane over the embryo to preserve the morphology. After a few minutes of fixation at room temperature, the embryo, together with a small amount of peripheral and underlying yolk, was excised from the bulk of the yolk and placed in Pannett-Compton saline solution (Alev et al., 2013). The vitelline membrane was then carefully removed and the embryo together with attached yolk was transferred to 4% PFA for further fixation and storage at 4°C. Out of 138 embryos obtained this way, 85 embryos were analyzed for stage distribution under a stereo microscope, and 73 out of 85 were within the cleavage stages according to the EGK staging system (13 EGK-I, 16 EGK-II, 15 EGK-III, 17 EGK-IV and 12 EGK-V). Pre-ovipositional GFP transgenic chick embryos and pre-ovipositional quail embryos were collected using the same abdominal massage method. Post-ovipositional zebra finch eggs were collected daily from locally maintained colonies and zebra finch embryos were processed for staining using the same protocol as for the chick embryos.

Scanning Electron Microscopy (SEM)

Samples for SEM analysis were prepared as described previously (Ikenouchi et al., 2013). Briefly, EGK stage chick embryos were fixed with 2.5% glutaraldehyde and 2% formaldehyde in 0.1 M sodium cacodylate buffer (pH 7.4) for 2 h at room temperature, followed by post-fixation with 1% OsO_4 in the same buffer for 2 h on ice. Post-fixed embryos were dehydrated in ethanol and then transferred to isoamyl acetate, followed by critical point drying (JCPD-5, JEOL). Specimen surface was coated using an osmium coater (Neoc-STB, Meiwafosis). For fracture surface images after whole-mount SEM, embryo specimen was settled on a small piece of adhesive tape, and fractured using a fine tungsten needle. All SEM images were acquired using scanning electron microscope JSM-5600LV (JEOL) with accelerating voltage set at 10 kV.

Immunofluorescence, Nuclear Staining, Imaging and Image Analysis

All steps except for the imaging were carried out at 4°C. For phosphorylated RNA polymerase II detection, fixed embryos were washed in PBTw (PBS with 0.1% Tween-20) for 3×20 min, followed by permeabilization in PBTr (PBS with 0.1% Triton X-100) for 3×30 min and by blocking in blocking solution (PBS with 1% DMSO, 0.1% Triton X-100 and 5% skimmed milk). Blocked embryos were incubated overnight with anti-RNA polymerase II CTD pSer5 (1:1000, Abcam Cat# ab5131), washed in PBTw, reblocked in blocking solution and incubated in the secondary antibody (Alexa Fluor 594 goat anti-rabbit IgG(H+L) (1:300, Molecular Probes Cat# a11012) before imaging. DAPI (Molecular Probes, D1306) and 36/E-Cadherin antibody (BD Transduction Laboratories, 610182) were used for whole-mount nuclear and E-cadherin staining, respectively. Yolk cell membrane was marked using rhodamine-labeled wheat germ agglutinin (WGA) (Vector Laboratories, Cat# RL-1022). Whole-mount fluorescence images were acquired using Olympus FV1000 microscope. Section fluorescence images were acquired using Olympus BX51W1 or Olympus FV1000 microscope after processing for 10 μm paraffin-embedded section or 12 μm cryosection followed by mounting in ProLong Gold antifade reagent with DAPI (Molecular Probes Cat# P36931). Mitotic division angles were measured using Fiji software and data were plotted as rose diagrams using GeoRose software.

ACKNOWLEDGEMENTS

We thank Anna Wrabel and Siu-Shan Mak for help with finch colony maintenance and Kanako Ota and members of the J.Y.H. laboratory for help and coordination during the collection of pre-ovipositional stage chicken embryos.

REFERENCES

1. Aanes, H., Winata, C. L., Lin, C. H., Chen, J. P., Srinivasan, K. G., Lee, S. G. P., Lim, A. Y. M., Hajan, H. S., Collas, P., Bourque, G. et al. (2011). Zebrafish mRNA sequencing deciphers novelties in transcriptome dynamics during maternal to zygotic transition. Genome Res. 21, 1328-1338.
2. Alev, C., Nakano, M., Wu, Y., Horiuchi, H. and Sheng, G. (2013). Manipulating the avian epiblast and epiblast-derived stem cells. Methods Mol. Biol. 1074, 151-173.
3. Baroux, C., Autran, D., Gillmor, C. S., Grimanelli, D. and Grossniklaus, U. (2008). The maternal to zygotic transition in animals and plants. Cold Spring Harb. Symp. Quant. Biol. 73, 89-100.
4. Bellairs, R., Lorenz, F. W. and Dunlap, T. (1978). Cleavage in the chick embryo. J. Embryol. Exp. Morphol. 43, 55-69.
5. Bellier, S., Chastant, S., Adenot, P., Vincent, M., Renard, J. P. and Bensaude, O. (1997). Nuclear translocation and carboxyl-terminal domain phosphorylation of RNA polymerase II delineate the two phases of zygotic gene activation in mammalian embryos. EMBO J. 16, 6250-6262.
6. Carvalho, L. and Heisenberg, C.-P. (2010). The yolk syncytial layer in early zebrafish development. Trends Cell Biol. 20, 586-592.
7. Danilchik, M. V., Bedrick, S. D., Brown, E. E. and Ray, K. (2003). Furrow microtubules and localized exocytosis in cleaving Xenopus laevis embryos. J. Cell Sci. 116, 273-283.
8. Eyal-Giladi, H. (1991). The early embryonic development of the chick, as an epigenetic process. Crit. Rev. Poult. Biol. 3, 143-166.
9. Eyal-Giladi, H. and Kochav, S. (1976). From cleavage to primitive streak formation: a complementary normal table and a new look at the first stages of the development of the chick. I. General morphology. Dev. Biol. 49, 321-337.
10. Foygel, K., Choi, B., Jun, S., Leong, D. E., Lee, A., Wong, C. C., Zuo, E., Eckart, M., Reijo Pera, R. A., Wong, W. H. et al. (2008). A novel and critical role for Oct4 as a regulator of the maternal-embryonic transition. PLoS ONE 3, e4109.
11. Gipson, I. (1974). Electron microscopy of early cleavage furrows in the chick blastodisc. J. Ultrastruct. Res. 49, 331-347.

12. Hackett, J. A. and Surani, M. A. (2013). Beyond DNA: programming and inheritance of parental methylomes. Cell 153, 737-739.

13. Harvey, S. A., Sealy, I., Kettleborough, R., Fenyes, F., White, R., Stemple, D. and Smith, J. C. (2013). Identification of the zebrafish maternal and paternal transcriptomes. Development 140, 2703-2710.

14. Hong, S.-K., Jang, M. K., Brown, J. L., McBride, A. A. and Feldman, B. (2011). Embryonic mesoderm and endoderm induction requires the actions of nonembryonic Nodal-related ligands and Mxtx2. Development 138, 787-795.

15. Ikenouchi, J., Hirata, M., Yonemura, S. and Umeda, M. (2013). Sphingomyelin clustering is essential for the formation of microvilli. J. Cell Sci. 126, 3585-3592.

16. Johnson, M. H. and Ziomek, C. A. (1981). The foundation of two distinct cell lineages within the mouse morula. Cell 24, 71-80.

17. Kochav, S. and Eyal-Giladi, H. (1971). Bilateral symmetry in chick embryo determination by gravity. Science 171, 1027-1029.

18. Kochav, S., Ginsburg, M. and Eyal-Giladi, H. (1980). From cleavage to primitive streak formation: a complementary normal table and a new look at the first stages of the development of the chick. II. Microscopic anatomy and cell population dynamics. Dev. Biol. 79, 296-308.

19. Lee, H. C., Choi, H. J., Park, T. S., Lee, S. I., Kim, Y. M., Rengaraj, D., Nagai, H., Sheng, G., Lim, J. M. and Han, J. Y. (2013a). Cleavage events and sperm dynamics in chick intrauterine embryos. PLoS ONE 8, e80631.

20. Lee, M. T., Bonneau, A. R., Takacs, C. M., Bazzini, A. A., DiVito, K. R., Fleming, E. S. and Giraldez, A. J. (2013b). Nanog, Pou5f1 and SoxB1 activate zygotic gene expression during the maternal-to-zygotic transition. Nature 503, 360-364.

21. Leichsenring, M., Maes, J., Mossner, R., Driever, W. and Onichtchouk, D. (2013). Pou5f1 transcription factor controls zygotic gene activation in vertebrates. Science 341, 1005-1009.

22. Li, W. M., Webb, S. E., Lee, K. W. and Miller, A. L. (2006). Recruitment and SNARE-mediated fusion of vesicles in furrow membrane remodeling during cytokinesis in zebrafish embryos. Exp. Cell Res. 312, 3260-3275.

23. Mathavan, S., Lee, S. G. P., Mak, A., Miller, L. D., Murthy, K. R. K., Govindarajan, K. R., Tong, Y., Wu, Y. L., Lam, S. H., Yang, H. et al. (2005). Transcriptome analysis of zebrafish embryogenesis using microarrays. PLoS Genet. 1, 260-276.

24. Parfitt, D.-E. and Zernicka-Goetz, M. (2010). Epigenetic modification affecting expression of cell polarity and cell fate genes to regulate lineage specification in the early mouse embryo. Mol. Biol. Cell 21, 2649-2660.

25. Park, T. S. and Han, J. Y. (2012). piggyBac transposition into primordial germ cells is an efficient tool for transgenesis in chickens. Proc. Natl. Acad. Sci. USA 109, 9337-9341.

26. Park, H. J., Park, T. S., Kim, T. M., Kim, J. N., Shin, S. S., Lim, J. M. and Han, J. Y. (2006). Establishment of an in vitro culture system for chicken preblastodermal cells. Mol. Reprod. Dev. 73, 452-461.

27. Rossant, J. and Tam, P. P. L. (2009). Blastocyst lineage formation, early embryonic asymmetries and axis patterning in the mouse. Development 136, 701-713.

28. Sasaki, H. (2010). Mechanisms of trophectoderm fate specification in preimplantation mouse development. Dev. Growth Differ. 52, 263-273.

29. Sheng, G. (2014). Day-1 chick development. Dev. Dyn. 243, 357-367.

30. Sun, Y., Tseng, W.-C., Fan, X., Ball, R. and Dougan, S. T. (2014). Extraembryonic signals under the control of MGA, Max, and Smad4 are required for dorsoventral patterning. Dev. Cell 28, 322-334.

31. Tadros, W. and Lipshitz, H. D. (2009). The maternal-to-zygotic transition: a play in two acts. Development 136, 3033-3042.

32. Tsunekawa, N., Naito, M., Sakai, Y., Nishida, T. and Noce, T. (2000). Isolation of chicken vasa homolog gene and tracing the origin of primordial germ cells. Development 127, 2741-2750.

33. Vassena, R., Boue, S., Gonzalez-Roca, E., Aran, B., Auer, H., Veiga, A. and Belmonte, J. C. I. (2011). Waves of early transcriptional activation and pluripotency program initiation during human preimplantation development. Development 138, 3699-3709.

34. Wang, H. and Dey, S. K. (2006). Roadmap to embryo implantation: clues from mouse models. Nat. Rev. Genet. 7, 185-199.

35. Watanabe, T., Biggins, J. S., Tannan, N. B. and Srinivas, S. (2014). Limited predictive value of blastomere angle of division in trophectoderm and inner cell mass specification. Development 141, 2279-2288.

36. Watt, J. M., Petitte, J. N. and Etches, R. J. (1993). Early development of the chick embryo. J. Morphol. 215, 165-182.

37. Xue, Z., Huang, K., Cai, C., Cai, L., Jiang, C.-y., Feng, Y., Liu, Z., Zeng, Q., Cheng, L., Sun, Y. E. et al. (2013). Genetic programs in human and mouse early embryos revealed by single-cell RNA sequencing. Nature 500, 593-597.

38. Yan, L., Yang, M., Guo, H., Yang, L., Wu, J., Li, R., Liu, P., Lian, Y., Zheng, X., Yan, J. et al. (2013). Single-cell RNA-Seq profiling of human preimplantation embryos and embryonic stem cells. Nat. Struct. Mol. Biol. 20, 1131-1139.

3

CHAPTER

THE MOLECULAR BASIS OF RADIAL INTERCALATION DURING TISSUE SPREADING IN EARLY DEVELOPMENT

Andra´ s Szabo´[1, 3] Isidoro Cobo[1, 3] Sharif Omara[1] Sophie McLachlan[1] Ray Keller[2] and Roberto Mayor[1]

[1]Deparment of Cell and Developmental Biology, University College London, London WC1E 6BT, UK

[2]Department of Biology, University of Virginia, Charlottesville, VA 22908, USA

[3]Co-first author

SUMMARY

Radial intercalation is a fundamental process responsible for the thinning of multilayered tissues during large-scale morphogenesis; however, its molecular mechanism has remained elusive. Using amphibian epiboly, the thinning and spreading of the animal hemisphere during gastrulation, here we provide evidence that radial intercalation is driven by chemotaxis of cells toward the external layer of the tissue. This role of chemotaxis in tissue spreading and thinning is unlike its typical role associated with largedistance directional movement of cells. We identify the chemoattractant as the complement component C3a, a factor normally linked with the immune system. The mechanism is explored by computational modeling and tested in vivo, ex vivo, and in vitro. This mechanism is robust against fluctuations of chemoattractant levels and expression patterns and explains expansion during epiboly. This study provides insight into the fundamental process of radial intercalation and could be applied to a wide range of morphogenetic events

Citation: Andra´ s Szabo´, Isidoro Cobo, Sharif Omara, Sophie McLachlan, Ray Keller, Roberto Mayor, The Molecular Basis of Radial Intercalation during Tissue Spreading in Early Development, DOI: http://dx.doi.org/10.1016/j.devcel.2016.04.008.

INTRODUCTION

Acquiring shape and form in multicellular organisms involves deformation of epithelial sheets through bending (invagination), extension through narrowing (convergent extension), and expansion via thinning (epiboly). During epiboly, the number of cell layers in a multilayered epithelium is reduced by cell intercalation, a process called radial intercalation (RI). RI was first described during the uniform expansion of the ectoderm in the animal pole region during amphibian gastrulation (Keller, 1980). Since then it has been recognized as a general morphogenetic process involved in a wide range of systems, including fish epiboly (Warga and Kimmel, 1990), fly gastrulation (Clark et al., 2011), amphibian and fish neural folding (Kee et al., 2008), regeneration of hydra (Kishimoto et al., 1996), and in mammalians during gastrulation (Yen et al., 2009), gut development (Yamada et al., 2010), and ear development (Chen et al., 2002). Two main mechanisms have been proposed so far to explain RI. Studies of amphibian epiboly proposed that RI is driven by adhesion to a fibronectin matrix accumulated on the basal surface of the ectoderm allowing protrusive activity only at the fibronectin-free cell surfaces (Marsden and DeSimone, 2001, Petridou et al., 2013, Sugrue and Hay, 1981). However, this would not explain the intercalation of cells that are not in direct contact with the fibronectin. Studies of teleost epiboly propose cell sorting via differential cell adhesion as the driving force behind RI (Kane et al., 2005, Málaga-Trillo et al., 2009, Schepis et al., 2012). A gradient of cell-cell adhesion molecules toward the superficial surface would drive ectodermal cells to move in the direction of the gradient, opposite to what the fibronectin hypothesis would predict (Kane et al., 2005, Málaga-Trillo et al., 2009, Schepis et al., 2012). Although both cell-matrix and cell-cell adhesions are likely to be involved in some capacity, it is beyond doubt that other mechanisms are required to fully explain this process.

Using *Xenopus laevis*, the original model system to study RI, here we propose a mechanism for epiboly in which the rearrangement of ectodermal cells is driven by an unexpected activity of complement component C3, a factor normally associated with the immune system (Ricklin et al., 2010, Leslie and Mayor, 2013). In immune cell homing, C3 is cleaved to produce C3a, a small anaphylatoxin peptide that binds to C3aR and triggers chemotaxis (Leslie and Mayor, 2013). We developed a computational model to test the hypothesis that RI is driven by deep cell (DC) chemotaxis toward superficial ectoderm cells (SCs). Several predictions of the model were tested in vivo, ex vivo, and in vitro, confirming the notion that short-range chemotaxis is required for ectoderm epiboly.

RESULTS

Radial Intercalation during*Xenopus* Epiboly Is Accompanied by C3 and C3aR Expression

Epiboly in *Xenopus* embryos takes place between developmental stages 8 and 11.5 during which the blastocoel roof (BCR) of the animal pole region expands and thins. At the onset of epiboly, the BCR consists of a tightly connected superficial epithelial monolayer (SL; Figures 1A and 1B , red), and a multilayered deep layer (DL; Figures 1A and 1B, green) comprising 2–4 tiers of loosely connected DCs (Keller, 1978, Keller, 1980). During epiboly, DCs undergo RI to form a single cell layer (Figure 1B; Keller, 1980), but they do not invade the SL (Keller, 1978). SCs change from cuboidal to squamous shape as the whole tissue expands.

In embryos undergoing epiboly, the complement component C3 is expressed at the expanding animal pole region (Figure 1C; in situ hybridization [ISH]) while absent in the ingressing vegetal pole (Figure 1D). Cross-sections along the BCR region indicate that C3 is specifically expressed in the SL (Figures 1E and 1F). Western blot analysis of separated SL and DL samples from stage 10 embryos shows the specificity of the C3a peptide to the SL and the presence of its cognate receptor C3aR in the DL (Figure 1G).

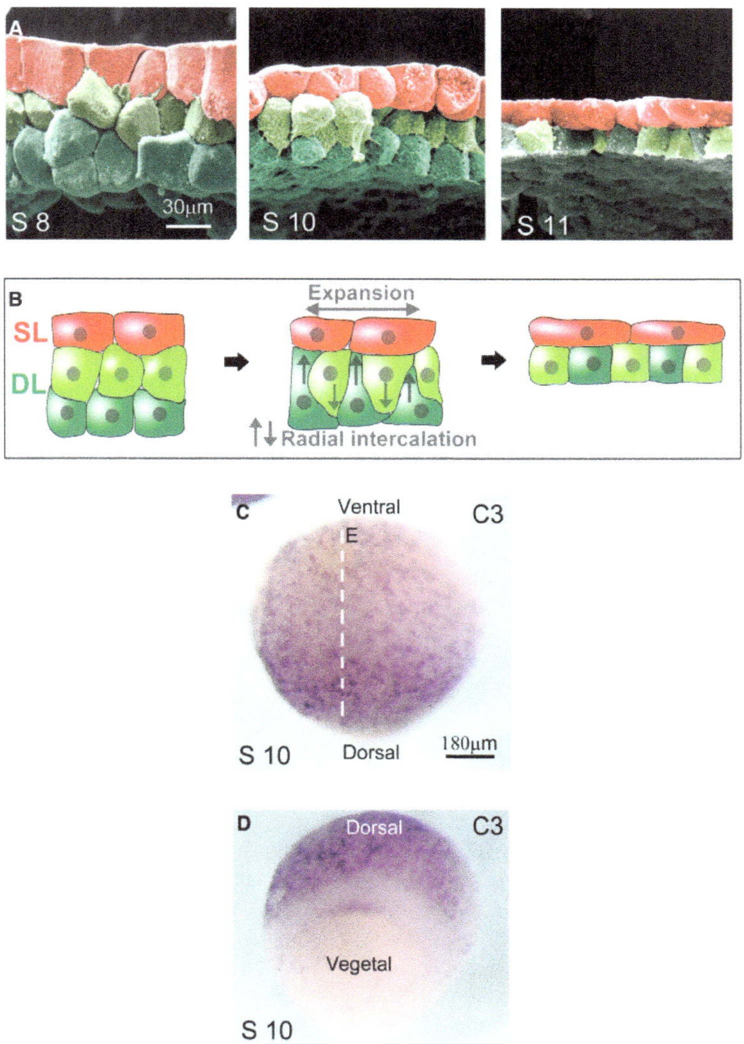

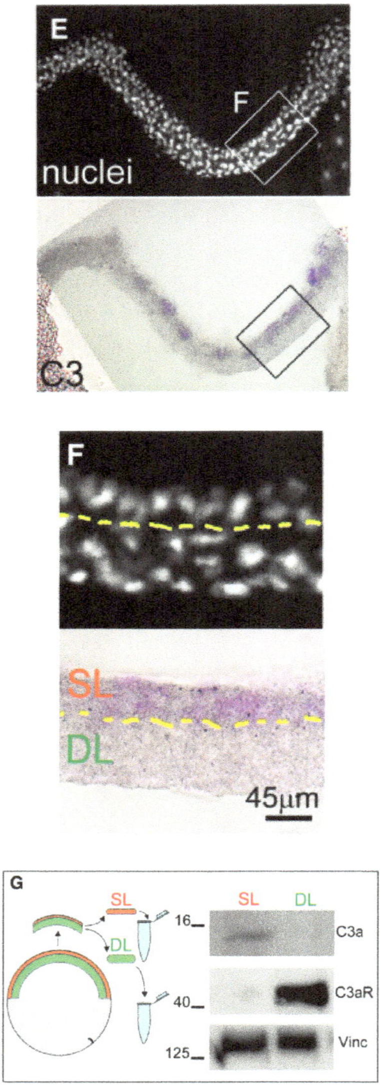

Figure 1: Radial Intercalation Is Accompanied by Expression of C3 and C3aR during Xenopus Epiboly (A) SEM images of blastocoel roof during epiboly in Xenopus embryos at stages 8, 10, and 11. False coloring indicates superficial cells (red), and the intercalation of outer-deep (light green) and innerdeep (dark-green) cells. (B) Schematic illustration of the process of epiboly in Xenopus shown in (A), including expansion of the superficial layer (SL) and the RI of the deep cell layer (DL). Note that the DL does not intercalate into the SL. (C and D) In situ hybridization (ISH) reveals C3 expression in the animal (C) but not in the vegetal (D) pole region during epiboly. (E and F) Cross-section (E) and zoom (F) of nuclei and ISH along the dashed line indicated in (C) show that C3 is expressed in the SL. (G) Western blot analysis showing differential expression of C3a and C3aR in the SL and DL of stage 10 embryos, respectively. Loading control, vinculin.

C3 and C3aR Are Required for Radial Intercalation

To investigate the role of C3 and C3aR on RI, we analyzed the number of cell layers of the BCR in loss-of-function experiments using antisense morpholino oligonucleotides (Mos) against C3aR (C3aRMo; Carmona-Fontaine et al., 2011), C3 (C3Mo; see Supplemental Information for controls) or antibodies against C3a (C3aAb; Carmona-Fontaine et al., 2011). Treatments left the tissue geometry unaffected prior to RI (Figure 2A) but led to a higher number of cell layers after epiboly when compared with control (Figures 2B and 2C) and eventually led to failure of blastopore closure and altered internal embryo structure. Importantly, control and rescue experiments and analysis of protein levels establish the specificity of these blocking reagents. Further characterization of these inhibitors shows that all of them produce the same phenotype, and therefore here we describe only one of them (C3aR depletion), except where mentioned otherwise.

It has been suggested that cell-cell adhesion (Kane et al., 2005, Málaga-Trillo et al., 2009, Schepis et al., 2012) and adhesion to fibronectin (Marsden and DeSimone, 2001, Petridou et al., 2013) are important for epiboly. To test if C3aR inhibition blocked epiboly by interfering with cell adhesion, we performed adhesion and sorting assays, finding no difference between control and C3aR-depleted cells. DC adhesion to fibronectin and fibronectin deposition in the BCR were found to be unaffected by the absence of C3aR. Germ layer specification, indicated by the expression of several known ectodermal and mesodermal markers, was normal after C3aRMo injection. In addition, C3aRMo injection had no observable effect on the cell size (17 ± 2 μm CoMo, 18 ± 6 μm C3aRMo) or number of cell divisions (1.3 ± 0.2 CoMo, 1.4 ± 0.2 C3aRMo, n = 200 cells). In conclusion, germ layer specification, cell adhesion, cell size, and proliferation are unaffected in embryos lacking C3aR.

To address whether the effect of C3aR inhibition on epiboly is independent of the SL, we blocked C3aR specifically in the DL by using a modified BCR explant culture (Marsden and DeSimone, 2001). Briefly, the DL from a mosaic labeled embryo was cultured on a dish underneath an SL from an unlabeled control embryo and imaged from above using time-lapse microscopy. The unlabeled SL allows imaging of the apical (top) portion of the DL in close vicinity of the SL (Figure 2D). In this ex vivo assay, intercalation of DCs is observed as newly emerging or disappearing cells, as these cells enter or leave the top portion of the DL (Figure 2E, asterisks mark intercalation, blue tracks). The ratio of such intercalating cells and cells that do not leave the visualized plane (intercalation percentage) in C3aRMo explants is severely reduced compared with the ratio found in CoMo explants, which indicates that C3aR activity is directly involved in RI at the BCR independently of the SL (Figure 2F).

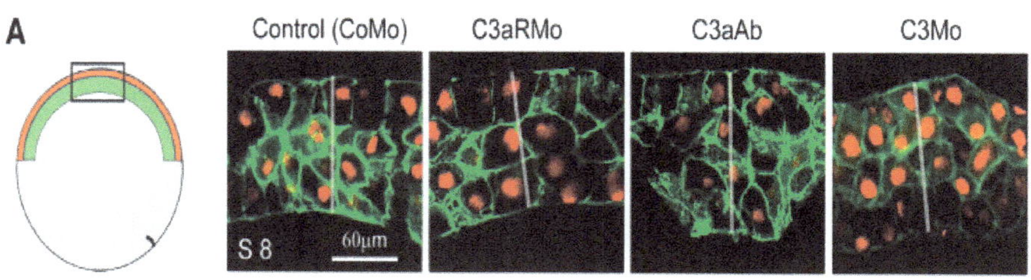

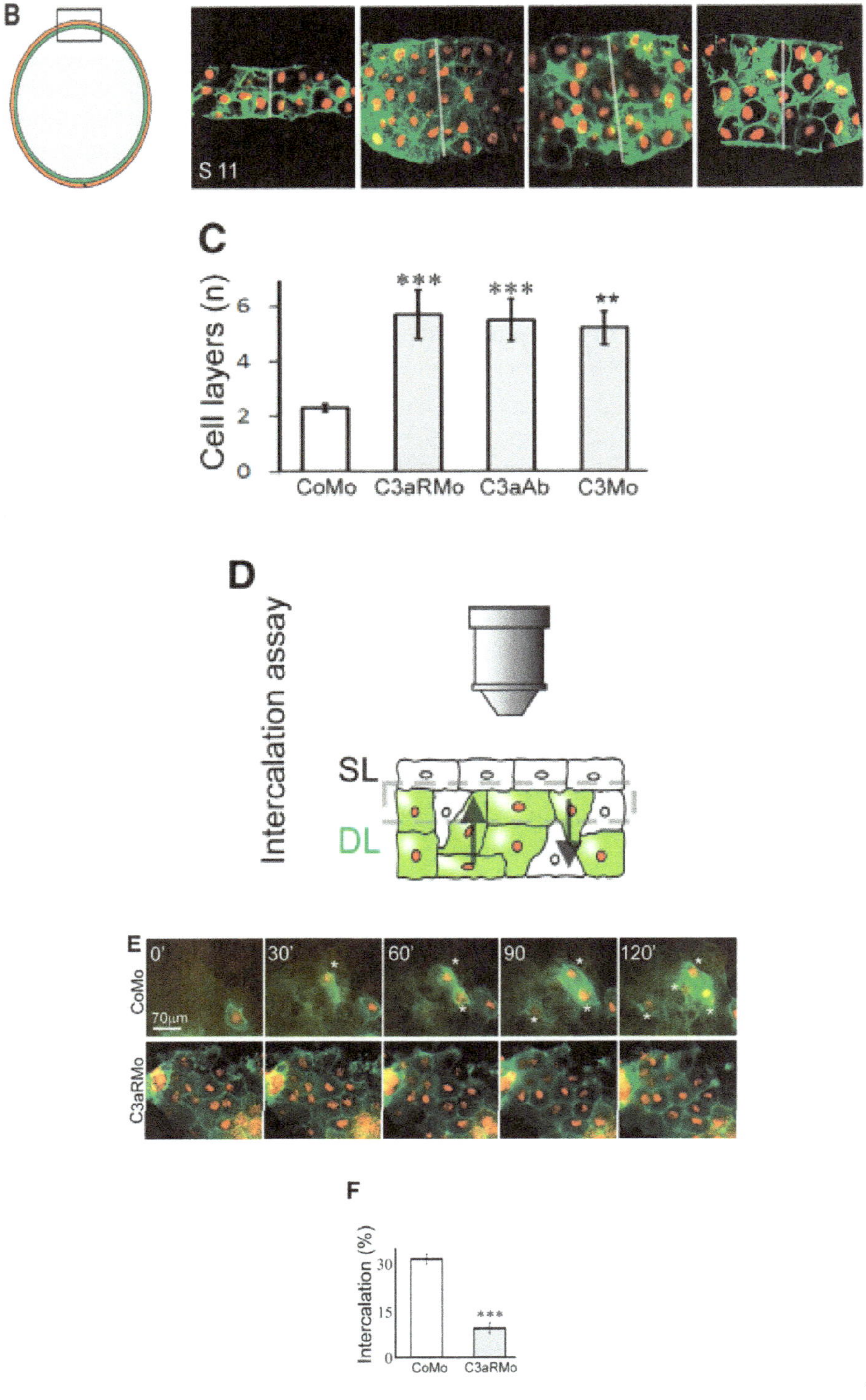
B
S 11
C
Cell layers (n)
0
2
4
6
***
***
**
CoMo
C3aRMo
C3aAb
C3Mo
D
Intercalation assay
SL
DL
E
0'
30'
60'
90
120'
CoMo
C3aRMo
70μm
F
Intercalation (%)
0
15
30
***
CoMo
C3aRMo

Figure 2: C3 and C3aR Are Required for Radial Intercalation (A and B) Sections showing the blastocoel roof before (A, stage 8) and after (B, stage 11) epiboly in control and treated embryos. Embryos deficient in C3aR (C3aRMo), treated with C3a antibody (C3aAb), or suppressed C3 expression (C3Mo) remain thick and multilayered by the end of epiboly (red, nucleus; green, membrane; white bars indicate tissue thickness). In contrast, the tissue thins into a dual-layered epithelium in control (CoMo) embryos. (C) The number of cell layers by the end of epiboly is increased with C3aRMo (n = 62), C3a antibody (C3aAb; n = 30), or C3Mo (n = 60) when compared with control embryos, indicating the lack of RI in the absence of C3 signaling. (D) Schematic of the ex vivo intercalation assay showing the explant from the side. A dashed rectangle indicates the focal range of imaging, which includes the top region of the DL that is in close vicinity of the SL. (E) Frames from time-lapse recording in the ex vivo assay show intercalating DCs (*) in a control (CoMo) and C3aR-deficient (C3aRMo) tissue explants. (F) The number of DCs that intercalate is significantly lower in C3aRMo tissues than in control tissues, showing that intercalation is hampered in the absence of C3aR signaling. Data are represented as means ± SD, t test significances are ***$p < 0.001$, **$p < 0.01$.

Radial Intercalation Driven by C3 Chemotaxis

A well-characterized activity of C3a is to promote chemotaxis during immune response through binding to its receptor C3aR (Leslie and Mayor, 2013). C3a/C3aR induce a similar function during development, whereby they promote cohesion through autocrine chemotaxis of the migrating neural crest cells (Carmona-Fontaine et al., 2011). Based on these observations, we hypothesized that RI is driven by close-range chemotaxis of the DCs toward the C3a producing SCs (Figure 3A).

We turned to computational modeling to test this hypothesis. We implemented a cellular Potts model (Graner and Glazier, 1992) with a quasi-2D cross-section of the BCR including an SL in which all SCs secrete a chemoattractant (Figures 3A and 3B), resulting in an emergent gradient. ISH shows a mosaic expression of C3 (Figure 1C), therefore we also tested an alternating production pattern in the SL and found that the emerging chemoattractant gradient is largely insensitive to the secretion pattern. As both patterns of expression generate the same result in the model, we only show a uniform secretion pattern for the rest of the study. DC chemotaxis is implemented by increasing the probability of cell displacements toward higher concentrations of the chemoattractant (Merks et al., 2004). Upon reaching the SL, DCs form an elastic bond with their SC neighbors with a probabilistic rule that allows for dynamic cell-cell adhesions. Such bonds have been shown to control tissue viscosity (Czirók et al., 2013) and may describe any mechanical resistance of the cells to change their distance, including but not limited to cell adhesion. During the course of the simulated 4 hr real time, all cells undergo one division without cell growth which corresponds to two divisions in the 2D tissue (Keller, 1980). For more details, see Supplemental Information.

Starting from two layers of DCs (Figure 3B, green; shades are only visual aid) and a single SC layer (red), the simulated tissue thins to produce a single layer of DCs via RI. In the absence of chemotaxis, intercalation is severely reduced and the average thickness of the tissue remains higher than with chemotaxis (Figures 3B and 3C). This change in thickness is comparable with the change observed in the BCR thickness in vivo at the end of epiboly in control (CoMo) or C3aR/C3-inhibited (C3aRMo, C3aAb, C3Mo) embryos (Figure 3D). Total tissue thickness in silico is largely insensitive to the cohesion between the DL and SL, which only influences the thickness ratio of the SL and DL.

Although chemotaxis actively drives DCs toward the SL, the model predicts passive cell movement in the opposite direction as well (Figure 3E). The percentage of DC movement toward the SL (Out, 51.3% ± 0.2%) is only slightly higher in silico than movement away from the SL (In, 48.1% ± 0.1%; Figure 3F). We tested this prediction using the ex vivo assay (Figure 2D), where the ratio of DCs intercalating toward the SL and DCs intercalating away from the SL showed a similar bidirectionality (Out, 50% ± 3%; In, 50% ± 2%; Figure 3F). Movement away from the SL is explained by volume exclusion in silico, but we cannot rule out other mechanisms ex vivo, such as repolarization upon contact (Höpker et al., 1999). These results, together with recent in vivo data on RI during zebrafish epiboly showing bidirectional intercalation (Bensch et al., 2013), confirm the bidirectional movement predicted by the computational model.

Taken together, our in silico results support the hypothesis that RI is driven by chemotaxis. Next, we explore the experimental foundation for the chemotactic interaction.

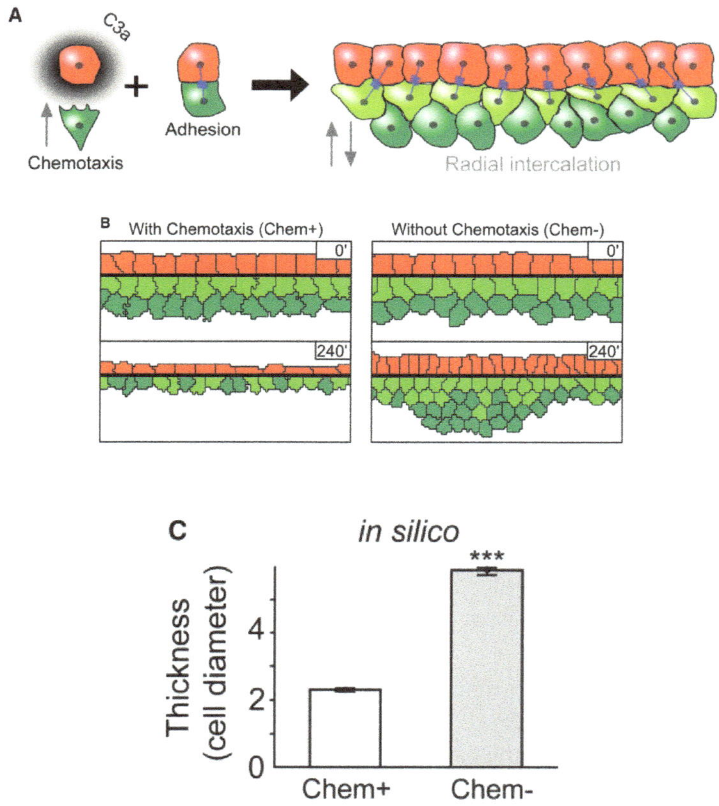

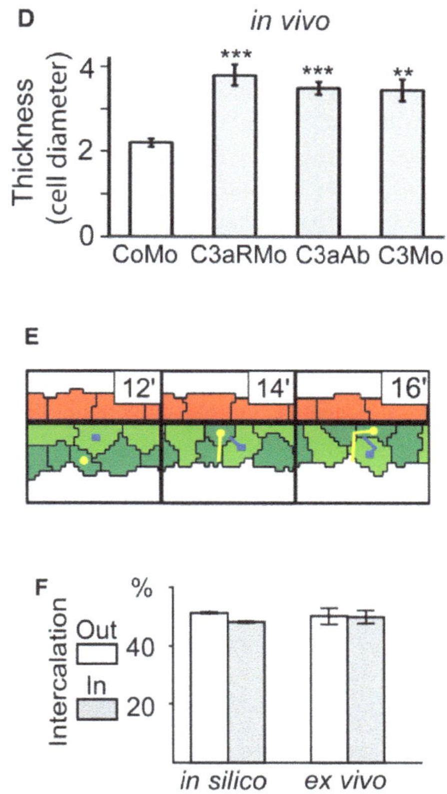

Figure 3: Chemotaxis-Based Radial Intercalation Hypothesis and Computer Simulation (A) Hypothesis: chemotaxis of DCs (green) toward SCs (red) via C3a drives RI and consequent tissue thinning. (B) Computer simulations of the hypothesis showing the initial (00) and final (2400) cell configurations with chemotaxis (Chem+) and without chemotaxis (Chem-). Coloring as in Figures 1A and 1B, for visual aid only. (C) Tissue thickness, measured in units of cell diameter after epiboly (20 mm), is significantly increased in simulations without chemotaxis (n = 30). (D) In vivo tissue thickness, measured in units of cell diameter after epiboly (20 mm), is significantly higher in embryos with either impaired C3 receptors (C3aRMo, n = 62) or C3 ligands (C3aAb, n = 35; C3Mo, n = 62) compared with control embryos (CoMo). (E) Prediction of bidirectional intercalation. In silico, DCs intercalate both toward the secreting SL, driven by chemotaxis, and against the chemotactic gradient away from the SL, driven by volume exclusion. (F) In silico DC movement toward the SL (Out) is only slightly more frequent than movement away from it (In). Direction of DC intercalation measured in the intercalation assay (ex vivo, see Figure 2) confirms the model's prediction. Error bars: SD, significance $^{**}p < 0.01$, $^{***}p < 0.001$.

Testing C3 Chemotaxis of the Deep Cells

To test chemotaxis, the basic assumption of our hypothesis, we utilized a classical chemotaxis assay by culturing deep and superficial explants at a distance (Figure 4A).

Our computational model predicts that, in the presence of chemotaxis, the DL explant will move toward the SL explant and, in the absence of chemotaxis, the clusters remain stationary. Next, we cultured and tracked different combinations of DC and SC explants in vitro. In line with the predictions, the otherwise non-motile DC explants exhibit directional movement toward non-motile SCs cultured nearby (Figures 4B–4D). This movement is compromised when C3 is inhibited in SCs or when C3aR is inhibited in DCs (Figures 4E and 4F), showing that chemotaxis of the DCs toward the SCs is dependent on the C3a/C3aR axis, and that the amount of C3a produced by the SCs is sufficient for chemotaxis. The specificity of C3a in this chemotaxis is shown by the directional movement of DC explants toward a localized source of purified C3a, but not toward a mutant form of C3a (C3aDesArg) that is unable to bind the receptor (Honczarenko et al., 2005) (Figures 4G–4I). As summarized by the chemotaxis indices (Figure 4J), we conclude that DC chemotaxis toward SCs is C3a/C3aR dependent, in line with the in silico measurements (Figure 4K). In vivo, the short distance between the SL and DL could allow a similar chemotaxis to occur in spite of the apparently low levels of C3a detected in the SCs by western blot (Figure 1G).

Such behavior in vivo is expected to generate a higher number of cell protrusions on the sides of DCs facing the SL. Direction of DC protrusions measured on scanning electron microscopy (SEM) images of fixed and fractured embryos during gastrulation revealed protrusions in all directions but with a very strong bias toward the SL (Figures 5A–5E). This finding suggests that chemotaxis is also present in vivo during the RI of the BCR.

Dynamic protrusive activity at the DL surface proximal to the SL was observed in our ex vivo assay, where the top portion of the DL explant is imaged from above through the SL (Figures 5F–5N). Protrusion formation was apparent in CoMo explants where stained DCs are adjacent to unlabeled DCs (Figure 5F, arrowheads). Protrusions were less frequent and smaller in C3aRMo explants (Figure 5G). Protrusions were quantified using the protrusion-retraction method (Scarpa et al., 2015). The area of these protrusions shows a significantly higher protrusive activity in control (CoMo) versus treated (C3aRMo) explants (Figures 5I and 5J). DL explants were imaged solely in the proximity of the SL, therefore any protrusion is indicative of protrusive activity toward or nearby the SL (Figure 5K). Protrusive activity between labeled DCs of CoMo explants is detected as increased levels of membrane staining caused by the overlap of protrusions at the cell surface (Figure 5L, red pseudocoloring). The threshold for the membrane signal to be considered as protrusive activity is set to match the level of a tissue with quasi-static boundaries. Activity is reduced in C3aRMo-treated DL explants (Figure 5M), also shown by the reduction of protrusive activity area per cell (Figure 5N). In conclusion, protrusive activity analyzed by SEM or by live imaging shows that most of the protrusions of DCs point toward the SCs, which is consistent with chemotaxis of DL cells toward the SL.

In the following, we explore an implication of the model, and then study its behavior while perturbing the level and the localization of the chemoattractant.

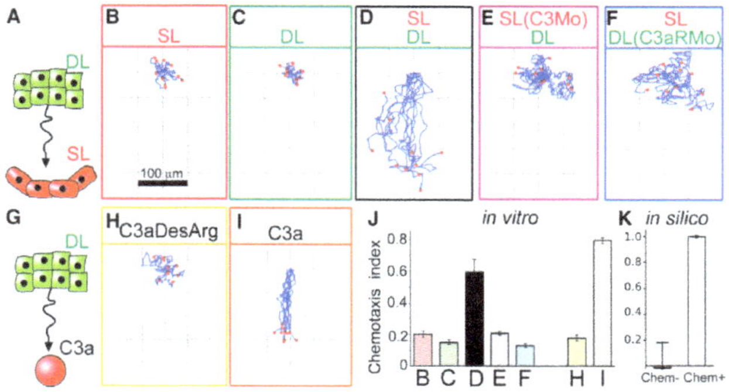

Figure 4: C3a-Based Chemotaxis between Cells of the Deep and Superficial Layers (A–K) Testing the basic assumption of the model. Classical chemotaxis assay using DL and SL explants (A), or DL explants with purified C3a protein (G). Trajectories of SL (B) and DL (C–F, H, and I) explants, and DL chemotaxis toward non-functional (C3aDesArg, H, n = 35) and functional (I, n = 38) C3a source with corresponding chemotaxis indices in vitro (J) and in silico (K). Error bars: SD.

Tissue Expansion Requires C3 Signaling

Our chemotaxis-based in silico model of RI predicts the expansion of the whole tissue. This expansion is concomitant with RI and depends on chemotaxis (Figures 6A–6C). DL expansion in our model is transmitted to the SL locally via an effective friction reducing the shear, relative movement between the SL and DL as a result of the elastic links between DCs and SCs. Consequently, the DL expands slightly more than the SL, which is in agreement with previous experimental observations (Keller, 1980, Bauer et al., 1994). We tested the prediction that expansion during epiboly depends on DC chemotaxis using time-lapse imaging of CoMo- and C3aRMo-treated embryos (Figure 6D). The expected expansion of the animal pole in CoMo embryos was observed through the increase in distance of SC pairs over time (Figures 6E). When RI was blocked using C3aRMo, SCs in the animal pole remained at an approximately constant distance from one another (Figure 6F), supporting the in silico prediction.

In order to test whether the interaction between the SL and DL is sufficient to drive this expansion independently of the rest of the embryo, we used our ex vivo assay where only the SL and DL are present (Figure 6G). We observed that the distance between DCs generally increased over time in CoMo explants but not in C3aRMo explants (Figures 6H and 6I), showing that the interaction between SL and DL is sufficient to drive expansion in a C3aR-dependent manner.

Expansion in all three systems (in silico, in vivo, and ex vivo) showed a significant reduction in the absence of chemotaxis or C3aR (Figure 6J). Expansion was quantified by measuring how the distance of tracked cell pairs changes during imaging. For a given cell pair, the expansion is $E = (d_f - d_i)/d_i$, where d_f is the final distance and d_i is the initial distance. Expansion of the tissue is the average of expansions for all measured cell pairs. We observed a slight expansion in the Mo-treated specimens, which is unexplained by our model. This expansion could result from any remaining functional C3aR due to incomplete blocking by the Mo in the experiments as opposed to perfect inhibition in the model. We cannot exclude the possibility of an alternative expansion mechanism acting independently of RI, such as cell-autonomous expansion in the SL. However, as the majority of expansion is lost upon C3aR inhibition, any alternative mechanism is expected to play a minor role during BCR epiboly.

Our simulations show that this mechanism could also drive thinning of tissues that have a higher number of cell layers than *Xenopus* ectoderm, such as zebrafish epiboly. Furthermore, we simulated experiments in which DCs were depleted, showing that there was no major effect on tissue expansion, indicating the robustness of our model. In conclusion, our combination of simulations and experiments show for the first time that RI is required for the expansion of an adjacent tissue that does not participate in intercalation, and that a minimal of two DC layers are required for this process.

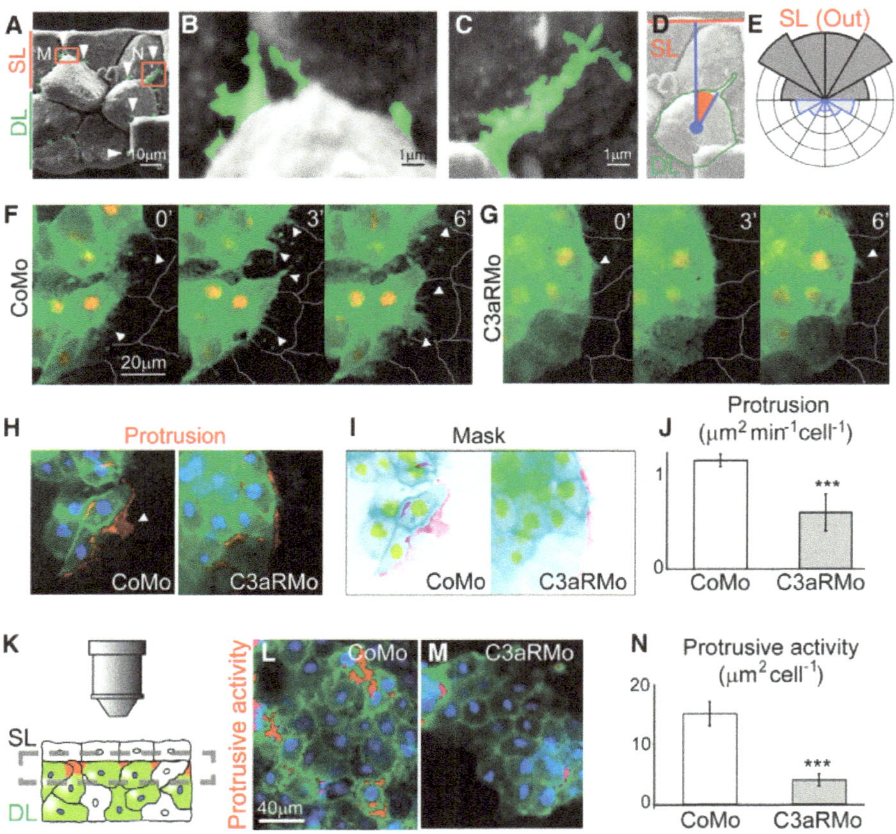

Figure 5: DC Protrusion Analysis (A–E) Direction of DC protrusions in vivo (arrowheads, A–C) was measured in relation to the SL external surface (D) to reveal a bias toward the SL (E, n = 64 embryos, 1445 protrusions). (F and G) Frames from time-lapse imaging showing DC protrusions (arrowheads) ex vivo (see Figure 2D). Protrusions are apparent in the vicinity of the SL in control

(CoMo) but not in the C3aRMotreated explants. Green, membrane; red, nuclei. (H–J) Protrusion activity in the ex vivo assay analyzed using the extension-retraction method. Red shows the difference in membrane signal (green) between frames 3 min apart (H). The difference reveals the extending protrusions (I, purple). (K–N) Ex vivo apical protrusive activity of DCs (L and M: green, membrane; red, protrusion; blue, nuclei) is decreased in C3aR-deficient explants (N: $n = 32$; ***$p < 0.01$; error, SD).

Chemoattractant Levels Modulate the Extent of Radial Intercalation and Extension

We investigated how the behavior of the model depends on the amount of chemoattractant in the system. Chemoattractant levels in the model are dynamically controlled by the secretion and decay rates, and the diffusion parameter, which together give rise to a quasi-steady concentration field. As these parameters are not directly accessible experimentally in vivo, here we measure the relative amount of chemoattractant produced in the SL in silico as a result of changing the secretion rate. A reduction of the produced chemoattractant levels to half did not impede RI, and tissue expansion was still visible in silico (Figure 7A). Increasing the chemoattractant levels by the same amount increased tissue expansion (Figure 7A). In silico tissue expansion increases rapidly even at low levels of chemoattractant but saturates at around triple concentration levels, where the tissue expands to slightly over double its linear size (Figure 7B, black). Tissue thickness similarly drops rapidly as the chemoattractant level is increased and reaches saturation (Figure 7B, red). Therefore the model predicts a slight reduction of epiboly at lower C3 expression levels but still noticeable epiboly even at levels as low as 10%.

To test this prediction, we analyzed C3 expression in subregions within the same embryo using ISH and found a considerable difference between the dorsal and ventral sides of the animal pole compared with the overall average in the whole-animal cap (Figures 7C and 7D). We found significantly higher expansion in the dorsal regions than in ventral regions as predicted by the model based on the same expression levels (Figures 7E and 7F). We conclude that this asymmetry in C3 expression might contribute to the known dorsoventral morphogenetic differences in gastrulation of *Xenopus* embryos (Bauer et al., 1994).

Localization of the Chemoattractant Source

Finally, we tested how the localization of the chemoattractant source within the tissue affects its morphogenesis. For this we compared simulations of normal chemoattractant expression (Control, Figures 8A and 8E) to simulations where no chemoattractant is present (Inhibition, Figures 8B and 8F), or is produced by all cells (Ubiquitous, Figures 8C and 8G), or is produced only in a restricted region of the SL (Localized, Figures 8D and 8H). As expected, inhibition resulted in thicker tissues than in controls. More surprisingly, a ubiquitous expression in the SL and DL resulted in tissue thinning similar to the control case in silico (Figure 8C). While a ubiquitous expression in uniformly packed tissues is expected to eliminate any chemotactic gradients, cell-free regions in our system take up a considerable volume. These regions represent the blastocoel cavity immediately below the DL and the external space above the SL. The chemoattractant diffuses from the cellular region into the cell-free area and is diluted in the comparatively large cavity of the blastocoel, giving rise to a gradient with decreasing concentration levels from the cellular region toward the blastocoel cavity (Figure 8G). Membranes of the DCs at the cavity edge fluctuate stochastically, extending into and retracting from the cavity. Membrane extensions into the cavity now occur against the generated chemotactic gradient, and therefore these are suppressed in the model. Through such a ratchet mechanism, this gradient directs cells from the edge of the blastocoel cavity toward the SL and promotes intercalation. In contrast, overexpression in uniformly packed tissues could not produce such a gradient due to the lack of cell-free regions. The mechanism is less efficient with ubiquitous expression than with the segregated expression of the control case. Finally, a localized source of chemoattractant in the SL (Figure 8D, red) is unable to rescue local tissue thinning in silico. The chemotactic gradient in this setting does not only contain a radial component but also has a significant tangential component that is parallel to the SL-DL interface (Figure 8H). This gradient is attracting DCs from lateral regions and paradoxically leads to tissue thickening at the region of the active tissue.

To test these predictions, we measured the thickness of the BCR cross-section in stage 11 embryos after epiboly in conditions related to the model simulations (Figures 8I–8L, white bar marks tissue thickness). Control and inhibition measurements were performed in CoMo-and C3Mo-treated embryos (Figures 8I and 8J). Ubiquitous expression was achieved by C3a mRNA injection leading to C3a overexpression in all cells and RI as predicted (Figure 8K). Localized C3 expression was achieved by grafting a piece of SL from a C3a mRNA-injected embryo (marked with RLDx in Figure 8L) into a C3Mo-treated host embryo. After healing, at stage 11 the tissue at the region of the graft was observed to be thicker than in either the control or the inhibited case indicating cell accumulation under the graft and confirming our prediction.

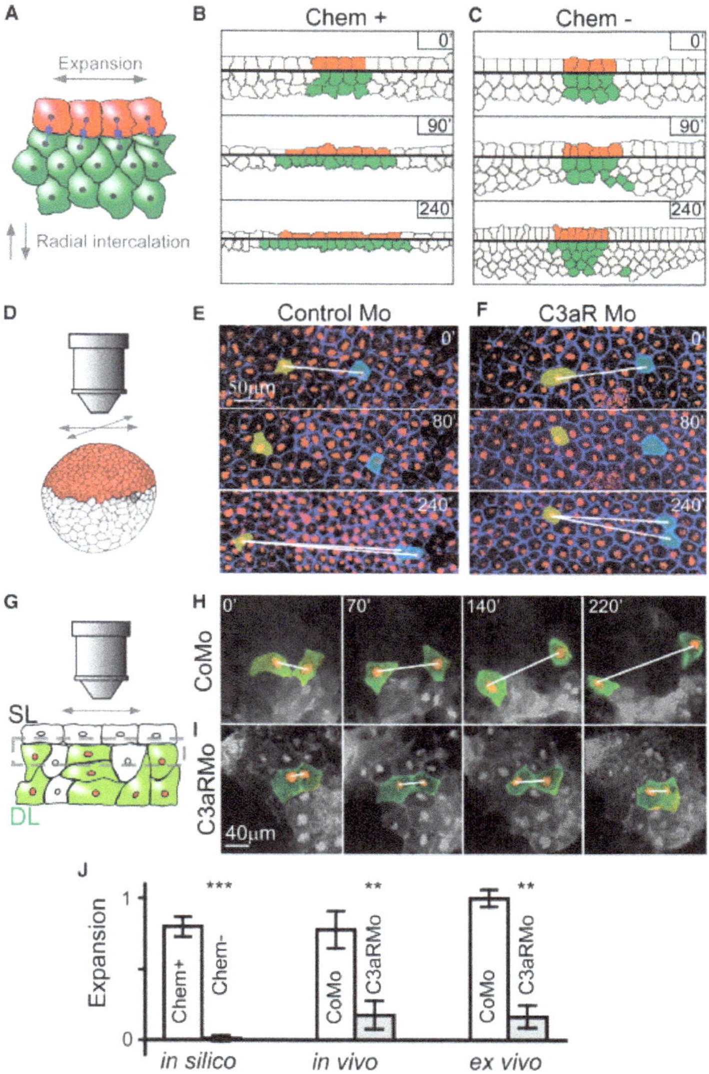

Figure 6: C3 Signaling Is Required for Epithelial Expansion (A–C) Model prediction of tissue expansion as a consequence of chemotaxis-driven RI. In the presence of chemotaxis, both the SL and DL expand simultaneously with RI (B, coloring for visual aid only), while no expansion is observed without chemotaxis (C). (D–F) Time-lapse imaging of epiboly in live embryos (D) reveals that while SCs drift away from one another following CoMo treatment (E), the distance of SCs in C3aR-deficient embryos does not increase (F) during the process. (G–I) Ex vivo study of tissue expansion using the intercalation assay (G). Cells in the DL separate as the isolated tissue undergoes expansion (H). This expansion is lost in tissues lacking C3aR (I). (J) Quantification of tissue expansions as the difference in the final and initial distances relative to the initial distance of tracked cell pairs in silico, in vivo, and ex vivo. Error, SD; significance, **p % 0.01, ***p < 0.001.

Quantification of tissue thicknesses shows the similarity of in silico predictions and in vivo validation (Figure 8M); while inhibition leads to significantly thicker tissues due to the lack of RI, ubiquitous overexpression shows no significant difference to the control. Note that, in silico, the phenotype is rescued with a 1.3-fold increase in expression levels, while the same expression levels result in a slightly but significantly thicker tissue (2.75 ± 0.08 cell diameters) than the control (2.3 ± 0.05 cell diameters). This shows that the ubiquitous expression pattern gives rise to a less efficient mechanism for thinning. A localized source of chemoattractant produces a significantly thicker tissue in the active region than the control, even surpassing the thickness of the passive, inhibited condition.

In summary, the counterintuitive predictions of the ubiquitous and localized sources and their experimental validation show that this chemotaxis-based morphogenetic mechanism is both robust and distinct from other chemotaxis processes where ubiquitous expression of the ligand impairs directional movement (e.g., chemotaxis of germ cells toward SDF-1; Doitsidou et al., 2002).

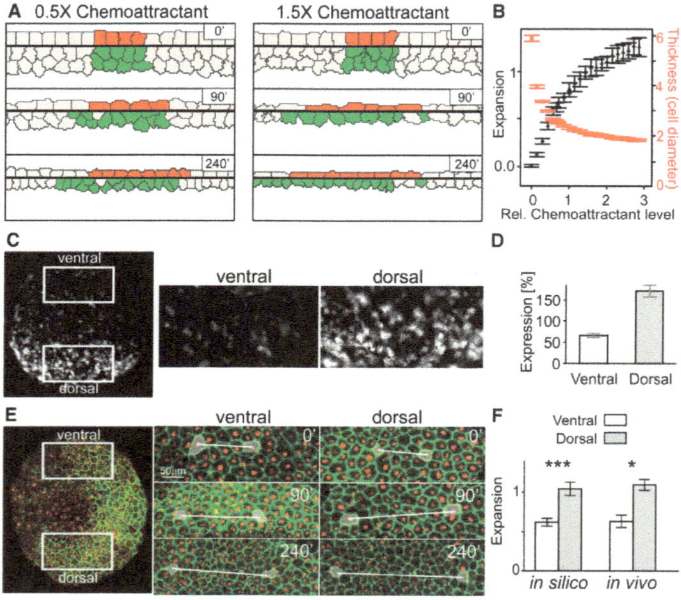

Figure 7: Dorsoventral Differences in Expansion Explained by Differential C3 Expression (A) The extent of thinning and expansion depends on the level of chemoattractant in the SL in silico. (B) Tissue expansion and thickness shown after 4 hr in simulations as a function of chemoattractant levels. (C and D) Approximation of C3 levels in vivo using intensity levels from ISH of a stage 10 embryo reveals differential expression in the animal cap with a reduced level in the ventral region (66% compared with the whole-animal cap) and an increased level in the dorsal region (172%). (E and F) Expansion in vivo in the ventral and dorsal areas of the animal pole. Expansion in the dorsal regions was significantly higher (n = 20) than in the ventral regions, as predicted by the model. Error bars in (B) and (F) show the SD; error bars in (D) show uncertainty of sampling. Significance ***$p < 0.001$, *$p < 0.05$.

DISCUSSION

Here we present a molecular mechanism for RI based on short-range chemotaxis that can facilitate thinning and expansion of multilayered epithelia. In silico modeling of epiboly enabled us to predict intercalation dynamics, and sensitivity to chemotactic levels and localization. These predictions were tested in support of our hypothesis using a combination of in vivo, in vitro, and ex vivo assays.

We chose to study RI during *Xenopus* epiboly, where it was first described (Keller, 1980). We show that in this system, chemotaxis and consequent RI are driven and directed by the complement component C3 and its receptor C3aR. Although these components are best known for their role in the immune system, an increasing body of evidence suggests that these and other parts of the immune system are involved in functions unrelated to their immune function (Denny et al., 2013, Bénard et al., 2004, Hawksworth et al., 2014, Leslie and Mayor, 2013). C3a, together with C5a, has been shown to play a role in regeneration of a vast variety of tissues through promoting cell survival, proliferation, differentiation, and chemotaxis (Schraufstatter et al., 2015). During early development, the presence of C3 and its receptor has been reported in the*Xenopus* gastrula (McLin et al., 2008) and mouse neurula (Jeanes et al., 2015), however, their function in these contexts remained unknown. The function of C3 signaling has been identified in the migrating neural crest (NC), where it promotes cohesion of the migrating collective (Carmona-Fontaine et al., 2011). Importantly, C3 and C3aR are expressed in the same cells during NC migration, and therefore the mechanism of chemotaxis utilized by the NC differs from the one proposed here, where expression of C3 and its receptor is segregated (Carmona-Fontaine et al., 2011, Woods et al., 2014).

Separation of ligand and receptor has been reported in many chemotaxis processes, such as during the in vivo chemotaxis of primordial germ cells (PGCs) in zebrafish (Doitsidou et al., 2002) or the short-range chemotaxis of leukocyte trafficking through the vascular endothelium (Zabel et al., 2015). In such systems, the localization of the source determines the behavior, which can be tested under at least three extreme conditions. First, a complete inhibition of the source leads to the lack of directional migration, as observed for PGCs (Doitsidou et al., 2002) and in our cells in vitro (Figures 4C, 4E, and 4H). In vivo, the absence of chemoattractants led to the severe reduction of intercalation and of thinning (Figures 2 and 8J), as predicted in silico (Figures 8B and 8M). Second, if the chemoattractant is expressed in all tissues ubiquitously, the chemoattractant gradient is lost and chemotaxis is expected to fail, as seen for the PGCs (Doitsidou et al., 2002). However, ubiquitous expression in our system does not share the phenotype with the inhibited case (Figure 8K) and is not predicted to do so (Figure 8C). In this specific geometry, the relatively thin tissue is bound by the vast blastocoel cavity, which acts as a sink for the chemoattractant and gives rise to the gradient (Figure 8G inset). Such source-sink systems are known to produce robust and steep gradients (Majumdar et al., 2014) and have been described in vitro (Volpe et al., 2012, Scherber et al., 2012) and have been suggested in vivo (Yu et al., 2009, Boldajipour et al., 2008, Donà et al., 2013). Nevertheless, our model demonstrates that this mechanism is less efficient than the control, which could explain the segregation. Finally, an ectopically localized source is expected to misdirect chemotaxing cells. As our chemotaxis-based hypothesis predicted, we found that introducing a local source of chemoattractant into a depleted embryo directs the cells toward this ectopic source to create a thick accumulation (Figure 8). Notably, this local re-establishment of the chemoattractant activity did not produce a local rescue. Taken together, our experiments of perturbed chemoattractant localization show that both our hypothesis and experimental setup behave as a chemotactic system.

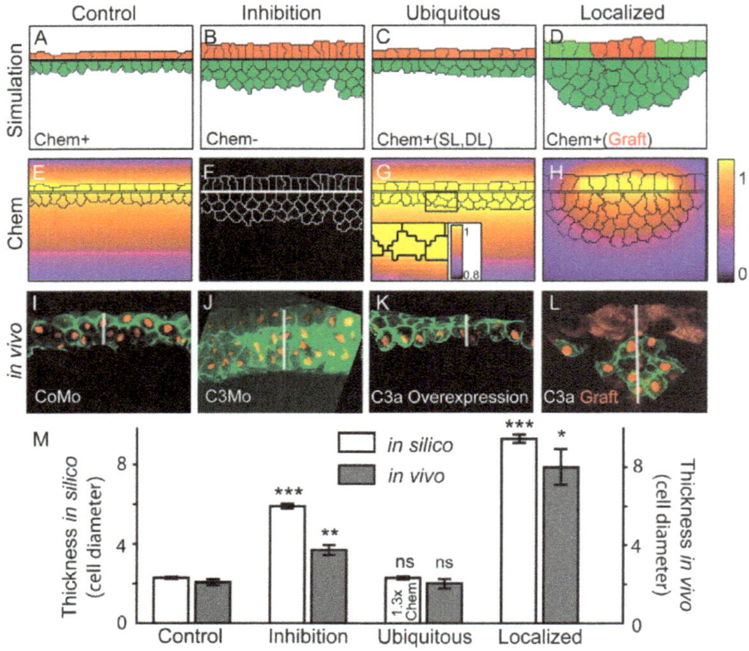

Figure 8: Localization of Chemoattractant Source Determines Radial Intercalation (A–H) Simulated cell configurations and corresponding chemoattractant levels at the end of epiboly with different localizations of chemoattractant production: only in the SL (Control, A, E), nowhere (Inhibition, B, F), in every cell (Ubiquitous, C, G), and in a restricted region of the SL (Localized, D, H, producing region shown in red). (I–L) Experimental validation of the predicted behavior showing sections of the blastocoel roof of stage 11 embryos. Red, nucleus; green, membrane; graft in (L) marked with cytosolic RLDx. White bar indicates tissue thickness. (M) Quantification of tissue thickness, measured in units of cell diameter after normal epiboly (20 mm), for the four settings in silico and in vivo. Error, SD. Significances compared with relevant control: *p < 0.05, **p < 0.01, ***p < 0.001; ns, p > 0.05. 1.33 Chem denotes 1.3-fold overexpression in both DCs and SCs.

Our results show that chemotaxis and DL intercalation are largely required for the expansion of the whole tissue occurring during epiboly, and that this expansion is independent of the rest of the embryo. Expansion of the non-intercalating SL is explained by our model through an effective friction between the DL and SL representing general cell adhesion or other independent mechanical interactions (Bergert et al., 2015). An alternative explanation for the SL-DL interaction with a similar expected result is that contact with the DCs could lower the basal surface tension of the SL in a C3/C3aR-dependent manner, allowing it to expand (Luu et al., 2011). As DL expansion in our model is driving SL expansion, the SL cannot expand beyond the DL. However, simulations with localized expression (Figure 8) reveal that the secreting region must be at least the same size as the receptive tissue; otherwise DCs outside the active region counteract expansion. While the boundary region of the SL and the neighboring mesoderm are beyond the scope of this study, we can contemplate that a yet unknown mechanism is at work to autonomously expand the SL beyond the DL. Such a mechanism is expected to act independently of C3 signaling, as its receptor is not expressed in the SL.

An apparently equivalent chemotaxis-based mechanism has been reported for the RI of the prechordal mesoderm (PCM) during *Xenopus* gastrulation (Damm and Winklbauer, 2011). These cells involute and subsequently migrate away from the blastopore against the expanding BCR. Cells of the PCM chemotax toward platelet-derived growth factor A (PDGF-A), produced by the nearby BCR, leading to the RI of the PCM, much like in the case of the expanding animal pole. Due to experimental limitations, expansion of the PCM is less accessible than the expansion analyzed in our system. It is tempting to speculate that the two processes share the same mechanism, and therefore any expansion resulting from the RI of the PCM may contribute to the expansion of the BCR, however this remains to be explored.

Our mechanism acts independently of the previously proposed adhesion-based mechanisms of epiboly and does not exclude them, as shown by the unaltered cell-cell adhesion and cell-fibronectin adhesion properties after blocking C3 signaling. The effect of C3 on DC protrusive activity is different from alternative adhesion-based hypotheses where a direct contact would be essential, e.g., DC protrusions promoted or stabilized on the surface of SCs, as supported by our observation that local C3 expression leads to accumulation of multiple layers of DCs around the active region, affecting even DCs that are not in direct contact with the SCs (Figure 8L). Nevertheless, our data do not exclude an essential role for fibronectin in epiboly, such as polarizing cells (Marsden and DeSimone, 2001, Petridou et al., 2013), preventing elevated tension in the animal cap (Petridou et al., 2013), regulating protrusions (Davidson et al., 2006), or even sequestering C3a to contribute to the gradient (Carmona-Fontaine et al., 2011). However, it is unlikely that these effects of fibronectin are sufficient to account for epiboly, as the matrix appears only hours after epiboly starts, and inhibition of cell-fibronectin binding is unable to affect the first half of epiboly (Marsden and DeSimone, 2001, Davidson et al., 2004). Therefore it is possible that C3 is required for the initiation and direction of RI and epibolic movements, and fibronectin is required for supporting and maintaining these movements during epiboly. Gastrulation is disrupted by blocking either C3 signaling or cell-fibronectin binding (Marsden and DeSimone, 2001,Petridou et al., 2013), showing the importance of epiboly for blastopore closure. Furthermore, epiboly also depends on the correct alignment of SC divisions through opposing apicobasal forces to keep the plane of division parallel to the SL (Woolner and Papalopulu, 2012). Upon disruption of these intracellular forces, SCs divide out of the epithelial plane and consequently epiboly is compromised (Woolner and Papalopulu, 2012). In-plane cell divisions have been shown to orient in order to relieve anisotropic tissue tension within the enveloping layer during zebrafish epiboly (Campinho et al., 2013). However, de-coupling of division orientation from tissue tension did not markedly hamper the progress of epiboly, although it led to abnormal fusion of the EVL cells (Campinho et al., 2013). While tissue organization in C3/C3aR-depleted embryos is affected, we observed no effect on the division of SCs in these embryos, making it unlikely that C3 signaling would be directly involved in the control of SC division.

Although the graded expression of E-cadherin required for the cell-cell adhesion-based mechanism is controversial (Song et al., 2013), there is strong evidence that the dynamics of E-cadherin is essential for epiboly in zebrafish (Babb and Marrs, 2004, Kane et al., 2005, Song et al., 2013, Shimizu et al., 2005, Arboleda-Estudillo et al., 2010). This dynamic cell adhesion is consistent with our model where DCs slide past the SL, similar to what has been proposed for the migration of prechordal plate mesendoderm, germ, and border cells (Ulrich et al., 2005,Kardash et al., 2010, Ulrich and Heisenberg, 2009, Cai et al., 2014), and also with recent data showing that guided cell migration in vivo requires dynamic filopodia (Meyen et al., 2015). Thus differential adhesion and chemotaxis could be acting in parallel. It would be interesting to know if an equivalent chemotactic mechanism of RI and epiboly based on the same C3a/C3aR molecules also operates in other animal models, such as zebrafish; however, the identification of the chemoattractant may be hampered by the presence of multiple C3 genes found in zebrafish (Forn-Cuní et al., 2014).

In summary, here we identify a molecular mechanism for RI based on short-range chemotaxis. We demonstrate that this mechanism is also involved in generating expansion during epiboly, both in the intercalating DL and non-intercalating SL. Chemotaxis in our particular study system is driven by the complement component C3, which has only been implicated in autocrine chemotaxis during development (Carmona-Fontaine et al., 2011). While the exact molecules may vary, this general mechanism of RI is likely to operate in other various epithelial morphogenetic events throughout developmental and pathological processes.

AUTHOR CONTRIBUTIONS

Conceptualization, R.M.; Methodology, R.M. and A.S.; Software, A.S.; Formal Analysis, A.S., I.C., and R.M.; Investigation, I.C., S.O., S.M., R.K., and R.M.; Writing – Original Draft, A.S. and R.M.; Writing – Review & Editing, A.S., S.M., R.K., and R.M.; Visualization, A.S. and R.M.; Supervision, R.M.; Project Administration, R.M.; Funding Acquisition, A.S., S.M., and R.M.

ACKNOWLEDGMENTS

We thank C. Stern, J. Leslie, and M. Tada for comments on the manuscript, and J. Lambris for providing us with the C3a peptide and for encouraging us to explore the non-immunological functions of the complement. Fibronectin antibody was kindly provided by D. DeSimone. This study was supported by grants from the MRC (J000655 and M010465), BBSRC (M008517), and the Wellcome Trust (084247/Z/07/Z) to R.M. A.S. was supported by a Marie Curie Fellowship (329968), S.M. by a Wellcome Trust PhD studentship. The authors acknowledge the use of the UCL Legion High

Performance Computing Facility (Legion@UCL), and associated support services, in the completion of this work.

REFERENCES

1. Arboleda-Estudillo, Y., Krieg, M., Stu¨ hmer, J., Licata, N.A., Muller, D.J., and Heisenberg, C.-P. (2010). Movement directionality in collective migration of germ layer progenitors. Curr. Biol. 20, 161–169.
2. Babb, S.G., and Marrs, J.A. (2004). E-cadherin regulates cell movements and tissue formation in early zebrafish embryos. Dev. Dyn. 230, 263–277.
3. Bauer, D.V., Huang, S., and Moody, S.A. (1994). The cleavage stage origin of Spemann's Organizer: analysis of the movements of blastomere clones before and during gastrulation in Xenopus. Development 120, 1179–1189.
4. Be´ nard, M., Gonzalez, B.J., Schouft, M.-T., Falluel-Morel, A., Vaudry, D., Chan, P., Vaudry, H., and Fontaine, M. (2004). Characterization of C3a and C5a receptors in rat cerebellar granule neurons during maturation. J. Biol. Chem. 279, 43487–43496.
5. Bensch, R., Song, S., Ronneberger, O., and Driever, W. (2013). Non-directional radial intercalation dominates deep cell behavior during zebrafish epiboly. Biol. Open 2, 845–854.
6. Bergert, M., Erzberger, A., Desai, R.A., Aspalter, I.M., Oates, A.C., Charras, G., Salbreux, G., and Paluch, E.K. (2015). Force transmission during adhesionindependent migration. Nat. Cell Biol. 17, 524–529.
7. Boldajipour, B., Mahabaleshwar, H., Kardash, E., Reichman-Fried, M., Blaser, H., Minina, S., Wilson, D., Xu, Q., and Raz, E. (2008). Control of chemokineguided cell migration by ligand sequestration. Cell 132, 463–473.
8. Cai, D., Chen, S.-C., Prasad, M., He, L., Wang, X., Choesmel-Cadamuro, V., Sawyer, J.K., Danuser, G., and Montell, D.J. (2014). Mechanical feedback through E-Cadherin promotes direction sensing during collective cell migration. Cell 157, 1146–1159.
9. Campinho, P., Behrndt, M., Ranft, J., Risler, T., Minc, N., and Heisenberg, C.-P. (2013). Tension-oriented cell divisions limit anisotropic tissue tension in epithelial spreading during zebrafish epiboly. Nat. Cell Biol. 15, 1405–1414.
10. Carmona-Fontaine, C., Theveneau, E., Tzekou, A., Tada, M., Woods, M.L., Page, K.M., Parsons, M., Lambris, J.D., and Mayor, R. (2011). Complement fragment C3a controls mutual cell attraction during collective cell migration. Dev. Cell 21, 1026–1037.
11. Chen, P., Johnson, J.E., Zoghbi, H.Y., and Segil, N. (2002). The role of Math1 in inner ear development: uncoupling the establishment of the sensory primordium from hair cell fate determination. Development 129, 2495–2505.
12. Clark, I.B.N., Muha, V., Klingseisen, A., Leptin, M., and Mu¨ ller, H.-A.J. (2011). Fibroblast growth factor signalling controls successive cell behaviours during mesoderm layer formation in Drosophila. Development 138, 2705–2715.

13. Cziro´ k, A., Varga, K., Me´ hes, E., and Szabo´ , A. (2013). Collective cell streams in epithelial monolayers depend on cell adhesion. New J. Phys. 15, 075006.

14. Damm, E.W., and Winklbauer, R. (2011). PDGF-A controls mesoderm cell orientation and radial intercalation during Xenopus gastrulation. Development 138, 565–575.

15. Davidson, L.A., Keller, R., and DeSimone, D.W. (2004). Assembly and remodeling of the fibrillar fibronectin extracellular matrix during gastrulation and neurulation in Xenopus laevis. Dev. Dyn. 231, 888–895.

16. Davidson, L.A., Marsden, M., Keller, R., and DeSimone, D.W. (2006). Integrin a5b1 and fibronectin regulate polarized cell protrusions required for Xenopus convergence and extension. Curr. Biol. 16, 833–844.

17. Denny, K.J., Coulthard, L.G., Jeanes, A., Lisgo, S., Simmons, D.G., Callaway, L.K., Wlodarczyk, B., Finnell, R.H., Woodruff, T.M., and Taylor, S.M. (2013). C5a receptor signaling prevents folate deficiency-induced neural tube defects in mice. J. Immunol. 190, 3493–3499.

18. Doitsidou, M., Reichman-Fried, M., Stebler, J., Ko¨ prunner, M., Do¨ rries, J., Meyer, D., Esguerra, C.V., Leung, T., and Raz, E. (2002). Guidance of primordial germ cell migration by the chemokine SDF-1. Cell 111, 647–659.

19. Dona` , E., Barry, J.D., Valentin, G., Quirin, C., Khmelinskii, A., Kunze, A., Durdu, S., Newton, L.R., Fernandez-Minan, A., Huber, W., et al. (2013). Directional tissue migration through a self-generated chemokine gradient. Nature 503, 285–289.

20. Forn-Cunı´, G., Reis, E.S., Dios, S., Posada, D., Lambris, J.D., Figueras, A., and Novoa, B. (2014). The evolution and appearance of C3 duplications in fish originate an exclusive teleost c3 gene form with anti-inflammatory activity. PLoS One 9, e99673.

21. Graner, F., and Glazier, J.A. (1992). Simulation of biological cell sorting using a two-dimensional extended Potts model. Phys. Rev. Lett. 69, 2013–2016. Hawksworth, O.A., Coulthard, L.G., Taylor, S.M., Wolvetang, E.J., and Woodruff, T.M. (2014). Brief report: complement C5a promotes human embryonic stem cell pluripotency in the absence of FGF2. Stem Cells 32, 3278–3284.

22. Honczarenko, M., Ratajczak, M.Z., Nicholson-Weller, A., and Silberstein, L.E. (2005). Complement C3a enhances CXCL12 (SDF-1)-mediated chemotaxis of bone marrow hematopoietic cells independently of C3a receptor. J. Immunol. 175, 3698–3706.

23. Ho¨ pker, V.H., Shewan, D., Tessier-Lavigne, M., Poo, M., and Holt, C. (1999). Growth-cone attraction to netrin-1 is converted to repulsion by laminin-1. Nature 401, 69–73.

24. Jeanes, A., Coulthard, L.G., Mantovani, S., Markham, K., and Woodruff, T.M. (2015). Co-ordinated expression of innate immune molecules during mouse neurulation. Mol. Immunol. 68, 253–260.

25. Kane, D.A., McFarland, K.N., and Warga, R.M. (2005). Mutations in half baked/ E-cadherin block cell behaviors that are necessary for teleost epiboly. Development 132, 1105–1116.
26. Kardash, E., Reichman-Fried, M., Maı^tre, J., Boldajipour, B., Papusheva, E., Messerschmidt, E.-M., Heisenberg, C.-P., and Raz, E. (2010). A role for Rho GTPases and cell-cell adhesion in single-cell motility in vivo. Nat. Cell Biol. 12, 11–47.
27. Kee, N., Wilson, N., De Vries, M., Bradford, D., Key, B., and Cooper, H.M. (2008). Neogenin and RGMa control neural tube closure and neuroepithelial morphology by regulating cell polarity. J. Neurosci. 28, 12643–12653.
28. Keller, R.E. (1978). Time-lapse cinemicrographic analysis of superficial cell behavior during and prior to gastrulation in Xenopus laevis. J. Morphol. 157, 223–247.
29. Keller, R.E. (1980). The cellular basis of epiboly: an SEM study of deep-cell rearrangement during gastrulation in Xenopus laevis. J. Embryol. Exp. Morphol. 60, 201–234.
30. Kishimoto, Y., Murate, M., and Sugiyama, T. (1996). Hydra regeneration from recombined ectodermal and endodermal tissue. I. Epibolic ectodermal spreading is driven by cell intercalation. J. Cell Sci. 109 (Pt 4), 763–772.
31. Kuriyama, S., and Mayor, R. (2009). A role for Syndecan-4 in neural induction involving ERK- and PKC-dependent pathways. Development 136, 575–584.
32. Leslie, J.D., and Mayor, R. (2013). Complement in animal development: unexpected roles of a highly conserved pathway. Semin. Immunol. 25, 39–46.
33. Longo, D., Peirce, S.M., Skalak, T.C., Davidson, L.A., Marsden, M., Dzamba, B., and DeSimone, D.W. (2004). Multicellular computer simulation of morphogenesis: blastocoel roof thinning and matrix assembly in Xenopus laevis. Dev. Biol. 271, 210–222.
34. Luu, O., David, R., Ninomiya, H., and Winklbauer, R. (2011). Large-scale mechanical properties of Xenopus embryonic epithelium. Proc. Natl. Acad. Sci. USA 108, 4000–4005.
35. Majumdar, R., Sixt, M., and Parent, C.A. (2014). New paradigms in the establishment and maintenance of gradients during directed cell migration. Curr. Opin. Cell Biol. 30, 33–40.
36. Ma′ laga-Trillo, E., Solis, G.P., Schrock, Y., Geiss, C., Luncz, L., Thomanetz, V., and Stuermer, C.A.O. (2009). Regulation of embryonic cell adhesion by the prion protein. PLoS Biol. 7, e55.
37. Marsden, M., and DeSimone, D.W. (2001). Regulation of cell polarity, radial intercalation and epiboly in Xenopus: novel roles for integrin and fibronectin. Development 128, 3635–3647.
38. McLin, V.A., Hu, C.-H., Shah, R., and Jamrich, M. (2008). Expression of complement components coincides with early patterning and organogenesis in Xenopus laevis. Int. J. Dev. Biol. 52, 1123–1133.

39. Merks, R.M.H., Newman, S.A., and Glazier, J.A. (2004). Cell-oriented modeling of in vitro capillary development. Lect. Notes Comput. Sci. 3305, 425–434.

40. Meyen, D., Tarbashevich, K., Banisch, T.U., Wittwer, C., Reichman-Fried, M., Maugis, B., Grimaldi, C., Messerschmidt, E.-M., and Raz, E. (2015). Dynamic filopodia are required for chemokine-dependent intracellular polarization during guided cell migration in vivo. Elife 4, 1–25.

41. Nieuwkoop, P., and Faber, J. (1967). Normal Table of Xenopus Laevis (Amsterdam: North-Holland).

42. Petridou, N.I., Stylianou, P., and Skourides, P.A. (2013). A dominant-negative provides new insights into FAK regulation and function in early embryonic morphogenesis. Development 140, 4266–4276.

43. Ploper, D., Lee, H.X., and De Robertis, E.M. (2011). Dorsal–ventral patterning: crescent is a dorsally secreted Frizzled-related protein that competitively inhibits Tolloid proteases. Dev. Biol. 352, 317–328.

44. Ricklin, D., Hajishengallis, G., Yang, K., and Lambris, J.D. (2010). Complement: a key system for immune surveillance and homeostasis. Nat. Immunol. 11, 785–797.

45. Scarpa, E., Szabo´ , A., Bibonne, A., Theveneau, E., Parsons, M., and Mayor, R. (2015). Cadherin switch during EMT in neural crest cells leads to contact inhibition of locomotion via repolarization of forces. Dev. Cell 34, 421–434.

46. Schepis, A., Sepich, D., and Nelson, W.J. (2012). aE-catenin regulates cell-cell adhesion and membrane blebbing during zebrafish epiboly. Development 139, 537–546.

47. Scherber, C., Aranyosi, A.J., Kulemann, B., Thayer, S.P., Toner, M., Iliopoulos, O., and Irimia, D. (2012). Epithelial cell guidance by self-generated EGF gradients. Integr. Biol. 4, 259.

48. Schraufstatter, I.U., Khaldoyanidi, S.K., and DiScipio, R.G. (2015). Complement activation in the context of stem cells and tissue repair. World J. Stem Cells 7, 1090.

49. Shimizu, T., Yabe, T., Muraoka, O., Yonemura, S., Aramaki, S., Hatta, K., Bae, Y.-K., Nojima, H., and Hibi, M. (2005). E-cadherin is required for gastrulation cell movements in zebrafish. Mech. Dev. 122, 747–763.

50. Song, S., Eckerle, S., Onichtchouk, D., Marrs, J.A., Nitschke, R., and Driever, W. (2013). Pou5f1-dependent EGF expression controls E-cadherin endocytosis, cell adhesion, and zebrafish epiboly movements. Dev. Cell 24, 486–501.

51. Steventon, B., Araya, C., Linker, C., Kuriyama, S., and Mayor, R. (2009). Differential requirements of BMP and Wnt signalling during gastrulation and neurulation define two steps in neural crest induction. Development 136, 771–779.

52. Sugrue, S.P., and Hay, E.D. (1981). Response of basal epithelial cell surface and cytoskeleton to solubilized extracellular matrix molecules. J. Cell Biol. 91, 45–54.

53. Swat, M.H., Thomas, G.L., Belmonte, J.M., Shirinifard, A., Hmeljak, D., and Glazier, J.A. (2012). Multi-scale modeling of tissues using compu cell 3D. In Computational Methods in Cell Biology, A.R. Asthagiri and A.P. Arkin, eds. (Academic Press), pp. 325–366.
54. Szabo´, A., Varga, K., Garay, T., Hegedus, B., and Cziro } ´ k, A. (2012). Invasion from a cell aggregate–the roles of active cell motion and mechanical equilibrium. Phys. Biol. 9, 016010.
55. Ulrich, F., and Heisenberg, C.-P. (2009). Trafficking and cell migration. Traffic 10, 811–818.
56. Ulrich, F., Krieg, M., Scho¨ tz, E.-M., Link, V., Castanon, I., Schnabel, V., Taubenberger, A., Mueller, D., Puech, P.-H., and Heisenberg, C.-P. (2005). Wnt11 functions in gastrulation by controlling cell cohesion through Rab5c and E-cadherin. Dev. Cell 9, 555–564.
57. Volpe, S., Cameroni, E., Moepps, B., Thelen, S., Apuzzo, T., and Thelen, M. (2012). CCR2 acts as scavenger for CCL2 during monocyte chemotaxis. PLoS One 7, e37208.
58. Warga, R.M., and Kimmel, C.B. (1990). Cell movements during epiboly and gastrulation in zebrafish. Development 108, 569–580.
59. Woods, M.L., Carmona-Fontaine, C., Barnes, C.P., Couzin, I.D., Mayor, R., and Page, K.M. (2014). Directional collective cell migration emerges as a property of cell interactions. PLoS One 9, e104969.
60. Woolner, S., and Papalopulu, N. (2012). Spindle position in symmetric cell divisions during epiboly is controlled by opposing and dynamic apicobasal forces. Dev. Cell 22, 775–787.
61. Yamada, M., Udagawa, J., Matsumoto, A., Hashimoto, R., Hatta, T., Nishita, M., Minami, Y., and Otani, H. (2010). Ror2 is required for midgut elongation during mouse development. Dev. Dyn. 239, 941–953.
62. Yen, W.W., Williams, M., Periasamy, A., Conaway, M., Burdsal, C., Keller, R.E., Lu, X., and Sutherland, A. (2009). PTK7 is essential for polarized cell motility and convergent extension during mouse gastrulation. Development 136, 2039–2048.
63. Yu, S.R., Burkhardt, M., Nowak, M., Ries, J., Petra´ sek, Z., Scholpp, S., Schwille, P., and Brand, M. (2009). Fgf8 morphogen gradient forms by a source-sink mechanism with freely diffusing molecules. Nature 461, 533–536.
64. Zabel, B.A., Rott, A., and Butcher, E.C. (2015). Leukocyte chemoattractant receptors in human disease pathogenesis. Annu. Rev. Pathol. 10, 51–81.

4

CHAPTER

ANALYSIS OF NEURAL PROGENITORS FROM EMBRYOGENESIS TO JUVENILE ADULT IN XENOPUS LAEVIS REVEALS BIPHASIC NEUROGENESIS AND CONTINUOUS LENGTHENING OF THE CELL CYCLE

Raphaël Thuret, Hélene Auger and Nancy Papalopulù

Faculty of Life Sciences, Michael Smith Building, University of Manchester, Oxford Road, Manchester M13 9PT, UK

ABSTRACT

Xenopus laevis is a prominent model system for studying neural development, but our understanding of the long-term temporal dynamics of neurogenesis remains incomplete. Here, we present the first continuous description of neurogenesis in X. laevis, covering the entire period of development from the specification of neural ectoderm during gastrulation to juvenile frog. We have used molecular markers to identify progenitors and neurons, short-term bromodeoxyuridine (BrdU) incorporation to map the generation of newborn neurons and dual pulse S-phase labelling to characterise changes in their cell cycle length. Our study revealed the persistence of Sox3-positive progenitor cells from the earliest stages of neural development through to the juvenile adult. Two periods of intense neuronal generation were observed, confirming the existence of primary and secondary waves of neurogenesis, punctuated by a period of quiescence before metamorphosis and culminating in another period of quiescence in the young adult. Analysis of multiple parameters indicates that neural progenitors alternate between global phases of differentiation and amplification and that, regardless of their behaviour, their cell cycle lengthens monotonically during development, at least at the population level.

Citation: Raphaël Thuret, Hélene Auger and Nancy Papalopulu, Analysis of neural progenitors from embryogenesis to juvenile adult in Xenopus laevis reveals biphasic neurogenesis and continuous lengthening of the cell cycle, doi: 10.1242/bio.013391.

INTRODUCTION

Neurogenesis refers to the generation of post-mitotic neurons and glia from dividing progenitor cells, and it is a process necessary for the establishment of a functional central nervous system (CNS). Neurogenesis starts very early in the formation of an organism and although the exact phylotypic stage may vary between species, the first neurons generally appear after gastrulation at, or around, the time of neural tube closure. A large part of neurogenesis takes place during embryogenesis but it continues in post-natal stages. In many organisms, including mammals, it continues in the adult, restricted to a few areas of the CNS (D'Amico et al., 2011; for review see Kaslin et al., 2008; Schmidt et al., 2013), where cells are mostly quiescent but can be reactivated (Goldshmit et al., 2012; Kroehne et al., 2011;Muñoz et al., 2015; Zupanc and Ott, 1999). Thus, neurogenesis is a prolonged process. However, there is evidence from many organisms that neurogenesis can be divided in more or less distinct phases, characterised by the behaviour of neural progenitor cells (defined here in a broad sense as proliferating cells which may undergo fate determining symmetric or asymmetric divisions). For example, a change in behaviour over developmental time can be observed during the development of the mammalian cortex, where the properties of neural progenitor cells change in terms of marker gene expression, mode of division and generation of distinct neuronal subtypes (reviewed in Merkle and Alvarez-Buylla, 2006). Thus, during development neural progenitors may alternate between periods of expansion and periods of neurogenesis.

Early neurogenesis in lower vertebrates has been an invaluable model system in understanding the molecular control of neurogenesis, aided by the accessibility of the early embryo. For example, the*Xenopus* embryo has been instrumental in studying neurogenesis at the neural plate stage. Neural inducers (Piccolo et al., 1996; Zimmerman et al., 1996), Notch-Delta mediated lateral inhibition (Chitnis et al., 1995), and vertebrate proneural genes (Lee et al., 1995; Ma et al., 1996) were first described in this species. The early zebrafish embryo has also been instrumental and recent work in this model has been invaluable in understanding the molecular control of adult neurogenesis (Adolf et al., 2006; Chapouton et al., 2006, 2010; März et al., 2010; Rothenaigner et al., 2011).

The recent development and adaptation of better genetic manipulation tools in *Xenopus* open up an opportunity to study neurogenesis beyond the early embryonic stages and as a continuous process. Indeed, *Xenopus* with its distinct embryonic/larval and juvenile stages, punctuated by the process of metamorphosis offers a tremendous opportunity to study how transitions between different phases of neurogenesis are controlled in vertebrates.

As a first step in this direction, a thorough understanding of the neurogenic phases from early to late developmental stages is necessary. It is widely believed that two distinct phases of neurogenesis (primary and secondary) exist. It is thought that the primary phase establishes the embryonic CNS and a secondary phase generates the adult nervous system by largely replacing the primary nervous system (Hughes, 1957; Lamborghini, 1987). However, the evidence of distinct phases is fragmented because it is collated from different studies (Schlosser et al., 2002). Moreover, the data are often indirect and the temporal borders unclear as previous studies relied on identifying neurons well after their birth date, by the onset of late differentiation markers (Hartenstein, 1993; Lamborghini, 1980; Thors et al., 1982a,b).

Here, we describe for the first time in a single study how neural progenitors proliferate, self-renew and give rise to neurons over the whole course of embryonic, larval and post-metamorphic development. We have used Sox3, as a molecular marker of neural progenitors (for review see Pevny and Placzek, 2005) and combined it with cell proliferation reagents [phosphoHistoneH3 (pH3), BrdU] and a pan-neuronal marker (xMyT1) (Bellefroid et al., 1996; Bonev et al., 2011; Hudson et al., 2011), to obtain accurate snapshots of neural progenitor behaviour and neuronal birth rate over this extensive time course. Our observations show that the neural progenitor population is globally increasing during development and is maintained at least as far as the young adult frog. Two phases of intense progenitor division coincide with high neuronal production and are interrupted by a long period of apparent quiescence, enabling neural progenitors to slowly expand. Finally, the cell cycle length is gradually increasing to reach a maximum duration of 40 h at larval stages.

In conclusion, our work provides a dynamic cellular description of neural progenitor behaviour during the whole course of *Xenopus* development and lays the foundation for future molecular studies.

RESULTS

Sox3+ Neural Progenitors are Maintained throughout Xenopus life

We studied a long period of development, covering almost 2 months (Fig. 1A) starting from mid-gastrula [Nieuwkoop and Faber stage (NF)10.5] to juvenile stage (NF66, Fig. 1A). Larval stages were chosen on the basis of a screen using pH3 antibody and looking for changes of mitotic activity on whole mount dissected CNS (data not shown). To facilitate comparison between developmental stages, we focused on one area of the CNS. We chose the anterior spinal cord/posterior hindbrain because it was reliably identifiable by morphological landmarks (position of the otic vesicle, tapering of hindbrain roof) during the whole course of development after neural tube closure.

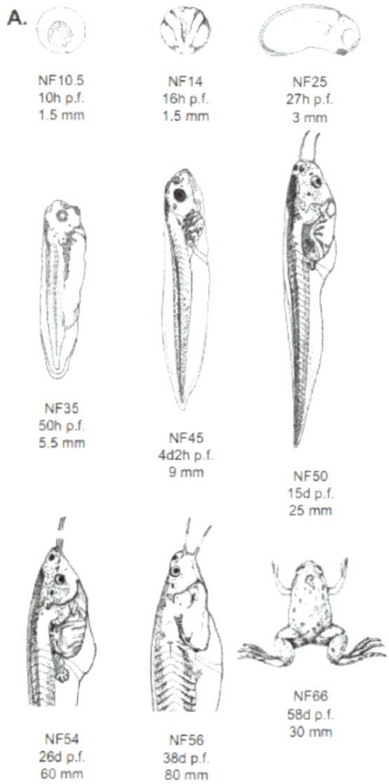
A.
NF10.5
10h p.f.
1.5 mm
NF14
16h p.f.
1.5 mm
NF25
27h p.f.
3 mm
NF35
50h p.f.
5.5 mm
NF45
4d2h p.f.
9 mm
NF50
15d p.f.
25 mm
NF54
26d p.f.
60 mm
NF56
38d p.f.
80 mm
NF66
58d p.f.
30 mm

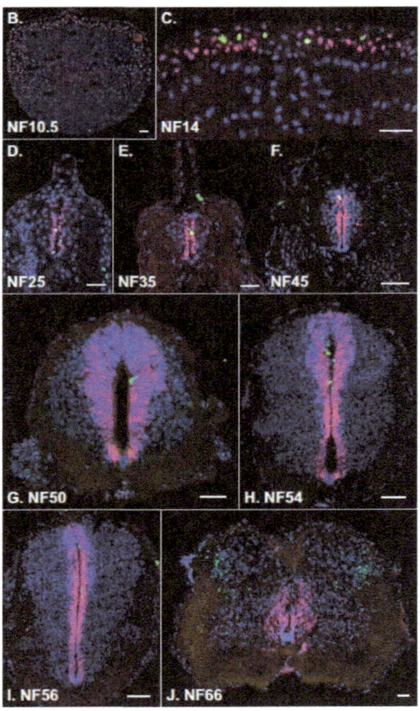
B.
NF10.5
C.
NF14
D.
NF25
E.
NF35
F.
NF45
G. NF50
H. NF54
I. NF56
J. NF66

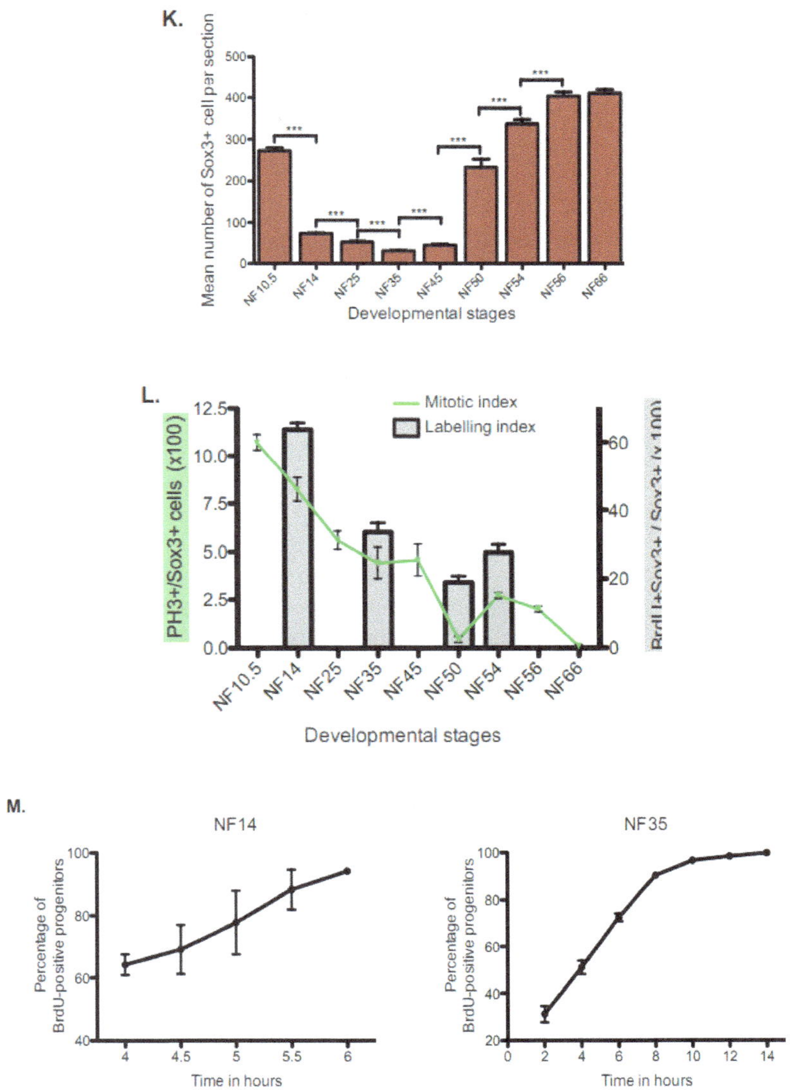

Figure 1: Sox3+ progenitors are maintained across *Xenopus* **development.** (A) Representation of *Xenopus laevis* stages used for the study. For each NF stage, the age in hours or days post-fertilisation (hpf or dpf) and the size of the animal (in mm) are provided. (B-J) Fluorescent immunodetection of Sox3 (red) and pH3 (green) at representative stages of *Xenopus*development. NF stages are indicated on each panel. Nuclei are counterstained with DAPI. Scale bars=50 μm. Note that during metamorphic stages, we identify a population of cells negative for Sox3 outside the ventricular zone but positive for PH3 in the posterior hindbrain/anterior spinal cord. These cells could represent a population of non-neurogenic glia such as astrocytes or oligodendrocytes (Maier and Miller, 1995; Yoshida et al., 1999). (K) Mean absolute number of Sox3-positive cells

per section during development. After the restriction of the expression of Sox3 to the neural plate, the number of positive cells decreases slightly until NF25 to augment and then reach a plateau by NF54. (L) Quantification of mitotic index (green curve) and labelling index (grey bars) of Sox3+ progenitors. The number of pH3-positive nuclei amongst Sox3-positive ones has been counted. Mitotic index is found to be high during neurulation (NF 14 to 25) and prometamorphosis (NF54-56). In between NF25 and NF54, the mitotic index decrease gradually to become nearly negligible at NF50. The labelling index represent the percentage of BrdU-positive Sox3-positive cells from animals injected with BrdU and fixed 3 h later. (M) Cumulative BrdU incorporation in neural progenitors at NF14 and NF35. More than 95% of the Sox3+ progenitors are positive for BrdU after 6 h at NF14 and after 10 to 12 h at NF35. Data represented as mean±s.e.m. ***$P<0.001$.

We analysed the immunoreactivity against Sox3 at mid-gastrula stage (NF10.5), neural plate (NF14), early tadpole (NF25), mid-tadpole (NF35), late tadpole (NF45), pre-metamorphic stage (NF50), metamorphic (NF54-56), and post-metamorphic juvenile stage (NF66; Fig. 1B-J). Prior to neural induction, Sox3 is expressed in all ectodermal cells of the gastrula (NF10.5, Fig. 1B) (Penzel et al., 1997;Zhang et al., 2003). From neural induction onwards, Sox3 becomes restricted and maintained to neural progenitors (Rogers et al., 2008). Analysis of pH3 labelling revealed that all mitotic cells are contained within the Sox3 domain (with the exception of some cells located at the lateral margins of the neural tube at NF66, Fig. 1B-J). Moreover the domain of expression of Sox3 does not overlap with immunoreactivity with neuronal markers such as xMyT1. Sox3 expression appears heterogeneous in intensity at the single cell level, therefore, it is likely that Sox3 labels neural progenitors, regardless of the possible existence of different subtypes or dynamic expression.

The data revealed that Sox3+ cells changes organization over the different developmental stages (Fig. 1B-J). At the neural plate stage (NF14), Sox3+ cells are organised into two neuroepithelial layers, a deep and a superficial one, as previously described (Schroeder, 1970, Fig. 1C). At neural tube closure (NF25, Fig. 1D), deep and superficial cells intercalate (Davidson and Keller, 1999) resulting in Sox3+ cells lining up the ventricle in a one-cell width layer (ventricular zone). This monolayer structure is maintained to early tadpole (NF35, Fig. 1E) but changes into a 2 to 4-cell width zone around the lumen of the spinal cord gradually in the stages leading to metamorphosis (from NF35 to NF50, Fig. 1E-G). During metamorphosis (NF54 to NF56, Fig. 1H-I), the progenitor zone thins again, although does not quite become a monolayer. After the climax of metamorphosis (NF66), the cytological structure of the spinal cord changes again compared to NF56, as it has been previously described (Thors et al., 1982b) (compare Fig. 1J to I). The length of the lumen of the spinal cord is reduced from approximately 800 μm to 400 μm. Sox3+ progenitors are redistributed in to a wider area around the ventricle, perhaps forming a ventricular and a subventricular zone. Overall, the cell density seems to be reduced in both the progenitors and the neurons compartments with a massive expansion of the white matter. Dividing cells (PH3+) outside the ventricular and subventricular zones appear towards the margins of the neural tube, coincident with the appearance of white matter.

The number of progenitor cells (number of Sox3+ cells per section, Fig. 1K) also changes during development. A sharp decrease is observed between gastrula (NF10.5) and neurulation (NF14); however, this is most likely due to neural induction that restricts the expression of Sox3 to neural progenitors. Between NF14 and 35, the number of Sox3+ cells decreases progressively from a mean of 70 to 30 Sox3+ cells per section. Some of this may be attributed to the elongation of the embryo between NF25 and NF35 (the total size of the embryo changes from 3 to 6 mm). A 1.5-fold increase is observed between NF35 and NF45 and a 6-fold increase between NF45 and NF50. This massive expansion may be responsible for the thickening of the progenitor zone at NF50 (Fig. 1G-I). In metamorphic stages (between NF50 and 56), the number of Sox3+ nuclei rises less sharply but steadily, reaches its maximum by NF56 and is maintained at NF66.

Sox3+ Cell Division Rate Peaks during Neurulation and Pro-Metamorphosis

To determine the rate of division of neural progenitors during development, we quantified the mitotic index, defined as the number of Sox3+ nuclei that were positive for the mitotic marker pH3 at any one time over the total number of Sox3+ progenitors (Fig. 1B-J and green line in L). This index enables us to observe if the population of neural progenitors is actively dividing at a particular stage of development and gives an indication of cell cycle length; assuming a constant length of M phase, a decline in the proportion of progenitors that are in M-phase shows that M-phase occupies a smaller portion of the cell cycle, indicative of lengthening of the cell cycle.

We observed that during gastrulation (NF10.5), about 10% of the cells in the ectoderm are found in mitosis. At early neurula (NF14), this percentage drops to 7.5% in the neuro-ectoderm and then decreases further to 5% at early tailbud stage (NF25). The mitotic index remains at this basal level of 5% up to larval stage (NF45). The biggest decrease is observed at pre-metamorphosis (NF50), when the mitotic index falls to just 0.4%, suggesting a really long cell cycle or a period of quiescence. Mitotic activity peaks again transiently during metamorphosis (NF54 and 56) to finally fade at juvenile stage (NF66).

We next estimated the labelling index on stages of development that are representative of the changes of the mitotic index to confirm this result by an independent method (Fig. 1L, charts bars). The labelling index is the quantification of BrdU+ cells amongst the Sox3+ population after a short pulse of BrdU (3 h). For all the stages analysed, the labelling index is in accordance with the mitotic index. At NF14, when 7.5% of the cells are found in mitosis, around 60% of the cells have incorporated BrdU indicative of fast cycling progenitors. At NF35, only 33% of the progenitors are positive for BrdU concomitant with the diminution of the mitotic index. At NF50, when the mitotic index in very low, only 19% of the progenitors are positive for BrdU and this proportion rises again to 28% at NF54.

Cumulative EdU incorporation of Sox3+ progenitors at NF14 and NF35 reveals that almost the entire Sox3+ population can be labelled within 6 to 12 h (depending on the stage; see Fig. 1M) confirming that Sox3-positive progenitors are actively dividing.

However, Sox3 continues to be expressed in progenitors even at stages when very few cells in mitosis are observed; therefore, Sox3 seems to label both actively (rapidly) dividing and quiescent (or slowly dividing) progenitors.

This data indicate that neural progenitors display two periods of active cell division, albeit with different cell cycle lengths (see also later); one from neurulation (around NF14) to larval stage (NF45) and a second one at early metamorphic stages (NF54 and 56). Neural progenitors also display two periods of relative quiescence, one before metamorphosis (NF50) and one when metamorphosis is complete (NF66). Notably, the distribution of mitotic cells at NF54 and NF56 seems different between the dorsal and the ventral part, suggesting some heterogeneity in progenitor behaviour at these stages (Fig. 1H,I).

Progenitors Alternate between Neurogenesis and Expansion

In order to understand the balance between neuronal production and progenitor expansion across development we looked both at the expansion of the neuronal population (neuronal birth) and the fate of dividing progenitors. We injected BrdU at various developmental stages ranging from NF10.5 to NF56 (Fig. 2A) and analysed BrdU incorporation 24 h later. In spite of the long time prior to analysis, BrdU is only available for 2.5 h (see Materials and Methods) therefore we were able to track progenitors that were in S-phase for the short period that BrdU was available and to identify them or their progeny 24 h later. BrdU persists at detectable levels in daughter cells after up to four divisions (Kiel et al., 2007). To assess the generation of neurons we used a pan-neuronal nuclear antibody against xMyT1, a transcription factor expressed from the early steps of neuronal differentiation pathway (Bellefroid et al., 1996). xMyt1 immunostaining is not detected either in progenitor cells that are Sox3-positive nor in neuronal precursors/prospective neurons that express X-Delta-1 (Chitnis et al., 1995;Henrique et al., 1995). The location of xMyt1 cells is more lateral than X-Delta-1 (Bellefroid et al., 1996; Schlosser et al., 2002), suggesting that xMyt1 is a slightly later marker and a good post-mitotic pan-neuronal marker.

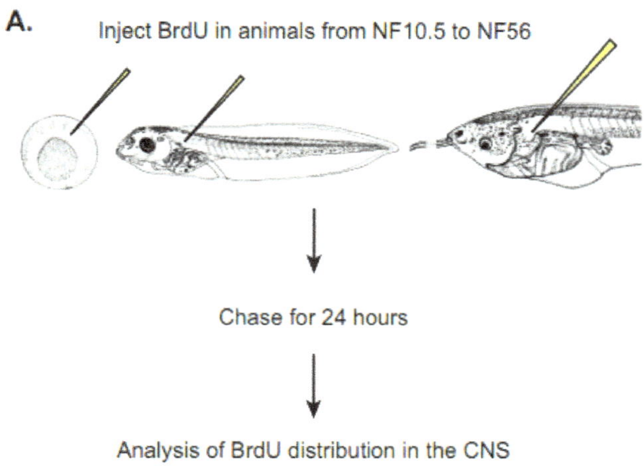

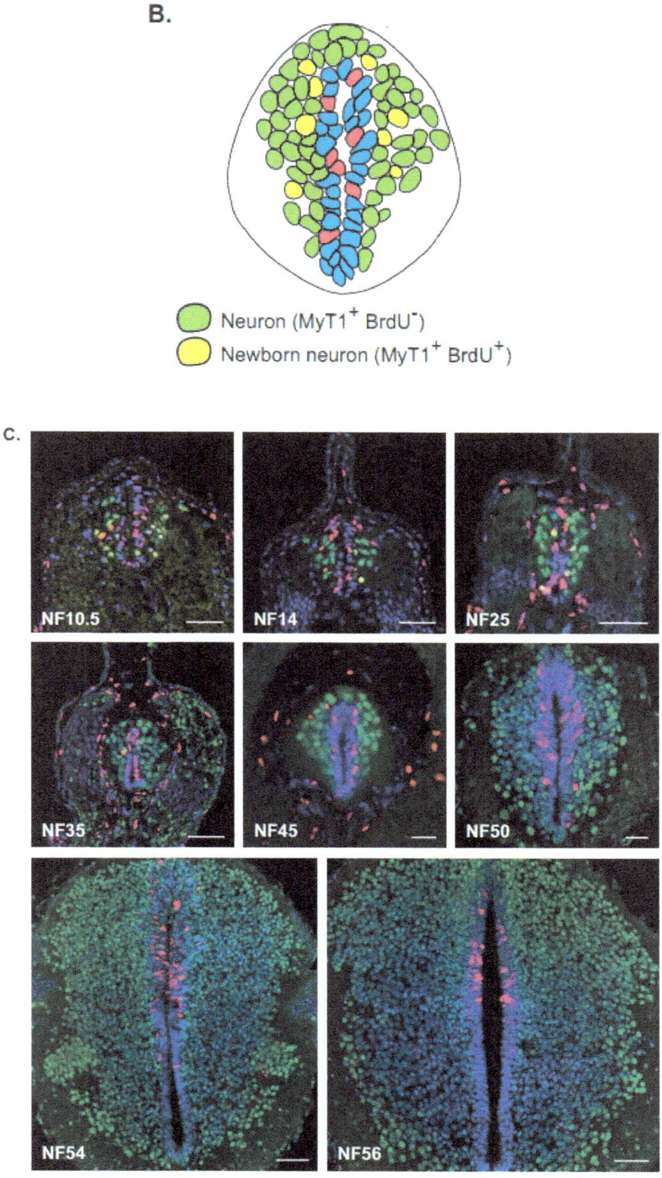

Figure 2: Neuronal birth during development. (A) Rationale of the birth dating experiment. Animals are injected with BrdU and kept for 24 h to allow BrdU incorporation and cell division to occur. (B) Schematic representation of spinal cord with identified cell populations. In a section of spinal cord, we can identify two categories of cells. (C) Fluorescent immunostaining against xMyT1 (green) and BrdU (red). Nuclei were counterstained with DAPI. Each developmental stage is indicated on the images. Scale bars=50 μm.

Two different categories of neurons were identified according to their immunoreactivity (Fig. 2B). (1) Neurons that were already differentiated at the time of labelling (BrdU− xMyT1+, Fig. 2B green nuclei). (2) Neurons that were born during the period of labelling from dividing progenitors (BrdU+ xMyT1+,Fig. 2B yellow nuclei).

We determined the rate of neuronal production, by quantifying the proportion of newly born neurons (BrdU+ MytT1+) during the time of labelling over the entire neuronal population (all MyT1+ nuclei) at the time of analysis (category 2 over 1+2, see above; Fig. 3A). We observed two main waves of neuronal birth. The first wave occurs during neurulation (primary neurogenesis, peaking at NF14) where 30% of the neurons found 24 h later in the spinal cord were born during this time period. The rate of neuronal birth rapidly decreases two-fold at NF25 to reach a low, basal level of neurogenesis of about 2% of newly born neurons in 24 h at NF35. This minimal level of neuronal birth is maintained throughout the end of embryogenesis up to pre-metamorphic stages. The second peak of neurogenesis is then happening during metamorphosis, at NF54 (Fig. 3A), where about 5% of the neuronal population is positive for BrdU, demonstrating their birth during the last 24 h. As evidenced by the mitotic index, most of the BrdU incorporation is observed in the dorsal part of the spinal at NF54 and 56, suggesting that Sox3+ progenitors are heterogeneous at these stages.

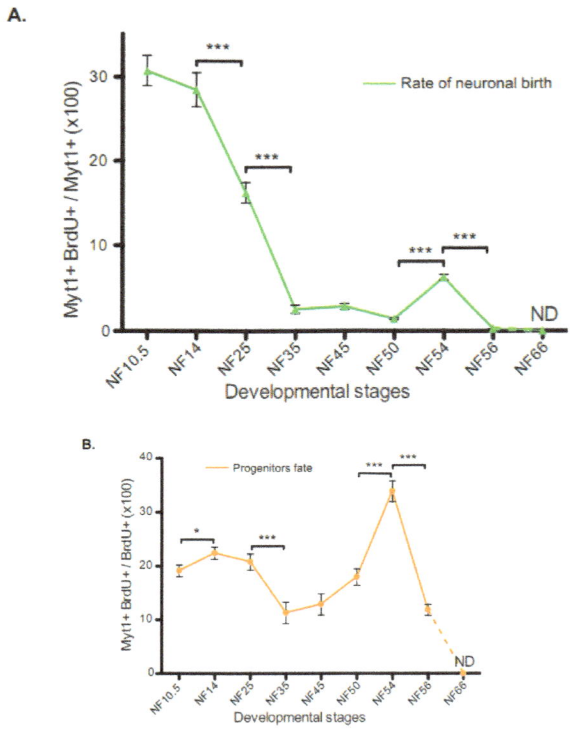

Figure 3: The behaviour of neural progenitors changes during development. The quantification of the birth dating results enables the estimation of two parameters according

to the populations considered. Since mitotic index was negligible at NF66, no birth dating data have been collected for this stage and values have been extrapolated to 0 (dashed lines). (A) Rate of neuronal birth. This graph represents the percentage of BrdU-positive cells among the xMyT1-positive cells and corresponds to the proportion of newly born neurons appeared during the last 24 h. Two main phases of neuronal birth can be identified at NF14 and NF54. In between, the rate of newly born neurons is very low. After NF54, the rate of newly born neurons decreases again. (B) Progenitors fate. This diagram represents the percentage of xMyT1-positive cells among the BrdU-positive population and is an indication of the choice made by progenitors to stay in the cell cycle by proliferating or to exit the cell cycle by differentiating into neurons. Data represented as mean±s.e.m. ***P<0.001, *P<0.05.

The analysis above estimates how many new neurons are added to the neuronal pool at any given point, but to gain insight into progenitor behaviour, we next looked at the outcome of progenitor division. We quantified the proportion of BrdU-positive cells that differentiated at different stages (xMyT1+ and BrdU+ over all BrdU+ cells, Fig. 3B). This shows that progenitors do not give rise to a homogeneously distributed progeny over development. On the contrary, we observe several changes in cell fate decision during the period studied. First, during early embryogenesis (from NF10.5 to NF 25), 20% of the BrdU- incorporating cells are becoming neurons. Between NF35 and 50, only 10% of BrdU+ cells become neurons (Fig. 3B), consistent with a low level of neuronal production (Fig. 3A). During this time, even though the mitotic and labelling index are steadily declining (Fig. 1L), the neural progenitors increase their number (Fig. 1K) more than giving rise to neurons (Fig. 3B), although a low level neuronal production persists (Fig. 3A). Finally, at metamorphosis, when neuronal production peaks again at NF54 (Fig. 3A), we observe that 30% of the cells incorporating BrdU give rise to neurons (Fig. 3B). Since the number of progenitor cells does not increase at this stage (Fig. 1K), we suggest that we are witnessing a switch in progenitor behaviour from self-renewal towards neuronal production.

The Cell Cycle Length Gradually Increases during *Xenopus* Development

Previous studies have shown that progenitors terminally dividing to give rise to neurons have a longer cell cycle (most particularly G1 phase) than progenitors that self-renew (Calegari, 2005; Salomoni and Calegari, 2010), although the S-phase is shorter (Arai et al., 2011; Cabochette et al., 2015). Since the majority of neuronal production in *Xenopus* takes place in two developmental phases, we wondered if this was linked to distinct changes of the cell cycle. In order to estimate the cell cycle length, we used the method of dual pulse S-phase labelling combining EdU, BrdU as well as Sox3 as a marker of the growth fraction (Martynoga et al., 2005; Peco et al., 2012; Sabherwal et al., 2014). We assumed that the growth fraction was identifiable to the Sox3+ progenitor population according to our cumulative BrdU incorporation results (Fig. 1M). We applied this method, during the first period of neuronal generation (neurulation, NF14 and mid-embryogenesis, NF35), the period of pre-metamorphic quiescence (larval stage, NF50) and the second period of neuronal generation (metamorphosis, NF54), adapting the protocol in a stage specific manner (Fig. 4A). This enabled us to determine the length of S-phase and the total duration of the cell cycle in the population of Sox3+ progenitors at these developmental stages (Fig. 4B,C). Our results show that the cell cycle gradually increases in length over developmental time (Fig. 4B). The average cell cycle is 5 h 52 m long at NF14, doubles in length at NF35 (12 h 24 m) and then increases again to a duration of 42 h at NF50. At NF54, the cell cycle shortens to a duration of 35 h, indicative of a reactivation of neural progenitors to produce new neurons. The same kind of trend is seen regarding the S-phase duration (Fig. 4C). We estimate that S-phase last for about 1 h 40 m at neural plate stage (NF14) and 2 h 20 m at mid-embryonic stage (NF35). Then, S-phase reaches 5 h at NF50 and remains of a comparable duration at NF54 (6 h 57 m). The lengthening of the cell cycle is consistent with the gradual decrease in the mitotic index between NF14 and NF50 (Fig. 1K). We subsequently observe a shortening of the cell cycle at NF54 in accordance with the increase of the mitotic index identified at this stage. This seems mostly due to the dorsal part of the spinal cord that is cycling much faster than the ventral part (Fig. 4D), confirming our observations of the distribution of aPH3 and BrdU staining at this stage (Fig. 1I, Fig. 2C). Our data also confirm and extend the conclusions of Frederick and Andrews that cell cycle is elongating, mostly by an increase of the G1-phase, which were based on flow cytometry of isolated nuclei from whole developing *Xenopus* embryos (Frederick and Andrews, 1994).

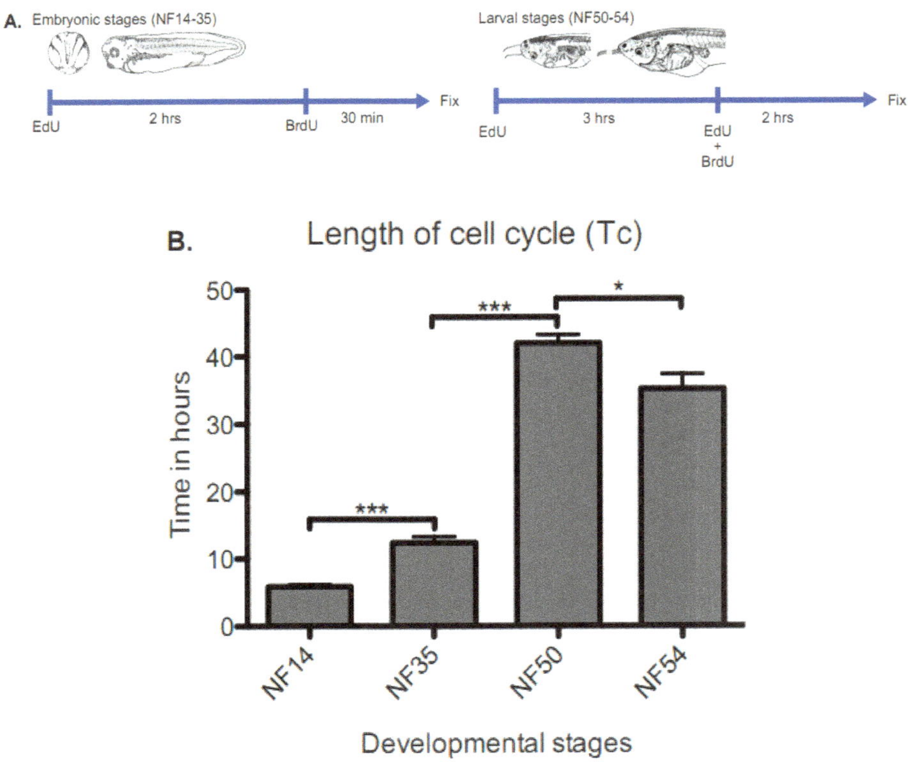

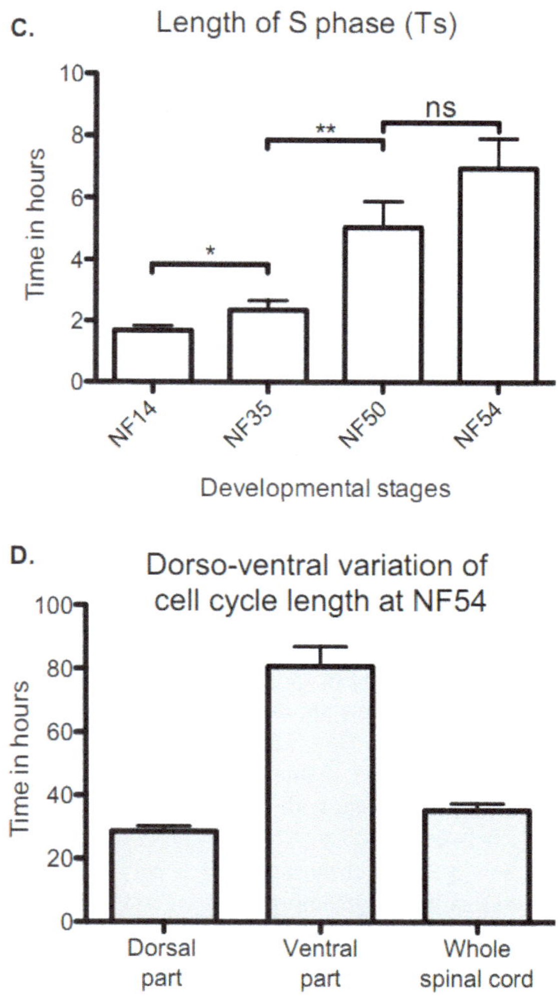

Figure 4: Estimation of cell cycle length during *Xenopus* **development.** (A) Protocols used for estimation of cell cycle length by dual pulse S-phase labelling. Different strategies have been used for stages corresponding to primary neurogenesis (NF14,*n*=13), progenitor amplification phase (NF35, *n*=9), prometamorphosis (NF50, *n*=5), and secondary neurogenesis during metamorphosis (NF54, *n*=5). (B) Total cell cycle length (Tc) and (C) S-phase length (Ts) at corresponding stages. Both Ts and Tc are increasing during development in a manner that seems independent of the neural progenitors behaviour. (D) Cell cycle length of the dorsal part versus the ventral part of the spinal cord at NF54. The overall cell cycle at this stage is 35 h represent a composite of a heterogeneous population of Sox3 progenitors where dorsal progenitors cycle faster (25 h) than the ventral ones (80 h). Data represented as mean±s.e.m. ***$P<0.001$, **$P<0.01$, *$P<0.05$.

DISCUSSION

We have presented a continuous cellular description of neurogenesis as a necessary first step to allow cellular behaviour to be understood from a molecular, mechanistic point of view. Our study revealed two distinct phases of neurogenesis, which is consistent with the previous view of primary and secondary neurogenesis based on compilation of cellular studies (Schlosser et al., 2002) or genes expression pattern of proneural genes in zebrafish and *Xenopus* (Mueller and Wullimann, 2003; Wullimann et al., 2005).

Summarising our data, we conclude that primary neurogenesis, corresponds to an initial neurogenic wave that takes place at neurulation (NF14), gradually declines until the mid-tadpole stage (NF35), and stays at a very low level until metamorphosis (NF54) when a second peak of neurogenesis is observed. During the second peak of neurogenesis, a smaller percentage of new neurons are added to the existing neuronal pool compared to the first (Fig. 3B). However, from the progenitors› point of view, secondary neurogenesis is a significant event; over 30% of progenitors present in the metamorphic neural tube (NF54, Fig. 3C) engage in neurogenic divisions, while only approx. 22% do so during primary neurogenesis (Fig. 3C). To reconcile the highly neurogenic progenitor fate (Fig. 3C) with the lower rate of generation of new neurons at NF54 (Fig. 3B), one has to consider that new neurons are added to a much larger population of existing neurons in NF54 than NF14.

Our data also revealed two phases of quiescence and allowed us to map these precisely on a temporal scale. The first period of quiescence takes place before metamorphosis (NF50), when we observed a very low number of cells in mitosis (pH3, Fig. 1L) as well as a very low rate of neuronal birth (Fig. 3A). The fact that there is a low but residual BrdU incorporation in the first period of quiescence indicates that some progenitors may cycle very slowly at NF50. Consistent with this, we showed a dramatic lengthening of the average cell cycle between NF35 to NF50 by direct measurements. Slow cycling is perhaps what one would expect if during the first period of quiescence progenitor cells are poised for reactivation at metamorphosis (NF54). Despite the maintenance of the expression of Sox3 in cells around the ventricle, the second period of quiescence observed in the post-metamorphic animal seems more definitive as the mitotic index falls to 0.1% between NF56 and NF66, consistent with other reports (D›Amico et al., 2011). It is also observed in other adult organisms, including mouse (Altman, 1969a,b; Grandel et al., 2006). We propose here the terms primary and secondary quiescence to describe these distinct events, in line with the terms primary and secondary neurogenesis. Both periods of quiescence are interesting in that they provide a reservoir of potentially re-activatable stem cells. Exit from secondary quiescence may be of particular interest in adult regeneration studies while exit from primary quiescence may be of particular interest in understanding the hormonal control of neuronal progenitor reactivation. Primary quiescence in *Xenopus* may indeed be a good model to study hormonally driven large-scale changes in neural progenitor behaviour in a vertebrate. In this perspective, our findings suggest similarities to the *Drosophila* model where it has been shown that neuroblasts enter quiescence during late embryogenesis, creating a break between embryonic and post-embryonic neurogenesis. Post-embryonic neuroblasts are then reactivated by insulin/IGF pathway (Chell and Brand, 2010). It is interesting to note that NF35 marks the onset of a period of change in the behaviour of progenitors, as evidenced by the elongation of the cell cycle length (also witnessed by the changes in mitotic and labelling indices) and rate of neuronal birth which intensifies at NF45 and lasts until NF50. At NF35, thyroid hormone receptors (TR) start to be expressed (Eliceiri and Brown, 1994) and act as repressors in the absence of thyroid hormone (TH). At NF54, TH starts to be produced (Leloup and Buscaglia, 1977) and binds to TR, converting repressive unbound TR to activating bound TR, enabling the metamorphic process. These changes in TR activity could be partially responsible for the change in progenitor behaviour. Notably, it has been shown before that TH treatment had an effect on neural progenitors, modifying their cycling properties (Denver et al., 2009).

By integrating information obtained by different methods, we also gained some insight into the behaviour of progenitors through developmental time, which would not have been possible with any of these methods alone. In principle, neural progenitors may divide symmetrically in terms of fate to produce 2 more progenitors (symmetric proliferative division) or 2 neurons (symmetric terminal division). Alternatively, they may also divide asymmetrically in terms of fate, to produce one neuron and one progenitor (asymmetric division). Of those, only symmetric proliferative divisions increase the number of progenitor cells, while terminal symmetric divisions decrease it and asymmetric divisions maintain it. Thus, the change in the number of progenitor cells (Fig. 1K) taken together with the rate of neuronal birth (Fig. 3A) and the fate of progenitor divisions (Fig. 3B) gave us a clue as to the mode of their division.

First, during early embryogenesis (from NF10.5 to NF 25), 20% of the BrdU incorporating cells are becoming neurons (Fig. 3B), which when taken together with the approximate stable number of progenitors between NF14 and 25 indicates that, at these stages, differentiated neurons are generated without depleting the progenitor pool, indicating asymmetric divisions. From neurula (NF14) to tadpole stage (NF35) the number of progenitors per section decreases by half. The most likely explanation to the reduction of progenitor numbers between these stages would be the antero-posterior elongation of the embryo, doubling its size from NF14 to NF25 (15 to 30 mm; Nieuwkoop and Faber, 1956). There is also an intense generation of neurons at this period; in fact, a higher percentage of neurons (30%) are generated in the 24 h following neurulation than at any other time in the formation of the animal (seeFig. 3A). Thus, we suggest that the data are consistent with an asymmetric neurogenic mode of division for the majority of progenitors at this stage, as this would tend to conserve rather than drastically reduce the number of progenitors, while allowing neuronal birth.

From NF35 to NF50, the rate of neuronal birth is very low and the number of progenitors increases, implying symmetric proliferative divisions between these stages. The massive increase of progenitor number observed between NF45 and NF 50 (Fig. 1K) may seem paradoxical if one considers the low level of BrdU incorporation at NF 45 and NF50 i.e. the stages preceding primary quiescence. However, it can be explained if one considers that cell division is integrated over a long period of time; for example, there is 2 days between NF35 to NF45 and 11 days between NF45 to NF 50 (see Fig. 1A). Alternatively, there may be a peak in proliferation, which takes place in a narrow window between NF45 and NF50 that we did not detect. Nonetheless, the fact that some divisions are neurogenic in fate (Fig. 3B) suggests that between these stages there is a mixture of behaviour where the majority of divisions are (slow) symmetric proliferative but some are asymmetric neurogenic.

Subsequently, during the period of secondary neurogenesis, which spans metamorphosis (between stages NF50 and NF56), the rate of progenitor pool expansion slows down and neuronal production picks up, implying a switch to an asymmetric neurogenic mode of division. The number of Sox3-positive progenitor cells stabilises between NF56 and NF 66 (Fig. 1J) and we no longer observed any mitotic cells. Of course, at all stages, the same outcome could be explained by more complicated scenarios, if there is a heterogeneity of behaviour in the progenitor pool. Nevertheless our interpretation of the data form a useful preliminary hypothesis that remains to be tested by more sophisticated methods.

Finally, our data revealed that neural progenitors display a seven-fold change in cell cycle length from neural plate to metamorphic stages in an apparently gradual increase. This was observed indirectly by the decrease in the mitotic index (Fig. 2L) and directly, by measurement of cell cycle and S phase length (Fig. 4). The short cell cycle length (6 h) observed during primary neurogenesis is similar with that observed, in the *Xenopus* retina (El Yakoubi et al., 2012), and is conserved at comparable development stages in zebrafish (He et al., 2012; Leung et al., 2011; Lyons, 2003), chicken (Peco et al., 2012) and mouse (Arai et al., 2011; Calegari, 2005; Caviness et al., 1995). In the mouse cortex, the transition from neuro-epithelial cells to progenitors to slow cycling stem cells during development have been shown to be accompanied by a gradual slowing down of the cell cycle to finally reach quiescence where cells can be reactivated (Lange and Calegari, 2010). Indeed, in our system, cell cycle gradually lengthens to reach a maximum length of about 40 h at NF50. At NF54, during secondary neurogenesis, we observe a shortening of the cell cycle, mainly coming from the dorsal part of the spinal cord indicative of the reactivation of neural progenitors and their subsequent differentiation.

In conclusion, our comprehensive analysis has revealed persistence of stem cells and population level changes during development. Specifically, there are two periods of neurogenesis and two periods of quiescence, accompanied by a continuous lengthening of the cell cycle and a possible alternation in division mode. These conclusions do not exclude the possibility of heterogeneity in progenitor behaviour (suggested by expression of TRα and TRβ; Denver et al., 2009) and remain to be confirmed by single cell analysis. In future, more comprehensive analyses at the single cell level will be needed to gain further insight into neural progenitor behaviour at different stages of development.

MATERIALS AND METHODS

Embryos, Tadpoles and Developmental Stages Choice

Animal experiments were approved by the University of Manchester Ethical Review Panel and were undertaken under UK Home Office project license PPL 70/7648. *Xenopus laevis* females were purchased from the *Xenopus* stock centre, Portsmouth, UK. Embryos were obtained by *in vitro*fertilization of hormone-induced laid eggs. Developmental stages were determined and are referenced in the text according to Nieuwkoop and Faber *Xenopus* table of development (Nieuwkoop and Faber, 1956). Once the animals reached feeding stage (NF45), they were kept at 16°C in filtered water and fed daily with spirulina until they reach the required NF stage for experimentations. The time interval between different stages is shown in Fig. 1A.

Bioavailability of Uridine Analogues

In order to set up our experiments, we first determined the time a uridine analogue was available for labelling after injection. To estimate this parameter, we performed a dual pulse S-phase labelling at NF35 using BrdU and 5-Ethynyl Uridine (EdU). After an initial injection of BrdU (10 mM, Roche), embryos were kept for an increasing amount of time (from 1 to 3.5 h) before being injected with EdU (10 mM, Life Technologies) and finally fixed 30 min later. The minimum amount of time necessary to see the appearance of EdU+ only cells corresponds to the time where BrdU is no longer available. We thus estimated that BrdU is available for a period of time ranging between 2 and 2.5 h since EdU+ only cells starts to be identified by 2.5 h (arrowheads) and become more numerous afterwards.

Birth Dating Experiments

In order to analyse the fate of dividing progenitors at different stages of development, we used the method of birth dating consisting of following the progeny of cells that incorporated a S-phase marker such as BrdU. Animals were injected with BrdU (10 mM, Roche) at the desired NF stage and kept for 24 h at 18°C (Fig. 2A) in order to allow neuronal differentiation to happen in the progeny of dividing cells. After fixation, samples were processed according to the method described in the immunofluorescence section. An additional incubation in 2 N HCl at 37°C for 30 min was performed before the blocking step of primary antibody incubation in order to enable accessibility to the BrdU antigen.

Cumulative EdU Incorporation

To estimate the growth fraction (i.e. percentage of cycling cells in the Sox3 population), we performed cumulative EdU incorporation at NF14 and NF35. Embryos were injected every 2 h with 10 nM EdU (4.2 nl at NF 14, 3×4.2 nl at NF35) and fixed every 30 min for NF14 or 2 h for NF35. Samples were then processed for Sox3 immunostaining and EdU detection and consecutively imaged. The percentage of Edu+Sox3+ was estimated from five consecutive sections taken from three different embryos for each time point.

Cell Cycle Length Analysis

Dual pulse S-phase labelling (Martynoga et al., 2005; Peco et al., 2012; Sabherwal et al., 2014) was used to estimate the length of the S-phase (Ts) and subsequently determine the cell cycle length (Tc) in the population of Sox3 progenitors. This method relies on the fact that the percentage of cells in a particular phase of the cell cycle is proportional to the length of this phase. The length of the S phase is estimated by two successive injections of different uridine analogues (in our case EdU and BrdU). Once the S phase length has been determined (see below), we can extrapolate the length of the whole cell cycle in the considered population. Practically, a first injection of EdU is provided and chased for a fixed amount of time (Texp) when BrdU is injected and chased for the minimal duration for reliable detection. Animals are then fixed and processed for EdU, BrdU and Sox3 detection. Three cell populations are identifiable:

1. Sox3+ cells corresponding the whole pool of cycling cells (proliferating fraction, P cells).
2. Sox3+ EdU+ only cells represent cells that left S phase before BrdU was available (leaving fraction, L cells).
3. Sox3+ EdU+ BrdU+ cells representing the cells that were in S phase at the time of fixation (replicative fraction, S cells).

According to Nowakowski (1989), the ratio of the duration of two phases of the cell cycle is equal to the ratio of the number cells in each of these phases. Hence, we can estimate the S phase length by the following relation:

$$\text{Texp}/\text{Ts} = \text{L cells}/\text{S cells}$$
$$\text{Ts} = \text{Texp}/(\text{L cells}/\text{S cells}).$$

Similarly, we can determine Tc by using a similar formula:

$$\text{Ts}/\text{Tc} = \text{S cells}/\text{P cells then Tc} = \text{Ts}/(\text{S cells}/\text{P cells})$$

We estimated that uridine analogues detection was robust and clear after 30 min at NF 14 and NF35 and after 2 h at NF 50 and NF54. Animals were isolated at the desired NF stage and injected first with EdU (10 mM in DMSO, Life Technologies). After 2 to 3 h incubation (depending on the developmental stage, Fig. 4A), an injection of BrdU (10 mM in DMSO, Roche, or EdU+BrdU at 10 mM each in case of 3 h incubation) was provided. Animals were then sacrificed 30 min to 2 h later according to their stage and processed as described in the immunofluorescence section. EdU detection was made using Click-iT EdU imaging kit from Life Technologies as described in Auger et al. (2012). BrdU antigen retrieval using 2 N HCl was performed after EdU staining and before primary antibody blocking step. The different cells categories described earlier were then counted on 10 consecutive sections from at least 5 different animals.

Immunofluorescence

Embryonic stages (up to NF45) were fixed in MEMFA for 1 h at room temperature, and then washed and kept in methanol at –20°C until embedding. Once anaesthetised in MS222, larval stages (from NF50) were fixed in 4% paraformaldehyde in 1× PBS overnight at 4°C. Central nervous systems were then dissected and kept in methanol at –20°C until further processing. Embedding, cryosectioning and immunofluorescence were performed as described in Auger et al. (2012). The following primary antibodies were used: rabbit anti-Sox3 (Eurogentec, custom made as described in Zhang et al.,

2003, 1:1000), rabbit anti-xMyt1 (Eurogentec, custom made as described in Sabherwal et al., 2009, 1 in 1000), mouse anti-pH3 (Abcam Ab14955, 1 in 300), mouse anti-BrdU (Roche Cat. No. 11 170 376 001, 1:500) and mouse anti-BrdU MoBU (Life Technologies B35128, 1:300). Corresponding secondary antibodies coupled to Alexa fluorophores (anti-mouse Alexa 488, anti-mouse Alexa 568, anti-rabbit Alexa 488, anti-rabbit Alexa 568, anti-rabbit Alexa 647) were purchased from Life Technology and used at 1:500 dilution. Double immunostaining with same species antibody was performed using Zenon Kit (Life Technologies). Once a regular antibody staining was performed for one of the antibodies, the other antibody was complexed with the Zenon IgG and incubated on the sections overnight at 4°C. Immunofluorescence combined with *in situ* hybridisation was performed as described in Auger et al. (2012). X-Delta-1 is a kind gift from Eric Bellefroid (used in Bellefroid et al., 1996)

Images Collection and Data Analysis

Images were collected on an Olympus FV1000 confocal microscope or a Nikon Widefield microscope. For each quantified value, at least 3 different animals were sampled on 10 consecutive 12 μm sections. Quantification was then manually performed using the Cell Counter plugin of ImageJ. Data shown are expressed as the mean±s.e.m. and are presented. Normality tests were applied on datasets and appropriate *t*-tests were used in order to assess the statistical significance of the data. Significance values are $***P<0.001$, $**P<0.01$, $*P<0.05$.

AUTHOR CONTRIBUTIONS

R.T. conceived designed, performed and analysed the experiments. H.A. performed and analysed part of the birth dating experiments. N.P. supervised the work; R.T. and N.P. wrote the manuscript.

ACKNOWLEDGEMENTS

We thank Dr Karel Dorey, Eamon Dubaissi, Shane Herbert, Nitin Sabherwal, Ximena Soto and Nick Phillips for their useful comments and discussions.

REFERENCES

1. Adolf, B., Chapouton, P., Lam, C. S., Topp, S., Tannhäuser, B., Strähle, U., Götz, M. and Bally-Cuif, L. (2006). Conserved and acquired features of adult neurogenesis in the zebrafish telencephalon. Dev. Biol. 295, 278-293.
2. Altman, J. (1969a). Autoradiographic and histological studies of postnatal neurogenesis. III. Dating the time of production and onset of differentiation of cerebellar microneurons in rats. J. Comp. Neurol. 136, 269-293.
3. Altman, J. (1969b). Autoradiographic and histological studies of postnatal neurogenesis. IV. Cell proliferation and migration in the anterior forebrain, with special reference to persisting neurogenesis in the olfactory bulb. J. Comp. Neurol. 137, 433-457.

4. Arai, Y., Pulvers, J. N., Haffner, C., Schilling, B., Nüsslein, I., Calegari, F. and Huttner, W. B. (2011). Neural stem and progenitor cells shorten S-phase on commitment to neuron production. Nat. Commun. 2, 154.
5. Auger, H., Thuret, R., El Yakoubi, W. and Papalopulu, N. (2012). A bromodeoxyuridine (BrdU) based protocol for characterizing proliferating progenitors in Xenopus embryos. Methods Mol. Biol. 917, 461-475.
6. Bellefroid, E. J., Bourguignon, C., Hollemann, T., Ma, Q., Anderson, D. J., Kintner, C. and Pieler, T. (1996). X-MyT1, a Xenopus C2HC-type zinc finger protein with a regulatory function in neuronal differentiation. Cell 87, 1191-1202.
7. Bonev, B., Pisco, A. and Papalopulu, N. (2011). MicroRNA-9 reveals regional diversity of neural progenitors along the anterior-posterior axis. Dev. Cell 20, 19-32.
8. Cabochette, P., Vega-Lopez, G., Bitard, J., Parain, K., Chemouny, R., Masson, C., Borday, C., Hedderich, M., Henningfeld, K. A., Locker, M. et al. (2015). YAP controls retinal stem cell DNA replication timing and genomic stability. Elife Sci. 4, e08488.
9. Calegari, F. (2005). Selective lengthening of the cell cycle in the neurogenic subpopulation of neural progenitor cells during mouse brain development. J. Neurosci. 25, 6533-6538.
10. Caviness, V. S., Takahashi, T. and Nowakowski, R. S. (1995). Numbers, time and neocortical neuronogenesis: a general developmental and evolutionary model. Trends Neurosci. 18, 379-383.
11. Chapouton, P., Adolf, B., Leucht, C., Tannhauser, B., Ryu, S., Driever, W. and Bally-Cuif, L. (2006). her5 expression reveals a pool of neural stem cells in the adult zebrafish midbrain. Development 133, 4293-4303.
12. Chapouton, P., Skupien, P., Hesl, B., Coolen, M., Moore, J. C., Madelaine, R., Kremmer, E., Faus-Kessler, T., Blader, P., Lawson, N. D. et al. (2010). Notch activity levels control the balance between quiescence and recruitment of adult neural stem cells. J. Neurosci. 30, 7961-7974.
13. Chell, J. M. and Brand, A. H. (2010). Nutrition-responsive glia control exit of neural stem cells from quiescence. Cell 143, 1161-1173.
14. Chitnis, A., Henrique, D., Lewis, J., Ish-Horowicz, D. and Kintner, C. (1995). Primary neurogenesis in Xenopus embryos regulated by a homologue of the Drosophila neurogenic gene Delta. Nature 375, 761-766.
15. D'Amico, L. A., Boujard, D. and Coumailleau, P. (2011). Proliferation, migration and differentiation in juvenile and adult Xenopus laevis brains. Brain Res. 1405, 31-48.
16. Davidson, L. A. and Keller, R. E. (1999). Neural tube closure in Xenopus laevis involves medial migration, directed protrusive activity, cell intercalation and convergent extension. Development 126, 4547-4556.

17. Denver, R. J., Hu, F., Scanlan, T. S. and Furlow, J. D. (2009). Thyroid hormone receptor subtype specificity for hormone-dependent neurogenesis in Xenopus laevis. Dev. Biol. 326, 155-168.
18. El Yakoubi, W., Borday, C., Hamdache, J., Parain, K., Tran, H. T., Vleminckx, K., Perron, M. and Locker, M. (2012). Hes4 controls proliferative properties of neural stem cells during retinal ontogenesis. Stem Cells 30, 2784-2795.
19. Eliceiri, B. P. and Brown, D. D. (1994). Quantitation of endogenous thyroid hormone receptors alpha and beta during embryogenesis and metamorphosis in Xenopus laevis. J. Biol. Chem. 269, 24459-24465.
20. Frederick, D. L. and Andrews, M. T. (1994). Cell cycle remodeling requires cell-cell interactions in developing Xenopus embryos. J. Exp. Zool. 270, 410-416.
21. Goldshmit, Y., Sztal, T. E., Jusuf, P. R., Hall, T. E., Nguyen-Chi, M. and Currie, P. D. (2012). Fgf-Dependent Glial cell bridges facilitate spinal cord regeneration in zebrafish. J. Neurosci. 32, 7477-7492.
22. Grandel, H., Kaslin, J., Ganz, J., Wenzel, I. and Brand, M. (2006). Neural stem cells and neurogenesis in the adult zebrafish brain: origin, proliferation dynamics, migration and cell fate. Dev. Biol. 295, 263-277.
23. Hartenstein, V. (1993). Early pattern of neuronal differentiation in the Xenopus embryonic brainstem and spinal cord. J. Comp. Neurol. 328, 213-231.
24. He, J., Zhang, G., Almeida, A. D., Cayouette, M., Simons, B. D. and Harris, W. A. (2012). How variable clones build an invariant retina. Neuron 75, 786-798.
25. Henrique, D., Adam, J., Myat, A., Chitnis, A., Lewis, J. and Ish-Horowicz, D. (1995). Expression of a Delta homologue in prospective neurons in the chick. Nature 375, 787-790.
26. Hudson, L. D., Romm, E., Berndt, J. A. and Nielsen, J. A. (2011). A tool for examining the role of the zinc finger myelin transcription factor 1 (Myt1) in neural development: Myt1 knock-in mice. Transgenic Res. 20, 951-961.
27. Hughes, A. (1957). The development of the primary sensory system in Xenopus laevis (Daudin). J. Anat. 91, 323-338.
28. Kaslin, J., Ganz, J. and Brand, M. (2008). Proliferation, neurogenesis and regeneration in the non-mammalian vertebrate brain. Philos. Trans. R. Soc. B Biol. Sci. 363, 101-122.
29. Kiel, M. J., He, S., Ashkenazi, R., Gentry, S. N., Teta, M., Kushner, J. A., Jackson, T. L. and Morrison, S. J. (2007). Haematopoietic stem cells do not asymmetrically segregate chromosomes or retain BrdU. Nature 449, 238-242.
30. Kroehne, V., Freudenreich, D., Hans, S., Kaslin, J. and Brand, M. (2011). Regeneration of the adult zebrafish brain from neurogenic radial glia-type progenitors. Development 138, 4831-4841.
31. Lamborghini, J. E. (1980). Rohon-beard cells and other large neurons in Xenopus embryos originate during gastrulation. J. Comp. Neurol. 189, 323-333.
32. Lamborghini, J. E. (1987). Disappearance of Rohon-Beard neurons from the spinal cord of larval Xenopus laevis. J. Comp. Neurol. 264, 47-55.

33. Lange, C. and Calegari, F. (2010). Cdks and cyclins link G1 length and differentiation of embryonic, neural and hematopoietic stem cells. Cell Cycle 9, 1893-1900.

34. Lee, J. E., Hollenberg, S. M., Snider, L., Turner, D. L., Lipnick, N. and Weintraub, H. (1995). Conversion of Xenopus ectoderm into neurons by NeuroD, a basic helix-loop-helix protein. Science 268, 836-844.

35. Leloup, J. and Buscaglia, M. (1977). Triiodothyronine, hormone of amphibian metamorphosis. Comptes Rendus Hebdomadaires Des Seances De L Academie Des Sciences Serie D 284, 2261-2263.

36. Leung, L., Klopper, A. V., Grill, S. W., Harris, W. A. and Norden, C. (2011). Apical migration of nuclei during G2 is a prerequisite for all nuclear motion in zebrafish neuroepithelia. Development 138, 5003-5013.

37. Lyons, D. A. (2003). Monitoring neural progenitor fate through multiple rounds of division in an intact vertebrate brain. Development 130, 3427-3436.

38. Ma, Q., Kintner, C. and Anderson, D. J. (1996). Identification of neurogenin, a vertebrate neuronal determination gene. Cell 87, 43-52.

39. Maier, C. E. and Miller, R. H. (1995). Development of Glial cytoarchitecture in the frog spinal cord. Dev. Neurosci. 17, 149-159.

40. Martynoga, B., Morrison, H., Price, D. J. and Mason, J. O. (2005). Foxg1 is required for specification of ventral telencephalon and region-specific regulation of dorsal telencephalic precursor proliferation and apoptosis. Dev. Biol. 283, 113-127.

41. März, M., Chapouton, P., Diotel, N., Vaillant, C., Hesl, B., Takamiya, M., Lam, C. S., Kah, O., Bally-Cuif, L. and Strähle, U. (2010). Heterogeneity in progenitor cell subtypes in the ventricular zone of the zebrafish adult telencephalon. Glia 58, 870-888.

42. Merkle, F. T. and Alvarez-Buylla, A. (2006). Neural stem cells in mammalian development. Curr. Opin. Cell Biol. 18, 704-709.

43. Mueller, T. and Wullimann, M. F. (2003). Anatomy of neurogenesis in the early zebrafish brain. Brain Res. Dev. Brain Res. 140, 137-155.

44. Muñ oz, R., Edwards-Faret, G., Moreno, M., Zuñ iga, N., Cline, H. and Larraın, J. (2015). Regeneration of Xenopus laevis spinal cord requires Sox2/3 expressing cells. Dev. Biol. [Epub ahead of print]

45. Nieuwkoop, P. D. and Faber, J. (1956). Normal table of Xenopus laevis (Daudin). A systematical and chronological survey of the development from the fertilized egg till the end of metamorphosis. Amsterdam: North-Holland.

46. Nowakowski, R. S., Lewin, S. B. and Miller, M. W. (1989). Bromodeoxyuridine immunohistochemical determination of the lengths of the cell cycle and the DNA-synthetic phase for an anatomically defined population. J. Neurocytol. 18, 311-318.

47. Peco, E., Escude, T., Agius, E., Sabado, V., Medevielle, F., Ducommun, B. and Pituello, F. (2012). The CDC25B phosphatase shortens the G2 phase of neural progenitors and promotes efficient neuron production. Development 139, 1095-1104.
48. Penzel, R., Oschwald, R., Chen, Y., Tacke, L. and Grunz, H. (1997). Characterization and early embryonic expression of a neural specific transcription factor xSOX3 in Xenopus laevis. Int. J. Dev. Biol. 41, 667-677.
49. Pevny, L. and Placzek, M. (2005). SOX genes and neural progenitor identity. Curr. Opin. Neurobiol. 15, 7-13.
50. Piccolo, S., Sasai, Y., Lu, B. and De Robertis, E. M. (1996). Dorsoventral patterning in Xenopus: inhibition of ventral signals by direct binding of chordin to BMP-4. Cell 86, 589-598.
51. Rogers, C. D., Archer, T. C., Cunningham, D. D., Grammer, T. C. and Silva Casey, E. M. (2008). Sox3 expression is maintained by FGF signaling and restricted to the neural plate by Vent proteins in the Xenopus embryo. Dev. Biol. 313, 307-319.
52. Rothenaigner, I., Krecsmarik, M., Hayes, J. A., Bahn, B., Lepier, A., Fortin, G., Gotz, M., Jagasia, R. and Bally-Cuif, L. (2011). Clonal analysis by distinct viral vectors identifies bona fide neural stem cells in the adult zebrafish telencephalon and characterizes their division properties and fate. Development 138, 1459-1469.
53. Sabherwal, N., Tsutsui, A., Hodge, S., Wei, J., Chalmers, A. D. and Papalopulu, N. (2009). The apicobasal polarity kinase aPKC functions as a nuclear determinant and regulates cell proliferation and fate during Xenopus primary neurogenesis. Development 136, 2767-2777.
54. Sabherwal, N., Thuret, R., Lea, R., Stanley, P. and Papalopulu, N. (2014). aPKC phosphorylates p27Xic1, providing a mechanistic link between apicobasal polarity and cell-cycle control. Dev. Cell 31, 559-571.
55. Salomoni, P. and Calegari, F. (2010). Cell cycle control of mammalian neural stem cells: putting a speed limit on G1. Trends Cell Biol. 20, 233-243.
56. Schlosser, G., Koyano-Nakagawa, N. and Kintner, C. (2002). Thyroid hormone promotes neurogenesis in theXenopus spinal cord. Dev. Dyn. 225, 485-498.
57. Schmidt, R., Strähle, U. and Scholpp, S. (2013). Neurogenesis in zebrafish - from embryo to adult. Neural Dev. 8, 3.
58. Schroeder, T. E. (1970). Neurulation in Xenopus laevis. An analysis and model based upon light and electron microscopy. J. Embryol. Exp. Morphol. 23, 427-462.
59. Thors, F., de Kort, E. J. M. and Nieuwenhuys, R. (1982a). On the development of the spinal cord of the clawed frog, Xenopus laevis. II. Experimental analysis of differentiation and migration. Anat. Embryol. 164, 443-454.
60. Thors, F., de Kort, E. J. M. and Nieuwenhuys, R. (1982b). On the development of the spinal cord of the clawed frog, Xenopus laevis. I. Morphogenesis and histogenesis. Anat. Embryol. 164, 427-441.

61. Wullimann, M. F., Rink, E., Vernier, P. and Schlosser, G. (2005). Secondary neurogenesis in the brain of the African clawed frog, Xenopus laevis, as revealed by PCNA, Delta-1, Neurogenin-related-1, and NeuroD expression. J. Comp. Neurol. 489, 387-402.

62. Yoshida, M., Shan, W. S. and Colman, D. R. (1999). Conserved and divergent expression patterns of the proteolipid protein gene family in the amphibian central nervous system. J. Neurosci. Res. 57, 13-22.

63. Zhang, C., Basta, T., Jensen, E. D. and Klymkowsky, M. W. (2003). The b-catenin/VegT-regulated early zygotic gene Xnr5 is a direct target of SOX3 regulation. Development 130, 5609-5624.

64. Zimmerman, L. B., De Jesús-Escobar, J. M. and Harland, R. M. (1996). The Spemann organizer signal noggin binds and inactivates bone morphogenetic protein 4. Cell 86, 599-606.

65. Zupanc, G. K. H. and Ott, R. (1999). Cell proliferation after lesions in the cerebellum of adult teleost fish: time course, origin, and type of new cells produced. Exp. Neurol. 160, 78-87.

5

CHAPTER

PAX7 LINEAGE CONTRIBUTIONS TO THE MAMMALIAN NEURAL CREST

Barbara Murdoch[1, 2], Casey DelConte[2], Martı´n I. Garcı´a-Castro[2]

[1]Biology Department, Eastern Connecticut State University, Willimantic, Connecticut, United States of America

[2]Department of Molecular, Cellular and Developmental Biology, Yale University, New Haven, Connecticut, United States of America

ABSTRACT

Background

Neural crest cells are vertebrate-specific multipotent cells that contribute to a variety of tissues including the peripheral nervous system, melanocytes, and craniofacial bones and cartilage. Abnormal development of the neural crest is associated with several human maladies including cleft/lip palate, aggressive cancers such as melanoma and neuroblastoma, and rare syndromes, like Waardenburg syndrome, a complex disorder involving hearing loss and pigment defects. We previously identified the transcription factor Pax7 as an early marker, and required component for neural crest development in chick embryos. In mammals, Pax7 is also thought to play a role in neural crest development, yet the precise contribution of Pax7 progenitors to the neural crest lineage has not been determined.

Methodology/Principal Findings

Here we use Cre/loxP technology in double transgenic mice to fate map the Pax7 lineage in neural crest derivates. We find that Pax7 descendants contribute to

Citation: Murdoch B, DelConte C, García-Castro MI (2012) Pax7 Lineage Contributions to the Mammalian Neural Crest. PLoS ONE 7(7): e41089. doi:10.1371/journal.pone.0041089.

multiple tissues including the cranial, cardiac and trunk neural crest, which in the cranial cartilage form a distinct regional pattern. The Pax7 lineage, like the Pax3 lineage, is additionally detected in some non-neural crest tissues, including a subset of the epithelial cells in specific organs.

Conclusions/Significance

These results demonstrate a previously unappreciated widespread distribution of Pax7 descendants within and beyond the neural crest. They shed light regarding the regionally distinct phenotypes observed in Pax3 and Pax7 mutants, and provide a unique perspective into the potential roles of Pax7 during disease and development.

INTRODUCTION

Unique to vertebrates, multipotent neural crest cells are found during early embryonic development, and in limited but specific regions in the adult [1]–[4]. The neural crest arises between the neural plate and non-neural ectoderm, and migrates extensively to numerous locations throughout the body, where they contribute to a wide range of cell and tissue derivatives.

Neural crest cells can be operationally defined based on their position within the anterior/posterior neural axis; cells at a given axial level follow distinct migratory routes and differentiate into a predictable array of progeny at their final destination [5]–[7]. From anterior to posterior, the axis can be roughly divided into three groups: the cranial, cardiac (also termed vagal), and trunk (including the sacral) neural crest. The cranial crest migrates dorsolaterally into the head region to produce cells including neurons, glia, and craniofacial mesenchyme such as various connective tissues (muscle, bones and cartilage) [6], [8]–[10]. Also produced by the cranial neural crest are thymic cells, the tooth primordia, and the bones of the middle ear and jaw. A forward migration of the cranial crest can also deposit cells into the frontonasal process. The cardiac neural crest cells migrate to the 3rd, 4th and 6th pharyngeal arches, which contribute to the muscle and connective tissue of the walls of the aortic arch arteries and the aorticopulmonary septum [11]. In the trunk, neural crest cells migrate either dorsolaterally to produce melanocytes, or ventrolaterally to produce cells of the peripheral nervous system (the dorsal root, sympathetic and parasympathetic (enteric) ganglia) [5]. The sacral neural crest, found in mice posterior to somite 28, also contributes to cells of the peripheral ganglia [7], [12].

The broad spectrum of neural crest tissue derivatives is equally matched by their excessive participation in disease. In humans developmental abnormalities attributed to the neural crest include defects of the heart (mostly of the outflow tract), craniofacial development (cleft lip/palate), peripheral nervous system (familial dysautonomia) and certain cancers including melanoma and neuroblastoma [13], [14].

During embryonic development, neural crest induction is initiated by a few signaling molecules such as BMP, FGF, and Wnt [15]–[17], which lead to the sequential up-regulation of transcription factors in neural crest precursors. The first cohort of transcription factors, including Pax3, Pax7, Zic1, Msx1 and Msx2, specify the neural plate border. This in turn initiates the expression of a second marker set, including Sox9, Sox10, FoxD3, AP-2, c-Myc, that specify the neural crest [18], [19]. These signals are further combined with additional downstream effectors to enable the proliferation, survival, detachment, migration and differentiation of neural crest cells. Even though mutations of several of these genes have been shown to alter specific neural crest derivatives, it is not yet clear what are the specific contributions of each transcription factor, nor their collaborative effects, downstream targets, or how these contribute to the production of neural crest-derived cells.

Our laboratory has identified the paired box domain transcription factor, Pax7, as one of the earliest markers of avian neural crest cells, and demonstrated its requirement for avian neural crest development [20]. However the lineage contributions of Pax7 precursors to mammalian neural crest development are unknown. In the E8 to E14 mouse embryo, Pax7 expression is detected in regions of the developing neural crest (the dorsal neural tube, cranial neural folds)[21]–[23]; regions not typical of the developing neural crest (brain, olfactory epithelium, frontonasal region, and somites) [22], [24], [25]; and in varied but restricted regions postnatally (frontonasal region, skeletal muscle satellite cells, brain and pituitary) [21], [26]–[28].

To date, transgenic mice have shown that the Pax7 lineage includes fat, dermis, muscle [29],[30]; olfactory neurons, ensheathing glia and supporting cells [25]. However, the Pax7 expression profile suggests a broader lineage potential for Pax7 descendants, like that seen for its closest Pax family member, Pax3 [31], [32]. Using transgenic mice, we expand the repertoire of lineages contributed by Pax7, with emphasis on the neural crest. In doing so, we identify some interesting similarities between the Pax3 and Pax7 lineages, within and outside of the typical neural crest.

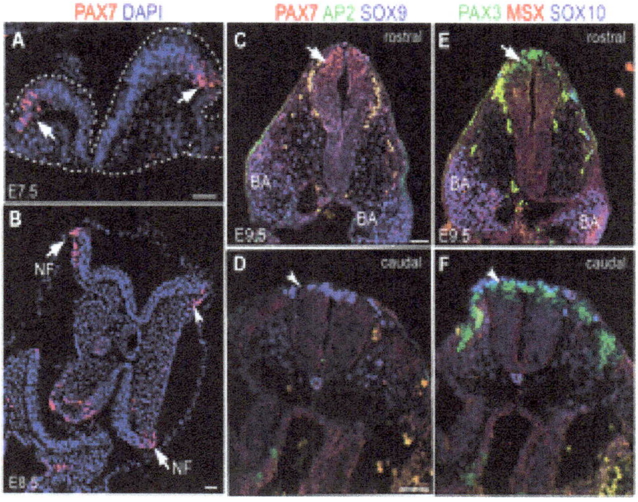

Figure 1: Spatiotemporal expression of neural crest markers including Pax7 in E7.5 to E9.5 mouse embryos.

In C57BL/6 embryos, immunohistochemistry detects Pax7 (red) in A,B) the E7.5 to E8.5 lateral neural folds (NF; arrows); B) the E8.5 cephalic mesenchyme (arrowhead). Dapi (blue) is a nuclear stain. C,E) and D,F) are serial sections from the same axial level at E9.5. C) Pax7 is detected in the rostral (arrow), but not D) caudal (arrowhead), dorsal neural tube, whereas F) Pax3 (green) is more highly expressed in the caudal (arrowhead) and not E) rostral (arrow) neural tube. C,D) AP2-α+ cells (green) are seen mostly in the ectoderm; rare AP2-α+ cells are seen together with Sox9+ (blue) and E,F) Sox10+ cells (blue) in the branchial arches (BA); Msx1+ cells (red) were undetected in these sections. Note that rounded cells adjacent to the neural tube are background, which appear also in the negative controls. Scale bars in A–D −100 um.

http://dx.doi.org/10.1371/journal.pone.0041089.g001

RESULTS

Spatiotemporal Expression of Neural Crest Markers Including Pax7 in the Early Mouse Embryo

Pax7 is one of the earliest markers of, and required for, neural crest development in the chick[20], [33]. However, the contribution of Pax7 precursors to mammalian neural crest tissues has not been determined. In the mouse, *in situ* showed Pax7 expression beginning at E8.0 through E14, in the closed neural tube, developing brain, nasal region and somites [21]–[23]. Using immunohistochemistry on E7.5 to E9.5 C57BL/6 embryos, we determined the spatiotemporal expression of Pax7 protein at the cellular level. We first detected Pax7 protein prior to neural tube closure at E7.5, at the tips of the lateral neural folds (Fig. 1A). At this early stage, Pax7+ cells form only a subpopulation of the larger expression domain of the neural crest, as indicated by the neural crest markers, AP2-α, Sox10 and Pax3 (Casey DelConte and Martín García-Castro, unpublished data). By E8.5 both the neural folds and cephalic mesenchyme contained small numbers of Pax7+ cells (Fig. 1B). Pax7 expression was more widely detected at E9.5 in the rostral neural tube (Fig. 1C), but markedly absent from the caudal neural tube (Fig. 1D), consistent with the suggestion that the Pax7 lineage contributes more highly to cranial regions [23]. In contrast to Pax7 expression, Pax3 expression was more readily detected in the caudal compared to rostral neural tube (Fig. 1E,F). Sox9+ (Fig. 1C,D) and Sox10+ cells (Fig. 1E,F) were seen mostly in the branchial arches (together with AP2-α+ cells), with rare cells in the dorsal neural tube. AP2-α+ cells were also detected in the developing ectoderm (Fig. 1C,D). From E10 to postnatally, Pax7 expression continues in the frontonasal region [21], [24], [25], skeletal muscle and in the pituitary gland [21], [26]–[28], confirming earlier studies (data not shown). These results indicate the early expression of Pax7, in more rostral regions of the mouse, which includes a subpopulation of presumptive neural crest precursors.

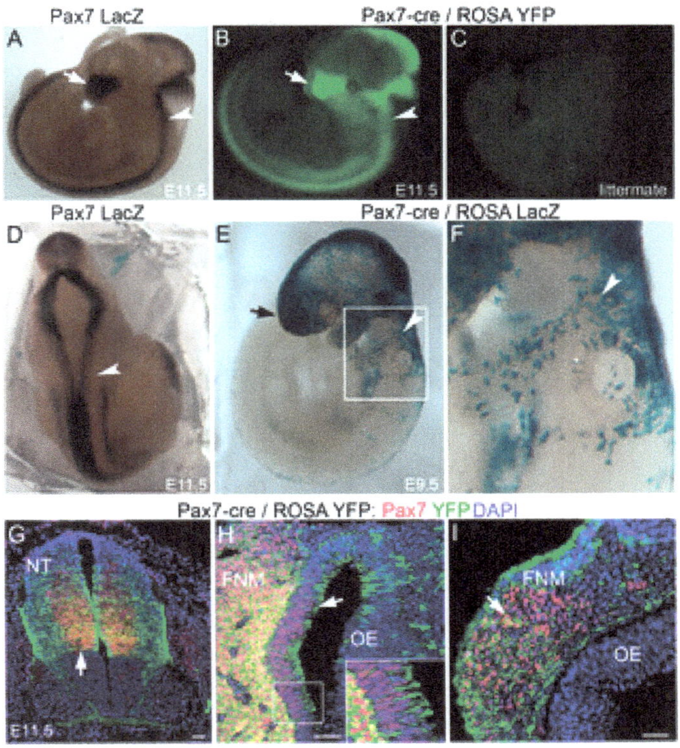

Figure 2: Pax7-cre/reporter mice for lineage tracing neural crest derivatives.

A,B,D) Equivalent Pax7 expression domains are seen in the frontonasal region (arrows) and neural tube (arrowheads) of E11.5 mouse embryos from different Pax7 transgenic lines: A,D) Pax7 LacZ (LacZ stain) and B) Pax7-cre/ROSA YFP (endogenous YFP expression). C) A littermate control, lacking YFP expression. E,F) E9.5 Pax7-cre/ROSA LacZ embryos show LacZ stain in neural crest regions, such as the frontonasal region (arrow) and cells leading to the pharyngeal arches (arrowheads). Box in (E) magnified in (F). G–I) In E11.5 Pax7-cre/ROSA YFP embryos, co-expression of Pax7 (red) and YFP reporter (green) can be detected in G) the neural tube (NT), H,I) olfactory epithelium (OE) and frontonasal mesenchyme (FNM; arrows). Inset highlights boxed region. Scale bars in G–I 100 um.

http://dx.doi.org/10.1371/journal.pone.0041089.g002

Pax7-cre/reporter Mice for Lineage Tracing Neural Crest Derivatives

To determine if Pax7 could identify precursors whose progeny contribute to neural crest tissues, we crossed Pax7-cre transgenic mice, driving Cre recombinase from the Pax7 locus, to *Gt(ROSA)26Sor*$^{tm(EYFP)Cos}$ reporter mice (hereafter termed ROSA YFP) [34], to fate map Pax7 descendants. Following Cre-mediated recombination in double transgenic Pax7-cre/ROSA YFP mice, a floxed stop cassette is excised allowing for the continued expression of YFP in Pax7-expressing precursors and their progeny, even after Pax7 expression ceases. The transgenic reporter line *Gtrosa26*^{tm1Sor} (hereafter termed ROSA LacZ), that expresses ß-galactosidase protein from the ROSA26 locus [35], produced a similar expression pattern after Pax7-cre mediated excision.

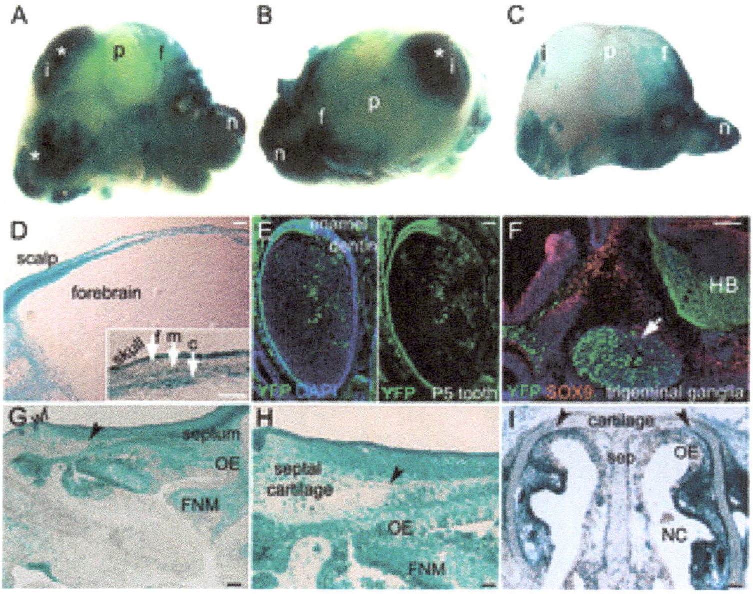

Figure 3: The Pax7 lineage contributions in the cranial region.

LacZ/YFP expression in E15.5 Pax7-cre/ROSA reporters is detected in multiple cranial regions that include: A,B) the brain (asterisks), skull, skin and nose; A–C) in the nasal (n) region, interparietal (i), parietal (p) and frontal (f) bones of the skull (viewed in (C) with brain removed); D) frontal bone (f), meninges (m) and connective tissue (c; arrows inset). E,F) YFP reporter (green) is detected in teeth (postnatal day 5 (P5)), and the trigeminal ganglia (arrow, F) near Sox9+ cells (red, F). DAPI nuclear stain in blue. G–I) Pax7-cre/ROSA LacZ mice express LacZ in the nasal cartilage, cells of the frontonasal mesenchyme (FNM), olfactory epithelium (OE) and whisker follicles (wf). Arrowheads in (G–I) show distinct border between dense and diffuse recombination+ cells in cartilage. Note (C) has been cropped in Photoshop for aesthetics. Sections are sagittal (D–H) and coronal (I). HB –hindbrain, NC –nasal cavity, sep -septum. Scale bars in D,F,G,I-500 um; E,H, inset -200 um.

http://dx.doi.org/10.1371/journal.pone.0041089.g003

We determined if Pax7-cre/ROSA YFP mice faithfully reported according to endogenous Pax7 expression, by comparing the YFP expression pattern, to that seen in multiple independent transgenic lines, including some whose reporter expression occurred independent of recombination. Similar expression patterns were seen in age-matched Pax7 LacZ (Fig. 2A,D), Pax7-cre/ROSA YFP (Fig. 2B), Pax7-cre/ROSA LacZ (data not shown and [36]) and Pax7-ZsGreen [37] transgenic mice, each recapitulating the expression domains of endogenous Pax7 [23], [38]. Additionally, in E11.5 Pax7-cre/ROSA YFP embryos, co-expression of Pax7 protein with YFP was detected in the dermomyotome (not shown), olfactory epithelium, neural tube, and frontonasal mesenchyme (Fig. 2G–I) [25]. Littermate controls of Pax7-cre/ROSA YFP embryos, having only a single or entirely lacking transgenic alleles, did not express YFP (Fig. 2C).

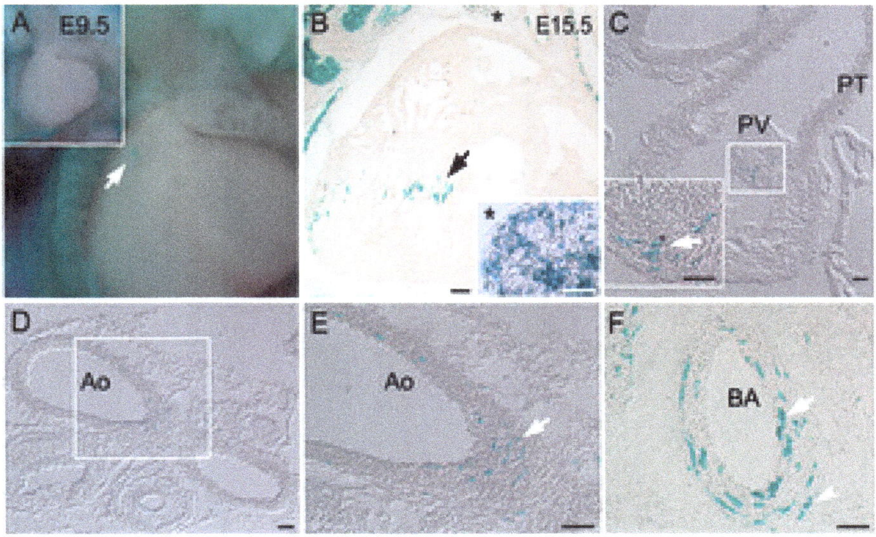

Figure 4: Limited cardiovascular tissue derived through the Pax7 lineage.

A–F) Recombination+ cells (LacZ+) are found in the heart and blood vessels of Pax7-cre/ROSA LacZ mice. A) Wholemount E9.5 embryos show faint LacZ+ cells in the heart (arrow). Inset shows lower magnification of panel. B–F) E15.5 sections show small populations of LacZ+ cells (arrows) in the B) heart, thymus (asterisk and inset), C) pulmonary valve (PV), D,E) aorta (Ao) and F) inner (arrow) through outer (arrowhead) layers of the brachiocephalic artery (BA). Boxes show magnified region in separate panel. PT –pulmonary trunk. Scale bars in B –500 um; insets, C–F –200 um.

http://dx.doi.org/10.1371/journal.pone.0041089.g004

Consistent with Pax7 expression, at E8.5 in Pax7-cre/reporter embryos, a few YFP+ cells were first detected in the cephalic mesenchyme and lateral neural folds [25] (and data not shown). By E9.5, Pax7 descendants (LacZ+) were detected in the midbrain, hindbrain, dorsal neural tube, frontonasal region, and in migratory neural crest cells emigrating from the dorsal neural tube to the branchial arches (Fig. 2E,F). Combined our results provide evidence validating the use of Pax7-cre/reporter mice to follow Pax7 descendants in the neural crest. Using these Pax7-cre/reporter mice we traced, the cranial, cardiac and trunk neural crest lineages.

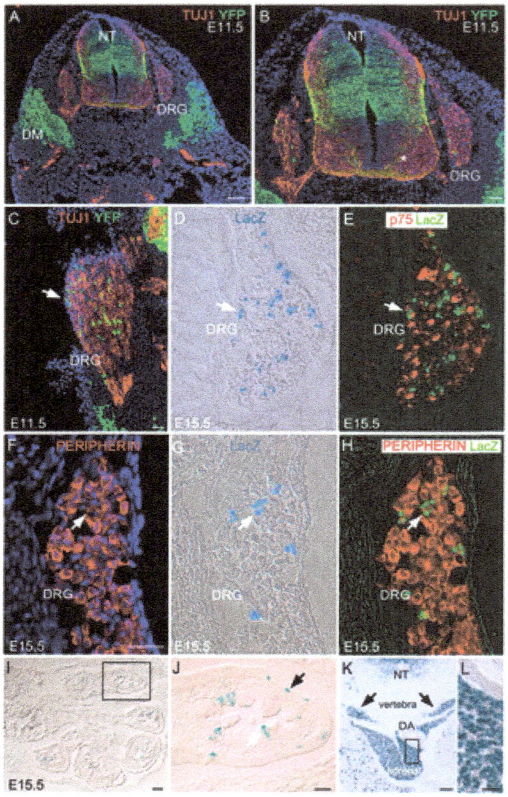

Figure 5: Pax7 contributions to trunk neural crest.

A–C) At E11.5 and D–H) E15.5 in Pax7-cre/ROSA reporter mice, Pax7 descendants (YFP+ or LacZ+, respectively) are detected in the neural tube (NT), including commissural axons (asterisk), dermomyotome (DM) and dorsal root ganglia (DRG). A–C) Only a subset of the total DRG neurons (TUJ1+, red) arise through the Pax7 lineage (YFP+, arrow). Neurons throughout the DRGs express E) P75 (red) and F,H) peripherin (red), including D–H) LacZ+ Pax7 descendants (arrows). In E,H), the LacZ signal (green) was pseudocolored in Photoshop. I,J) Within the physiological umbilical hernia, subsets of Pax7 descendants (LacZ+) are detected in the lining of the midgut (white arrow) and presumptive enteric neurons (black arrow, J), K,L) sympathetic tissue (arrows) and adrenal gland (all E15.5). Boxes show magnified region in adjacent panel. DA-dorsal aorta. Scale bars in A,I,K -500 um; B,C,F,J,K magnified region –200 um.

http://dx.doi.org/10.1371/journal.pone.0041089.g005

The Pax7 Lineage Contributions in the Cranial Region

In the cranial region of E15.5 Pax7-cre/reporter mice, both neural crest and non-neural crest recombinants were detected (Fig. 3). LacZ+ cells were seen in the brain, skin and nose (Fig. 3A,B). In the skull, cranial crest derivatives arising via the Pax7 lineage included subpopulations of cells of the frontal, parietal and interparietal bones (Fig. 3A–D). Below the skull, recombinants were detected in the connective tissue and meninges (Fig. 3D). In Pax7-cre/ROSA YFP mice, YFP+ cells were seen in postnatal day 5 teeth (Fig. 3E), and at E15.5, in the trigeminal ganglia and hindbrain, independent of Sox9+ precursors (Fig. 3F). Sagittal sections of the head showed LacZ+ cells in the presumptive neural crest tissues of the whisker follicles, the frontonasal mesenchyme, olfactory epithelium and cartilage of the septum (Fig. 3G–I). Interestingly, recombinant cells were more highly detected in the lateral cartilage (93.5±0.8%, $p<0.001$) compared to medial cartilage of the septum (14.9±1.3%) (Fig. 3G–I). These results show the widespread contributions of the Pax7 lineage to numerous cranial tissues, some with distinct spatial patterns.

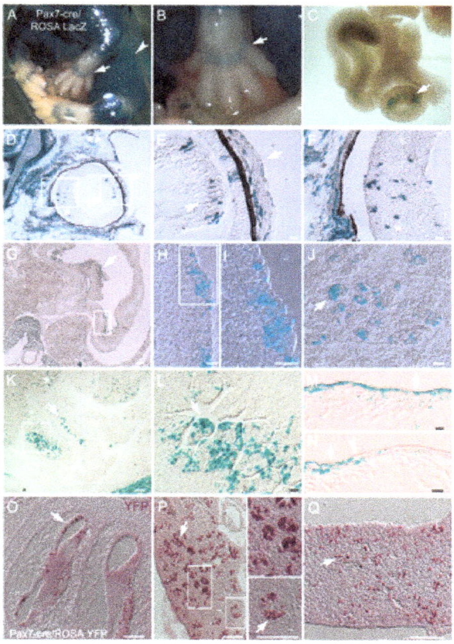

Figure 6: The Pax7 lineage outside of the neural crest.

In E15.5 Pax7-cre/reporter embryos, LacZ+ Pax7 descendants are seen in: A) the skin (arrowhead) and skeletal muscle (asterisk); A,B) restricted regions of the carpal bone primordium (arrows) and C) portions of the wholemount gut (arrow). In sections, a subset of LacZ+ cells are detected in: D–F) the eye; the presumptive eyelid muscle (double asterisk), in the lens (arrowhead) and its surrounding tissue (arrow), and neural retina (asterisk); G–I) the lining of the stomach (arrow); J) the pancreas (arrow identifies presumptive acinar cells); K,L) the lining of the gut (arrow; asterisk denotes the liver); and M,N) the skin (arrow), where a distinct border distinguishes the Pax7 lineage from non-lineage Pax7 (arrowhead). O–Q) YFP+ cells visualized with non-fluorescent VIP are detected in: O) the cochlea of the ear; P) lining of lung; and Q) liver (arrows for all). Both YFP and LacZ Pax7-cre/reporters gave the same results, as did fluorescent and non-fluorescent detection methods. Boxes show magnified region in adjacent panel. Scale bars in D,G,K,O–Q are 500 um; E,F,H–J, L–N are 100 um.

http://dx.doi.org/10.1371/journal.pone.0041089.g006

Limited Cardiovascular Tissue Derived through the Pax7 Lineage

Compared to the cranial neural crest, and in contrast to the Pax3 lineage [31], there were limited contributions of the Pax7 lineage to cardiovascular tissues. Faint LacZ+ cells were seen in wholemount hearts from E9.5 Pax7-cre/ROSA LacZ embryos (Fig. 4A). Although in sections of E15.5 samples, few LacZ+ cells were readily detected in the heart and thymus (Fig. 4B), the pulmonary valve (Fig. 4C), the arch of the aorta just proximal to the entrance site for the ductus arteriosus (Fig. 4D,E), and in the brachiocephalic (Fig. 4F) and carotid arteries (data not shown). These results indicate a limited contribution of the Pax7 lineage to cardiovascular tissues and the cardiac neural crest.

Pax7 Contributions to Trunk Neural Crest

Trunk sections of Pax7cre/ROSA YFP and LacZ embryos showed recombinant+ cells (YFP+ or LacZ+) in the dermomyotome, neural tube, commissural axons, and dorsal root ganglia (DRGs; Fig. 5A–H). At E11.5 (Fig. 5A–C) and E15.5 (Fig. 5D–H), the neuronal antigens TUJ1 (Fig. 5A–C), P75 (Fig. 5E) and peripherin (Fig. 5F,H) were widely expressed throughout DRG neurons. However, only a subset of DRG neurons ($4.5 \pm 0.7\%$; 32/702 TUJ1+ neurons counted), arises via the Pax7 lineage (Fig. 5A–H). Pax7 progeny also included subsets of presumptive enteric neurons surrounding the outer portion of the gut, and epithelial cells lining the gut (Fig. 5I,J), sympathetic tissue and cells of the adrenal gland (Fig. 5K,L). These results indicate a broad, albeit restricted, contribution to the trunk neural crest by the Pax7 lineage.

The Pax7 Lineage in Non-neural Crest Regions

Our focus was primarily on neural crest derivatives, however we examined tissues not commonly associated with the neural crest, as some were previously reported in the Pax3 lineage [31], the closest Pax family member to Pax7. After histochemical staining, E15.5 Pax7-cre/ROSA LacZ embryos appeared mostly LacZ+ (Fig. 6A). In wholemount, Pax7 descendants were detected in the lung, liver, thymus (data not shown), skeletal muscle (Fig. 6A) and gut (Fig. 6C). Unlike the Pax3 lineage, Pax7 descendants were not detected in the kidney. In the paws, LacZ+ cells were restricted to the presumptive carpal bone primordium (Fig. 6A,B).

Sections revealed a more detailed analysis of the cells and tissues to which Pax7 descendants contribute. A striking finding for the Pax7 lineage was their contribution to only a subset of seemingly homogeneous cell populations within certain tissues. In the eye LacZ+ Pax7 descendants contributed to subsets of cells in and around the lens, neural retina, presumptive ocular muscles and eyelids (Fig. 6D–F). A similar trend was found in the stomach lining (Fig. 6G–I), pancreas (including acinar cells, Fig. 6J), gut lining (Fig. 6K,L) and skin (Fig. 6M,N), where each contained a population of cells formed in part by Pax7 derivatives. In addition to the underlying connective tissue, the skin contained Pax7 derivatives in the epidermis and dermis that showed a distinct regional boundary, with positive cells in the back skin that went midway around to the belly (Fig. 6M,N). Non-fluorescent detection of YFP revealed Pax7 lineage contributions to lung bronchioles (Fig. 6P), liver (Fig. 6Q) and, consistent with the detection of lineage+ cells in the E9.5 otic vesicle (Fig. 2E,F), cells in the cochlea of the ear (Fig. 6O). Similarly, Wnt-1-cre and Pax3-cre lineages contribute to the otic vesicle and inner ear [32]. These results show a previously unappreciated diversity for the Pax7 lineage throughout the body, including overlap with some of the Pax3 and Wnt-1 lineages[31].

DISCUSSION

Using transgenic mice, we mapped the Pax7 lineage focusing on the neural crest. Consistent with the predominance of craniofacial defects seen in Pax7 mutants [23], we found a higher contribution of Pax7 descendants in the cranial compared to either cardiac or trunk neural crest (Fig. 3–5). More globally, Pax7 descendants showed two interesting traits: 1. a regional pattern especially evident in the head cartilage (Fig. 3), and 2. within specific tissues, contributions to only a subset of cells in a presumed homogeneous population (Fig. 3–6). Our results reveal a previously unappreciated widespread distribution of Pax7 descendants, within and beyond the neural crest.

Although neural crest derivatives were prevalent in the Pax7 lineage, many Pax7 descendants were found outside of the ectoderm-derived neural crest tissues. The lining of the stomach, pancreas, gut and lungs each contained Pax7 descendants, as did cells in the liver (Fig. 6), indicating endoderm derivatives. Thus, together with the mesoderm derivatives, muscle and fat[29], [30], Pax7 precursors contribute to tissues derived from all three germ layers.

Pax7 in the Neural Crest Lineage: Comparisons to Wnt1-cre Mice

Previous studies have fate mapped the neural crest lineage using transgenic mice [39], [40], with the most common and extensive neural crest labeling detected with the Wnt1-cre driver[41]–[43]. Neural crest cells targeted by Wnt1-cre label descendants in craniofacial structures, olfactory, cardiac, endocrine, peripheral and enteric nervous system derivatives [41]–[44]. Here, in neural crest derivatives, we highlight the most pronounced distinctions between Wnt1 and Pax7 descendants.

Like Wnt1-cre, Pax7-cre mice targeted neural crest cells at all axial levels, but in contrast labeled only relatively small subpopulations of certain neural crest derivatives (Fig. 3–5). Wnt1-cre mice label more cells in the frontal and interparietal bones of the skull [42] main arteries surrounding the heart [43], and in the DRGs [45]. Pax7, but not Wnt1 descendants, contribute to the parietal bone in a mosaic pattern, although these are likely derived from the mesoderm[42], [46]. Such a mosaic pattern is evident in multiple tissues (Fig. 5,6; olfactory and gut epithelia, inner ear, lung, pancreas, etc.), and consistent with patterns seen using several different Cre drivers (Pax7-, Wnt-1-, Pax3-, Bmi-, Id2- cre) [25]. These patterns suggest the input of multiple progenitors to certain tissues, thus allowing for compensation by alternate progenitors after progenitor loss, to maintain tissue integrity. This is supported by the only mild defects seen in Pax7 mutants at birth [23].

Despite these differences, in certain cranial crest derivatives, like the frontonasal mesenchyme and olfactory ensheathing glia, similar proportions of labeled cells are seen in both Wnt1-cre and Pax7-cre/reporter mice [25], [44], [46], [47]. This pattern suggests a convergence of the Pax7 and Wnt1-derived neural crest lineages, at least in some regions.

Regional Restrictions in the Pax7 Lineage

Of particular interest regarding the Pax7 neural crest lineage is the distinct regional restriction of labeled cells where approximately 6 times more Pax7 descendants populated the lateral compared to the central head cartilage (Fig. 3; 93.5% vs 14.9%). This pattern could arise either through the seeding of more Pax7 derivatives in the lateral compared to medial septum or by the preferential expansion and/or survival of the lateral Pax7 derivatives. A more detailed examination using BrdU labeling and apoptosis assays would provide a better understanding of the development of this head cartilage.

The Pax7 Compared to Pax3 Lineage

Of the 9 mammalian Pax genes, Pax7 is most structurally similar to Pax3, each having a paired domain, octapeptide region and homeodomain, and together form the Pax subgroup III[48], [49]. Since Pax7 and Pax3 have only partially overlapping expression domains, one would predict both common and distinct descendants. When *Cre* is inserted into exon 1 of Pax3, and thus faithfully recapitulates the endogenous Pax3 expression domain, Pax3 descendants can be found in the dorsal neural tube, somites, skeletal muscle, adrenal gland and facial structures [31]. Here we report a very similar pattern of derivatives generated by Pax7 progenitors (Figs 2, 3, 4, and 5). Additional derivatives common to both Pax7 and Pax3 lineages include the ear, colon epithelia, stomach (Fig. 5,6) and olfactory epithelium [31], [50],[51]. Unlike Pax3 descendants [31], Pax7 descendants were not detected in the kidney.

In certain tissues, like the enteric plexus and cardiac neural crest, the Pax7 lineage was less prevalent than the Pax3 lineage [31]. This is reflective of the loss of enteric ganglia and severe cardiovascular defects seen in homozygous Pax3 *"Splotch"* mutants [51]–[53], but absent from Pax7 mutants [23].

The Pax7 lineage was detected in some tissues where Pax3 descendants have not been reported - the presumptive carpal cartilage, lens and neuroepithelium of the eye, cells in the liver, pancreas, skin epithelium, and lung bronchioles (Fig. 6, and data not shown). The differences between the Pax7 and Pax3 derivatives reflect the divergent regulation of Pax7 compared to Pax3 expression. Given this diverse range of Pax7 derivatives it is intriguing to speculate that defects in survival and/or turnover of cells in the pituitary [26], lung, liver, gut, etc., may similarly, and in addition to muscle stem cell defects, contribute to the stunted neonatal size and death of Pax7 mutants.

Pax7 Progenitors -Multipotent or Restricted?

Pax7 progenitors contribute to many different tissues and cell types, representing derivatives that are distributed throughout the body from each of the three germ layers. Are Pax7 progenitors then multipotent, or do they instead represent varying yet more restricted progenitors? The Cre-LoxP lineage-tracing approach used in this study loyally marks progenitors that expressed Pax7, however it does not allow one to match the precise time and region of emergence of a Pax7+ precursor with its contributions to specific tissues and cell types. Timely inducible promoters and/or enhancer specific strategies offer better-fitted alternatives for this purpose. A previous study induced Pax7-cre expression with tamoxifen at varying developmental stages, and tested for reporter expression one day later [30]. Injections on E7.5 (the first time point to reveal Pax7 derivatives), showed a few labeled cells near the dorsal neural tube, indicative of neural crest, whereas cells labeled on E8.5 included the cephalic neural crest and craniofacial region (in addition to parts of the brain, spinal cord, and somites). After injections between E9.5 to E13.5, labeling persisted in the craniofacial region, spinal cord and somitic tissues. Our detection of Pax7 protein at E7.5 to E9.5 in the neural folds, cephalic mesenchyme and rostral dorsal neural tube (Fig. 1), and later in the frontonasal mesenchyme and olfactory epithelium (Fig. 2 and [25]), together with our lineage tracing results, are in agreement with and compliment this study. In most of the tissues with Pax7 lineage contributions, we did not detect Pax7 expression and hence could not identify a tissue-specific Pax7 precursor. These findings are also supported by the induction experiments [30]and demonstrate the dynamic nature of Pax7 progenitors during different stages of development. That these labeled cells are indeed Pax7 descendants is likely, since our labeling pattern is consistent with multiple transgenic lines, even those whose reporter expression is independent of recombination (Fig. 2) [23], [36]–[38]. Additionally, YFP could be detected coincident with endogenous Pax7 expression (Fig. 2 and data not shown) [25]. Future studies using clonal analyses of individual Pax7+ progenitors will reveal their potency and help to dissect the molecular pathways leading to the differentiation of specific cell types.

CONCLUSION

Here we used Pax7-cre/reporter double transgenic mice to identify the Pax7 lineage in the neural crest. We found the Pax7 lineage contributed more highly to the cranial compared to either cardiac or trunk neural crest. Two patterns formed by Pax7 descendants were notable: 1. a regionally restricted pattern in the cranial cartilage and 2. contributions to only a subset of seemingly identical cells, within a given tissue. Outside of the neural crest, the Pax7 lineage was widespread, contributing to sensory organs, the lining of organs, and parts of the digestive system. We thus uncover a previously unappreciated broad tissue and cell diversity for the Pax7 lineage, which may help to better understand the development and function of these structures. Our results additionally, have implications for regenerative medicine, as future studies dissect the molecular regulation required to direct the differentiation of Pax7 precursors to specific cell types.

MATERIALS AND METHODS

Ethics Statement

All experimental procedures were specifically approved (protocol # 2009-11115) and performed in accordance with the Yale Animal Resources Center and Institutional Animal Care and Use Committee policies.

Transgenic Mice

Pax7 ^{ICNm} mice (C57Bl/6; termed *Pax7-cre* throughout the paper), obtained from Mario Capecchi [36], were crossed with either C57Bl/6 female *Gt(ROSA)26Sor*$^{tm(EYFP)Cos}$ mice [34], expressing enhanced yellow fluorescent protein from the ROSA26 locus (termed ROSA YFP throughout the paper) or *Gtrosa26*^{tm1Sor} [35], expressing ß-galactosidase protein from the ROSA26 locus (termed ROSA LacZ throughout the paper). Mouse reporter lines were gifts from Diane Krause. Both reporter lines produced a similar reporter expression pattern after Pax7-cre mediated excision. Both *Pax7-cre/ROSA YFP* and *Pax7-cre/ROSA LacZ* double transgenic mice, where Cre recombinase expression is under the control of the endogenous Pax7 regulatory elements, express the reporters without disrupting Pax7 function [36]. Mice were genotyped using PCR for Pax7, the Cre knock-in allele of Pax7 [36], and the modified ROSA26 allele containing either YFP [34], or LacZ [35]. Pax7 primers located within Pax7 exon 10 5′-GCTCTGGATACACCTGAGTCT-3′; 5′-TCGGCCTTCTTCTAGGTTCTGCTC-3′ (ck118 and ck256 respectively; wildtype 465 bp product) were combined with an *IRES-Cre* primer 5′-GGATAGTGAAACAGGGGCAA-3′ (ck172; 340 bp product) [36]; ROSA YFP primers: Rosa1 5′-AAAGTCGCTCTGAGTTGTTAT-3′ and Rosa3 5′-GGAGCGGGAGAAATGGATATG-3′ (wildtype ROSA locus, 650 bp product); Rosa3 and YRF 5′-CGACCACTACCAGCAGAACA-3′ (ROSA26 YFP, 850 bp product); ROSA LacZ primers: Rosa1 5′-AAAGTCGCTCTGAGTTGTTAT-3′ and Rosa2 5′-GCGAAGAGTTTGTCCTCAACC-3′ (wildtype ROSA locus 650 bp, ROSA26 LacZ 350 bp product). PCR parameters: 35 cycles of 95°C 30 seconds, 58°C 30 seconds, 72°C 60 seconds. Efficiency of Cre-lox recombination was verified by the specific co-localization of Pax7 with YFP proteins in *Pax7-cre/ROSA YFP* embryos and the similar reporter expression patterns detected in Pax7-LacZ [38], Pax7-Zs-green [37] and Pax7-cre/*Gtrosa26*^{tm1Sor} mice[36]. For each time point, at least three Pax7-cre/reporter positive embryos or mice were analyzed.

Tissue Preparation

Mice were sacrificed in a CO_2 chamber, perfused with cold PBS and 4% paraformaldehyde (PFA) in PBS and post-fixed in 4% PFA at 4°C [54]. Embryos were immersion-fixed in 4% PFA overnight. The day of vaginal plug was defined as E0.5. Tissues were cryoprotected in sucrose, embedded in either Tissue-Tek medium (OCT; Sakura Finetek, Torrance, CA) or gelatin and frozen in liquid nitrogen. 12 μm sections were stored at –20°C for subsequent analysis.

Immunohistochemistry

Sections were immersed in PBS, permeabilized in 0.1% Triton-X-100/PBS and blocked with 4% normal serum prior to primary antibody incubation. Secondary antibodies (1:200) used were of specific isotypes conjugated to biotin (Vector labs), Alexa 568 or Alexa 488 (Invitrogen Molecular Probes). Primary antibodies used: mouse anti-rat βIII tubulin (neuron-specific tubulin-TUJ1, 1:500) Covance; mouse anti -Pax7, Developmental Studies Hybridoma Bank (developed by Kawakami, A), rabbit anti-GFP (1:400, to detect YFP), Millipore. Note: all YFP panels show anti-GFP immunofluorescence that was confirmed using non-fluorescent VIP immunohistochemistry with anti-GFP antibodies (as seen in Fig. 6O–Q), with the exception of the YFP-containing panels in Fig. 2B,C which detect endogenous YFP fluorescence. Nuclei were stained with 0.5 ug/ml diaminopyridine imidazole (DAPI) and sections coverslipped in Vectashield (Vector Laboratories, Burlingame, CA) for fluorescent antigens or 50% glycerol for VIP. All images were visualized with either a Nikon dissecting scope or Nikon Eclipse 80 i microscope using a SPOT camera (SPOT Diagnostic Instruments Inc.) with SPOT software (v4.5) and compiled using Adobe Photoshop CS4.

Histochemistry

Lac Z histochemistry was performed on wholemounts or cryosections postfixed in 4% PFA, permeabilized in 0.1% Triton-X-100 and washed in PBS before adding staining buffer (2 mM $MgCl_2$, 0.01% deoxycholate, 0.02% Nonidet-P40, and 100 mM $NaPO_4$, pH 7.3) containing 1 mg/mL X-gal, 5 mM potassium ferrocyanide, and 5 mM potassium ferricyanide. Staining proceeded either on ice for wholemounts or at 37°C for sections, protected from light for 1–10 hours. Negative controls did not demonstrate β-galactosidase staining and included C57BL/6 non-transgenic mice and Pax7-cre transgenic mice without the ROSA reporter transgene. Results were confirmed using antibodies to β-galactosidase.

ACKNOWLEDGMENTS

We thank the following colleagues for sharing the reagents that made this project possible – Vivian Lee for Sox9, ß-galactosidase antibodies; Mario Capecchi (Pax7-cre mice); Diane Krause (ROSA -YFP and –LacZ mice).

AUTHOR CONTRIBUTIONS

Conceived and designed the experiments: BM CD MIGC. Performed the experiments: BM CD. Wrote the paper: BM MIGC.

REFERENCES

1. Wong CE, Paratore C, Dours-Zimmermann MT, Rochat A, Pietri T, et al. (2006) Neural crest-derived cells with stem cell features can be traced back to multiple lineages in the adult skin. J Cell Biol 175: 1005–1015.

2. Clewes O, Narytnyk A, Gillinder K, Loughney A, Murdoch A, et al. (2011) Human Epidermal Neural Crest Stem Cells (hEPI-NCSC)–Characterization and Directed Differentiation into Osteocytes and Melanocytes. Stem Cell Reviews and Reports. pp. 1–16.
3. Jinno H, Morozova O, Jones KL, Biernaskie JA, Paris M, et al. (2010) Convergent genesis of an adult neural crest-like dermal stem cell from distinct developmental origins. Stem Cells 28: 2027–2040.
4. Nagoshi N, Shibata S, Nakamura M, Matsuzaki Y, Toyama Y, et al. (2009) Neural crest-derived stem cells display a wide variety of characteristics. Journal of Cellular Biochemistry 107: 1046–1052.
5. Serbedzija GN, Fraser SE, Bronner-Fraser M (1990) Pathways of trunk neural crest cell migration in the mouse embryo as revealed by vital dye labelling. Development 108: 605–612.
6. Serbedzija GN, Bronner-Fraser M, Fraser SE (1992) Vital dye analysis of cranial neural crest cell migration in the mouse embryo. Development 116: 297–307.
7. Serbedzija GN, Burgan S, Fraser SE, Bronner-Fraser M (1991) Vital dye labelling demonstrates a sacral neural crest contribution to the enteric nervous system of chick and mouse embryos. Development 111: 857–866.
8. Couly GF, Le Douarin NM (1987) Mapping of the early neural primordium in quail-chick chimeras. II. The prosencephalic neural plate and neural folds: implications for the genesis of cephalic human congenital abnormalities. Dev Biol 120: 198–214.
9. Lumsden A, Sprawson N, Graham A (1991) Segmental origin and migration of neural crest cells in the hindbrain region of the chick embryo. Development 113: 1281–1291.
10. Noden DM (1975) An analysis of migratory behavior of avian cephalic neural crest cells. Dev Biol 42: 106–130.
11. Kirby ML, Waldo KL (1995) Neural crest and cardiovascular patterning. Circ Res 77: 211–215.
12. Kapur RP (2000) Colonization of the murine hindgut by sacral crest-derived neural precursors: experimental support for an evolutionarily conserved model. Dev Biol 227: 146–155.
13. Etchevers HC, Amiel J, Lyonnet S (2006) Molecular bases of human neurocristopathies. Adv Exp Med Biol 589: 213–234.
14. Lee G, Papapetrou EP, Kim H, Chambers SM, Tomishima MJ, et al. (2009) Modelling pathogenesis and treatment of familial dysautonomia using patient-specific iPSCs. Nature 461: 402–406.
15. Basch ML, Garcia-Castro MI, Bronner-Fraser M (2004) Molecular mechanisms of neural crest induction. Birth Defects Res C Embryo Today 72: 109–123.
16. Raible DW (2006) Development of the neural crest: achieving specificity in regulatory pathways. Curr Opin Cell Biol 18: 698–703.

17. Sakai D, Wakamatsu Y (2005) Regulatory mechanisms for neural crest formation. Cells Tissues Organs 179: 24–35.

18. Betancur P, Bronner-Fraser M, Sauka-Spengler T (2010) Assembling neural crest regulatory circuits into a gene regulatory network. Annu Rev Cell Dev Biol 26: 581–603.

19. Sauka-Spengler T, Bronner-Fraser M (2008) A gene regulatory network orchestrates neural crest formation. Nat Rev Mol Cell Biol 9: 557–568.

20. Basch ML, Bronner-Fraser M, Garcia-Castro MI (2006) Specification of the neural crest occurs during gastrulation and requires Pax7. Nature 441: 218–222.

21. Jostes B, Walther C, Gruss P (1990) The murine paired box gene, Pax7, is expressed specifically during the development of the nervous and muscular system. Mech Dev 33: 27–37.

22. Lang D, Brown CB, Milewski R, Jiang YQ, Lu MM, et al. (2003) Distinct enhancers regulate neural expression of Pax7. Genomics 82: 553–560.

23. Mansouri A, Stoykova A, Torres M, Gruss P (1996) Dysgenesis of cephalic neural crest derivatives in Pax7−/− mutant mice. Development 122: 831–838.

24. LaMantia AS, Bhasin N, Rhodes K, Heemskerk J (2000) Mesenchymal/epithelial induction mediates olfactory pathway formation. Neuron 28: 411–425.

25. Murdoch B, DelConte C, Garcia-Castro MI (2010) Embryonic Pax7-expressing progenitors contribute multiple cell types to the postnatal olfactory epithelium. J Neurosci 30: 9523–9532.

26. Hosoyama T, Nishijo K, Garcia MM, Schaffer BS, Ohshima-Hosoyama S, et al. (2010) A Postnatal Pax7 Progenitor Gives Rise to Pituitary Adenomas. Genes Cancer 1: 388–402.

27. Stoykova A, Gruss P (1994) Roles of Pax-genes in developing and adult brain as suggested by expression patterns. J Neurosci 14: 1395–1412.

28. Ziman MR, Fletcher S, Kay PH (1997) Alternate Pax7 transcripts are expressed specifically in skeletal muscle, brain and other organs of adult mice. Int J Biochem Cell Biol 29: 1029–1036.

29. Lepper C, Conway SJ, Fan C-M (2009) Adult satellite cells and embryonic muscle progenitors have distinct genetic requirements. Nature 460: 627–631.

30. Lepper C, Fan CM (2010) Inducible lineage tracing of Pax7-descendant cells reveals embryonic origin of adult satellite cells. Genesis 48: 424–436.

31. Engleka KA, Gitler AD, Zhang M, Zhou DD, High FA, et al. (2005) Insertion of Cre into the Pax3 locus creates a new allele of Splotch and identifies unexpected Pax3 derivatives. Dev Biol 280: 396–406.

32. Freyer L, Aggarwal V, Morrow BE (2011) Dual embryonic origin of the mammalian otic vesicle forming the inner ear. Development 138: 5403–5414.

33. Otto A, Schmidt C, Patel K (2006) Pax3 and Pax7 expression and regulation in the avian embryo. Anat Embryol (Berl) 211: 293–310.

34. Srinivas S, Watanabe T, Lin CS, William CM, Tanabe Y, et al. (2001) Cre reporter strains produced by targeted insertion of EYFP and ECFP into the ROSA26 locus. BMC Dev Biol 1: 4.

35. Soriano P (1999) Generalized lacZ expression with the ROSA26 Cre reporter strain. Nat Genet 21: 70–71.

36. Keller C, Hansen MS, Coffin CM, Capecchi MR (2004) Pax3:Fkhr interferes with embryonic Pax3 and Pax7 function: implications for alveolar rhabdomyosarcoma cell of origin. Genes Dev 18: 2608–2613.

37. Bosnakovski D, Xu Z, Li W, Thet S, Cleaver O, et al. (2008) Prospective isolation of skeletal muscle stem cells with a Pax7 reporter. Stem Cells 26: 3194–3204.

38. Relaix F, Rocancourt D, Mansouri A, Buckingham M (2004) Divergent functions of murine Pax3 and Pax7 in limb muscle development. Genes Dev 18: 1088–1105.

39. Pietri T, Eder O, Blanche M, Thiery JP, Dufour S (2003) The human tissue plasminogen activator-Cre mouse: a new tool for targeting specifically neural crest cells and their derivatives in vivo. Dev Biol 259: 176–187.

40. Yamauchi Y, Abe K, Mantani A, Hitoshi Y, Suzuki M, et al. (1999) A novel transgenic technique that allows specific marking of the neural crest cell lineage in mice. Dev Biol 212: 191–203.

41. Chai Y, Jiang X, Ito Y, Bringas P Jr, Han J, et al. (2000) Fate of the mammalian cranial neural crest during tooth and mandibular morphogenesis. Development 127: 1671–1679.

42. Jiang X, Iseki S, Maxson RE, Sucov HM, Morriss-Kay GM (2002) Tissue origins and interactions in the mammalian skull vault. Dev Biol 241: 106–116.

43. Jiang X, Rowitch DH, Soriano P, McMahon AP, Sucov HM (2000) Fate of the mammalian cardiac neural crest. Development 127: 1607–1616.

44. Forni PE, Taylor-Burds C, Melvin VS, Williams T, Wray S (2011) Neural crest and ectodermal cells intermix in the nasal placode to give rise to GnRH-1 neurons, sensory neurons, and olfactory ensheathing cells. J Neurosci 31: 6915–6927.

45. Joseph NM, Mukouyama YS, Mosher JT, Jaegle M, Crone SA, et al. (2004) Neural crest stem cells undergo multilineage differentiation in developing peripheral nerves to generate endoneurial fibroblasts in addition to Schwann cells. Development 131: 5599–5612.

46. Yoshida T, Vivatbutsiri P, Morriss-Kay G, Saga Y, Iseki S (2008) Cell lineage in mammalian craniofacial mesenchyme. Mech Dev 125: 797–808.

47. Barraud P, Seferiadis AA, Tyson LD, Zwart MF, Szabo-Rogers HL, et al. (2010) Neural crest origin of olfactory ensheathing glia. Proceedings of the National Academy of Sciences 107: 21040–21045.

48. Robson EJ, He SJ, Eccles MR (2006) A PANorama of PAX genes in cancer and development. Nat Rev Cancer 6: 52–62.

49. Noll M (1993) Evolution and role of Pax genes. Curr Opin Genet Dev 3: 595–605.
50. Buckiova D, Syka J (2004) Development of the inner ear in Splotch mutant mice. Neuroreport 15: 2001–2005.
51. Lang D, Chen F, Milewski R, Li J, Lu MM, et al. (2000) Pax3 is required for enteric ganglia formation and functions with Sox10 to modulate expression of c-ret. J Clin Invest 106: 963–971.
52. Auerbach R (1954) Analysis of the developmental effects of a lethal mutation in the house mouse. J Exp Zool 127: 305–329.
53. Epstein DJ, Vogan KJ, Trasler DG, Gros P (1993) A mutation within intron 3 of the Pax-3 gene produces aberrantly spliced mRNA transcripts in the splotch (Sp) mouse mutant. Proc Natl Acad Sci U S A 90: 532–536.
54. Murdoch B, Roskams AJ (2008) A novel embryonic nestin-expressing radial glia-like progenitor gives rise to zonally restricted olfactory and vomeronasal neurons. J Neurosci 28: 4271–4282.

6

CHAPTER

SOX2 IS SEQUENTIALLY REQUIRED FOR PROGENITOR PROLIFERATION AND LINEAGE SPECIFICATION IN THE DEVELOPING PITUITARY

Sam Goldsmith, Robin Lovell-Badge and Karine Rizzoti

The Francis Crick Institute, Mill Hill Laboratory, The Ridgeway, Mill Hill, London NW7 1AA, UK.

ABSTRACT

SOX2/Sox2 mutations are associated with pituitary hormone deficiencies and the protein is required for pituitary progenitor proliferation, but its function has not been well characterized in this context. SOX2 is known to activate expression of Six6, encoding a homeodomain transcription factor, in the ventral diencephalon. Here, we find that the same relationship likely exists in the pituitary. Moreover, because Six6 deletion is associated with a similar phenotype as described here for loss of Sox2, Six6 appears as an essential downstream target of SOX2 in the gland. We also uncover a second role for SOX2. While cell differentiation is reduced in Sox2 mutants, some endocrine cells are generated such as POMC positive cells in the intermediate lobe. However, loss of SOX2 here results in complete downregulation of the melanotroph pioneer factor PAX7, and subsequently a switch of identity from melanotrophs to ectopic corticotrophs. Rescuing proliferation by ablating the cell cycle negative regulator p27 in Sox2 mutants does not restore melanotroph emergence. Therefore SOX2 has two independent roles during pituitary morphogenesis, first promotion of progenitor proliferation, and subsequently acquisition of melanotroph identity.

Citation: Sam Goldsmith, Robin Lovell-Badge and Karine Rizzoti, SOX2 is sequentially required for progenitor proliferation and lineage specification in the developing pituitary, doi: 10.1242/dev.137984.

INTRODUCTION

Pituitary endocrine secretions are regulated by the hypothalamus, located within the ventral diencephalon (VD). The two organs are connected through the pituitary stalk, allowing transmission of information between the two components of the axis. The hypothalamus integrates peripheral information and regulates pituitary secretions accordingly. The gland comprises three lobes, anterior, intermediate and posterior; endocrine cells types are located in the anterior (AL) and intermediate (IL) lobes while the posterior lobe comprises hypothalamic axon termini and glial cells. Secreted hormones control the function of other endocrine glands, and also different tissues. They are required for normal function of the organism as they intervene in many different physiological processes, including growth, reproduction, metabolism, and response to stress. Consequently deficits are associated with significant morbidity (Kelberman et al., 2009). It is therefore important to understand the role of genes and their proteins whose mutations are associated with pituitary hormone deficiencies to characterize the aetiology of such deficiencies, improve diagnosis and, in consequence, treatments. Mutations in the gene encoding the HMG box transcription factor SOX2 are associated with pituitary hormone deficiencies, or hypopituitarism, both in humans and mice (Kelberman et al., 2008; Kelberman et al., 2006). SOX2 is present from the earliest stages of pituitary development, in cells of the pituitary anlagen or Rathke pouch (RP) and is maintained in the post-natal and adult gland (Fauquier et al., 2008). Its expression defines a population of pituitary progenitor/stem cells (SC) throughout all these stages (Andoniadou et al., 2013; Rizzoti et al., 2013). However, the function of SOX2 within such cells is poorly defined.

We, and others, have shown that SOX2 is involved at several levels for hypothalamopituitary axis development and function. It is required in the VD for specification of the future hypothalamus and for the infundibulum, which is essential for both induction and maintenance of RP and will later give rise to the pituitary stalk and posterior lobe (Lee et al., 2012; Zhao et al., 2012) (Trowe et al., 2013). SOX2 is also required for the development of IL lobe tumors in mice mutant for p27KIP (Li et al., 2012)). In the embryo, the IL is specified in the dorsal part of RP. PAX7, which is expressed from 15.5dpc, acts as a pioneer transcription factor for emergence of melanotrophs, the sole endocrine cell type populating this lobe (Budry et al., 2012). These secrete MSH (melanocyte stimulating hormone) to regulate pigmentation. MSH is proteolytically cleaved from POMC (pro-opiomelanocortin), which also gives rise to ACTH (adrenocorticotropic hormone), secreted by AL corticotrophs. In the embryonic gland, P27KIP, which is mostly known as a cell cycle negative regulator but also performs several other functions ((Godin and Nguyen, 2014), is required to prevent cell cycle re-entry in differentiated cells (Bilodeau et al., 2009), while postnatal pituitary tumors develop in p27 null mice, exclusively within the IL (Fero et al., 1996; Nakayama et al., 1996). P27 has been shown to recruit co-repressors to downregulate Sox2 expression (Li et al., 2012). We demonstrated the relevance of this genetic interaction and the role of SOX2 in tumor development, by showing that deletion of one allele of Sox2 in p27-/- mice prevented occurrence of IL tumors (Li et al., 2012).

In this report, we characterize the role of SOX2 during pituitary morphogenesis. Because of its important role in the VD, conditional loss-of-function approaches are necessary to study specific functions of SOX2 during pituitary development. Here, we have used four Cre drivers to conditionally delete the gene in RP, while maintaining its expression in the VD, allowing phenotypic analysis of early and late phenotypes. We first demonstrate that SOX2 is required for normal levels of cell proliferation in RP. This is in agreement with (Jayakody et al., 2012), but we go on to reveal that Sox2 deletion results in a complete down-regulation of SIX6, known for its role in RP progenitor proliferation (Li et al., 2002). We then demonstrate a second role for SOX2. Deletion of the gene results in a reduction in endocrine cell differentiation, but we still observe some hormone secreting cells. In particular, some POMC positive cells are present in the developing IL of Sox2 mutants. However we show here that these are not melanotrophs, but ectopic corticotrophs, and that this can be explained by a complete downregulation of the melanotroph cell fate factor PAX7 in the absence of Sox2. We further demonstrate that this phenotype is independent of the rate of cell division. These results therefore demonstrate a sequential requirement for SOX2 during pituitary morphogenesis, first to promote progenitor proliferation and secondly for IL cell fate acquisition. This may also underlie its role in IL tumor development.

RESULT

Deletion of Sox2 in RP Results in Reduction of Progenitor Proliferation.

SOX2 is expressed throughout RP at 10.5dpc, becoming gradually restricted to the cells lining the cleft as development progresses (Fauquier et al., 2008). To understand the role of the protein during pituitary development, we deleted the gene using two different Cre drivers, FOXG1Cre (Hebert and McConnell, 2000) and NKX3.1Cre (Y.P.H., S. M. Price, Z. Chen, W. A. Banach-Petrosky, C. Abate-Shen and M.M.S., unpublished). Foxg1 is ubiquitously expressed in RP (Xuan et al., 1995). Accordingly, lineage tracing experiment using the R26ReYFP allele reveals eYFP expression throughout RP in Foxg1Cre/+;R26ReYFP/+ embryos at 10.5dpc (Figure.1A). By 18.5dpc, all cells in the pituitary appear eYFP positive (Figure.S1A). Nkx3.1 is expressed in RP from 10.5dpc until at least 14.5dpc, but is restricted to the dorsal region (Treier et al., 1998). In Nkx3.1Cre/+;R26ReYFP/+ embryos, eYFP is first detected in a few cells in RP at 10.5dpc (Figure.S1B), becoming substantially upregulated in the dorsal RP at 12.5dpc, where SOX2 is also predominantly present (Figure.1A). By the end of gestation, reporter activity is mostly observed in the IL, but there is also a significant contribution of eYFP positive cells in AL (Figure.S1B).

When we used either FOXG1Cre or NKX3.1Cre to delete Sox2, down-regulation of the protein closely matches the activity profile of the Cre drivers (Figure.1B, C). Expression of SOX2 is almost completely extinguished in RP at 10.5dpc in Sox2fl/fl;Foxg1Cre/+ embryos. Notably, morphological abnormalities, including hypoplasia and failure to separate from the underlying oral ectoderm, are observed at 12.5dpc (Figure.1B). In Sox2fl/fl;Nkx3.1Cre/+ embryos, SOX2 expression is down-regulated later, at 12.5dpc (Figure. 1C). Pituitary hypoplasia is observed as a consequence at 14.5dpc, mostly in the dorsal region of RP (Figure.1C,S3A). Most Sox2fl/fl;Nkx3.1Cre/+ and all Sox2fl/fl;Foxg1Cre/+animals die shortly after birth. This could be consecutive to hypopituitarism, and notably to a deficiency in ACTH, which would be expected in the in Sox2fl/fl;Nkx3.1Cre/+ animals because the pituitary is one essential organ where this driver is active and Sox2 expressed (Schneider et al., 2000); we did not however examine other tissues where the driver is active and SOX2 is essential. When using Foxg1Cre, post-natal lethality could be due to hypopituitarism, but there are also severe forebrain defects (Ferri et al., 2013).

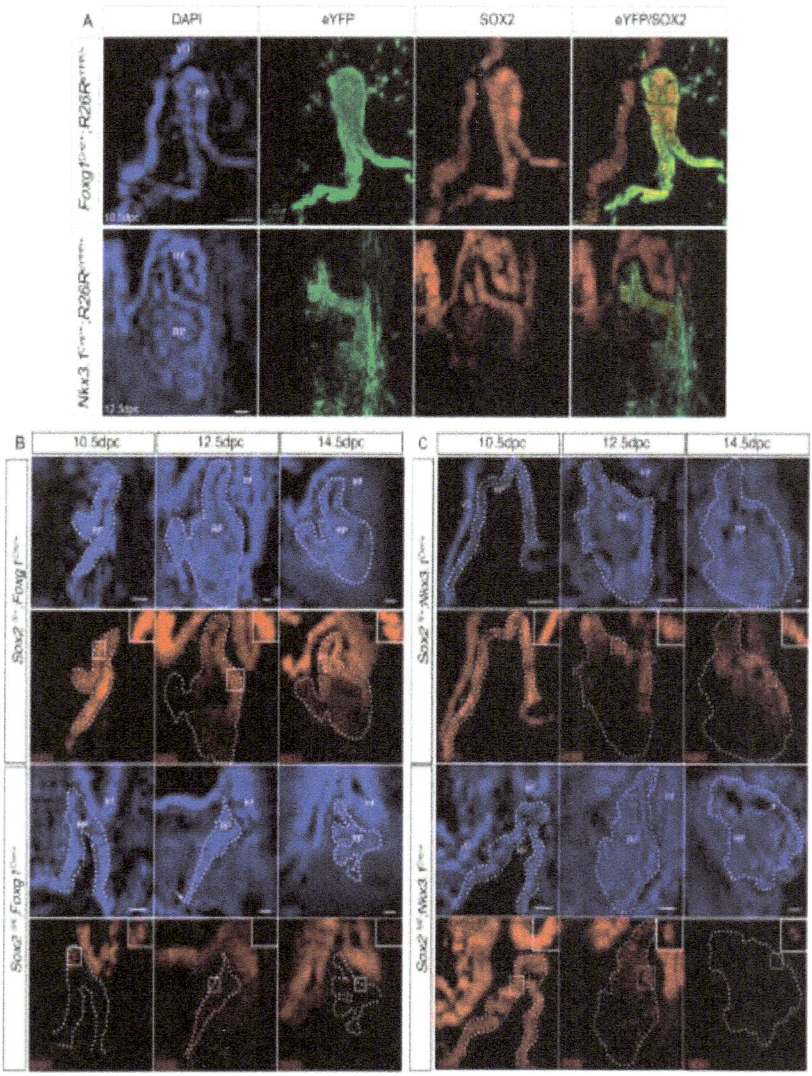

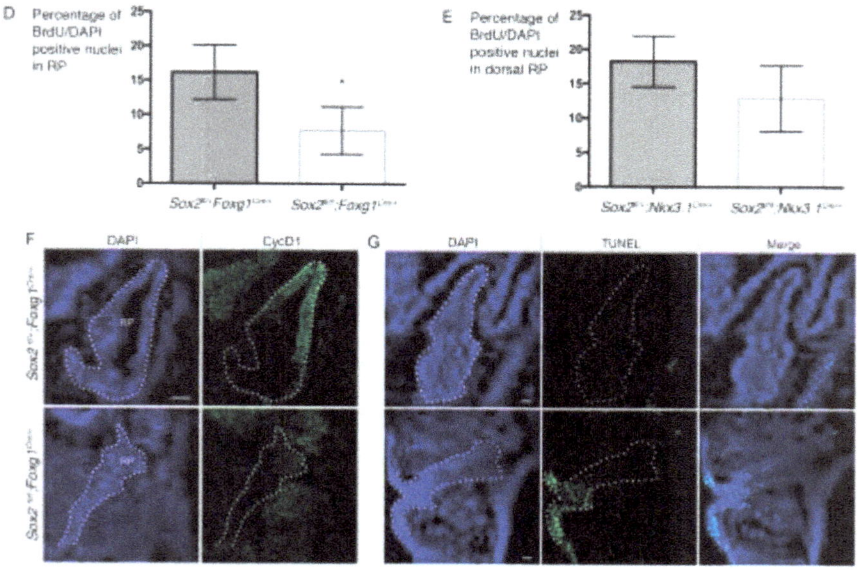

Figure 1: Loss of SOX2 results in a reduction of RP progenitors proliferation.

The RP hypoplasia, observed when both Nkx3.1Cre and Foxg1Cre are used to delete Sox2 could be caused by a defect in proliferation, an increase in apoptosis, or both. Progenitor proliferation was examined by performing a one-hour BrdU incorporation pulse at 12.5dpc. A reduction in the percentage of dorsal proliferating progenitors was observed in homozygous mutant Sox2fl/fl;Nkx3.1Cre/+ RPs, compared to Sox2fl/+;Nkx3.1Cre/+ embryos but this did not reach statistical significance (Figure.1E). When Sox2 was deleted using FOXG1Cre, a further significant reduction in BrdU incorporation was observed throughout RP in embryos homozygous for the conditional allele versus those that were heterozygous (Figure.1D). We then examined expression of the cell cycle promoter cyclin D1 and observed a clear downregulation of the protein in Sox2fl/fl;Foxg1Cre/+ embryos (Figure.1F) suggesting that the cell cycle was stopped at the G1/S checkpoint.

We examined apoptosis by performing TUNEL assays at 12.5dpc. In $Sox2^{fl/fl};Nkx3.1^{Cre/+}$ embryos, we did not observe any significant difference with controls (data not shown). In contrast, in $Sox2^{fl/fl};Foxg1^{Cre/+}$ embryos there is a ventral zone in the area of the hypoplastic RP that is still attached to the oral ectoderm where cells are undergoing apoptosis (Figure. 1G). At 11.5dpc, in wild type embryos, apoptosis occurs in the same region and this is thought to result in separation of RP from the underlying oral ectoderm (Charles et al., 2005). In $Sox2^{fl/fl};Foxg1^{Cre/+}$ embryos at 12.5dpc, we observe more apoptosis in this ventral domain. However, the hypoplastic pouch does not appear to separate from the ectoderm, even at a later stage (Figure.2D). This suggests that ventral apoptosis is not the only process required from separation of RP and that hypoplasia perturbs this event. In conclusion, Sox2 deletion in RP is associated with reduced proliferation and downregulation of CyclinD1 expression, in agreement with a previous study (Jayakody et al., 2012). Earlier and more efficient deletion of the gene, using Foxg1Cre instead of Nkx3.1Cre, results in a more severe phenotype, with formation of a hypoplastic RP that fails to separate from the underlying ectoderm.

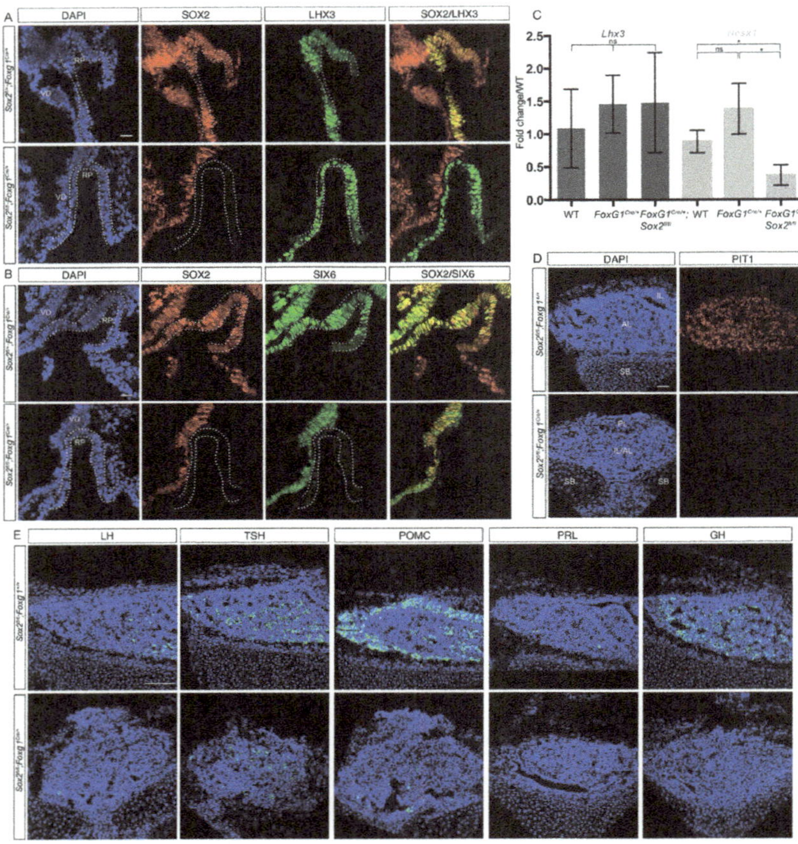

Figure 2: Loss of SOX2 is associated with down-regulation of the transcription factors SIX6 and HESX1, and a later dramatic reduction in cell numbers of all hormonal lineages.

Ubiquitous Sox2 Deletion in RP Results in the Complete Loss of SIX6 Expression, Down-Regulation of Hesx1 Transcripts, and a Deficiency in Most Pituitary Endocrine Cell Types

To better characterize the pathways and factors involved downstream of SOX2 in RP progenitor proliferation, we performed a marker analysis in $Sox2^{fl/fl}$;$Foxg1^{Cre/+}$ mutants. We first examined expression of LHX3, a LIM homeodomain transcription factor necessary for progenitor maintenance (Sheng et al., 1996). Its expression in Sox2fl/fl;Foxg1Cre/+ RP appears unaffected at 10.5dpc, and is similar to that observed in

$Sox^{2fl/+}$;$Foxg^{1Cre/+}$ embryos (Figure.2A). Transcripts for PITX2, another homeodomain transcription factor required for maintenance of RP progenitors (Charles et al., 2005; Gage et al., 1999), also appear unaffected by the loss of SOX2 in $Sox2^{fl/fl}$;$Foxg1^{Cre/+}$ embryos at 12.5dpc (Figure.S2A). Pituitary hypoplasia and reduced proliferation are phenotypic features observed in embryos null for the homeobox gene Six6 (Li et al., 2002). These two features, with varying degrees of severity, are also seen in $Sox2^{fl/fl}$;$Nkx3.1^{Cre/+}$ and $Sox^{2fl/fl}$;$Foxg1^{Cre/+}$ mutants (Figure.1B, C). At 10.5dpc, SOX2 and SIX6 are co-localised throughout RP in Sox2fl/+;$Foxg1^{Cre/+}$ embryos (Figure.2B). However, in mutant $Sox2^{fl/fl}$;$Foxg1^{Cre/+}$ embryos SIX6 expression is dramatically downregulated at 10.5dpc, coincident with loss of SOX2 (Figure.2B).

is required for forebrain and pituitary development (Andoniadou et al., 2007; Dasen et al., 2001). It has been proposed that SOX2 could directly participate in Hesx1 regulation because it can bind the Hesx1 promoter and induce its transcriptional activation in vitro (Kelberman et al., 2006). We therefore quantified Hesx1 expression using real time PCR with RNA extracted from whole heads at 11.5dpc, as the gene is exclusively present in RP at this stage, before any significant morphological defect appears in $Sox2^{fl/fl}$;$Foxg1^{Cre/+}$ embryos. In $Sox2^{fl/fl}$;$Foxg1^{Cre/+}$ embryos we observe a significant reduction of more than 50% of Hesx1 transcript levels compared to wildtype embryos (Figure.2C). To assess the specificity of this result we examined Lhx3 transcripts levels in the same samples. We observed no significant difference in Lhx3 expression levels, in agreement with our previous results showing that expression of its protein product is not affected (Figure.2A).

We also examined the expression of the paired-like homeobox transcription factor PROP1, which is up-regulated as HESX1 expression fades in RP and is necessary for emergence of the PIT1 lineage and gonadotrophs (Gage et al., 1996; Sornson et al., 1996). Expression of Prop1 is clearly downregulated in mutant embryos at 12.5dpc (Figure. S2A) as previously observed (Jayakody et al., 2012). At 15.5dpc, as endocrine cell differentiation takes place, we observe a dramatic reduction of PIT1 protein expression (Figure.2D), and a loss of most hormonal cell types at 16.5dpc (Figure.2E), while $Sox2^{fl/fl}$;$Nkx3.1^{Cre/+}$embryos show a less dramatic but clear reduction in differentiated cell types (Figure.S2B). In contrast, we found proportionally more thyrotrophs in $Sox2^{fl/fl}$;$Foxg1^{Cre/+}$ than any other endocrine cell types. As PIT1 is completely down regulated, we hypothesize that these represent PIT1 independent, transient thyrotrophs emerging at 12.5dpc, before PIT1 dependent definitive thyrotrophs arise at 15.5dpc (Kelberman et al., 2009). These are normally localized rostrally however in $Sox2^{fl/fl}$;$Foxg1^{Cre/+}$ embryos the morphology of RP is severely disrupted; to the extent that any delineation between the IL and AL lobes is no longer discernable. As thyrotrophs are the primary endocrine cell type remaining in RP, loss of tissue organization results in these cells becoming spread both rostrally and caudally throughout the gland. Moreover we observe in a proportion of mutants a disruption of the basisphenoid cartilage, probably because of the maintenance of an abnormal connection with the oral ectoderm (Figure. 1G). This leads to ectopic location of pituitary tissue in continuity with the oral ectoderm (Figure.2D) (Jayakody et al., 2012).

In summary, downregulation of Six6, Hesx1 and Prop1 in mutants suggests that SOX2 participates in regulation of the expression of these three factors, which are normally co-expressed with SOX2 and are necessary for correct development of RP (Dasen et al., 2001; Gage et al., 1996; Li et al., 2002; Sornson et al., 1996; Yoshida et al., 2009). Consequently, very little endocrine cell differentiation takes place in $Sox2^{fl/fl}$;$Foxg1^{Cre/+}$ hypoplastic pituitaries, where essentially only some early differentiating thyrotrophs are observed. The three genes could be direct or indirect targets of SOX2, but we hypothesize that Six6 regulation is likely to be direct, as we previously showed this to be the case in the ventral diencephalon (Lee et al., 2012).

SOX2 is Expressed at Low Levels in IL Melanotrophs and is Not Required for SIX6 Expression in These Cells

By the end of gestation, at 18.5dpc, SOX2 positive cells are mainly found surrounding the lumen of the IL and AL, although some are also scattered in the AL parenchyma (Figure 3A,B) (Fauquier et al., 2008). These cells, highly positive for SOX2 ($SOX2^{Hi}$), do not express any hormones and at least a proportion of them represent SCs (Andoniadou et al., 2013; Rizzoti et al., 2013).

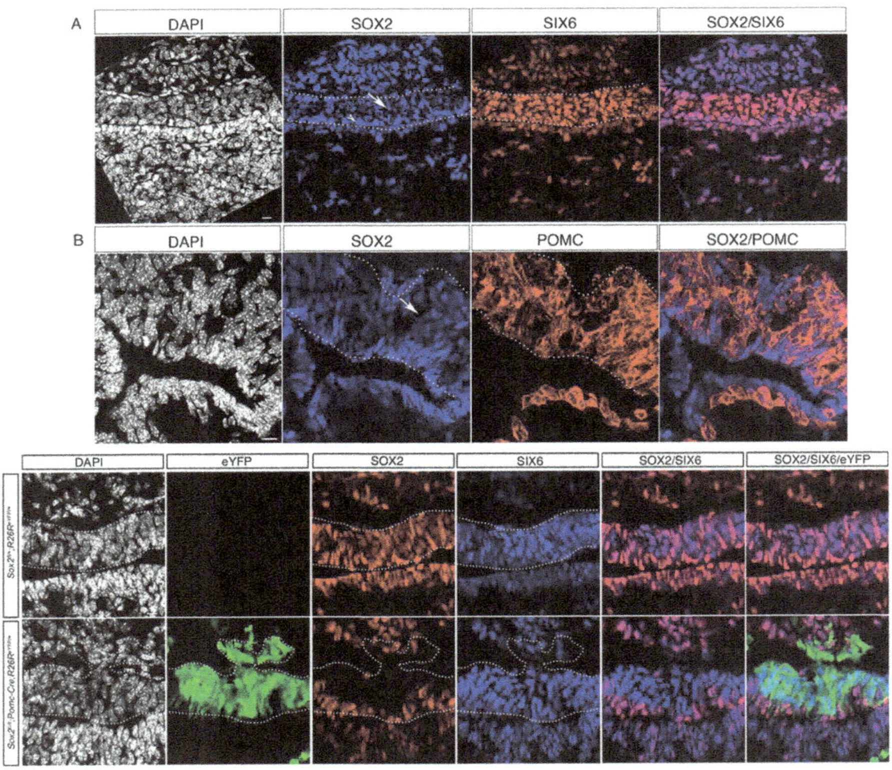

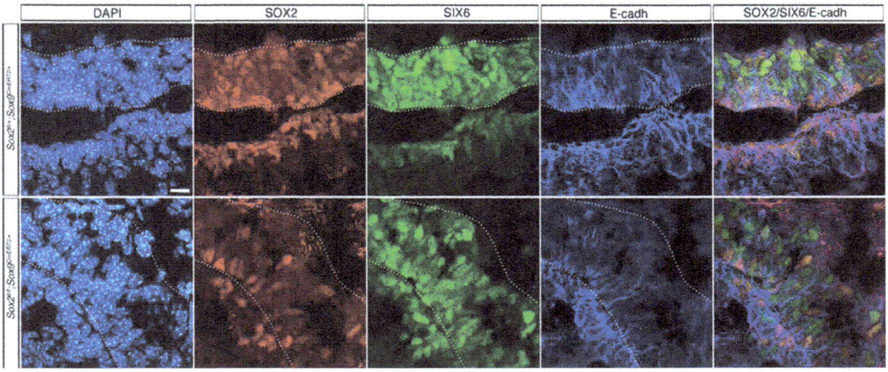

Figure 3: SOX2 regulates SIX6 expression exclusively in the SOX2Hi progenitor population.

In IL, SOX2 is maintained throughout the lobe, albeit at lower levels than in the epithelial stem/progenitor cell layer lining the cleft (Figure.3A). Cells expressing low levels of SOX2 in the IL ($SOX2^{Low}$) are POMC positive melanotrophs (Figure.3B). These are the only differentiated cells in the pituitary to maintain nuclear expression of SOX2. This appears to break with the usual assumption that SOX2 is associated with stem cells, even though there are many exceptions (Ekonomou et al., 2005; Hoefflin and Carter, 2014). It is also a finding that has received little attention. We therefore decided to focus our analysis on the role of SOX2 in IL.

At 18.5dpc, SIX6 and SOX2 remain co-expressed but, in contrast with SOX2, the levels of SIX6 expression appear uniform in both the cells lining the cleft and in melanotrophs (Figure.3A). We set out to investigate whether, as we observed in RP, SOX2 is required for SIX6 expression at this later stage. We first examined whether SOX2 is necessary for SIX6 expression within terminally differentiated melanotrophs (SOX2Low; SIX6 positive; POMC positive cells). We used Pomc-Cre, which is expressed in all melanotrophs and in a proportion of corticotrophs (Langlais et al., 2013), to delete Sox2 (Figure.3C). In $Sox2^{fl/fl}$;Pomc-Cre; $R26R^{eYFP/+}$ embryos at 18.5dpc, SOX2 expression is efficiently lost in eYFP positive cells in the body of the IL, while expression is maintained in the eYFP negative, $SOX2^{Hi}$ cells that line the lumen. SIX6 expression appears unaffected by the loss of SOX2 in IL eYFP positive cells (Figure.3C).

We further investigated the SOX2-SIX6 interaction by deleting SOX2 in undifferentiated cells, using Sox9CreErT2, which, like Sox9 itself, is expressed from 14.5dpc in pituitary stem/progenitor cells (Rizzoti et al., 2013). At 18.5dpc, Ecadherin expression is enriched in progenitors lining the pituitary cleft (Chauvet et al., 2009). We therefore assessed deletion of SOX2 by examining expression of the protein in E-cadherin positive cells. We did not observe any obvious difference in Ecadherin expression between $Sox2^{fl/+}$;$Sox9^{CreErT2}$ and $Sox2^{fl/fl}$;$Sox9^{CreErT2}$ at 18.5dpc (Figure.3D). However, following CreER2 induction, SOX2 is absent in a proportion of E-cadherin positive cells lining the cleft in $Sox2^{fl/fl}$;$Sox9^{CreErT2}$ pituitaries, while SIX6 expression appears unaffected by the loss of SOX2 (Figure.3D). This could simply mean that SIX6 expression does not rely on SOX2 at this stage, in progenitors/SCs. Nevertheless, the identity of SCs may be altered as Sox2 is deleted. Therefore SIX6 positive; SOX2 negative cells may not be progenitors/SCs anymore but differentiated/differentiating cells in which maintenance of SIX6 is independent of SOX2, as we show in IL POMC positive cells (Figure.3C).

Loss of SOX2 in IL Results in Downregulation of PAX7 and a Switch in IL cell fate from Melanotrophs to Corticotrophs.

To better characterize the IL phenotype in Sox2 mutants, we focused our analysis on Sox2fl/fl;Nkx3.1$^{Cre/+}$ embryos where this lobe is still discernable, in contrast with Sox2$^{fl/fl}$;Foxg1$^{Cre/+}$ embryos where the phenotype is more severe. Deletion of Sox2 in the dorsal region of RP at 12.5dpc in Sox2$^{fl/fl}$;Nkx3.1$^{Cre/+}$ embryos results in reduced proliferation, a severe reduction in the size of the dorsal RP, and subsequently IL at 18.5dpc (Figure.1C, E, 4A and Figure.S2). This is associated with a significant reduction in the percentage of POMC positive cells in IL (Figure.4B).

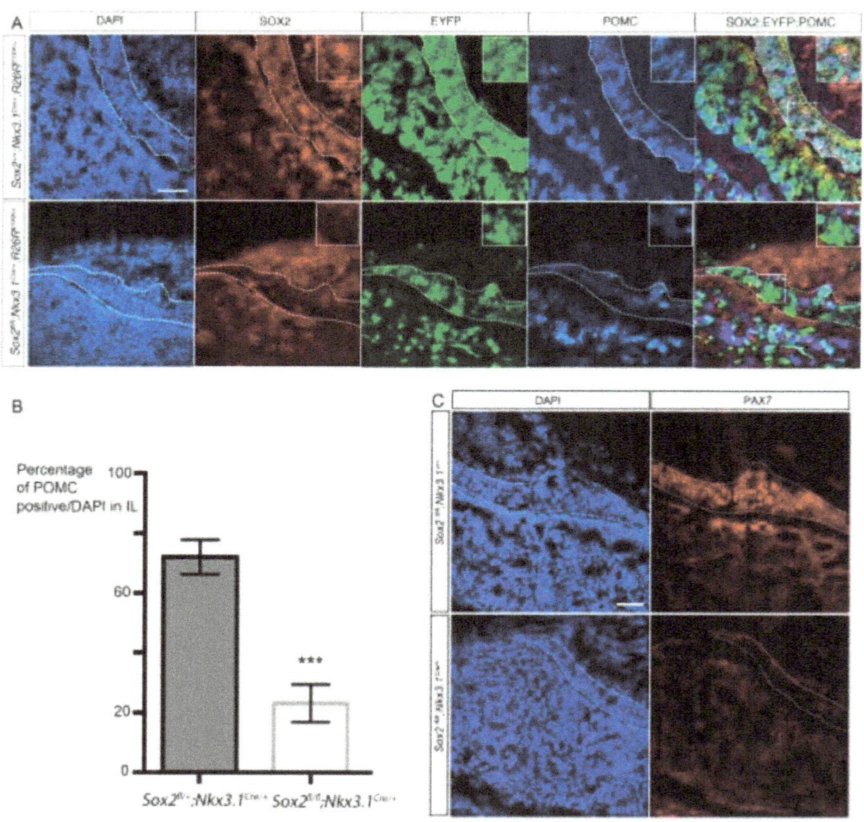

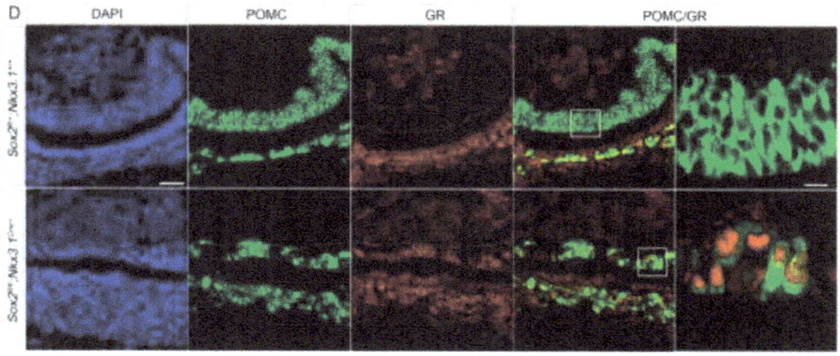

Figure 4: Sox2 deletion in IL progenitors results in reduction of POMC positive cells, complete downregulation of PAX7, and ectopic differentiation of corticotrophs.

To investigate the deficit in POMC positive cells in IL, we examined the expression of factors required for the emergence of melanotrophs. We first examined the expression of TPIT, which is required for POMC activation in both melanotrophs and corticotrophs (Pulichino et al., 2003). In the IL of $Sox2^{fl/fl}$;$Nkx3.1^{Cre/+}$ embryos, its expression is dramatically down-regulated, mirroring the loss of POMC (Figure.S3). In Tpit null pituitaries, IL cells can change fate and become ectopic PIT1 independent thyrotrophs or SF1 positive gonadotrophs (Pulichino et al., 2003). We did not observe any ectopic expression of TSH or SF1 in the $Sox2^{fl/fl}$;$Nkx3.1^{Cre/+}$ IL, demonstrating that IL cells did not adopt an alternative thyrotroph or gonadotroph fate in mutants (Figure.S2,S3). The paired homeodomain protein PAX7 is a pioneer transcription factor required for melanotroph fate (Budry et al., 2012). It is expressed exclusively in the IL, becoming upregulated just prior to TPIT at 15.5dpc. In $Sox2^{fl/fl}$;$Nkx3.1^{Cre/+}$ embryos at 18.5dpc we observe a complete loss of PAX7 expression (Figure. 4C). This result suggests that the Sox2 deleted, POMC negative cells in the mutant IL fail to commit to the melanotroph fate. In addition, the complete loss of PAX7 raises questions about the identity of the POMC positive cells present in the mutant IL.

A switch in IL cell fate from melanotrophs to corticotrophs has been observed in Pax7 null pituitaries (Budry et al., 2012). This phenotype is characterized by ectopic expression of the Glucocorticoid Receptor (GR) in POMC positive Pax7 null IL cells. GR is normally excluded from melanotrophs and confined to glucocorticoid responding cells of the AP, including POMC positive corticotrophs (Budry et al., 2012). We therefore examined expression of GR and observed that it is ectopically expressed in the IL of $Sox2^{fl/fl}$;$Nkx3.1^{Cre/+}$ embryos, co-localising with POMC (Figure 4D). This strongly suggests that the SOX2 deleted, POMC positive IL cells are in fact ectopic corticotrophs that have switched fate in the absence of SOX2, and consequently PAX7. This implies a specific requirement for SOX2 in IL cell fate.

SOX2 is not Required for Maintenance of PAX7 Expressio

To better characterize the interaction between SOX2 and PAX7, we examined PomcCre;$Sox2^{fl/fl}$;$R26R^{eYFP/+}$ embryos, where Sox2 is deleted exclusively in differentiated melanotrophs and corticotrophs (Langlais et al., 2013) (Figure.3B). We did not observe a loss of PAX7, or ectopic GR expression in these mutants (Figure.5), suggesting that SOX2 is required for initiation rather than maintenance of PAX7 expression.

However, we still detect a small number of ectopic corticotrophs, negative for PAX7 and positive for GR (Figure.5, arrow). We therefore analyzed in detail the activity of the Pomc-Cre transgene by examining Pomc-Cre;$R26R^{eYFP/+}$embryos. At 14.5dpc, which is before the initiation of endogenous POMC expression at 16.5dpc, we observe a few eYFP positive cells in the Pomc-Cre;$R26R^{eYFP/+}$ IL, demonstrating early ectopic expression of the Pomc-Cre transgene (Figure.S4). Sox2 is therefore deleted ectopically in these cells before PAX7 is expressed at 15.5dpc, resulting in the differentiation of ectopic corticotrophs, as observed in $Sox2^{fl/fl}$;$Nkx3.1^{Cre/+}$ mutants. All together these data further reinforce the suggestion that SOX2 is required for initiation and not maintenance of PAX7 expression.

SOX2 Regulates Progenitor Proliferation and Melanotroph Cell Lineage Specification Independently.

The adoption by a proportion of cells of an alternative corticotroph cell fate in the absence of SOX2 in IL, where the protein is normally maintained in differentiated cells, led us to hypothesize that SOX2 has two independent functions. Firstly SOX2 is required for the generation of a sufficient number of progenitors (Figure.1), and secondly for the specification of melanotroph fate by inducing PAX7 expression (Figure. 4). To verify this hypothesis, we decided to restore proliferation in $Sox2^{fl/fl}$;Nkx3.1Cre/+ mutants and examine melanotroph differentiation.

We observed, as reported previously (Jayakody et al., 2012), a slight up-regulation of the cell cycle negative regulator P27KIP in Sox2 null mutants (data not shown). In the developing pituitary P27KIP has been demonstrated to prevent re-entry into the cell cycle of differentiated cells. In its absence, cell differentiation is not perturbed but there is ectopic proliferation (Bilodeau et al., 2009). We therefore examined whether proliferation and melanotroph cell differentiation were restored in $p27^{-/-}$; $Sox2^{fl/fl}$;$Nkx3.1^{Cre/+}$ embryos.

We first explored proliferation levels in $p27^{-/-}$;$Sox2^{fl/fl}$;$Nkx3.1^{Cre/+}$ pituitaries following a one hour pulse of EdU at 18.5dpc (Figure.6A). We observed a significant reduction in EdU incorporation in $Sox2^{fl/fl}$;$Nkx3.1^{Cre/+}$ pituitaries compared to wildtype. In contrast, there was no significant reduction in EdU incorporation in p27-/-; $Sox2^{fl/fl}$;$Nkx3.1^{Cre/+}$ embryos compared to wild-type, demonstrating a rescue of proliferation in double mutants (Figure.6A). In agreement with these results, we observe a thicker IL in $p27^{-/-}$;$Sox2^{fl/fl}$;$Nkx3.1^{Cre/+}$ compared to $Sox2^{fl/fl}$;$Nkx3.1^{Cre/+}$ embryos (Figure.6C).

We then quantified the percentage of POMC positive cells in IL (Figure.6B). We found a significant reduction in the percentage of POMC positive cells in the IL of both $Sox2^{fl/fl}$;$Nkx3.1^{Cre/+}$ and $p27^{-/-}$;$Sox2^{fl/fl}$;$Nkx3.1^{Cre/+}$ IL's compared to wild-type embryos.

However there is a slight, but significant increase in the percentage of POMC positive cells in the IL of p27$^{-/-}$;Sox2$^{fl/fl}$;Nkx3.1$^{Cre/+}$ embryos compared to Sox2$^{fl/fl}$;Nkx3.1$^{Cre/+}$ embryos, demonstrating a partial rescue of the differentiation defect.

We then examined PAX7 and POMC expression. PAX7 and POMC expression is unaffected in p27 null embryos, and there is no re-activation of PAX7 in p27-/-; Sox2fl/fl;Nkx3.1Cre/+ pituitaries (Figure.6C). As a consequence, ectopic expression of GR is observed, which demonstrates the presence of ectopic corticotrophs, (Figure. 6D). Therefore, despite restoring proliferation and improving the proportion of differentiated cells, the melanotroph lineage specification defect is still present. These results further argue for an independent role of SOX2 in RP progenitor proliferation and subsequently in acquisition of IL cell fate.

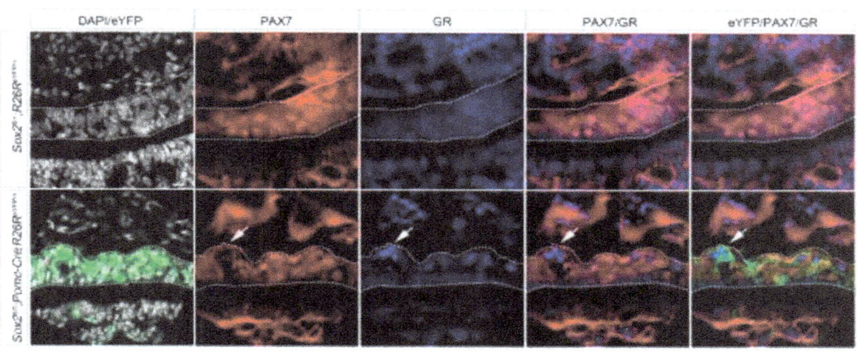

Figure.5: Sox2 deletion in IL after cell commitment does not affect melanotroph cell fate.

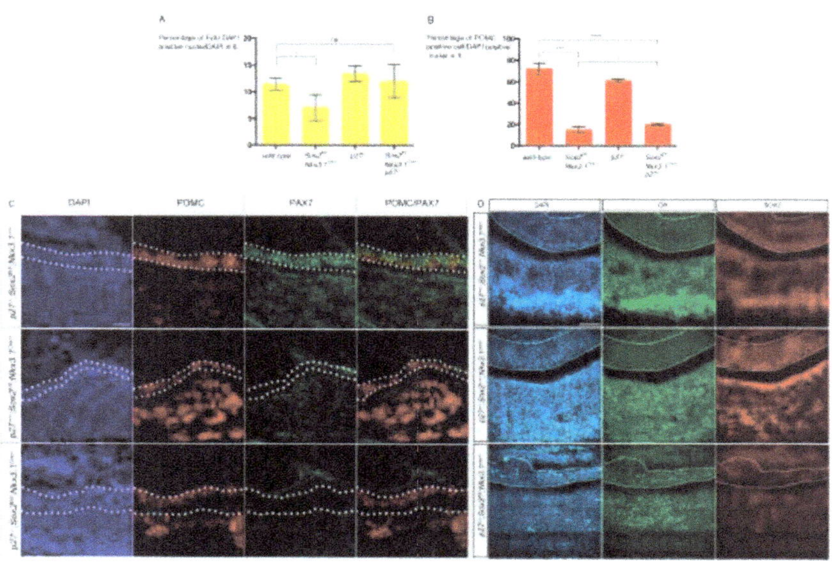

Figure 6: Deletion of p27 in Sox2$^{fl/fl}$;Nkx3.1$^{Cre/+}$ embryos result in a rescue of IL cell proliferation, but melanotroph cell fate acquisition is not restored.

DISCUSSION

Congenital defects in pituitary development and function can be associated with substantial morbidity and/or deficiencies that compromise quality of life. These can pose significant challenges for treatment, especially when linked to other clinical problems. For example, heterozygous loss-of-function mutations in SOX2 in humans are associated with severe eye defects, ranging from micropthalmia to anopthalmia, but the patients also exhibit other congenital defects including hypopituitarism that can affect everything from height to puberty (Macchiaroli et al., 2014) (Suzuki et al., 2014). However, the role of SOX2 in the developing pituitary is not well understood. Here we have explored this in the mouse, revealing the role of the protein in cell proliferation and differentiation and defining several critical downstream genes.

To characterize the role of SOX2 in the developing pituitary we have conditionally deleted the gene using two different Cre drivers. Deletion using $Foxg1^{Cre}$, expressed early and ubiquitously in RP (Wang et al., 2010), showed that SOX2 is required for normal levels of proliferation in RP progenitors. This is in agreement with (Jayakody et al., 2012) who deleted Sox2 using $Hesx1^{Cre}$, which displays a pattern of activity very similar to that of Foxg1Cre. The generation of a reduced pool of progenitors severely impairs pituitary development and emergence of endocrine cell types is consequently dramatically reduced. This is supported by the slight but significant improvement in IL cell differentiation observed after rescue of cell proliferation by removing the cell cycle inhibitor p27 (see below). Cre driven by Nkx3.1 is active later than when it is driven by Foxg1, moreover, it is mostly restricted to the dorsal part of RP, the future IL. RP development is less perturbed in $Sox2^{fl/fl}$;$Nkx3.1^{Cre/+}$ compared to $Sox2^{fl/fl}$;$Foxg1^{Cre/+}$ mutants and we observe a later phenotype, mostly affecting the IL. As found in the AL, the proportion of differentiated cells is reduced, but strikingly there is a complete down-regulation of the melanotroph fate pioneer factor PAX7. This results in a switch of identity of the IL POMC positive cells to ectopic corticotrophs. When we rescue the proliferation defect in Sox2 mutants by removing p27, we do not observe a rescue of IL melanotroph identity. These results demonstrate that SOX2 has two independent functions during pituitary morphogenesis, proliferation of progenitors and later cell fate acquisition in the IL.

We have examined the expression of key transcription factors required in RP progenitors, and are co-expressed with SOX2, to better understand its early role. Despite a severe hypoplastic phenotype in $Sox2^{fl/fl}$;$Foxg1^{Cre/+}$ mutants, the early pattern and expression levels of LHX3/Lhx3 and Pitx2 are not affected by the loss of SOX2. In contrast, we observe a clear down-regulation of Prop1, as reported previously (Jayakody et al., 2012), and of Hesx1, which was suspected of being a SOX2 target (Kelberman et al., 2006). Jayakody et al used Hesx1Cre to delete Sox2. This could have given misleading results given that (i) the Cre driver depends on the product of its intended target gene (Sox2) for its own expression and (ii) $Hesx1^{Cre}$;$Sox2^{fl/fl}$ embryos are already lacking one functional allele of Hesx1. While we both obtain a similar hypoplasia in the anterior lobe, they did not report any IL phenotype. HESX1 and PROP1 are sequentially expressed in RP progenitors, with a slight overlap. In contrast with Sox2 mutant phenotypes, deletion of Hesx1 is mostly associated with hyperproliferation in RP (Dasen et al., 2001) while Prop1 loss results in accumulation of progenitors near the cleft because these are unable to differentiate and populate the developing AL. PROP1, the first exclusive marker of pituitary identity, therefore promotes progenitor commitment (Ward et al., 2005). The significance of a direct or indirect regulatory role for SOX2 in HESX1 and PROP1 expression is unclear, but it may reflect an involvement of SOX2 in establishment/maintenance of pituitary identity. SOX2 and PROP1 remain coexpressed in rat pituitary progenitors/SC until PROP1 expression fades in adults (Yoshida et al., 2009; Yoshida et al., 2011) and it would be of interest to determine whether SOX2 is continuously required to maintain normal levels of PROP1 expression.

In contrast, the significance of Six6 regulation by SOX2 is very clear, but the molecular mechanisms appear complex. SOX2 directly regulates Six6 in the VD (Lee et al., 2012), but the enhancer bound by SOX2 in the VD is not active in RP (Lee et al., 2012). The complexity of the regulatory network is further revealed by the maintenance of SIX6 expression after Sox2 deletion in IL melanotrophs using PomcCre (Langlais et al., 2013) and in late progenitors/SCs, using $Sox9^{CreERT2}$ (Rizzoti et al., 2013). We expect the properties of progenitors to be affected by the loss of SOX2. Therefore the maintenance of SIX6 expression after Sox2 deletion in progenitors may rather be linked to their progression toward commitment/differentiation once SOX2 is downregulated, rather than simply reflect SOX2 independent regulation of SIX6 in late versus early progenitors. The specific roles of SOX2 in the variety of cell types where it is expressed rely on its interaction with different partners, and the nature of the complexes that bind DNA. Therefore these results suggest that SOX2 has a specific partner in RP progenitors for Six6 transactivation, but that SOX2 then becomes redundant in differentiated and differentiating cells.

IX6 associates with DACH corepressors to downregulate transcription of p27 in the eye, and hence indirectly promote proliferation (Li et al., 2002). In RP, SIX6 has been proposed to positively regulate progenitor proliferation by the same mechanism (Li et al., 2002). This fits well with the observed up-regulation of p27 in Sox2 mutants (data not shown; (Jayakody et al., 2012)), because SIX6 is essentially absent (Figure.7). Therefore SOX2 may promote RP progenitor proliferation indirectly, through upregulation of SIX6, which in turn represses p27. SIX6 is probably not the only relevant target of SOX2 in RP progenitors, but we expect it to play a significant role because Six6 (Li et al., 2002) and Sox2 mutant RP phenotypes are relatively similar. Once p27 is up-regulated in MSH secreting cells, we hypothezise that it can recruit co-repressors to down-regulate expression of Sox2, as shown in vitro (Li et al., 2012). However low levels of SOX2 persist in melanotrophs, and elevated levels of SOX2 in these and/or SC, as observed in p27 null mice, result in IL tumor development (Li et al., 2012). It is important to determine which of these two cell types is most sensitive to alteration in SOX2 expression to understand mechanisms of tumor formation.

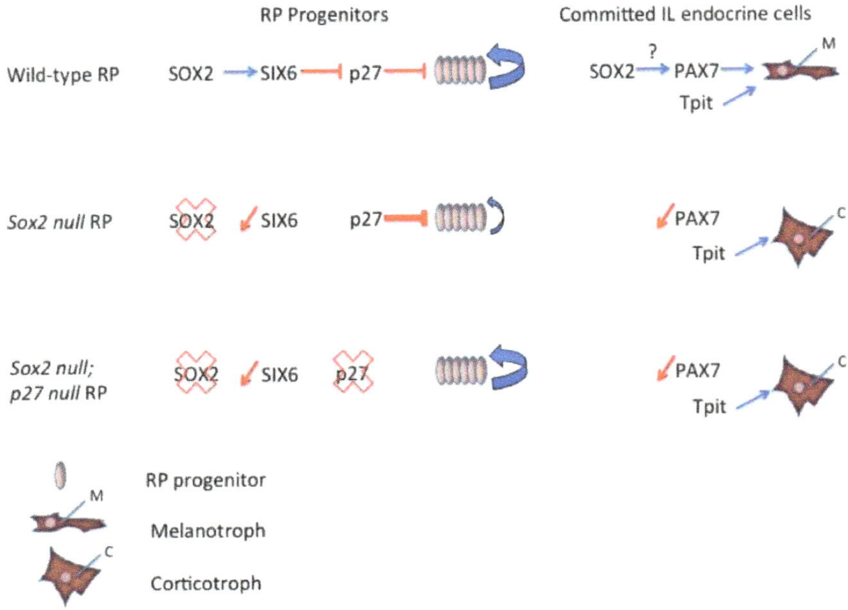

Figure 7: Model representing the roles of SOX2 during pituitary morphogenesis.

In RP progenitors, SOX2 is likely to directly activate Six6 expression. SIX6 has been shown to repress p27 expression, hence promoting progenitors proliferation (Li et al., 2002). In the absence of SOX2, we show that SIX6 is downregulated while p27 appears to be up-regulated (data not show, (Jayakody et al., 2012)); as a result, cell proliferation is reduced. In support of this hypothesis, we show that ablating p27 in Sox2 mutants results in a rescue of proliferation. In melanotroph precursors, we show that SOX2 is required for PAX7 up-regulation, but not T-pit expression as Sox2 mutant develop ectopic Tpit; POMC positive corticotrophs in IL. This is independent of SOX2 role in cell proliferation as ablating pss27 has no effect on the emergence of ectopic IL corticotrophs in Sox2 mutants.

In an attempt to restore proliferation in Sox2 mutants, we decided to delete p27. In agreement with our hypothesis we observe a clear rescue of proliferation in double mutants. However, the differentiated cells in the IL of Sox2;p27 double mutants are still ectopic corticotrophs. This is a strong argument in favor of a second role for SOX2 in acquisition of melanotroph identity. Our results further suggest that SOX2 is required for up-regulation, but not maintenance, of the melanotroph pioneer transcription factor PAX7. The molecular mechanisms underlying this requirement are unknown. We analyzed 143kb surrounding Pax7 for ECR (evolutionary conserved regulatory elements); we could not find any SOX2 consensus binding sites within these. Moreover Chip Seq experiments in an altered AtT20 corticotroph cell line, engineered to overexpress SOX2, did not reveal any binding of SOX2 at the Pax7 locus (J. Drouin, personal communication). Further investigations are required to determine whether SOX2 requires a specific partner to up-regulate expression of Pax7, perhaps via variant binding sites, or alternatively to demonstrate that regulation is indirect. In conclusion our work has uncovered a new role for SOX2 for melanotroph cell fate acquisition, independently of its early role in promoting progenitor proliferation (Figure.7). We also reveal that SOX2 is maintained at low levels in melanotrophs where its expression is likely regulated by p27 (Li et al., 2012). It is important to understand the significance of this, because it may be relevant for development of tumors in the IL. Finally, we have better placed SOX2 is the hierarchy of transcription factors involved in early development of the pituitary. This may aid clinical diagnosis of pituitary defects and be relevant for regenerative medicine approaches for their treatment (Suga et al., 2011) (Dincer et al., 2013).

ACKNOWLEDGEMENTS

We are grateful to Jacques Drouin (Institut de Recherches Cliniques de Montreal) for the Tpit antibody, Pomc-Cre mice and sharing non-published data. We would like to thank Michael M. Shen (Columbia University College of Physicians and Surgeons) for the $Nkx3.1^{Cre/+}$ mice. We are indebted to the Confocal microscopy platform and Biological Services for animal husbandry. We would like to thank members of the Lovell-Badge's lab for help and support. This work was supported by the UK Medical Research Council (U117512772), the NIH (R01EB016629-03) and Worldwide Cancer Research (13-1270)

REFERENCES

1. Andoniadou, C.L., Matsushima, D., Mousavy Gharavy, S.N., Signore, M., Mackintosh, A.I., Schaeffer, M., Gaston-Massuet, C., Mollard, P., Jacques, T.S., Le Tissier, P., et al. (2013). Sox2(+) stem/progenitor cells in the adult mouse pituitary support organ homeostasis and have tumor-inducing potential. Cell Stem Cell 13, 433-445.
2. Andoniadou, C.L., Signore, M., Sajedi, E., Gaston-Massuet, C., Kelberman, D., Burns, A.J., Itasaki, N., Dattani, M., and Martinez-Barbera, J.P. (2007). Lack of the murine homeobox gene Hesx1 leads to a posterior transformation of the anterior forebrain. Development 134, 1499-1508.
3. Bilodeau, S., Roussel-Gervais, A., and Drouin, J. (2009). Distinct developmental roles of cell cycle inhibitors p57Kip2 and p27Kip1 distinguish pituitary progenitor cell cycle exit from cell cycle reentry of differentiated cells. Mol Cell Biol 29, 1895-1908.
4. Budry, L., Balsalobre, A., Gauthier, Y., Khetchoumian, K., L'Honore, A., Vallette, S., Brue, T., Figarella-Branger, D., Meij, B., and Drouin, J. (2012). The selector gene Pax7 dictates alternate pituitary cell fates through its pioneer action on chromatin remodeling. Genes Dev 26, 2299-2310.
5. Charles, M.A., Suh, H., Hjalt, T.A., Drouin, J., Camper, S.A., and Gage, P.J. (2005). PITX genes are required for cell survival and Lhx3 activation. Mol

Endocrinol 19, 1893-1903.

6. Chauvet, N., El-Yandouzi, T., Mathieu, M.N., Schlernitzauer, A., Galibert, E., Lafont, C., Le Tissier, P., Robinson, I.C., Mollard, P., and Coutry, N. (2009). Characterization of adherens junction protein expression and localization in pituitary cell networks. J Endocrinol 202, 375-387.
7. Dasen, J.S., Barbera, J.P., Herman, T.S., Connell, S.O., Olson, L., Ju, B., Tollkuhn, J., Baek, S.H., Rose, D.W., and Rosenfeld, M.G. (2001). Temporal regulation of a paired-like homeodomain repressor/TLE corepressor complex and a related activator is required for pituitary organogenesis. Genes Dev 15, 3193-3207.
8. Dincer, Z., Piao, J., Niu, L., Ganat, Y., Kriks, S., Zimmer, B., Shi, S.H., Tabar, V., and Studer, L. (2013). Specification of functional cranial placode derivatives from human pluripotent stem cells. Cell reports 5, 1387-1402.
9. Ekonomou, A., Kazanis, I., Malas, S., Wood, H., Alifragis, P., Denaxa, M., Karagogeos, D., Constanti, A., Lovell-Badge, R., and Episkopou, V. (2005). Neuronal migration and ventral subtype identity in the telencephalon depend on SOX1. PLoS Biol 3, e186.
10. Fauquier, T., Rizzoti, K., Dattani, M., Lovell-Badge, R., and Robinson, I.C. (2008). SOX2-expressing progenitor cells generate all of the major cell types in the adult mouse pituitary gland. Proc Natl Acad Sci U S A 105, 2907-2912.
11. Fero, M.L., Rivkin, M., Tasch, M., Porter, P., Carow, C.E., Firpo, E., Polyak, K., Tsai, L.H., Broudy, V., Perlmutter, R.M., et al. (1996). A syndrome of multiorgan hyperplasia with features of gigantism, tumorigenesis, and female sterility in p27(Kip1)-deficient mice. Cell 85, 733-744.
12. Ferri, A., Favaro, R., Beccari, L., Bertolini, J., Mercurio, S., Nieto-Lopez, F., Verzeroli, C., La Regina, F., De Pietri Tonelli, D., Ottolenghi, S., et al. (2013). Sox2 is required for embryonic development of the ventral telencephalon through the activation of the ventral determinants Nkx2.1 and Shh. Development 140, 1250- 1261.
13. Furuyama, K., Kawaguchi, Y., Akiyama, H., Horiguchi, M., Kodama, S., Kuhara, T., Hosokawa, S., Elbahrawy, A., Soeda, T., Koizumi, M., et al. (2010). Continuous cell supply from a Sox9-expressing progenitor zone in adult liver, exocrine pancreas and intestine. Nat Genet 43, 34-41.
14. Gage, P.J., Brinkmeier, M.L., Scarlett, L.M., Knapp, L.T., Camper, S.A., and Mahon, K.A. (1996). The Ames dwarf gene, df, is required early in pituitary ontogeny for the extinction of Rpx transcription and initiation of lineage-specific cell proliferation. Mol Endocrinol 10, 1570-1581.
15. Gage, P.J., Suh, H., and Camper, S.A. (1999). Dosage requirement of Pitx2 for development of multiple organs. Development 126, 4643-4651.
16. Godin, J.D., and Nguyen, L. (2014). Novel functions of core cell cycle regulators in neuronal migration. Adv Exp Med Biol 800, 59-74.
17. Hebert, J.M., and McConnell, S.K. (2000). Targeting of cre to the Foxg1 (BF-1)

locus mediates loxP recombination in the telencephalon and other developing head structures. Dev Biol 222, 296-306.

18. Hoefflin, S., and Carter, D.A. (2014). Neuronal expression of SOX2 is enriched in specific hypothalamic cell groups. Journal of chemical neuroanatomy 61-62, 153- 160.
19. Jayakody, S.A., Andoniadou, C.L., Gaston-Massuet, C., Signore, M., Cariboni, A., Bouloux, P.M., Le Tissier, P., Pevny, L.H., Dattani, M.T., and Martinez-Barbera, J.P. (2012). SOX2 regulates the hypothalamic-pituitary axis at multiple levels. The Journal of clinical investigation 122, 3635-3646.
20. Kelberman, D., de Castro, S.C., Huang, S., Crolla, J.A., Palmer, R., Gregory, J.W., Taylor, D., Cavallo, L., Faienza, M.F., Fischetto, R., et al. (2008). SOX2 plays a critical role in the pituitary, forebrain, and eye during human embryonic development. J Clin Endocrinol Metab 93, 1865-1873.
21. Kelberman, D., Rizzoti, K., Avilion, A., Bitner-Glindzicz, M., Cianfarani, S., Collins, J., Chong, W.K., Kirk, J.M., Achermann, J.C., Ross, R., et al. (2006). Mutations within Sox2/SOX2 are associated with abnormalities in the hypothalamo-pituitarygonadal axis in mice and humans. J Clin Invest 116, 2442-2455.
22. Kelberman, D., Rizzoti, K., Lovell-Badge, R., Robinson, I.C., and Dattani, M.T. (2009). Genetic regulation of pituitary gland development in human and mouse. Endocr Rev 30, 790-829.
23. Langlais, D., Couture, C., Kmita, M., and Drouin, J. (2013). Adult pituitary cell maintenance: lineage-specific contribution of self-duplication. Mol Endocrinol 27, 1103-1112.
24. Lee, B., Rizzoti, K., Kwon, D.S., Kim, S.Y., Oh, S., Epstein, D.J., Son, Y., Yoon, J., Baek, K., and Jeong, Y. (2012). Direct transcriptional regulation of Six6 is controlled by SoxB1 binding to a remote forebrain enhancer. Dev Biol 366, 393-403.
25. Li, H., Collado, M., Villasante, A., Matheu, A., Lynch, C.J., Canamero, M., Rizzoti, K., Carneiro, C., Martinez, G., Vidal, A., et al. (2012). p27(Kip1) directly represses Sox2 during embryonic stem cell differentiation. Cell Stem Cell 11, 845-852.
26. Li, X., Perissi, V., Liu, F., Rose, D.W., and Rosenfeld, M.G. (2002). Tissue-specific regulation of retinal and pituitary precursor cell proliferation. Science 297, 1180- 1183.
27. Livak, K.J., and Schmittgen, T.D. (2001). Analysis of relative gene expression data using real-time quantitative PCR and the 2(-Delta Delta C(T)) Method. Methods 25, 402-408.
28. Macchiaroli, A., Kelberman, D., Auriemma, R.S., Drury, S., Islam, L., Giangiobbe, S., Ironi, G., Lench, N., Sowden, J.C., Colao, A., et al. (2014). A novel heterozygous SOX2 mutation causing congenital bilateral anophthalmia, hypogonadotropic hypogonadism and growth hormone deficiency. Gene 534, 282-285.

29. Meyers, E.N., and Martin, G.R. (1999). Differences in left-right axis pathways in mouse and chick: functions of FGF8 and SHH. Science 285, 403-406.

30. Nakayama, K., Ishida, N., Shirane, M., Inomata, A., Inoue, T., Shishido, N., Horii, I., and Loh, D.Y. (1996). Mice lacking p27(Kip1) display increased body size, multiple organ hyperplasia, retinal dysplasia, and pituitary tumors. Cell 85, 707- 720.

31. Pulichino, A.M., Vallette-Kasic, S., Tsai, J.P., Couture, C., Gauthier, Y., and Drouin, J. (2003). Tpit determines alternate fates during pituitary cell differentiation. Genes Dev 17, 738-747.

32. Rizzoti, K., Akiyama, H., and Lovell-Badge, R. (2013). Mobilized adult pituitary stem cells contribute to endocrine regeneration in response to physiological demand. Cell Stem Cell 13, 419-432.

33. Rizzoti, K., Brunelli, S., Carmignac, D., Thomas, P.Q., Robinson, I.C., and LovellBadge, R. (2004). SOX3 is required during the formation of the hypothalamopituitary axis. Nat Genet 36, 247-255.

34. Sajedi, E., Gaston-Massuet, C., Signore, M., Andoniadou, C.L., Kelberman, D., Castro, S., Etchevers, H.C., Gerrelli, D., Dattani, M.T., and Martinez-Barbera, J.P. (2008). Analysis of mouse models carrying the I26T and R160C substitutions in the transcriptional repressor HESX1 as models for septo-optic dysplasia and hypopituitarism. Dis Model Mech 1, 241-254.

35. Schneider, A., Brand, T., Zweigerdt, R., and Arnold, H. (2000). Targeted disruption of the Nkx3.1 gene in mice results in morphogenetic defects of minor salivary glands: parallels to glandular duct morphogenesis in prostate. Mech Dev 95, 163- 174.

36. Sheng, H.Z., Zhadanov, A.B., Mosinger, B., Jr., Fujii, T., Bertuzzi, S., Grinberg, A., Lee, E.J., Huang, S.P., Mahon, K.A., and Westphal, H. (1996). Specification of pituitary cell lineages by the LIM homeobox gene Lhx3. Science 272, 1004-1007.

37. Sornson, M.W., Wu, W., Dasen, J.S., Flynn, S.E., Norman, D.J., O'Connell, S.M., Gukovsky, I., Carriere, C., Ryan, A.K., Miller, A.P., et al. (1996). Pituitary lineage determination by the Prophet of Pit-1 homeodomain factor defective in Ames dwarfism. Nature 384, 327-333.

38. Srinivas, S., Watanabe, T., Lin, C.S., William, C.M., Tanabe, Y., Jessell, T.M., and Costantini, F. (2001). Cre reporter strains produced by targeted insertion of EYFP and ECFP into the ROSA26 locus. BMC Dev Biol 1, 4.

39. Suga, H., Kadoshima, T., Minaguchi, M., Ohgushi, M., Soen, M., Nakano, T., Takata, N., Wataya, T., Muguruma, K., Miyoshi, H., et al. (2011). Self-formation of functional adenohypophysis in three-dimensional culture. Nature 480, 57-62.

40. Suzuki, J., Azuma, N., Dateki, S., Soneda, S., Muroya, K., Yamamoto, Y., Saito, R., Sano, S., Nagai, T., Wada, H., et al. (2014). Mutation spectrum and phenotypic variation in nine patients with SOX2 abnormalities. J Hum Genet 59, 353-356.

41. Taranova, O.V., Magness, S.T., Fagan, B.M., Wu, Y., Surzenko, N., Hutton, S.R.,

and Pevny, L.H. (2006). SOX2 is a dose-dependent regulator of retinal neural progenitor competence. Genes Dev 20, 1187-1202.

42. Treier, M., Gleiberman, A.S., O'Connell, S.M., Szeto, D.P., McMahon, J.A., McMahon, A.P., and Rosenfeld, M.G. (1998). Multistep signaling requirements for pituitary organogenesis in vivo. Genes Dev 12, 1691-1704.
43. Trowe, M.O., Zhao, L., Weiss, A.C., Christoffels, V., Epstein, D.J., and Kispert, A. (2013). Inhibition of Sox2-dependent activation of Shh in the ventral diencephalon by Tbx3 is required for formation of the neurohypophysis. Development 140, 2299-2309.
44. Wang, Y., Martin, J.F., and Bai, C.B. (2010). Direct and indirect requirements of Shh/Gli signaling in early pituitary development. Developmental biology 348, 199-209.
45. Ward, R.D., Raetzman, L.T., Suh, H., Stone, B.M., Nasonkin, I.O., and Camper, S.A. (2005). Role of PROP1 in pituitary gland growth. Mol Endocrinol 19, 698-710.
46. Xuan, S., Baptista, C.A., Balas, G., Tao, W., Soares, V.C., and Lai, E. (1995). Winged helix transcription factor BF-1 is essential for the development of the cerebral hemispheres. Neuron 14, 1141-1152.
47. Yoshida, S., Kato, T., Susa, T., Cai, L.Y., Nakayama, M., and Kato, Y. (2009). PROP1 coexists with SOX2 and induces PIT1-commitment cells. Biochem Biophys Res Commun 385, 11-15.
48. Yoshida, S., Kato, T., Yako, H., Susa, T., Cai, L.Y., Osuna, M., Inoue, K., and Kato, Y. (2011). Significant quantitative and qualitative transition in pituitary stem / progenitor cells occurs during the postnatal development of the rat anterior pituitary. J Neuroendocrinol 23, 933-943.
49. Zhadanov, A.B., Bertuzzi, S., Taira, M., Dawid, I.B., and Westphal, H. (1995). Expression pattern of the murine LIM class homeobox gene Lhx3 in subsets of neural and neuroendocrine tissues. Dev Dyn 202, 354-364.
50. Zhao, L., Zevallos, S.E., Rizzoti, K., Jeong, Y., Lovell-Badge, R., and Epstein, D.J. (2012). Disruption of SoxB1-dependent Sonic hedgehog expression in the hypothalamus causes septo-optic dysplasia. Dev Cell 22, 585-596.

SECTION II. GENETIC AND EPIGENETIC REGULATION

7

CHAPTER

GATA6-DEPENDENT GLI3 REPRESSOR FUNCTION IS ESSENTIAL IN ANTERIOR LIMB PROGENITOR CELLS FOR PROPER LIMB DEVELOPMENT

Shinichi Hayashi[1,2], Ryutaro Akiyama[1,2], Julia Wong[1], Naoyuki Tahara[1,2], Hiroko Kawakami[1,2], Yasuhiko Kawakami[1,2]

[1]Department of Genetics, Cell Biology and Development, University of Minnesota, Minneapolis, Minnesota, United States of America

[2]Stem Cell Institute, University of Minnesota, Minneapolis, Minnesota, United States of America

ABSTRACT

Gli3 is a major regulator of Hedgehog signaling during limb development. In the anterior mesenchyme, GLI3 is proteolytically processed into GLI3R, a truncated repressor form that inhibits Hedgehog signaling. Although numerous studies have identified mechanisms that regulate Gli3 function in vitro, it is not completely understood how Gli3 function is regulated in vivo. In this study, we show a novel mechanism of regulation of GLI3R activities in limb buds by Gata6, a member of the GATA transcription factor family. We show that conditional inactivation of Gata6 prior to limb outgrowth by the Tcre deleter causes preaxial polydactyly, the formation of an anterior extra digit, in hindlimbs. A recent study suggested that Gata6represses Shh transcription in hindlimb buds. However, we found that ectopic Hedgehog signaling precedes ectopic Shh expression. In conjunction, we observed Gata6 and Gli3genetically interact, and compound heterozygous mutants develop preaxial polydactyly without ectopic Shh expression, indicating an additional prior mechanism to prevent polydactyly. These results support the idea that Gata6 possesses dual roles during limb development: enhancement of Gli3 repressor function to repress Hedgehog signaling in the anterior limb bud,

Citation: Hayashi S, Akiyama R, Wong J, Tahara N, Kawakami H, Kawakami Y (2016)Gata6-Dependent GLI3 Repressor Function is Essential in Anterior Limb Progenitor Cells for Proper Limb Development. PLoS Genet 12(6): e1006138. doi:10.1371/journal.pgen.1006138.

and negative regulation of Shh expression. Our in vitro and in vivo studies identified that GATA6 physically interacts with GLI3R to facilitate nuclear localization of GLI3R and repressor activities of GLI3R. Both the genetic and biochemical data elucidates a novel mechanism byGata6 to regulate GLI3R activities in the anterior limb progenitor cells to prevent polydactyly and attain proper development of the mammalian autopod.

INTRODUCTION

Understanding the developmental mechanisms that regulate progenitor cells to generate organs with specific morphology and function is a central topic in developmental biology. The vertebrate limb has been serving as an excellent system for such studies. In particular, mesenchymal progenitor cells in limb buds are specified, patterned and expanded to generate each skeletal element with a distinct morphology at each defined position to create the stereotypical limb skeletal system. The mammalian autopod possesses five digits, termed as d1-d5, in an anterior to posterior order. The number and identity of digits have been used as a readout of specification, patterning, and proliferative expansion of progenitor cells [1].

Sonic Hedgehog (*Shh*) is expressed in the zone of polarizing activity (ZPA), located at the posterior mesenchyme of the limb bud, and acts as a major regulatory molecule for limb development [1, 2]. Anterior-posterior specification of digit progenitors is regulated by the concentration and duration of progenitor exposure to SHH [3–6]. SHH also regulates the proliferative expansion of mesenchymal progenitor cells to generate a sufficient number of cells to develop into cartilage condensations [7, 8]. Accordingly, ectopic expression of *Shh* in the anterior portion is associated with preaxial polydactyly, which is characterized by the formation of ectopic digits in the anterior of the limb [9]. By contrast, the most anterior digit (d1) develops in a SHH-independent manner [10, 11]. Recent studies have shown that anterior genetic programs, such as *Irx3-Irx5* and *Sall4*, are required for development of d1, at least in part, by excluding SHH signaling from the anterior mesenchyme [12, 13].

The glioma-associated oncogene family (GLI) proteins are zinc finger DNA binding proteins, which play diverse roles in animal development and diseases [14]. Among the three *Gli* genes,*Gli3* encodes a bi-functional molecule, acting as both an activator (GLI3A) and a repressor (GLI3R), whose balance depends on Hedgehog signaling [14]. In the presence of Hedgehog ligands, its signal transduction at primary cilia causes inhibition of proteolytic processing of GLI3 [15]. This results in the accumulation of a full-length activator form of GLI3 (GLI3A) in the posterior mesenchyme. In contrast, in the absence of Hedgehog signaling, GLI3 is subjected to proteolysis, generating a truncated repressor form (GLI3R), which accumulates in the anterior mesenchyme. Because GLI1 lacks a repressor domain and GLI2 predominantly functions as an activator [16, 17], GLI3R is the major GLI repressor in the limb [18].

Consistent with the important function of *Gli3* in limb development, its mutations cause developmental defects in mice and humans [19–21]. In particular, *Gli3*$^{-/-}$ mice develop polydactyly [21]. Genetic studies in mice demonstrated that a predominant function of *Gli3* is to repress Hedgehog signaling target genes [22, 23]. Furthermore, it has been shown that the balance of GLI3A and GLI3R regulates digit number and identity [24–26]. Numerous studies have shown that multiple mechanisms regulate GLI3 functions in vitro, such as posttranslational modifications, degradation, cytoplasmic retention, and primary cilium-mediated processing (reviewed in [14, 27, 28]). In vivo studies in mice demonstrated that *Gli3*genetically interacts with *Hox* genes, *Zic3* and *Alx4* during limb development [29–31]. Despite these studies, the in vivo control of *Gli3* function during proper limb development is still to be elucidated.

The *Gata* family of zinc finger transcription factors is an important regulator of tissue and organ development. The *Gata* family is subdivided into the *Gata1/2/3* subfamily and the *Gata4/5/6*subfamily, which show expression in hematopoietic cell lineages and meso-endoderm lineages, respectively [32, 33]. In particular, *Gata6* is essential for endoderm formation and is also involved in the development of various mesoderm- and endoderm-derived organs, such as the cardiovascular system and pancreas [34–37]. Moreover, a recent study suggested that*Gata6* functions as a negative regulator of *Shh* expression in limb buds by binding to its limb bud-specific cis-regulatory element, ZRS [38].

In this study, we found that broad deletion of *Gata6* in the limb mesenchymal progenitors caused hindlimb-specific preaxial polydactyly, which is associated with ectopic SHH signaling in the anterior hindlimb bud. We discovered that *Gata6* and *Gli3* genetically interact to regulate normal patterning of the hindlimb. Furthermore, we show that direct association of GATA6 with GLI3R promoted nuclear localization and transcriptional repressor activity of GLI3R. Our work identified that genetic and biochemical interactions between *Gata6* and *Gli3* act as essential mechanisms to regulate GLI3R activity for proper autopod patterning.

RESULTS

Inactivation of *Gata6* in early Mesoderm caused Hindlimb Specific Preaxial Polydactyly

Prior studies have identified expression of *Gata6* in developing limb buds [38–40]. *Gata6* null embryos die during gastrulation [34, 35]; therefore, we inactivated *Gata6* in the meso-endoderm by using the conditional allele of *Gata6* (*Gata6*^{fl}) [41] and the *Tcre* line, which recombines in the early meso-endoderm [42]. We found that *Tcre; Gata6*$^{fl/fl}$ mutants (hereafter referred to as *Gata6* cKO) die around E12.5–14.5 with broad hemorrhage (Fig 1A and 1E). This result is consistent with a former study, demonstrating a role of proper dosage of *Gata4* and*Gata6* for vessel integrity [43]. We found that *Gata6* cKO embryos exhibited polydactyly in the hindlimb, while forelimbs seem to be unaffected (Fig 1A–1C, 1E and 1G). Alcian blue staining demonstrated that the mutant hindlimbs possess patterned digits, d1-d5, and an extra digit on the anterior edge, which morphologically resembles d1. Based on the position and morphology, tarsal and metatarsal elements were also patterned. Two ectopic tarsal elements, likely the navicular and medial cuneiform, were present proximally to the ectopic 1st metatarsal (Fig 1D and 1H). These observations indicate that the autopod is patterned along the anterior-posterior axis, and the absence of *Gata6* induces the formation of an extra anterior digit with the associated tarsal and metatarsal elements.

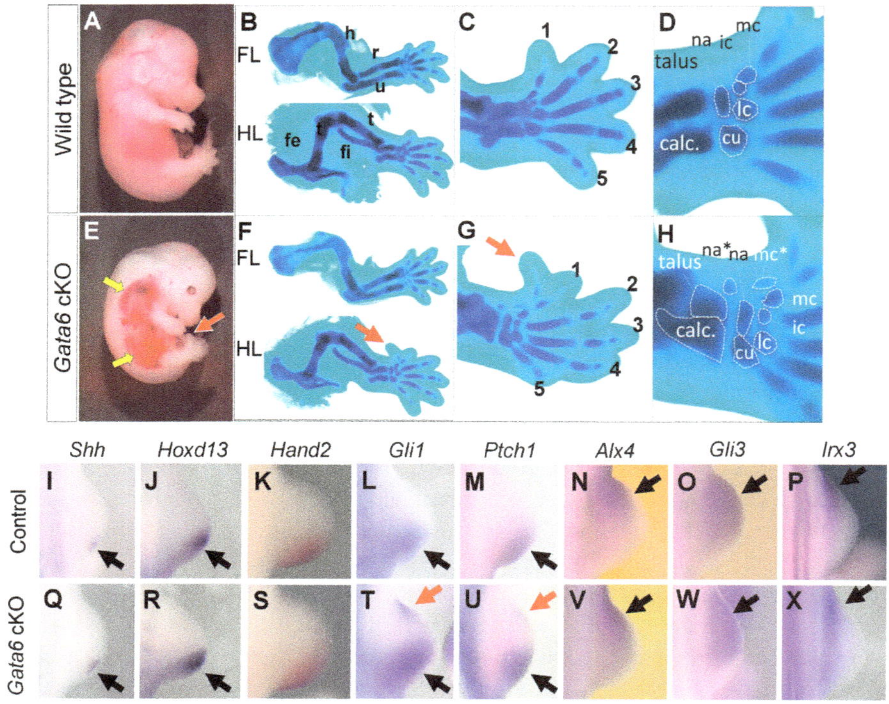

Figure 1: Loss of Gata6 causes preaxial polydactyly in hindlimbs.

A-H: Lateral views (A, E) of whole E14.5 embryos, and Alcian blue-stained cartilage (B-D, F-H) of wild type (A-D) and *Gata6* cKO (F-H) embryos at E14.5. C and G show hindlimb autopod, and D and H show tarsal and metatarsal elements. Red arrows in E-Gpoint to the anterior ectopic digit. Yellow arrows point to hemorrhage in *Gata6* cKO embryos. Digits are numbered with 1–5 in C and G. Asterisks in H indicates ectopic elements. calc: calcaneus, cu: cuboid, fe: femur, fi: fibula, ic: intermediate cuneiform, lc: lateral cuneiform, mc: medial cuneiform, na: navicular ti: tibia. I-X: in situ hybridization of wild type (I-P) and *Gata6* cKO (Q-X) hindlimb buds at E10.5 with indicated probes. Black and red arrows point to normal and ectopic signals, respectively.

http://dx.doi.org/10.1371/journal.pgen.1006138.g001

Ectopic Hedgehog Signaling Activation in *Gata6* cKO Hindlimbs

Preaxial polydactyly is known to be associated with ectopic *Sonic Hedgehog* (*Shh*) expression in the anterior margin. At E10.5, we detected posteriorly-localized *Shh* expression without ectopic anterior expression (n = 4, 39–40 somite stage, Fig 1I and 1Q). Consistent with this normal expression, *Hoxd13* (n = 3) and *Hand2* (n = 6), upstream regulators of limb bud *Shh*expression [44], were normally expressed in the posterior mesenchyme (Fig 1J, 1K, 1R and 1S). However, *Gli1* (n = 3) and *Patch1* (n = 3), targets of Hedgehog signaling, were detected in the anterior margin of *Gata6* cKO hindlimb buds (Fig 1L, 1M, 1T and 1U). Expression of anterior marker genes, such as *Alx4* (n = 3), *Gli3* (n = 4) and *Irx3* (n = 3), were not significantly affected in *Gata6* cKO hindlimb buds (Fig 1N–1P and 1V–1X).

We also examined gene expression at a later stage. At E11.5, we detected ectopic *Shh*expression in the anterior border of *Gata6* cKO hindlimbs (n = 4). Consistent with evident ectopic *Shh* expression, expression of *Hoxd13* (n = 3), *Gli1* (n = 6), *Ptch1* (n = 6) and*Gremlin1* (n = 3) was also detected in the anterior margin. This data indicates that ectopic Hedgehog signaling became evident at E10.5 in *Gata6* cKO hindlimb buds, although ectopic*Shh* expression was undetectable. At a later stage (E11.5), ectopic *Shh* expression became evident and all SHH targets, examined in this study, were detected in the anterior margin.

Shh expression is negatively regulated in the anterior margin by various genes. Thus, we examined expression of negative regulators of *Shh* expression. In addition to *Alx4* and *Gli3*(Fig 1) [23, 45], expression of *Etv4* (n = 3), *Etv5* (n = 5), *Tulp3* (n = 3), *Twist1* (n = 3), whose loss can cause ectopic *Shh* expression in the anterior margin [46–52], did not show evident alteration. Therefore, it is unlikely that these genes account for the preaxial polydactyly phenotype in *Gata6* cKO hindlimbs.

Reduction of *Shh* Dosage Rescued Ectopic SHH Signaling but not Ectopic Anterior Digit Formation

If ectopic *Shh* expression accounts for the preaxial polydactyly in *Gata6* cKO hindlimbs, we would expect that reducing *Shh* dosage might rescue the phenotype. Therefore, we genetically reduced *Shh* dosage from the *Gata6* cKO background using the *Shh* null allele [2]. *Gata6* cKO;$Shh^{+/-}$ mutants did not survive beyond E12.5, thus, we examined expression of SHH target genes (*Gli1* and *Ptch1*) and expression of *Sox9*, an early marker of chondrogenic condensation [53].

Removing one allele of *Shh* from the *Gata6* cKO background resulted in posteriorly restricted expression of *Gli1* and *Ptch1*, and the ectopic anterior expression became undetectable (n = 4,Fig 2A–2C and 2E–2G). However, ectopic chondrogenic condensation in the anterior portion was still detected by *Sox9* expression at E12.5 (n = 3, Fig 2I–2K). Removing both alleles of*Shh* from the *Gata6* cKO background resulted in the loss of *Gli1* and *Ptch1* expression and single digit condensation, the same phenotype as $Shh^{-/-}$ limbs (n = 3, Fig 2D, 2H and 2L) [10,11]. These results indicate that *Shh* functions downstream of *Gata6* during preaxial polydactyly development. However, ectopic chondrogenic condensation in the anterior portion of *Gata6*cKO; $Shh^{+/-}$ hindlimbs suggests that additional mechanisms could be involved in the preaxial polydactyly in *Gata6* cKO hindlimbs.

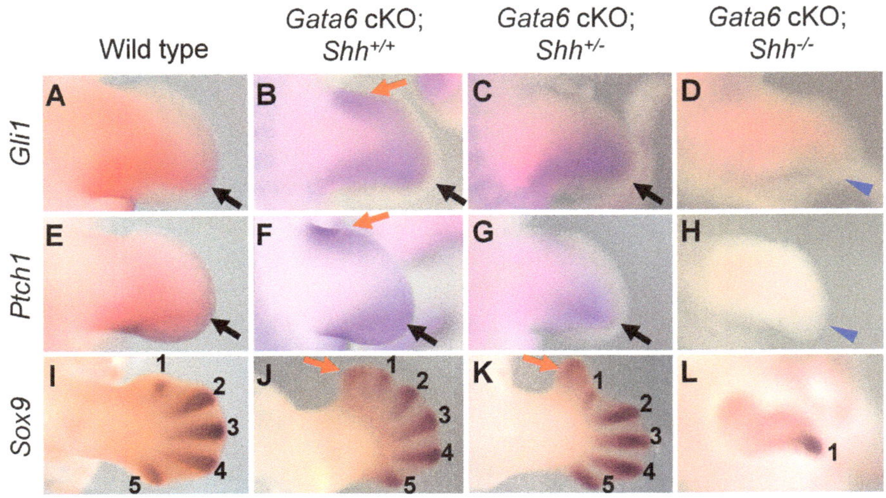

Figure 2: Expression pattern of SHH targets and digit condensation in *Gata6* cKO;*Shh* **allelic series.**

Expression pattern of *Gli1* (A-D), *Ptch1* (E-H) and *Sox9* (I-L) of wild type (A, E, I),*Gata6* cKO (B, F, J), *Gata6* cKO; *Shh*$^{+/-}$ (C, G, K) and *Gata6* cKO; *Shh*$^{-/-}$ (D, H, L) hindlimb buds. A-H: E11.5, I-L: E12.5. In A-H, black arrows and red arrows point to normal and ectopic signals, respectively. Blue arrowheads indicate loss of expression inD and H. In I-L, digit condensations are labeled as 1–5, and ectopic condensation is marked with red arrows.

http://dx.doi.org/10.1371/journal.pgen.1006138.g002

Gli3 Genetically Interacts with *Gata6* in Forelimbs and Hindlimbs

GLI3 is a major regulator of Hedgehog signaling, and thus, *Gli3* might be involved in preaxial polydactyly in *Gata6* cKO hindlimbs. To test this hypothesis, we genetically removed *Gli3* from the *Gata6* cKO background. *Gli3*$^{+/-}$ hindlimbs developed a small spike in the anterior region [21, 54], while most of the *Tcre; Gata6*$^{+/fl}$ hindlimbs were indistinguishable from the wild-type hindlimbs at E14.5–15.5 (Fig 3F–3H, Table 1). *Tcre; Gata6*$^{+/fl}$; *Gli3*$^{+/-}$ compound heterozygous hindlimbs developed an extra digit in the anterior region (Fig 3I). Unexpectedly, we also found that this interaction operates in forelimbs. *Gli3*$^{+/-}$ forelimbs developed d1, which was associated with small ectopic cartilage condensation at the distal tip. Contrary to this, *Tcre; Gata6*$^{+/fl}$; *Gli3*$^{+/-}$ compound heterozygous forelimbs developed an evident extra digit with incomplete penetrance (Fig 3A–3D) or an extra digit that partially fused with endogenous d1 with incomplete penetrance. These results demonstrate a genetic interaction between *Gli3* and *Gata6* in fore- and hind-limbs.

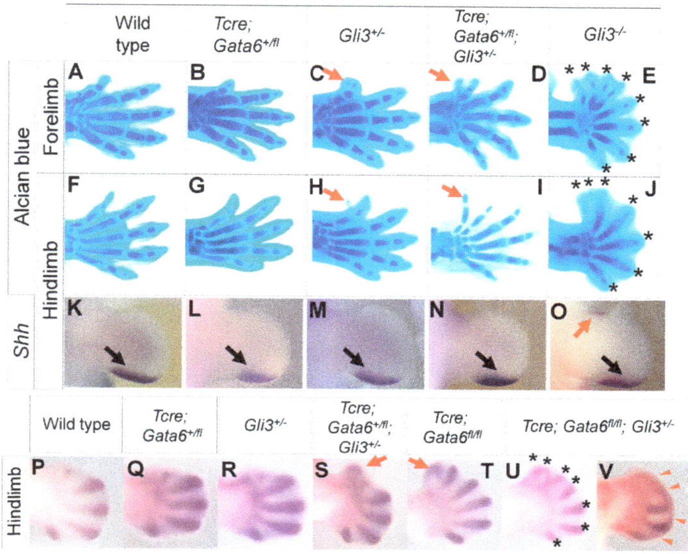

Figure 3: Genetic interaction between *Gata6* **and** *Gli3* **in** preaxial polydactyly development.

A-J: Alcian blue-stained autopod of indicated genotypes at E15.5. A-E: forelimbs, F-J: hindlimbs. Thin red arrows point to bifurcated d1 (C) and small projection (H) in fore- and hind-limbs, respectively in $Gli3^{+/-}$ mutants. Thick red arrows in D and I point to anterior ectopic digits. Asterisks in E and J indicate digit tips of $Gli3^{-/-}$ autopod. K-O: Expression pattern of *Shh* in hindlimb buds of indicated genotypes at E11.5. Black and red arrows point to normal and ectopic signals, respectively. P-V: *Sox9* in situ hybridization in hindlimbs of indicated genotypes at E12.5. Red arrows in S and T point to anterior ectopic digit condensation. Asterisks in U indicate distal tips of digit condensation. Red arrowheads in V point to distally-fused condensation.

http://dx.doi.org/10.1371/journal.pgen.1006138.g003

Table 1: Number of hindlimbs with indicated phenotypes at E14.5–16.5.

http://dx.doi.org/10.1371/journal.pgen.1006138.t001

Genotype	Number of hindlimbs with normal digits	Number of hindlimbs with small projection	Number of hindlimbs with anterior extra digit
Wild type	140/140 (100%)	0/140	0/140
$Gli3^{+/-}$	2/18 (11.1%)	16/18 (88.9%)	0/18 (0%)
Tcre; Gata6$^{+/fl}$	61/66 (92.4%)	0/66 (0%)	5/66 (7.6%)
Tcre; Gata6$^{+/fl}$; $Gli3^{+/-}$	3/54 (5.6%)	0/54 (0%)	51/54 (94.4%)

doi:10.1371/journal.pgen.1006138.t001

Because the *Gata6* cKO limb phenotype was evident in hindlimbs, we focused the following analysis on hindlimbs. Ectopic *Shh* expression can cause preaxial polydactyly, therefore, we examined *Shh* expression at E11.5. We detected a small domain of anterior ectopic *Shh*expression in $Gli3^{-/-}$ hindlimbs (n = 3/6, Fig 3O), as previously reported [23]. By contrast, *Tcre; Gata6*$^{+/fl}$; *Gli3*$^{+/-}$ compound heterozygous hindlimbs did not exhibit anterior ectopic *Shh*expression (n = 6), similar to wild-type, *Tcre; Gata6*$^{+/fl}$ (n = 6) and *Gli3*$^{+/-}$ (n = 6) hindlimb buds (Fig 3K–3N). Therefore, preaxial polydactyly in *Tcre; Gata6*$^{+/fl}$; *Gli3*$^{+/-}$ compound heterozygous limbs were unlikely to be caused by ectopic *Shh* expression. Given that GLI3R prevents ectopic digit formation in the anterior portion [55], these results suggest that an interaction between *Gata6* and *Gli3* contributes to GLI3R activities.

Gata6 cKO; *Gli3*$^{+/-}$ embryos do not survive beyond E12.5, therefore, we further examined the interaction between *Gata6* and *Gli3* by visualizing digit condensation by *Sox9* in situ hybridization. Both *Gli3*$^{+/-}$ and *Tcre; Gata6*$^{+/fl}$ hindlimbs exhibited similar expression patterns to wild-type hindlimbs at E12.5 (Fig 3P–3R). Correlating with preaxial polydactyly at E15.5, *Gata6*cKO and *Tcre; Gata6*$^{+/fl}$; *Gli3*$^{+/-}$ compound heterozygous hindlimbs exhibited ectopic anterior digit condensation (Fig 3S and 3T). *Gata6* cKO; *Gli3*$^{+/-}$ hindlimbs were slightly underdeveloped and exhibited seven digit condensations (n = 2/6, Fig 3U), distally-fused condensation (n = 2/6,Fig 3V) or one extra anterior condensation, similar to *Gata6* cKO hindlimbs (n = 2/6). Formation of multiple extra digits and distal fusion of cartilage condensation are characteristics of *Gli3*$^{-/-}$ limbs [21]. Therefore, we speculate that the *Gata6* cKO; *Gli3*$^{+/-}$ genotype may be in conditions similar to the *Gli3*$^{-/-}$ genotype in hindlimbs. These results further support the idea that loss of *Gata6* leads to reduction of GLI3R activities.

In order to further characterize the *Gata6-Gli3* interaction, we examined gene expression at E11.5. Expression of *Gli1* and *Patch1* was posteriorly restricted in wild-type, *Tcre; Gata6*$^{+/fl}$and *Gli3*$^{+/-}$ hindlimbs (Fig 4A–4C and 4H–4J). Hindlimbs with the Tcre; Gata6$^{+/fl}$; Gli3$^{+/-}$, Gata6 cKO, Gata6 cKO; Gli3$^{+/-}$ or Gli3$^{-/-}$ genotypes exhibited anterior ectopic expression of these genes (Fig 4D–4G and 4K–4N). The ectopic expression domain was larger in *Gata6* cKO and*Gata6* cKO; *Gli3*$^{+/-}$ hindlimb buds than that in *Tcre; Gata6*$^{+/fl}$; *Gli3*$^{+/-}$ and *Gli3*$^{-/-}$ hindlimbs, likely due to ectopic *Shh* expression in the *Gata6* cKO background.

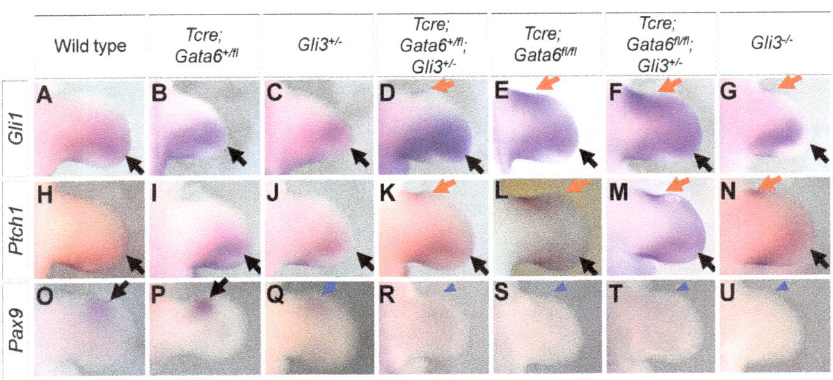

Figure 4: Expression pattern of *Gli1*, *Ptch1* **and** *Pax9* **in** *Gata6*; *Gli3* **allelic series.**

In situ hybridization of *Gli1* (A-G), *Ptch1* (H-N) and *Pax9* (O-U) of hindlimb buds of indicated genotypes at E11.5. Black and red arrows point to normal and ectopic signals, respectively. Blue arrows and arrowheads indicate reduced and loss of *Pax9* signals, respectively.

http://dx.doi.org/10.1371/journal.pgen.1006138.g004

Pax9, whose expression requires high levels of GLI3R activities [56], was detected in the anterior of wild-type and *Tcre; Gata6*$^{+/fl}$ hindlimbs, and was reduced in *Gli3*$^{+/-}$ hindlimb buds (Fig 4O–4Q). In *Tcre; Gata6*$^{+/fl}$; *Gli3*$^{+/-}$, *Gata6* cKO, *Gata6* cKO; *Gli3*$^{+/-}$ hindlimbs, *Pax9*expression was undetectable, similar to *Gli3*$^{-/-}$ hindlimbs (Fig 4R–4U).

These alterations of gene expression at E11.5 are consistent with the idea that GLI3R activities were reduced in hindlimbs with the *Tcre; Gata6*$^{+/fl}$; *Gli3*$^{+/-}$, *Gata6* cKO and *Gata6*cKO; *Gli3*$^{+/-}$ genotypes.

GATA6 and GLI3 Functionally and Physically Interact in Vitro

Ectopic *Shh* expression in the *Gata6* cKO background could affect gene expression patterns in hindlimb buds. Therefore, we set up in vitro experiments to further investigate how *Gata6*regulates *Gli3* function. We first set up luciferase reporter assays using 12xGLI-binding site luciferase [31]. Transfecting a C-terminally truncated form of human *GLI3* that could function as GLI3R caused significant reduction of the reporter activities, while transfecting human*GATA6* did not affect the reporter activities. Co-transfecting *GLI3R* and *GATA6* caused further reduction of the reporter activities (Fig 5A). These results are consistent with the in vivo data and support the idea that *GATA6* functionally interacts with and contributes to GLI3R activities.

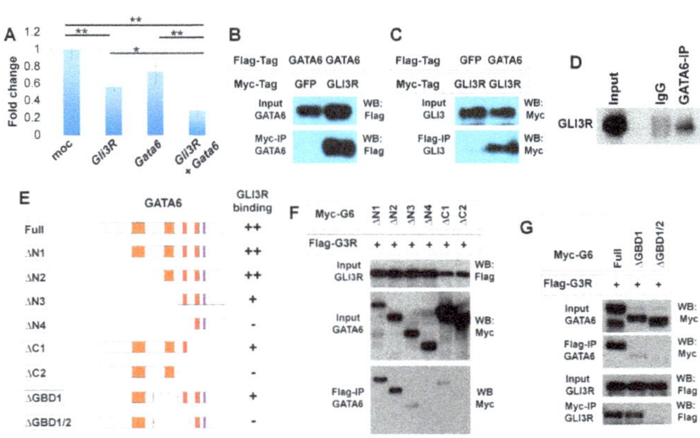

Figure 5: Physical and functional interaction between GATA6 and GLI3R.

A: GLI-specific luciferase reporter assay with indicated expression constructs. *: p<0.01, **: p<0.001. B, C: Co-immunoprecipitation assay of Flag-GATA6 and Myc-GLI3R. (B) Pulldown with anti-Myc, detection by anti-Flag. (C) Pulldown with anti-Flag, detection by anti-Myc. D: Co-immunoprecipitation of GATA6 and GLI3R from wild-type hindlimb buds. E: Schematic presentation of deletion mutants of GATA6. Binding with GLI3R in F and G is summarized in the right side of the panel. Orange bars represent transactivation domains. Red and blue bars represent zinc finger DNA binding domains and the nuclear localization signal, respectively. F, G: Co-immunoprecipitation assay of Flag-GLI3R and GATA6 mutants.

http://dx.doi.org/10.1371/journal.pgen.1006138.g005

Next, we tested whether GATA6 and GLI3R physically interact by co-immunoprecipitation assays. HEK293T cells were transfected with Flag-tagged GATA6, Myc-tagged GLI3R or GFP. Flag-GATA6 and Myc-GLI3R were co-immunoprecipitated, demonstrating that GATA6 and GLI3R can interact (Fig 5B and 5C). We also confirmed that the interaction occurs in vivo. GLI3R was detected in immunoprecipitated complex from E10.25–10.5 wild-type hindlimb buds using ant-GATA6 (Fig 5D). To further characterize their interaction, we mapped the GLI3R interaction domain in GATA6. For this purpose, we generated serial deletion mutants (Fig 5E), and performed co-immunoprecipitation assays with Flag-GLI3R. The ΔN1 and ΔN2 mutants showed a strong interaction with Flag-GLI3R. The ΔN3 and ΔC1 mutants exhibited weak interaction, and we did not detect any interactions of Flag-GLI3R with ΔN4 and ΔC2 (Fig 5F).

We also generated intra-molecular deletion mutants. These mutants lack the GLI3R-binding domain (GBD) 1, which includes the second putative transactivation domain (ΔGBD1), or both GBD1 and GBD2 (ΔGBD1/2). We did not detect any interaction of ΔGBD1/2 with GLI3R, although ΔGBD1 exhibited a weak interaction with GLI3R (Fig 5G). These results suggest that the zinc finger domain 1 (ZFD1) is critical to interact with GLI3R. The weak interaction of ΔN3, ΔC1 and ΔGBD1, which possess the ZFD1, also suggests that both the N- and C-terminal regions around the ZFD1 contribute to the interaction with GLI3R, in collaboration with the ZFD1.

Interaction between GATA6 and GLI3R Regulates Subcellular Localization of GLI3R

Our analyses indicated the presence of genetic and physical interactions between *Gata6* and*Gli3*. Given that both GATA6 and GLI3R act as transcription factors, we next examined subcellular localization of these proteins after co-transfecting HEK293T cells with Flag-GLI3R and either full length or mutant forms of Myc-GATA6.

We observed three patterns of localization (Fig 6A and 6B). First, co-transfection of either full length GATA6, ΔN1-GATA6 or ΔN2-GATA6, which can interact with GLI3R and possess the nuclear localization signal (NLS), resulted in predominant nuclear localization of both GLI3R and GATA6. Second, we co-transfected ΔN3-GATA6 or ΔN4-GATA6, which possess the NLS, but have either very weak or undetectable interactions with GLI3R. In these transfection assays, GLI3R localization became either predominantly cytoplasmic or localized similarly in both the cytoplasm and nucleus, although GATA6 was predominantly detected in the nucleus. Third, we co-transfected ΔC1-GATA6 or ΔC2-GATA6, which lack the NLS and have very weak or undetectable interactions with GLI3R. We detected GATA6 predominantly in the cytoplasm, consistent with the lack of NLS. GLI3R was also predominantly located in the cytoplasm or located similarly in the nucleus and cytoplasm.

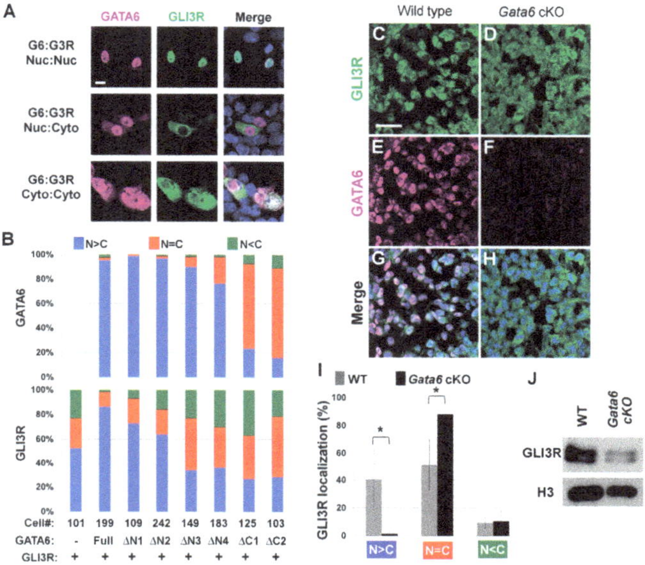

Figure 6: GATA6 Regulates Subcellular Localization of GLI3R.

A: Representative in vitro images of nuclear GATA6+nuclear GLI3R (upper), nuclear GATA6+cytosolic GLI3R (middle) and cytosolic GATA6+cytosolic GLI3R (bottom). B: Quantitation of subcellular localization of GATA6 and GLI3R. N<C: predominantly cytoplasmic, N = C: similarly in cytoplasm and in nucleus, N>C: predominantly nuclear localized. GATA6 mutants, indicated at the bottom, are shown in Fig 5E. The number of cells examined for each set of transfection is indicated in the panel. C-H: Representative images of the anterior-proximal mesenchyme of hindlimb buds at E10.25. C, E, G: wild type, D, F, H: *Gata6* cKO. I: Quantitation of subcellular localization of GLI3R in the anterior-proximal mesenchyme of hindlimb buds at E10.25. Gray and black bars represent wild-type and *Gata6* cKO samples, respectively. The graph shows percentage of GLI3R localization patterns, such as predominantly nuclear (N>C), similarly in the nucleus and cytoplasm (N = C), or predominantly cytoplasmic (N<C). A total of 597 cells from three wild-type embryos and a total of 528 cells from three *Gata6* cKO embryos were examined. * indicates P<0.05. J: Western blot of nuclear fractions from anterior part of wild-type and *Gata6* cKO hindlimb buds at E10.25–10.5. Histone H3 (H3) is included as a loading control.

http://dx.doi.org/10.1371/journal.pgen.1006138.g006

These results indicate a correlation between GLI3R nuclear localization and nuclear GATA6 that possesses a GLI3R-interaction ability. This correlation suggests that physical association between GATA6 and GLI3R contributes to nuclear localization and the repressor activities of GLI3R. We next tested this idea in vivo by examining GLI3R nuclear localization. The earliest molecular alteration in *Gata6* cKO hindlimb buds in our study is ectopic *Gli1* and *Ptch1*expression at E10.5 (Fig 1). Therefore, we re-examined *Gata6*/GATA6 expression, although their mRNA expression patterns were examined in previous studies [38–40]. *Gata6* mRNA was detected in the anterior-proximal part of hindlimb buds at E10.25 (34 somite stage), but the strong signals in endoderm -derived tissues seem to mask the limb bud signals. Therefore, we also performed immunofluorescence of GATA6 in combination with limb bud mesenchyme markers, such as Fibroblast growth factor10 (FGF10) [57] or Dual specificity phosphatase6 (DUSP6) [58–60]. Co-staining with these markers on transverse sections indicates that GATA6 is present in the ventral side of the proximal region in anterior hindlimb buds at E10.25. The GATA6 signal was undetectable in limb buds in the middle-posterior region.

In the anterior proximal region of limb buds at E10.25, we detected GLI3R predominantly in the nucleus or similarly in the nucleus and cytoplasm (Fig 6C, 6E, 6G and 6I). By contrast, *Gata6*cKO hindlimb buds showed a reduced percentage of cells with predominant nuclear GLI3R signals. Accordingly, we detected an increased percentage of cells with nuclear/cytoplasmic GLI3R (Fig 6D, 6H and 6I). Western blot analysis of nuclear extracts from the anterior part of hindlimb buds showed reduced GLI3R levels in *Gata6* cKO, compared to wild-type embryos (Fig 6J). Although the presence of nuclear GLI3R in *Gata6* cKO hindlimb buds indicates *Gata6*-independent GLI3R nuclear localization mechanisms in the anterior mesenchyme, reduced GLI3R levels provide evidence that *Gata6* contributes to GLI3R nuclear localization. These results are consistent with the in vitro data, and further support the idea that *Gata6* regulation of GLI3R nuclear localization contributes to GLI3R activities during normal limb development.

DISCUSSION

In this study, we found hindlimb-specific preaxial polydactyly in *Gata6* mutants. The skeletal phenotype of *Gata6* mutants was restricted to hindlimbs, and the forelimbs developed normally. Several possibilities would account for such limb type-specific phenotypes. For instance, a recent study showed that *Gata4* is differentially expressed in forelimb buds (high) and hindlimb buds (low) [38]. *Gata4* and *Gata6* are functionally redundant during heart development and for vascular integrity [36, 43]; therefore, *Gata4* might compensate for loss of*Gata6* in forelimb buds [38]. Another possibility is that differences in the sensitivity to Hedgehog signaling contribute to different phenotypes in fore- and hind-limbs. It is suggested that levels of Hedgehog signaling are higher in hindlimb mesenchyme than forelimb mesenchyme [12], and that hindlimbs are more sensitive to changes in the levels of Hedgehog signaling. Higher Hedgehog signaling, in combination with reduced GLI3R, might have contributed to hindlimb-specific polydactyly in *Gata6* cKO. This idea is consistent with ectopic digit formation in *Tcre; Gata6*$^{+/fl}$; *Gli3*$^{+/-}$ forelimbs, in which GLI3R activities would be lower than and SHH signaling levels would be higher than *Gli3*$^{+/-}$ forelimbs. These two scenarios are not mutually exclusive, and they might cooperate together to ensure proper Hedgehog signaling and pentadactyly in mammalian limbs.

Our study proposes two mechanisms by which *Gata6* regulates proper autopod patterning. One mechanism is by enhancing GLI3R activities to repress Hedgehog signaling in the anterior mesenchyme, and the other is by negative regulation of *Shh* expression in the anterior mesenchyme.

Genetic studies have shown that preaxial polydactyly is associated with ectopic expression of*Shh* in the anterior mesenchyme [9]. Expression of *Shh* is positively and negatively regulated in the posterior and anterior mesenchyme, respectively. *Twist1*, *Alx4*, *Gli3*, *Tulp3* and *Etv4-Etv5* act as negative regulators, for their loss of function caused ectopic *Shh* expression [23,47, 49, 51, 61]. Genetic and biochemical studies have shown that *Hand2* and *Hoxd13*positively regulate *Shh* expression through the limb bud-specific cis-regulatory element, ZRS [44, 62]. Anterior *Shh* expression could be induced by loss of negative regulators or ectopic expression of positive regulators [63]. Given that these regulators did not exhibit significant alteration in *Gata6* cKO hindlimb buds, the preaxial polydactyly phenotype in *Gata6* cKO limbs is unlikely to be induced through these genes. A recent study suggested that *Gata6* represses*Shh* in the limb through binding to ZRS [38]. Our data is consistent with this report, and demonstrated that *Shh* and its targets are ectopically expressed in *Gata6* cKO hindlimb buds at E11.5. Restoration of normal expression pattern of *Gli1* and *Ptch1* in *Gata6* cKO; *Shh*$^{+/-}$hindlimbs also supports the idea that *Gata6* is upstream of *Shh*.

The second role is repressing ectopic Hedgehog signaling by enhancing repressor function of*Gli3*. Ectopic *Shh* expression in the *Gata6* cKO background affects data interpretations; however, compound heterozygous mutant analyses could enable separate analysis of the two mechanisms and support the second mechanism. Previous studies have shown *Gli3* to genetically interact with other genes during limb development. Studies on *Hox* genes suggested that the *Gli3*$^{-/-}$ polydactyly phenotype is mediated by *Hoxd9* and *Hoxd10* [29, 64]. In addition, it has been shown that polydactyly of *Gli3*$^{-/-}$ limbs becomes milder on the *Alx4*$^{-/-}$ or*Zic3*$^{-/-}$ background [30, 31], which suggested that the *Gli3*$^{-/-}$ polydactyly phenotype requires*Alx4* or *Zic3*. In contrast to these reports, loss of one allele of *Gata6* enhanced the polydactyly phenotype of *Gli3*$^{+/-}$ hindlimbs. Therefore, unlike previous genetic studies, our study identified*Gata6* as a negative factor for polydactyly development. Given that GLI3R prevents extra-digit formation in the anterior mesenchyme [55], our results suggest that *Gata6* cooperates with GLI3R activities.

It is believed that d1 develops in a *Shh*-independent manner, while development of d2-d5 requires *Shh* [5, 6, 10, 11]. Genetic manipulation of *Gli3* in mice provided evidence that high levels of GLI3R in the anterior of limb buds is necessary for proper d1 development and ensuring pentadactyly [24, 55, 65]. Expression pattern of *Pax9*, which requires high levels of GLI3R [56], indicates that *Gata6* contributes to GLI3R activities in the anterior of hindlimb buds. In particular, *Pax9* was undetectable in *Tcre; Gata6*$^{+/fl}$; *Gli3*$^{+/-}$ hindlimb buds, similar to*Gata6* cKO and *Gli3*$^{-/-}$ hindlimb buds. These altered expression pattern of *Pax9* correlates with ectopic digit condensation and preaxial polydactyly, and further supports the idea that *Gata6*cooperate with *Gli3* for proper GLI3R activities in the anterior of hindlimb buds.

How does *Gata6* cooperate with *Gli3*? Our data support the idea that GATA6 physically interacts with GLI3R, facilitates the nuclear localization of GLI3R, and enhances the repressor activities of GLI3R. Reduced nuclear GLI3R localization in *Gata6* cKO hindlimb supports the idea that this interaction-mediated nuclear GLI3R localization would also occur in vivo. A recent study showed that *Gata4, 5, and 6* can repress *Gli*-dependent reporter activation in vitro [66]. This study suggested that GATA inhibits SHH-dependent GLI activator function by protein interaction in the chick presomitic mesoderm. Based on this report and our study, GATA might modulate both GLI3R (this study) and SHH-dependent GLI activator [66] in a context-dependent manner. Since expression of *Gata* genes is reported in other *Gli3*-positive developing tissues, such as the branchial arch, somite and central nervous system [16, 67,68], *Gata* regulation of GLI3R might be a shared mechanism during the development of other organs.

MATERIALS AND METHODS

Ethics Statement

Animal breeding was performed according to the approval by the Institutional Animal Care and Use Committee of the University of Minnesota. Compressed CO_2 gas from a cylinder followed by cervical dislocation was the methods of euthanasia for mice. All efforts were made to minimize suffering.

Mouse Lines and Embryo

The mouse lines for $Gata6^{fl}$ [41], $Gli3^{-}$ [69] and *Tcre* [42] were maintained on a mixed genetic background. Skeletal preparation was done as previously published [70]. Whole mount in situ hybridization was done as previously published [13].

Expression Constructs

The full-length human GATA6 construct and the human GLI3 construct were published [31,71]. The GLI3R construct was generated by deleting the 3' part of full-length cDNA, and cloned into 3xFlag CMV7. GATA6 deletion constructs were generated by PCR-based cloning and cloned in pcDNA3.1 or pCS2.

Immunofluorescence and Confocal Imaging for GLI3R Localization

For in vitro analysis, cells were fixed with 4% PFA for two hours at room temperature, washed with PBS and stained with anti-Flag (Sigma, M2, F3165, dilution 1:500) and anti-Myc tag (Abcam, ab9106, dilution 1:500) antibodies. For in vivo analysis, embryos were fixed for two hours in 4% PFA at 4C, washed with cold PBS, and cryosectioned with the OCT compound at 14 μm thickness. Sections were stained according to a standard procedure [13] without heat-induced epitope retrieval. Anti-GATA6 (R&D Systems, AF1700, dilution 1:400) and anti-GLI3R (Clone 6F5, dilution 1:200) [15, 72] were used. Alexa fluorophore-labelled secondary antibodies were obtained from Invitrogen (1:1000 dilution). Fluorescent confocal images were obtained by using Zeiss LSM 710 laser scanning microscope system (Carl Zeiss Microscopy), and analyzed using ZEN2009 software (Carl Zeiss Microscopy).

For subcellular localization analysis in vitro, images were acquired form six arbitrary areas from two plates. Nuclear/cytoplasmic localization of GLI3R and GATA6 was blindly evaluated in cells that were doubly transfected with GLI3R and GATA6 (or its mutants) except for samples that are transfected with GLI3R alone. For in vivo samples, nuclear/cytoplasmic localization of GLI3R was evaluated similarly in the anterior-proximal domain where GATA6 signals in wild-type hindlimb buds were detected. In *Gata6* cKO embryos, the anterior-proximal domain, similar to wild-type embryos, was selected for GLI3R subcellular localization. The quantification was performed similar to in vitro samples.

GATA6 Localization in Hindlimb Buds

In order to clarify GATA6 localization in hindlimb bud mesenchyme, GATA6 was simultaneously detected with limb bud mesenchyme markers, such as FGF10 or DUSP6. Wild-type embryos were fixed, washed and cryosectioned as described above. Sections were simultaneously stained by anti-GATA6 (R&D AF1700 or Cell Signaling #5851, dilution 1:1,600) and anti-FGF10 (Santa Cruz, sc-7917, dilution 1:100) or anti-DUSP6 (Sigma, Clone 3G2, dilution 1:200). Sections were reacted with Alexa fluorophore-labelled secondary antibodies, and fluorescent signals were detected by Zeiss LSM 710 according to a standard procedure [13].

Luciferase Reporter Assay

NIH3T3 cells in 48-well plates were transfected with the 12xGLI-binding site-TK minimum promoter-luciferase [31] with pRL-TK, *GATA6* and/or GLI3R expression constructs by using Fugene6 (Promega). Forty hours after transfection, cells were subjected to analysis using the Dual-Luciferase Reporter Assay System (Promega). Experiments were performed in triplicate, and statistical significance was analyzed by One-way ANOVA followed by the Tukey's comparison.

Co-Immunoprecipitation Assay and Nuclear GLI3R Detection

HEK293T cells were transfected with expression constructs by using the standard calcium phosphate method. Cell lysates, prepared after two days, were passed through 25 gauge syringes to ensure protein extraction from the nucleus, and co-immunoprecipitation assays were performed by using Dynabeads protein G (Invitrogen) and anti-Flag (Sigma, M2, F3165, 2μg) or anti-Myc tag (Abcam, ab9106, 1 μg) antibodies. Proteins were resolved by SDS-PAGE, transferred to PVDF membranes (Millipore, MA, USA), reacted with anti-Myc tag or anti-Flag antibodies, followed by HRP goat anti-mouse or rabbit IgG, and a chemiluminescence detection.

For co-immunoprecipitation assays with in vivo samples, hindlimb buds were collected from wild-type embryos at E10.25–10.5. After pooling, the samples were lysed and subjected to co-immunoprecipitation procedures [73] using anti-GATA6 (Cell Signaling, #5851) and Dynabeads protein G. The protein complex was eluted, and detected by Western using anti-GLI3 (R&D Systems, AF3690, dilution 1:100) and the

PicoLUCENT PLUS HRP detection kit (G-Bioscience) according to the manufacturer's instructions.

For nuclear GLI3R detection by Western, anterior one third of hindlimb buds at E10.25–10.5 were collected, and the nuclear fraction was prepared after dissociating cells by using the NE-PER kit (Thermo Fischer) according to the manufacturer's instructions. The nuclear extracts were analyzed by Western using anti-GLI3 (R&D Systems, AF3690) and anti-Histone H3 (Abcam, ab-1791).

ACKNOWLEDGMENTS

We are grateful to Drs. Christine Iacobuzio-Donahue, Juan Carlos Izpisua Belmonte, Mark Lewandoski, Xin Sun, Stephanie Ware, Rolf Zeller and Yi Zhong for sharing plasmids and/or mouse lines. We are also grateful to Dr. Susan Scales for anti-GLI3R antibodies, to Dr. Michael O'Connor for the use of his LSM710, to Dr. Laura Gammill for suggestions and reagents, to Dr. Naoyuki Wada for critical reading, and to Asha Elgonda, Malina Peterson and Samantha Young for their excellent technical support. We thank Malina Peterson and Austin Johnson for editorial assistance.

AUTHOR CONTRIBUTIONS

Conceived and designed the experiments: SH YK. Performed the experiments: SH RA JW NT HK YK. Analyzed the data: SH RA NT YK. Wrote the paper: SH YK.

REFERENCES

1. Zeller R, Lopez-Rios J, Zuniga A. Vertebrate limb bud development: moving towards integrative analysis of organogenesis. Nat Rev Genet. 2009;10(12):845–58. pmid:19920852. doi: 10. 1038/nrg2681.
2. Chiang C, Litingtung Y, Lee E, Young KE, Corden JL, Westphal H, et al. Cyclopia and defective axial patterning in mice lacking Sonic hedgehog gene function. Nature. 1996;383(6599):407–13. pmid:8837770. doi: 10. 1038/383407a0
3. Riddle RD, Johnson RL, Laufer E, Tabin C. Sonic hedgehog mediates the polarizing activity of the ZPA. Cell. 1993;75(7):1401–16. pmid:8269518. doi: 10. 1016/0092-8674(93)90626-2
4. Yang Y, Drossopoulou G, Chuang PT, Duprez D, Marti E, Bumcrot D, et al. Relationship between dose, distance and time in Sonic Hedgehog-mediated regulation of anteroposterior polarity in the chick limb. Development (Cambridge, England). 1997;124(21):4393–404. pmid:9334287.
5. Harfe BD, Scherz PJ, Nissim S, Tian H, McMahon AP, Tabin CJ. Evidence for an expansion-based temporal Shh gradient in specifying vertebrate digit identities. Cell. 2004;118(4):517–28. pmid:15315763. doi: 10. 1016/j. cell. 2004. 07. 024
6. Ahn S, Joyner AL. Dynamic changes in the response of cells to positive hedgehog signaling during mouse limb patterning. Cell. 2004;118(4):505–16. pmid:15315762. doi: 10. 1016/j. cell. 2004. 07. 023

7. Towers M, Mahood R, Yin Y, Tickle C. Integration of growth and specification in chick wing digit-patterning. Nature. 2008;452(7189):882–6. pmid:18354396. doi: 10. 1038/nature06718.

8. Zhu J, Nakamura E, Nguyen MT, Bao X, Akiyama H, Mackem S. Uncoupling Sonic hedgehog control of pattern and expansion of the developing limb bud. Developmental cell. 2008;14(4):624–32. pmid:18410737. doi: 10. 1016/j. devcel. 2008. 01. 008.

9. Masuya H, Sagai T, Wakana S, Moriwaki K, Shiroishi T. A duplicated zone of polarizing activity in polydactylous mouse mutants. Genes & development. 1995;9(13):1645–53. Epub 1995/07/01. pmid:7628698. doi: 10. 1101/gad. 9. 13. 1645

10. Chiang C, Litingtung Y, Harris MP, Simandl BK, Li Y, Beachy PA, et al. Manifestation of the limb prepattern: limb development in the absence of sonic hedgehog function. Developmental biology. 2001;236(2):421–35. pmid:11476582. doi: 10. 1006/dbio. 2001. 0346

11. Kraus P, Fraidenraich D, Loomis CA. Some distal limb structures develop in mice lacking Sonic hedgehog signaling. Mechanisms of development. 2001;100(1):45–58. pmid:11118883. doi: 10. 1016/s0925-4773(00)00492-5

12. Li D, Sakuma R, Vakili NA, Mo R, Puviindran V, Deimling S, et al. Formation of proximal and anterior limb skeleton requires early function of Irx3 and Irx5 and is negatively regulated by Shh signaling. Developmental cell. 2014;29(2):233–40. pmid:24726282. doi: 10. 1016/j. devcel. 2014. 03. 001.

13. Akiyama R, Kawakami H, Wong J, Oishi I, Nishinakamura R, Kawakami Y. Sall4-Gli3 system in early limb progenitors is essential for the development of limb skeletal elements. Proceedings of the National Academy of Sciences of the United States of America. 2015;112(16):5075–80. doi: 10. 1073/pnas. 1421949112 pmid:25848055; PubMed Central PMCID: PMCPMC4413345.

14. Hui CC, Angers S. Gli proteins in development and disease. Annu Rev Cell Dev Biol. 2011;27:513–37. Epub 2011/08/02. doi: 10. 1146/annurev-cellbio-092910-154048 pmid:21801010.

15. Wen X, Lai CK, Evangelista M, Hongo JA, de Sauvage FJ, Scales SJ. Kinetics of hedgehog-dependent full-length Gli3 accumulation in primary cilia and subsequent degradation. Molecular and cellular biology. 2010;30(8):1910–22. doi: 10. 1128/MCB. 01089-09 pmid:20154143; PubMed Central PMCID: PMCPMC2849461.

16. McDermott A, Gustafsson M, Elsam T, Hui CC, Emerson CP Jr. , Borycki AG. Gli2 and Gli3 have redundant and context-dependent function in skeletal muscle formation. Development (Cambridge, England). 2005;132(2):345–57. doi: 10. 1242/dev. 01537 pmid:15604102.

17. Pan Y, Bai CB, Joyner AL, Wang B. Sonic hedgehog signaling regulates Gli2 transcriptional activity by suppressing its processing and degradation. Molecular and cellular biology. 2006;26(9):3365–77. doi: 10. 1128/MCB. 26. 9. 3365–3377. 2006 pmid:16611981; PubMed Central PMCID: PMCPMC1447407.

18. Wang B, Fallon JF, Beachy PA. Hedgehog-regulated processing of Gli3 produces an anterior/posterior repressor gradient in the developing vertebrate limb. Cell. 2000;100(4):423–34. pmid:10693759. doi: 10. 1016/s0092-8674(00)80678-9

19. Winter RM, Huson SM. Greig cephalopolysyndactyly syndrome: a possible mouse homologue (Xt-extra toes). Am J Med Genet. 1988;31(4):793–8. doi: 10. 1002/ajmg. 1320310411 pmid:3239570.

20. Pettigrew AL, Greenberg F, Caskey CT, Ledbetter DH. Greig syndrome associated with an interstitial deletion of 7p: confirmation of the localization of Greig syndrome to 7p13. Hum Genet. 1991;87(4):452–6. pmid:1879832. doi: 10. 1007/bf00197167

21. Hui CC, Joyner AL. A mouse model of greig cephalopolysyndactyly syndrome: the extra-toesJ mutation contains an intragenic deletion of the Gli3 gene. Nature genetics. 1993;3(3):241–6. Epub 1993/03/01. doi: 10. 1038/ng0393-241 pmid:8387379.

22. Hu MC, Mo R, Bhella S, Wilson CW, Chuang PT, Hui CC, et al. GLI3-dependent transcriptional repression of Gli1, Gli2 and kidney patterning genes disrupts renal morphogenesis. Development (Cambridge, England). 2006;133(3):569–78. doi: 10. 1242/dev. 02220 pmid:16396903.

23. Buscher D, Bosse B, Heymer J, Ruther U. Evidence for genetic control of Sonic hedgehog by Gli3 in mouse limb development. Mechanisms of development. 1997;62(2):175–82. Epub 1997/03/01. S0925-4773(97)00656-4 [pii]. pmid:9152009. doi: 10. 1016/s0925-4773(97)00656-4

24. Wang C, Ruther U, Wang B. The Shh-independent activator function of the full-length Gli3 protein and its role in vertebrate limb digit patterning. Developmental biology. 2007;305(2):460–9. pmid:17400206. doi: 10. 1016/j. ydbio. 2007. 02. 029

25. te Welscher P, Zuniga A, Kuijper S, Drenth T, Goedemans HJ, Meijlink F, et al. Progression of vertebrate limb development through SHH-mediated counteraction of GLI3. Science (New York, NY. 2002;298(5594):827–30. pmid:12215652. doi: 10. 1126/science. 1075620

26. Litingtung Y, Dahn RD, Li Y, Fallon JF, Chiang C. Shh and Gli3 are dispensable for limb skeleton formation but regulate digit number and identity. Nature. 2002;418(6901):979–83. pmid:12198547. doi: 10. 1038/nature01033

27. Aberger F, Ruiz IAA. Context-dependent signal integration by the GLI code: the oncogenic load, pathways, modifiers and implications for cancer therapy. Seminars in cell & developmental biology. 2014;33:93–104. doi: 10. 1016/j. semcdb. 2014. 05. 003 pmid:24852887; PubMed Central PMCID: PMCPMC4151135.

28. Briscoe J, Therond PP. The mechanisms of Hedgehog signalling and its roles in development and disease. Nature reviews. 2013;14(7):416–29. doi: 10. 1038/ nrm3598 pmid:23719536.

29. Sheth R, Bastida MF, Ros M. Hoxd and Gli3 interactions modulate digit number in the amniote limb. Developmental biology. 2007;310(2):430–41. doi: 10. 1016/j. ydbio. 2007. 07. 023 pmid:17714700.

30. Panman L, Drenth T, Tewelscher P, Zuniga A, Zeller R. Genetic interaction of Gli3 and Alx4 during limb development. The International journal of developmental biology. 2005;49(4):443–8. Epub 2005/06/22. 051984lp [pii] doi: 10. 1387/ijdb. 051984lp pmid:15968591.

31. Quinn ME, Haaning A, Ware SM. Preaxial polydactyly caused by Gli3 haploinsufficiency is rescued by Zic3 loss of function in mice. Human molecular genetics. 2012;21(8):1888–96. Epub 2012/01/12. doi: 10. 1093/hmg/dds002 [pii]. pmid:22234993; PubMed Central PMCID: PMC3313802.

32. Chlon TM, Crispino JD. Combinatorial regulation of tissue specification by GATA and FOG factors. Development (Cambridge, England). 2012;139(21):3905–16. doi: 10. 1242/dev. 080440 pmid:23048181; PubMed Central PMCID: PMCPMC3472596.

33. Molkentin JD. The zinc finger-containing transcription factors GATA-4, -5, and -6. Ubiquitously expressed regulators of tissue-specific gene expression. The Journal of biological chemistry. 2000;275(50):38949–52. doi: 10. 1074/jbc. R000029200 pmid:11042222.

34. Morrisey EE, Tang Z, Sigrist K, Lu MM, Jiang F, Ip HS, et al. GATA6 regulates HNF4 and is required for differentiation of visceral endoderm in the mouse embryo. Genes & development. 1998;12(22):3579–90. pmid:9832509; PubMed Central PMCID: PMCPMC317242. doi: 10. 1101/gad. 12. 22. 3579

35. Koutsourakis M, Langeveld A, Patient R, Beddington R, Grosveld F. The transcription factor GATA6 is essential for early extraembryonic development. Development (Cambridge, England). 1999;126(9):723–32. pmid:10383242.

36. Zhao R, Watt AJ, Battle MA, Li J, Bondow BJ, Duncan SA. Loss of both GATA4 and GATA6 blocks cardiac myocyte differentiation and results in acardia in mice. Developmental biology. 2008;317(2):614–9. doi: 10. 1016/j. ydbio. 2008. 03. 013 pmid:18400219; PubMed Central PMCID: PMCPMC2423416.

37. Decker K, Goldman DC, Grasch CL, Sussel L. Gata6 is an important regulator of mouse pancreas development. Developmental biology. 2006;298(2):415–29. doi: 10. 1016/j. ydbio. 2006. 06. 046 pmid:16887115; PubMed Central PMCID: PMCPMC2824170.

38. Kozhemyakina E, Ionescu A, Lassar AB. GATA6 is a crucial regulator of Shh in the limb bud. PLoS Genet. 2014;10(1):e1004072. Epub 2014/01/15. doi: 10. 1371/journal. pgen. 1004072 PGENETICS-D-13-01628 [pii]. pmid:24415953; PubMed Central PMCID: PMC3886911.

39. Karamboulas K, Dranse HJ, Underhill TM. Regulation of BMP-dependent chondrogenesis in early limb mesenchyme by TGFbeta signals. Journal of cell science. 2010;123(Pt 12):2068–76. Epub 2010/05/27. doi: 10. 1242/jcs. 062901 [pii]. pmid:20501701.

40. Alexandrovich A, Qureishi A, Coudert AE, Zhang L, Grigoriadis AE, Shah AM, et al. A role for GATA-6 in vertebrate chondrogenesis. Developmental biology. 2008;314(2):457–70. Epub 2008/01/15. doi: 10. 1016/j. ydbio. 2007. 12. 001 S0012-1606(07)01570-9 [pii]. pmid:18191120.

41. Sodhi CP, Li J, Duncan SA. Generation of mice harbouring a conditional loss-of-function allele of Gata6. BMC developmental biology. 2006;6:19. pmid:16611361.

42. Perantoni AO, Timofeeva O, Naillat F, Richman C, Pajni-Underwood S, Wilson C, et al. Inactivation of FGF8 in early mesoderm reveals an essential role in kidney development. Development (Cambridge, England). 2005;132(17):3859–71. pmid:16049111. doi: 10. 1242/dev. 01945

43. Xin M, Davis CA, Molkentin JD, Lien CL, Duncan SA, Richardson JA, et al. A threshold of GATA4 and GATA6 expression is required for cardiovascular development. Proceedings of the National Academy of Sciences of the United States of America. 2006;103(30):11189–94. Epub 2006/07/19. 0604604103 [pii] doi: 10. 1073/pnas. 0604604103 pmid:16847256; PubMed Central PMCID: PMC1544063.

44. Galli A, Robay D, Osterwalder M, Bao X, Benazet JD, Tariq M, et al. Distinct roles of Hand2 in initiating polarity and posterior Shh expression during the onset of mouse limb bud development. PLoS Genet. 2010;6(4):e1000901. Epub 2010/04/14. doi: 10. 1371/journal. pgen. 1000901 pmid:20386744; PubMed Central PMCID: PMC2851570.

45. Qu S, Tucker SC, Ehrlich JS, Levorse JM, Flaherty LA, Wisdom R, et al. Mutations in mouse Aristaless-like4 cause Strong's luxoid polydactyly. Development (Cambridge, England). 1998;125(14):2711–21. pmid:9636085.

46. Zhang Z, Verheyden JM, Hassell JA, Sun X. FGF-regulated Etv genes are essential for repressing Shh expression in mouse limb buds. Developmental cell. 2009;16(4):607–13. Epub 2009/04/24. S1534-5807(09)00082-3 [pii] doi: 10. 1016/j. devcel. 2009. 02. 008 pmid:19386269.

47. Mao J, McGlinn E, Huang P, Tabin CJ, McMahon AP. Fgf-dependent Etv4/5 activity is required for posterior restriction of Sonic Hedgehog and promoting outgrowth of the vertebrate limb. Developmental cell. 2009;16(4):600–6. Epub 2009/04/24. S1534-5807(09)00079-3 [pii] doi: 10. 1016/j. devcel. 2009. 02. 005 pmid:19386268; PubMed Central PMCID: PMC3164484.

48. Bourgeois P, Bolcato-Bellemin AL, Danse JM, Bloch-Zupan A, Yoshiba K, Stoetzel C, et al. The variable expressivity and incomplete penetrance of the twist-null heterozygous mouse phenotype resemble those of human Saethre-Chotzen syndrome. Human molecular genetics. 1998;7(6):945–57. pmid:9580658. doi: 10. 1093/hmg/7. 6. 945

49. Zhang Z, Sui P, Dong A, Hassell J, Cserjesi P, Chen YT, et al. Preaxial polydactyly: interactions among ETV, TWIST1 and HAND2 control anterior-posterior patterning of the limb. Development (Cambridge, England). 2010;137(20):3417–26. Epub 2010/09/10. dev. 051789 [pii] doi: 10. 1242/dev. 051789 pmid:20826535; PubMed Central PMCID: PMC2947755.

50. Patterson VL, Damrau C, Paudyal A, Reeve B, Grimes DT, Stewart ME, et al. Mouse hitchhiker mutants have spina bifida, dorso-ventral patterning defects and polydactyly: identification of Tulp3 as a novel negative regulator of the Sonic hedgehog pathway. Human molecular genetics. 2009;18(10):1719–39. Epub

2009/02/19. ddp075 [pii] doi: 10. 1093/hmg/ddp075 pmid:19223390; PubMed Central PMCID: PMC2671985.

51. Norman RX, Ko HW, Huang V, Eun CM, Abler LL, Zhang Z, et al. Tubby-like protein 3 (TULP3) regulates patterning in the mouse embryo through inhibition of Hedgehog signaling. Human molecular genetics. 2009;18(10):1740–54. Epub 2009/03/17. ddp113 [pii] doi: 10. 1093/hmg/ddp113 pmid:19286674; PubMed Central PMCID: PMC2671991.

52. Cameron DA, Pennimpede T, Petkovich M. Tulp3 is a critical repressor of mouse hedgehog signaling. Dev Dyn. 2009;238(5):1140–9. Epub 2009/04/01. doi: 10. 1002/dvdy. 21926 pmid:19334287.

53. Wright E, Hargrave MR, Christiansen J, Cooper L, Kun J, Evans T, et al. The Sry-related gene Sox9 is expressed during chondrogenesis in mouse embryos. Nature genetics. 1995;9(1):15–20. pmid:7704017. doi: 10. 1038/ng0195-15

54. Mo R, Freer AM, Zinyk DL, Crackower MA, Michaud J, Heng HH, et al. Specific and redundant functions of Gli2 and Gli3 zinc finger genes in skeletal patterning and development. Development (Cambridge, England). 1997;124(1):113–23. Epub 1997/01/01. pmid:9006072.

55. Lopez-Rios J, Speziale D, Robay D, Scotti M, Osterwalder M, Nusspaumer G, et al. GLI3 constrains digit number by controlling both progenitor proliferation and BMP-dependent exit to chondrogenesis. Developmental cell. 2012;22(4):837–48. Epub 2012/04/03. doi: 10. 1016/j. devcel. 2012. 01. 006 S1534-5807(12)00040-8 [pii]. pmid:22465667.

56. McGlinn E, van Bueren KL, Fiorenza S, Mo R, Poh AM, Forrest A, et al. Pax9 and Jagged1 act downstream of Gli3 in vertebrate limb development. Mechanisms of development. 2005;122(11):1218–33. pmid:16169709. doi: 10. 1016/j. mod. 2005. 06. 012

57. Min H, Danilenko DM, Scully SA, Bolon B, Ring BD, Tarpley JE, et al. Fgf-10 is required for both limb and lung development and exhibits striking functional similarity to Drosophila branchless. Genes & development. 1998;12(20):3156–61. pmid:9784490. doi: 10. 1101/gad. 12. 20. 3156

58. Dickinson RJ, Eblaghie MC, Keyse SM, Morriss-Kay GM. Expression of the ERK-specific MAP kinase phosphatase PYST1/MKP3 in mouse embryos during morphogenesis and early organogenesis. Mechanisms of development. 2002;113(2):193–6. pmid:11960712 doi: 10. 1016/s0925-4773(02)00024-2

59. Klock A, Herrmann BG. Cloning and expression of the mouse dual-specificity mitogen-activated protein (MAP) kinase phosphatase Mkp3 during mouse embryogenesis. Mechanisms of development. 2002;116(1–2):243–7. pmid:12128234 doi: 10. 1016/s0925-4773(02)00153-3

60. Kawakami Y, Rodriguez-Leon J, Koth CM, Buscher D, Itoh T, Raya A, et al. MKP3 mediates the cellular response to FGF8 signalling in the vertebrate limb. Nature cell biology. 2003;5(6):513–9. pmid:12766772. doi: 10. 1038/ncb989

61. Qu S, Niswender KD, Ji Q, van der Meer R, Keeney D, Magnuson MA, et al. Polydactyly and ectopic ZPA formation in Alx-4 mutant mice. Development

(Cambridge, England). 1997;124(20):3999–4008. Epub 1997/11/28. pmid:9374397.

62. Lettice LA, Horikoshi T, Heaney SJ, van Baren MJ, van der Linde HC, Breedveld GJ, et al. Disruption of a long-range cis-acting regulator for Shh causes preaxial polydactyly. Proceedings of the National Academy of Sciences of the United States of America. 2002;99(11):7548–53. pmid:12032320. doi: 10. 1073/pnas. 112212199

63. Charite J, McFadden DG, Olson EN. The bHLH transcription factor dHAND controls Sonic hedgehog expression and establishment of the zone of polarizing activity during limb development. Development (Cambridge, England). 2000;127(11):2461–70. Epub 2000/05/11. pmid:10804186.

64. Zakany J, Zacchetti G, Duboule D. Interactions between HOXD and Gli3 genes control the limb apical ectodermal ridge via Fgf10. Developmental biology. 2007;306(2):883–93. pmid:17467687. doi: 10. 1016/j. ydbio. 2007. 03. 517

65. Hill P, Gotz K, Ruther U. A SHH-independent regulation of Gli3 is a significant determinant of anteroposterior patterning of the limb bud. Developmental biology. 2009;328(2):506–16. doi: 10. 1016/j. ydbio. 2009. 02. 017 pmid:19248778.

66. Daoud G, Kempf H, Kumar D, Kozhemyakina E, Holowacz T, Kim DW, et al. BMP-mediated induction of GATA4/5/6 blocks somitic responsiveness to SHH. Development (Cambridge, England). 2014;141(20):3978–87. doi: 10. 1242/dev. 111906 pmid:25294942; PubMed Central PMCID: PMCPMC4197703.

67. Nardelli J, Thiesson D, Fujiwara Y, Tsai FY, Orkin SH. Expression and genetic interaction of transcription factors GATA-2 and GATA-3 during development of the mouse central nervous system. Developmental biology. 1999;210(2):305–21. doi: 10. 1006/dbio. 1999. 9278 pmid:10357893.

68. Ruest LB, Xiang X, Lim KC, Levi G, Clouthier DE. Endothelin-A receptor-dependent and -independent signaling pathways in establishing mandibular identity. Development (Cambridge, England). 2004;131(18):4413–23. doi: 10. 1242/dev. 01291 pmid:15306564; PubMed Central PMCID: PMCPMC2818681.

69. Buscher D, Grotewold L, Ruther U. The XtJ allele generates a Gli3 fusion transcript. Mamm Genome. 1998;9(8):676–8. pmid:9680393. doi: 10. 1007/ s003359900845

70. Kawakami Y, Tsuda M, Takahashi S, Taniguchi N, Esteban CR, Zemmyo M, et al. Transcriptional coactivator PGC-1alpha regulates chondrogenesis via association with Sox9. Proceedings of the National Academy of Sciences of the United States of America. 2005;102(7):2414–9. pmid:15699338. doi: 10. 1073/pnas. 0407510102

71. Zhong Y, Wang Z, Fu B, Pan F, Yachida S, Dhara M, et al. GATA6 activates Wnt signaling in pancreatic cancer by negatively regulating the Wnt antagonist Dickkopf-1. PloS one. 2011;6(7):e22129. doi: 10. 1371/journal. pone. 0022129 pmid:21811562; PubMed Central PMCID: PMCPMC3139620.

72. Osterwalder M, Speziale D, Shoukry M, Mohan R, Ivanek R, Kohler M, et al. HAND2 targets define a network of transcriptional regulators that compartmentalize

the early limb bud mesenchyme. Developmental cell. 2014;31(3):345–57. Epub 2014/12/03. doi: 10. 1016/j. devcel. 2014. 09. 018 S1534-5807(14)00624-8 [pii]. pmid:25453830.

73. Suzuki A, Raya A, Kawakami Y, Morita M, Matsui T, Nakashima K, et al. Nanog binds to Smad1 and blocks bone morphogenetic protein-induced differentiation of embryonic stem cells. Proceedings of the National Academy of Sciences of the United States of America. 2006;103(27):10294–9. pmid:16801560. doi: 10. 1073/pnas. 0506945103

8

CHAPTER

EMBRYONIC TRANSCRIPTION IS CONTROLLED BY MATERNALLY DEFINED CHROMATIN STATE

Saartje Hontelez[1], Ila van Kruijsbergen[1], Georgios Georgiou[1], Simon J. van Heeringen[1], Ozren Bogdanovic[2], Ryan Lister[2,3] & Gert Jan C. Veenstra[1]

[1]Department of Molecular Developmental Biology, Radboud Institute for Molecular Life Sciences, Faculty of Science, Radboud University, PO Box 9101, 6500 HB Nijmegen, The Netherlands

[2]ARC Center of Excellence in Plant Energy Biology, The University of Western Australia, Perth, Western Australia 6009, Australia

[3] The Harry Perkins Institute of Medical Research, Perth, Western Australia 6009, Australia

ABSTRACT

Histone-modifying enzymes are required for cell identity and lineage commitment, however little is known about the regulatory origins of the epigenome during embryonic development. Here we generate a comprehensive set of epigenome reference maps, which we use to determine the extent to which maternal factors shape chromatin state in *Xenopus* embryos. Using α-amanitin to inhibit zygotic transcription, we find that the majority of H3K4me3- and H3K27me3-enriched regions form a maternally defined epigenetic regulatory space with an underlying logic of hypomethylated islands. This maternal regulatory space extends to a substantial proportion of neurula stage-activated promoters. In contrast, p300 recruitment to distal regulatory regions requires embryonic transcription at most loci. The results show that H3K4me3 and H3K27me3 are part of a regulatory space that exerts an extended maternal control well into post-gastrulation development, and highlight the combinatorial action of maternal and zygotic factors through proximal and distal regulatory sequences.

Citation: Saartje Hontelez, Ila van Kruijsbergen,Georgios Georgiou,Simon J. van Heeringen,Ozren Bogdanovic,Ryan Lister& Gert Jan C. Veenstra, Embryonic transcription is controlled by maternally defined chromatin state, doi:10.1038/ncomms10148.

INTRODUCTION

During early embryonic development cells differentiate, acquiring specific transcription and protein expression profiles. Histone modifications can control the activity of genes through regulatory elements in a cell-type-specific manner[1, 2, 3, 4]. Recent advances have been made in the annotation of functional genomic elements of mammalian cells, *Drosophila* and*Caenorhabditis* through genome-wide profiling of chromatin marks[5, 6]. Immediately after fertilization, the embryonic genome is transcriptionally silent, and zygotic genome activation (ZGA) occurs after a number of mitotic cycles[7]. In *Drosophila* and zebrafish (*Danio rerio*) ZGA starts after 8 and 9 mitotic cycles, respectively, in mammals transcription starts at the two-cell stage[8, 9], whereas in *Xenopus* this happens after the first 12 cleavages at the mid-blastula transition (MBT)[10, 11, 12]. Permissive H3K4me3 and repressive H3K27me3 histone modifications emerge during blastula and gastrula stages[13, 14, 15, 16]. To date, little is known about the origin and specification of the epigenome in embryonic development of vertebrates, which is essential for understanding physiological cell lineage commitment and differentiation. To explore the developmental origins of epigenetic regulation we have generated epigenome reference maps during early development of *Xenopus tropicalis* embryos and assessed the need for embryonic transcription in their acquisition. We find a hierarchical appearance of histone modifications, with a priority for promoter marks which are deposited hours before transcription activation on regions with hypomethylated DNA. Surprisingly, the promoter H3K4me3 and the Polycomb H3K27me3 modifications are largely maternally defined (MaD), providing maternal epigenetic control of gene activation that extends well into neurula and tailbud stages. By contrast, p300 recruitment to distal regulatory elements is largely under the control of zygotic factors. Moreover, this maternal-proximal and zygotic-distal dichotomy of gene regulatory sequences also differentiates between early and late Wnt signalling target genes, suggesting that different levels of permissiveness are involved in temporal target gene selection.

RESULTS

Progressive Specification of Chromatin State

We have performed chromatin immunoprecipitation (ChIP) sequencing of eight histone modifications, RNA polymerase II (RNAPII) and the enhancer protein p300 at five stages of development: blastula (st. 9), gastrula (st. 10.5, 12.5), neurula (st. 16) and tailbud (st. 30). These experiments allow identification of enhancers (H3K4me1, p300)[17, 18, 19, 20], promoters (H3K4me3, H3K9ac)[14, 21, 22, 23], transcribed regions (H3K36me3, RNAPII)[22] and repressed and heterochromatic domains (H3K27me3, H3K9me2, H3K9me3 and H4K20me3)[1, 14, 24, 25]. In addition we generated pre-MBT (st. 8) maps for three histone modifications (H3K4me3, H3K9ac and H3K27me3) and single-base resolution DNA methylome maps using whole-genome bisulfite sequencing of blastula and gastrula (st. 9 and 10.5) embryos (Fig. 1; Supplementary Fig. 1). Our data set consists of 2.7 billion aligned sequence reads representing the most comprehensive set of epigenome reference maps of vertebrate embryos to date. Using a

Hidden Markov Model approach[26] we have identified 19 chromatin states based on co-occurring ChIP signals (Fig. 2a). This analysis identifies combinations of ChIP signals at specific genomic sequences without distinguishing between overlapping histone modifications that result from regional or cell-type specificity and co-occurrence in the same cells[14]. Seven main groups were recognized, namely (i) Polycomb (H3K27me3, deposited by Polycomb Repressive Complex 2 (PRC2)), (ii) poised enhancers, (iii) p300-bound enhancers, (iv) transcribed regions, (v) promoters, (vi) heterochromatin and (vii) unmodified regions (Fig. 2a; Supplementary Fig. 2). Alluvial plots of state coverage per stage show that all states increase in coverage during development, except for the unmodified state (Fig. 2b; Supplementary Fig. 2a). Unmodified regions decrease in coverage during development, however, even at tailbud stage 67% of the total epigenome remains naive for the modifications and bound proteins in our data set (Supplementary Fig. 2b). Promoter coverage remains relatively constant during development from blastula to tailbud stages, in contrast to the Polycomb state which increases in coverage during gastrulation. P300-bound enhancers are highly dynamic during development (Fig. 2b). Global enrichment levels of modified regions show similar dynamics, and reveal a priority for promoter marking at or before the blastula stage, followed by enhancer activation and heterochromatic repression during late blastula and gastrulation stages (Supplementary Fig. 3a,b). A detailed time course between fertilization and early gastrulation shows that both H3K4me3 and H3K9ac emerge hours before the start of embryonic transcription (Supplementary Fig. 3c). We and others have previously reported that H3K4me3 is acquired during blastula stages[14]. Indeed, H3K4me3 and H3K9ac levels increase strongly before the MBT, well before embryonic transcription starts. This however raises the question to what extent histone modifications are regulated by maternal or embryonic factors.

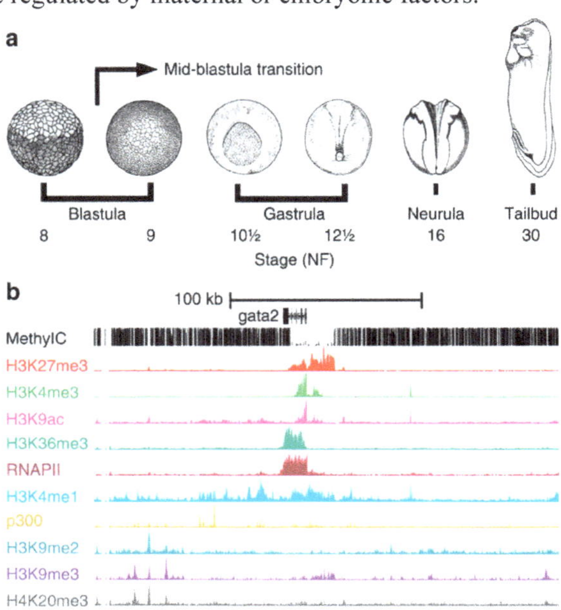

Figure 1: Reference epigenome maps of *Xenopus tropicalis* **development.**

(**a**) Genome-wide profiles were generated for stages 8 and 9 (blastula, before and after MBT), 10.5 and 12.5 (gastrula), 16 (neurula) and 30 (tailbud). Adapted from Tan, M.H. *et al. Genome Res.* **23**, 201–216 (2013), under a Creative Commons License (Attribution-NonCommercial 3.0 Unported License), as described at http://creativecommons.org/licenses/by/3.0/. (**b**) *Gata2* locus with late gastrula (stage 10.5) methylC-seq, ChIP-seq enrichment of histone modifications, RNAPII and p300 (cf. Supplementary Figs 1 and 2).

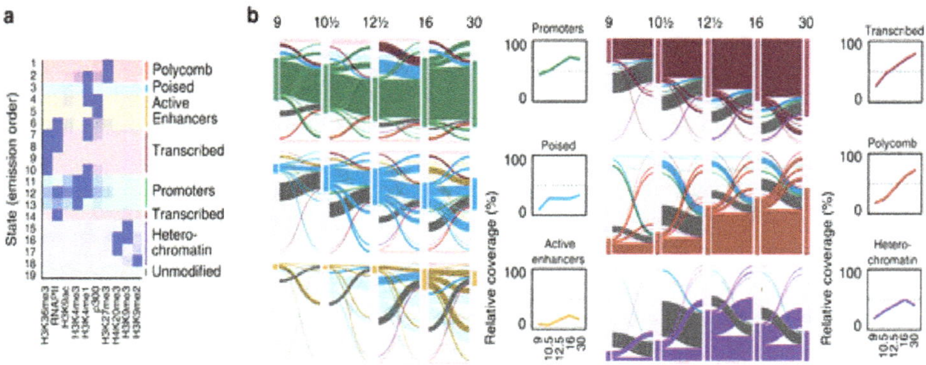

Figure 2: Chromatin state dynamics.

(**a**) Emission states (same for all developmental stages) of the hidden Markov model, identifying the 19 most prevalent combinations of histone modifications and bound proteins. From top to bottom: Polycomb (red), Poised enhancers and promoters (blue), Active Enhancers (gold), Transcribed (dark magenta), Promoter (green), Heterochromatin (purple) and unmodified (grey). (**b**) Alluvial plots of chromatin state coverage during development. Each plot shows the transitions (to and from the highlighted group of chromatin states) across developmental stages (stages 9–30). The height represents the base pair coverage of the chromatin state relative to the modified genome. The 'modified genome' has a chromatin state other than unmodified in any of the stages 9–30. From top to bottom left: promoters (green), poised (blue), p300-bound enhancers (gold). From top to bottom right: transcribed (dark magenta), Polycomb (red) and heterochromatin (purple). Line plots: Chromatin state coverage per stage as a percentage of the modified genome.

Maternal and Zygotic Epigenetic Regulation

To determine the maternal and zygotic contributions to chromatin state, we used α-amanitin to block embryonic transcription (Fig. 3a). α-Amanitin blocks the translocation of RNA polymerase II (RNAPII) on DNA, thereby preventing transcript elongation[27]. It is therefore expected that injection of α-amanitin into embryos will stall RNAPII, immobilizing it on DNA after its recruitment to pre-initiation complexes. Indeed, both RNAPII elongation and embryonic transcription were effectively blocked in α-amanitin-injected embryos (Fig. 3b,c; Supplementary Fig. 4a). New transcription is necessary for gastrulation[11, 28, 29], but α-amanitin-injected embryos survive to the equivalent of stage 11 control embryos. ChIP sequencing of replicates of α-amanitin-injected and control embryos (stage 11) revealed that the majority of H3K4me3 (86%) and H3K27me3 (90%) regions are consistently modified with these modifications independently of embryonic transcription (Fig. 3d; Supplementary Fig. 4b,c). This is especially surprising given the temporal hierarchy of H3K27me3 and H3K4me3, and the relatively late acquisition of H3K27me3 (Fig. 2b). By contrast, only 15% of the p300-bound regions recruit p300 independently of active transcription (Fig. 3d). This suggests that the promoter-permissive H3K4me3 mark and the Polycomb-repressive H3K27me3 mark are mostly controlled by maternal factors (maternally defined, MaD), whereas p300 binding to regulatory regions is largely zygotically defined (ZyD). Regions with MaD H3K4me3 and H3K27me3 acquire these modifications more robustly and also earlier during development compared with ZyD regions (Supplementary Fig. 4d). By contrast, ZyD p300-bound regions show more robust p300 recruitment during gastrulation compared with p300 MaD regions. These data show a pervasive maternal influence on the developmental acquisition of key histone modifications.

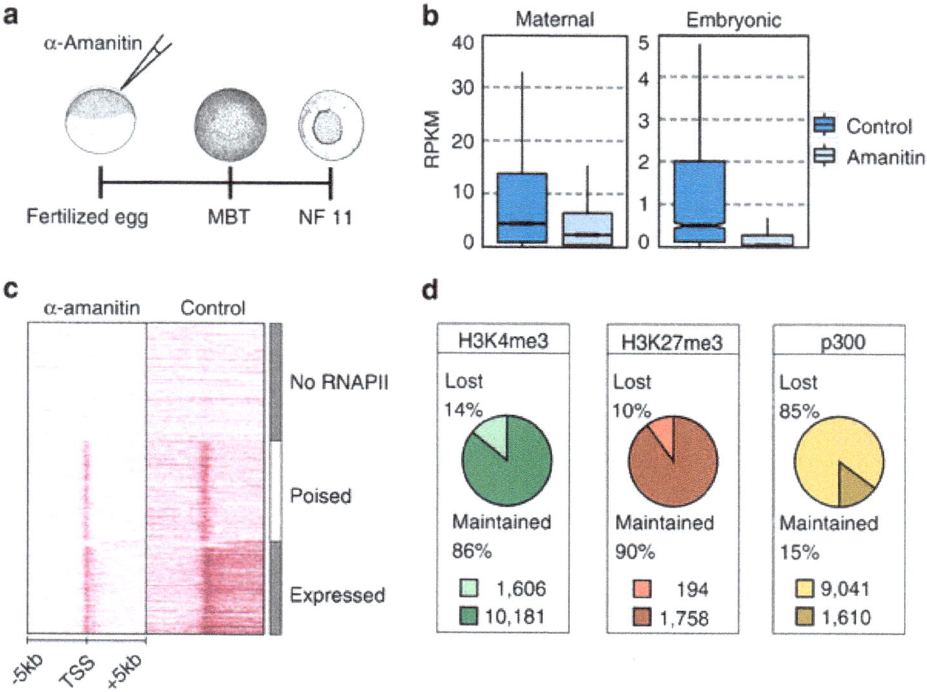

Figure 3: Developmental acquisition of chromatin states.

(**a**) Inhibition of embryonic transcription with α-amanitin, adapted from Tan, M.H. *et al. Genome Res.* **23**, 201–216 (2013), under a Creative Commons License (Attribution-

NonCommercial 3.0 Unported License), as described at http://creativecommons.org/licenses/by/3.0/. (**b**) RNAPII on the TSS of genes in control and α-amanitin-injected embryos (stage 11). (**c**) Box plots showing RNA expression levels (RPKM) of maternal and embryonic transcribed genes in control and α-amanitin-injected embryos (stage 11). Box: 25th (bottom), 50th (internal band), 75th (top) percentiles. Whiskers: 1.5 × interquartile range of the lower and upper quartiles, respectively. (**d**) ChIP-sequencing on chromatin of α-amanitin-injected and control embryos reveals maternal and zygotic origins of H3K4me3, H3K27me3 or p300 binding. Data from two biological replicates, see Supplementary fig. 4.

DNA Methylation Logic of Maternal Control

Trimethylation of H3K4 and H3K27 has been associated with CpG density and a lack of DNA methylation. The Set1 and related MLL complexes are responsible for H3K4me3 (ref. 10). Set1 is recruited to hypomethylated CpG domains via the Cxxc1 protein (Cfp1)[30, 31, 32]. In the absence of H3K4me3, PRC2 binding to hypomethylated CpGs results in H3K27me3 and inhibition of gene activation[13, 33]. Using our whole-genome bisulfite sequencing data we determined that MaD H3K4me3 promoters are predominantly hypomethylated (Fig. 4a; Supplementary Fig. 5a;Supplementary Data 1). Conversely, promoters decorated with ZyD H3K4me3 almost exclusively have highly methylated promoters. Demethylation of ZyD promoters was not detected, and methylation levels of MaD and ZyD regions were similar in stage 9 and stage 10.5 (Supplementary Fig. 5a,b). In addition, H3K4me3 often extends asymmetrically from promoters into gene bodies (+1–2 kb from transcription start site (TSS); Supplementary Fig. 5c), likely representing the second and third nucleosomes that are trimethylated via RNAPII-recruited Set1 in actively transcribed genes[34]. Concordantly, α-amanitin reduces H3K4me3 at downstream positions. Interestingly, we also find poised enhancers that gain H3K4me3 in α-amanitin-injected embryos and which exhibit intermediate to high levels of DNA methylation (Supplementary Fig. 5d,e).

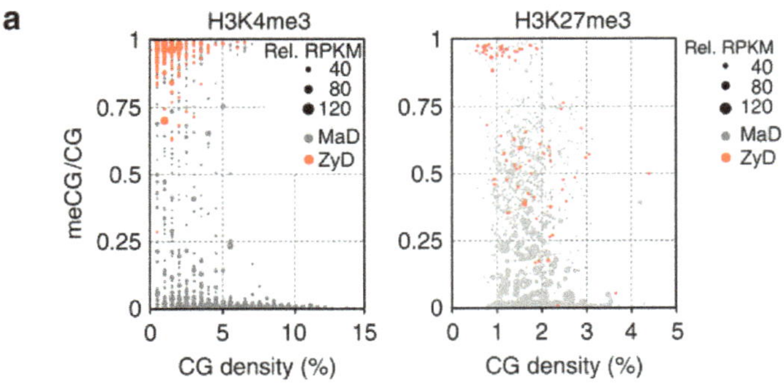

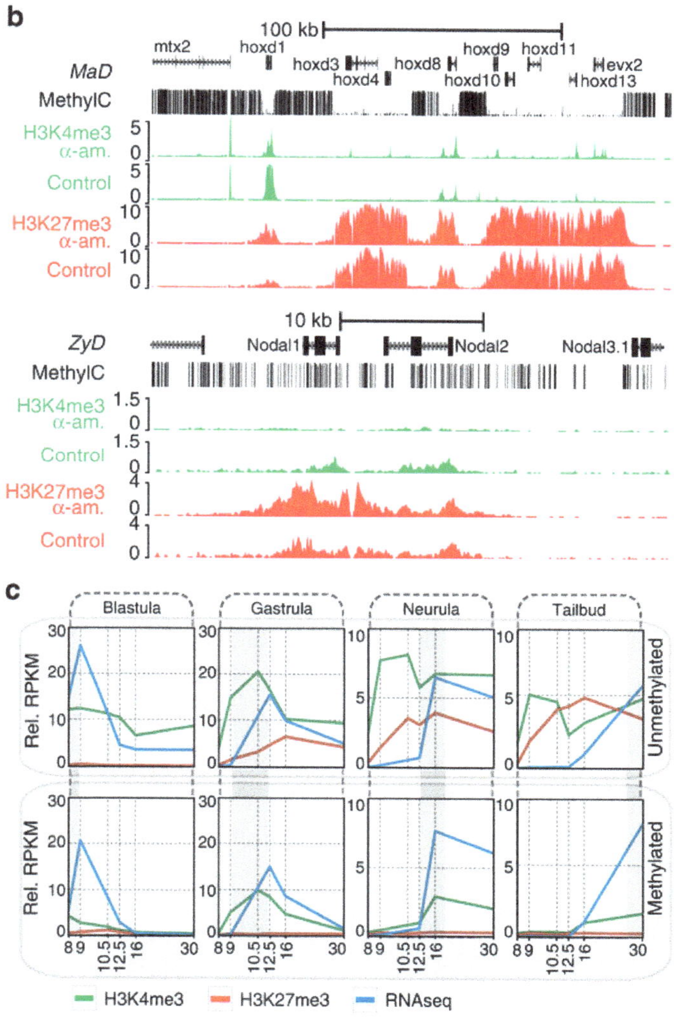

Figure 4: DNA methylation logic of maternally versus zygotically defined H3K4me3 and H3K27me3.

(**a**) CpG density and methylation at stage 9 of promoters (H3K4me3: ±100 bp from TSS; H3K27me3: ±2.5 kb from TSS) that contain a zygotic defined (ZyD, lost in α-amanitin treated embryos, red) or maternal defined (MaD, maintained in α-amanitin treated embryos, grey) peak for H3K4me3 (left) or H3K27me3 (right) after inhibition of embryonic transcription. The size of the dot indicates the relative RPKM of the histone modification (background corrected). (**b**) *Hoxd* (MaD) and *nodal1, -2* (ZyD) loci with stage 9 methylC-seq, H3K4me3 and H3K27me3 in control and α-amanitin-injected embryos. (**c**) Developmental profiles of H3K4me3 and H3K27me3 (median background corrected RPKM) at genes without detectable maternal mRNA do correlate with activation for methylated promoters (lower panels) but not for hypomethylated CpG island promoters (upper panels).

The majority of promoters with ZyD H3K27me3 shows intermediate to high levels of DNA methylation (Fig. 4a; Supplementary Fig. 5a; Supplementary Data 1). Some of the MaD H3K27me3 regions are methylated, but the highly enriched H3K27me3 domains (larger dots) are almost exclusively both maternally defined and hypomethylated. This is illustrated by the *hoxd*cluster which harbours a large hypomethylated domain with MaD H3K4me3 and H3K27me3 (Fig. 4b). There are also examples of reciprocal changes of H3K4 and H3K27 methylation, for example at the hypermethylated promoters of *nodal1* and *nodal2*.

ZyD p300-bound regions are generally hypermethylated, whereas MaD p300-bound regions show a variable degree of DNA methylation (Supplementary Fig. 5e). However, promoters that overlap with MaD p300 peaks are hypomethylated in 77% of the cases, whereas 96% of the promoters that are associated with ZyD p300 peaks are hypermethylated (Supplementary Fig. 5f), showing that p300-recruiting hypomethylated promoters tend to be under complete maternal control, for both H3K4 methylation and p300 recruitment.

To further explore the relationships between DNA methylation, histone modifications and developmental activation of transcription we determined correlations with different measures of gene activity such as RNA-seq and ChIP-seq of RNAPII and H3K36me3 (Supplementary Fig. 6). We find that H3K36me3 and RNAPII in gene bodies correlate well with each other but less with transcript levels (RNA-seq), presumably due to the effects of RNA stability. A much lower correlation was found between either measure of gene activity and the promoter marks H3K4me3 and H3K9ac, especially at early stages. In part this may be caused by time delays of transcriptional activation relative to acquisition of permissive histone modifications[14, 15]. It raises the question to what extent a lack of DNA methylation at promoters, which is associated with MaD H3K4me3, uncouples promoter marking and transcriptional activation. Therefore, we grouped transcribed genes without detectable maternal messenger RNA[35] based on the stage of maximum expression and DNA methylation (Fig. 4c). We find that developmentally activated promoters with hypomethylated CpG islands are trimethylated at H3K4 or H3K27 early on, irrespective of the time of transcriptional activation. By contrast, methylated promoters show a much closer relation between H3K4me3 and gene expression. Although H3K4me3 is known to stabilize the transcription initiation factor Taf3 (a subunit of TFIID) and can also interact with the chromatin remodeller Chd1 (refs 36, 37, 38), hypomethylated promoters gain H3K4me3 autonomously with their hypomethylated CpG island status, independent of embryonic transcription.

ZyD p300-Bound Domains Shape Enhancer Clusters

P300 can be recruited by transcription factors that bind to regulatory elements. We therefore modelled transcription factor motif contributions to p300 binding across multiple developmental stages (see Methods). The results predict specific transcription factors to recruit p300 in a stage-specific manner (Fig. 5a). Clustering of MaD and ZyD p300-bound regions with H3K4me3, H3K4me1 and RNAPII data revealed that ZyD p300 is recruited to distal regulatory sequences that lose both p300 and RNAPII binding in the presence of α-amanitin, whereas MaD p300 binding mostly includes promoter-proximal regions that are H3K4me3-decorated and recruit RNAPII in the presence of α-amanitin but without elongating (Fig. 5b). Indeed, MaD p300 regions are enriched for promoter-related motifs (Supplementary Fig. 7). Although some ZyD p300-bound regions overlap with annotated transcription start sites (Supplementary Fig. 5f), most of these sequences are decorated with H3K4me1 in the absence of H3K4me3, suggesting they correspond to distal regulatory sequences (Fig. 5b). Both MaD- and ZyD p300-bound regulatory regions recruit embryonically regulated transcription factors such as Otx2, Gsc, Smad2/3, Foxh1, T (Xbra), Vegt and Eomes (Supplementary Fig. 8)[39, 40, 41], suggesting that multiple transcription factors contribute to p300 recruitment.

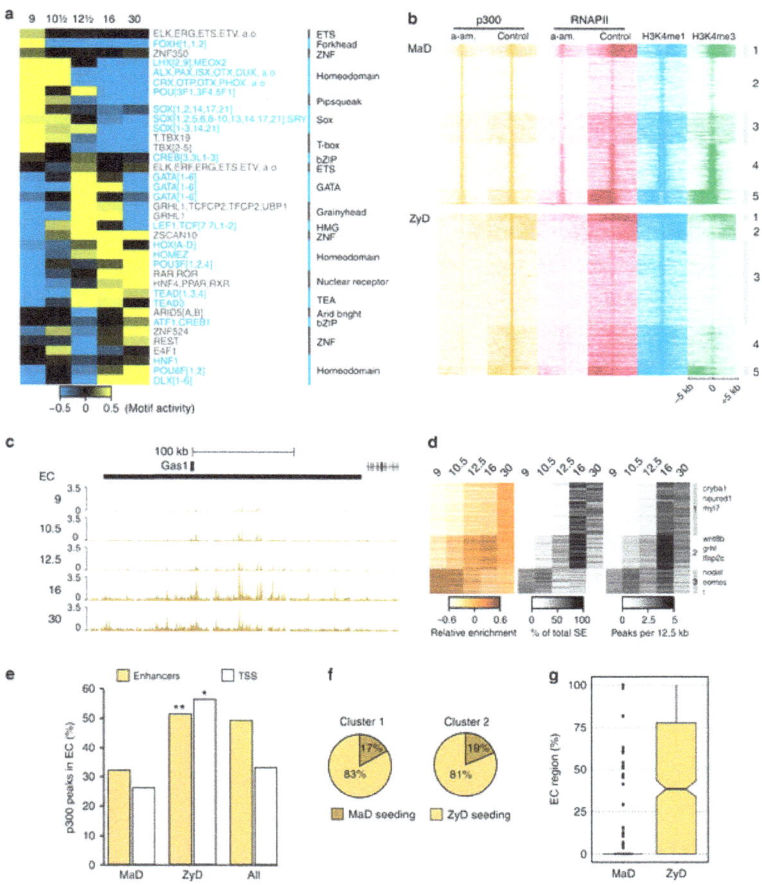

Figure 5: Zygotically controlled p300 recruitment shapes enhancer clusters (EC) domains.

(**a**) Modelled transcription factor motif activity to p300 enrichment (see Methods). Activity reflects modelled contributions in p300 peak RPKM. (**b**) Heatmaps of MaD (upper panel) and ZyD (lower panel) p300 binding sites in α-amanitin treated and control embryos. (**c**) Developmental increase in genomic coverage of the *gas1* EC by acquisition of p300 binding at enhancers. (**d**) EC dynamics of p300 enrichment (left panel), percentage of total EC region identified in each stage based on stage-dependent p300 binding (middle panel) and number of p300 peaks (per 12.5 kb) in EC. (**e**) Percentage of zygotic defined (ZyD, lost in α-amanitin treated embryos) and maternal defined (MaD, maintained in α-amanitin treated embryos) p300 peaks that map to ECs. Asterisks indicate significance as more or less p300 peaks than expected by chance calculated using cumulative hypergeometric test: $*P=6E-14$; $**P=5E-29$ (**f**) Percentage of ECs that have a MaD or ZyD seeding peak at stage 9. (**g**) Box plot showing the percentage of the EC region that is defined by MaD or ZyD p300 peaks. Box: 25th (bottom), 50th (internal band), 75th (top) percentiles. Whiskers: 1.5 × interquartile range of the lower and upper quartiles, respectively. Outliers are indicated with black dots.

Large enhancer clusters (ECs) are thought to improve the stability of enhancer–promoter interactions, are associated with genes coding for developmental regulators, and have been implicated in cell differentiation[42, 43, 44]. During development the cluster size of p300-bound enhancers grows dynamically by p300 seeding of individual enhancers (Fig. 5c,d, see Methods). Histone modifications and transcript levels of EC-associated genes are developmental stage specific, confirming the association of ECs with developmental genes (Supplementary Fig. 9;Supplementary Data 2). Analysis of the percentage of the total EC regions identified in each stage show that most p300-bound ECs increase in genomic coverage during development by newly gained p300 binding at enhancers (EC clusters 1 and 2), whereas a group of early ECs (EC cluster 3) decrease in coverage as a result of the decreasing number of p300 peaks that contribute to the EC.

We next examined how MaD and ZyD p300-bound regions contribute to p300-bound ECs. Approximately 50% of all ZyD p300-bound enhancers are located in ECs at stage 11. Among MaD p300-bound enhancers this fraction is much reduced (Fig. 5e). Similarly, a much larger fraction of ZyD p300-bound promoters is found in ECs compared with MaD p300-bound promoters. Up to 20% of the developmental ECs that are seeded at stage 9 have a MaD p300 seeding site (Fig. 5f). However, very few ECs can be called based on MaD p300, showing that formation of p300-bound enhancer clusters requires embryonic transcription (Fig. 5g).

Extended Maternal Epigenetic Control

We next examined the extent to which the MaD epigenome is maintained during development. Genes were grouped based on MaD or ZyD trimethylation of H3K4 and H3K27 in the promoter (Supplementary Data 3, see Methods). For p300 we counted the total number of MaD and ZyD peaks in the *cis*-regulatory landscapes of genes (Fig. 6a). Remarkably, MaD H3K4me3-regulated genes represent the majority of all H3K4me3-enriched genes in both early and late developmental stages. Even at neurula and tailbud stages only a small fraction of the H3K4me3-decorated genes are ZyD. Similarly, maternal control of H3K27me3 also extends late into development, albeit to a smaller degree. After gastrulation, the number of MaD H3K27me3 regulated genes slightly decreases, whereas ZyD increases. However, also at neurula stage more than 50% of the Polycomb (PRC2)-regulated genes are under MaD H3K27me3 control. By contrast, p300 in *cis*-regulatory regions of genes is almost exclusively ZyD in all stages (Fig. 6a).

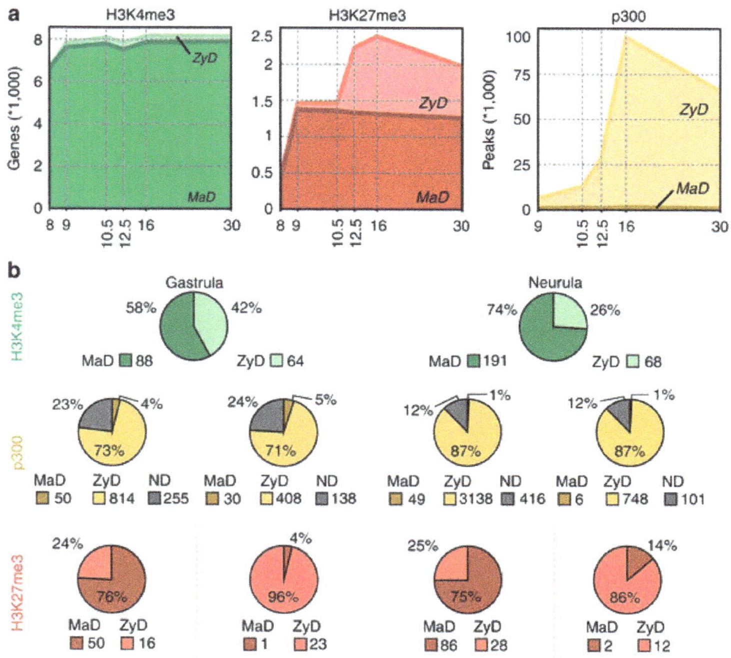

Figure 6: Maternal epigenetic control extends beyond gastrulation.

Maternally defined (MaD) peaks emerge at or before stage 11 independent of embryonic transcription. Zygotically defined (ZyD) peaks appear before stage 11 and are lost in α-amanitin treated embryos, or emerge at or after stage 12. Not determined (ND) peaks are not consistently detected in replicates 1 and 2 and generally have low enrichment values. (**a**) Total number of genes with a MaD or ZyD peak in their promoter (H3K4me3 and H3K27me3), or total number of MaD and ZyD peaks per GREAT region (p300). ND peaks are not shown. (**b**) MaD and ZyD regulation of gastrula and neurula expressed genes. The pie charts show the number genes with a MaD or ZyD peak in their promoter (H3K4me3 and H3K27me3) or the number of MaD, ZyD and ND peaks per *cis*-regulatory region (p300). The H3K27me3 and p300 pie charts represent: Gastrula expressed genes with a MaD (far left) or ZyD (middle left) H3K4me3 peak; neurula expressed genes with a MaD (middle right) or ZyD (far right) H3K4me3 peak.

Many genes may maintain MaD H3K4me3 because they are constitutively expressed throughout development. We therefore analysed the regulation of genes that are exclusively embryonically transcribed. We find that 487 of 983 (49.5%) genes which are expressed between blastula and tailbud stages but not expressed in oocytes or before the MBT, feature a MaD H3K4me3 promoter (Supplementary Fig. 10a). Most of the MaD H3K4me3 genes that are modified by PRC2 exhibit MaD H3K27me3. When separating embryonic transcripts based on developmental activation, we find MaD

H3K4me3 for 58% of the gastrula genes and up to 74% of the neurula expressed genes (Fig. 6b; Supplementary Fig. 10b). In most cases MaD H3K4me3-regulated genes also have MaD H3K27me3 control. This indicates an important role for the MaD epigenome in the regulation of embryonic transcripts.

To explore the distinctions between expression inside and outside the maternal regulatory space, we analysed Wnt signalling targets. Early Wnt/beta-catenin signalling serves to specify dorsal fates following fertilization, leading to organizer gene expression. This has been shown to depend on Prmt2-mediated promoter poising before the MBT[45]. Indeed, we find that seven of eight early Wnt/beta-catenin targets have a hypomethylated island promoter marked with MaD H3K4me3 (Fig. 7a; Supplementary Fig. 10c). Wnt signalling also plays an important role after the MBT, when it ventralises and patterns mesoderm. The majority of these later targets turn out to have a methylated promoter with ZyD H3K4me3. Notably, these ZyD H3K4me3 late Wnt targets are associated with high binding of p300 in their locus; many of the p300 binding events happen at distal regulatory regions. In contrast, MaD H3K4me3 Wnt targets have less p300 binding but are marked with H3K27me3 (Fig. 7a,b). These results illustrate the dichotomy in proximal and distal regulation that is associated with transcriptional activation of maternal and zygotic Wnt target genes, which is paradigmatic of the distinctive maternal and zygotic epigenetic programs that are orchestrated by DNA methylation and exert a long-lasting influence in development (Fig. 8).

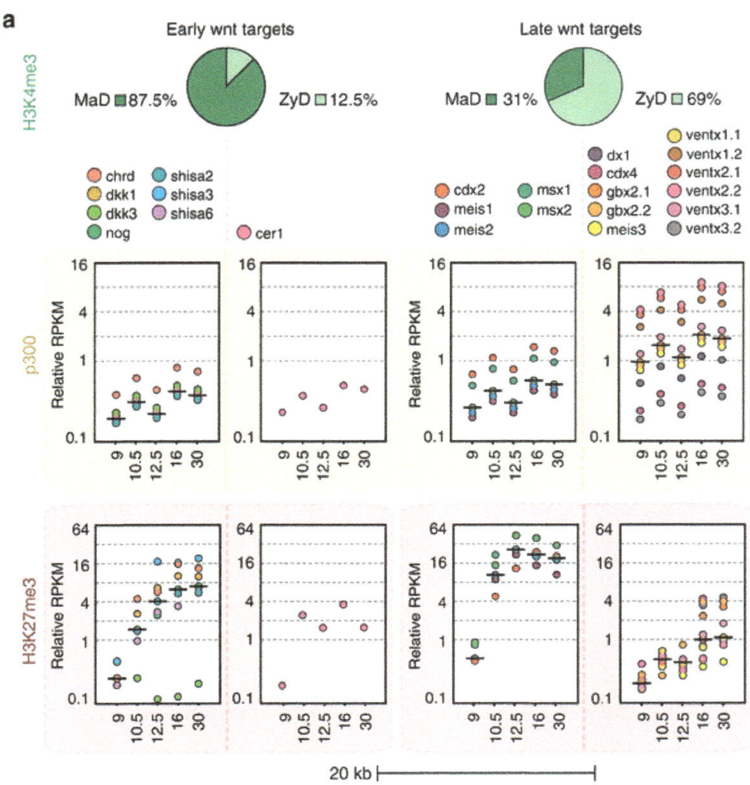

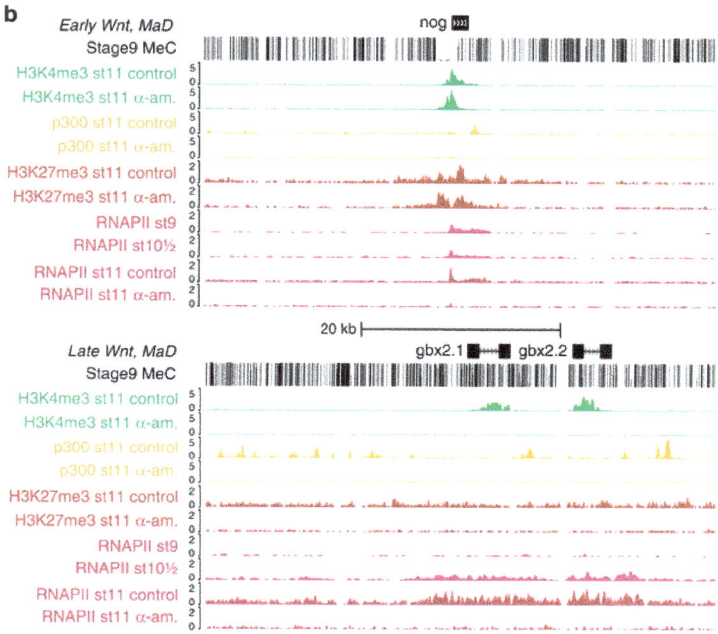

Figure 7: Maternal and zygotic regulatory space separates early and late Wnt target genes.

(**a**) The number of genes with MaD or ZyD H3K4me3 (pie charts) and relative RPKM (dot plots, horizontal line: median) of p300 in *cis*-regulatory regions of genes and H3K27me3 on promoters (±2.5 kb from TSS) at different developmental stages that have maternally or zygotically defined H3K4me3 at the promoter. Early targets sia1 and sia2 are not included, these genes lose H3K4me3 after stage 9 and cannot be assigned to MaD or ZyD space based on our stage 11 α-amanitin data. H3K4me3 on these genes is acquired at stage 8, before embryonic transcription. (**b**) Browser views of the early Wnt target *nog (noggin)* and the late Wnt targets *gbx2.1* and *gbx2.2* with ChIP-seq enrichment of H3K4me3, p300 and RNAPII on control and α-amanitin-injected embryos and RNAPII on stages 9 and 10.5.

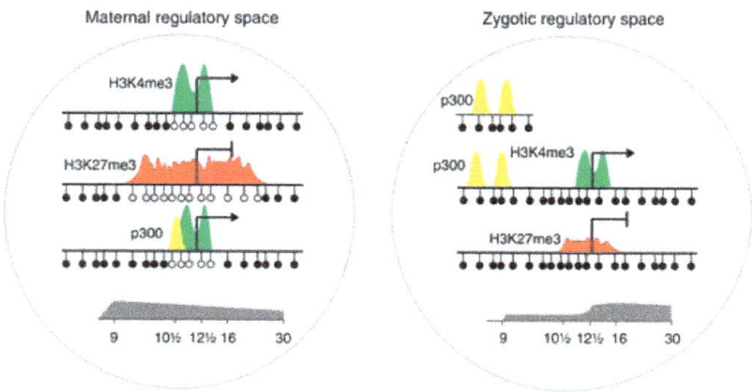

Figure 8: Model of maternal and zygotic regulatory space.

This shows the segregation of maternal regulatory space, which contains hypomethylated promoters that are mainly controlled by maternal factors, and zygotic regulatory space, which includes methylated promoters and enhancers that are under zygotic control. Most p300-bound enhancers are in zygotic space, however, they can regulate promoters in both maternal and zygotic space, crossing the regulatory space border. This may contribute to varying degrees of permissiveness to transcriptional activation. Maternal regulatory space extends well into neurula and tailbud stages and includes many embryonic genes which are activated at specific stages of development. Zygotic regulatory space requires zygotic transcription, is established from the mid-blastula stage onwards but increases in relative contribution during development.

DISCUSSION

The H3K4me3 modification poises promoters for transcription initiation by stabilizing Taf3/TFIID binding[36, 37]. Promoter H3K4 methylation based on an underlying DNA methylation logic driven by maternal factors at the blastula stage sets the stage for a default programme of gene expression. Most constitutively expressed housekeeping genes are within this maternal regulatory space, as well as a subset of developmentally regulated genes. Remarkably, many late expressed genes have hypomethylated promoters and are already poised for activation by H3K4me3 during early blastula stages. H3K4me3 is not sufficient for gene transcription and additional embryonic factors are required for activation in many cases. Genes with MaD H3K4me3 generally have fewer p300-bound enhancers associated with them, suggesting they are regulated by promoter-proximal elements. This further underscores the permissive nature of this regulation, as opposed to zygotically regulated events at both promoters (H3K4me3) and enhancers (recruitment of p300). The H3K27me3 modification is gradually acquired between blastula and gastrula stages on spatially regulated genes, repressing lineage-specific genes in other lineages[13, 14]. The acquisition of this modification in the absence of transcription indicates that it is uncoupled from the inductive events of the early embryo, suggesting a default maternal response to a lack of transcriptional activation. The results indicate that maternal factors set permissions and time-dependent constraints on a subset of genes with reduced CpG methylation at their promoter. These permissions and constraints are likely to channel embryonic cell fates into a limited number of directions by controlling hierarchical developmental progression by master regulators. Previously we observed that DNA methylation does not lead to transcriptional repression in early embryos, whereas it does in oocytes and late embryos[46]. The observations described here suggest a new role of DNA methylation in defining a maternal-embryonic programme of gene expression. In zebrafish, the maternal methylome is reprogrammed between fertilization and ZGA, to match the paternal methylome. This also occurs in maternal haploid fish, and appears to align with CG content[47, 48], suggesting an intrinsic maternal mechanism that sets the stage for the MaD epigenome.

Gene expression outside maternal regulatory space could be mediated by p300-associated enhancers, most of which require new transcription for recruitment of p300. Promoter and enhancer activation in the ZyD regulatory space likely involves binding of specific factors. Indeed, we find that both MaD- and ZyD p300-bound regulatory regions recruit embryonically regulated transcription factors. Enhancers often contain binding sites for many different proteins, which can play different roles in opening up chromatin, recruitment of co-activators and establishing looping interactions with promoters. Future experiments will shed light on the maternal–zygotic hierarchy and the regulatory transitions underlying these events and the roles of maternal and zygotic pioneer factors. We find that ZyD p300-bound enhancers shape enhancer clusters. These form dense hubs of regulatory activity, and EC p300 binding is generally correlated with the expression of the associated genes. The work reported here suggests that recruitment of p300 to 'seeding' enhancers precedes establishing cluster-wide activity of the local enhancer landscape. Future work will also need to address to which extent seeding causes relaxation and opening of the local chromatin and activity of neighbouring enhancers.

Key proteins of the molecular machinery involved in DNA methylation (Dnmt3a, Tet2), H3K4me3 (Mll1-4, Kdm5b/c), H3K27me3 (Ezh2, Eed, Kdm6a/b) and enhancer histone acetylation (p300) are not only highly conserved between species but also frequently mutated in cancer[49, 50, 51]. Moreover cancer-specific hypermethylated regions tend to correspond to Polycomb-regulated loci in embryonic stem cells and DNA methylation may restrict H3K27 methylation globally[52, 53]. In addition, the sequence signatures of hypomethylated regions that acquire H3K4me3 or H3K27me3 are conserved between fish, frogs and humans[13]. These observations suggest that the molecular mechanisms that orchestrate the maternal and zygotic regulatory space are conserved. One key difference between mammals and non-mammalian vertebrates is the specification of extra-embryonic lineages between zygotic genome activation and the blastocyst stage in mammals[10], so it is likely that the way this plays out for specific genes differs between species. In summary, our results provide an unprecedented view of the far reach of maternal factors in zygotic life through chromatin state. The dichotomy of maternal promoter-based and embryonic enhancer regulation demarcates an epigenetic maternal-to-zygotic transition that is maternal permissive to the expression of some embryonic genes and restrictive to others. This highlights the combinatorial interplay of maternal and zygotic factors through distinct mechanisms.

METHODS

Animal Procedures

X. tropicalis embryos were obtained by *in vitro* fertilization, dejellied in 3% cysteine and collected at the indicated stage. Fertilized eggs were injected with 2.3 nl of 2.67 ng μl^{-1} α-amanitin and developed until the control embryos reached mid-gastrulation (stage 11). Animal use was conducted under the DEC permission (Dutch Animal Experimentation Committee) RU-DEC 2012–116 and 2014–122 to G.J.C.V.

ChIP Sequencing and RNA Sequencing

Chromatin for ChIP was prepared as previously described[54, 55], with minor modifications. Antibody was incubated with chromatin overnight, followed by incubation with

Dynabeads Protein G for 1 h. The following antibodies were used: anti-H3K4me1 (Abcam ab8895, 1 μg per 15 embryo equivalents (Eeq)), anti-H3K4me3 (Abcam ab8580, 1 μg per 15 Eeq), anti-H3K9ac (Upstate/Millipore 06-942, 1 μg per 15 Eeq), anti-H3K36me3 (Abcam ab9050, 1 μg per 15 Eeq), anti-H3K27me3 (Upstate/Millipore 07-449, 1 μg per 15 Eeq), anti-H3K9me2 (Diagenode C15410060, 1 μg per 15 Eeq), anti-H3K9me3 (Abcam ab8898, 2 μg per 15 Eeq), anti-H4K20me3 (Abcam ab9053, 2 μg per 15 Eeq), anti-p300 (Santa Cruz sc-585, 1 μg per 15 Eeq) and anti-RNAPII (Diagenode C15200004, 1 μg per 15 Eeq). For all ChIP-seq samples of the epigenome reference maps and RNAPII ChIP-seq samples of the α-amanitin experiments three biological replicates of different chromatin isolations of 45 embryos were pooled. Two biological replicates for H3K4me3 (α-amanitin injected: 90 and 56 Eeq; control: 45 and 67 Eeq), H3K27me3 (α-amanitin injected: 90 and 180 Eeq; control: 45 and 202 Eeq) and p300 (α-amanitin injected: 112 and 56 Eeq; control: 112 and 67 Eeq) ChIP-seq samples of the α-amanitin experiments were generated. For RNA-seq samples of the α-amanitin experiments RNA from five embryos from one biological replicate was isolated and depleted of ribosomal RNA as previously described[35]. Samples were subjected to a qPCR quality check pre- and post preparation. Libraries were prepared with the Kapa Hyper Prep kit (Kapa Biosystems), and sequencing was done on the Illumina HiSeq2000 platform. Reads were mapped to the reference *X. tropicalis* genome JGI7.1, using STAR (RNA-seq) or BWA (ChIP-seq) allowing one mismatch.

MethylC-seq

Genomic DNA from *Xenopus* embryos stages 9 and 10.5 was obtained as described before[56]. MethylC-seq library generation was performed as described previously[57]. Library amplification was performed with KAPA HiFi HotStart Uracil+ DNA polymerase (Kapa Biosystems, Woburn, MA, USA), using six cycles of amplification. Single-read MethylC-seq libraries were processed and aligned as described previously[58].

Quantitative PCR

PCR reactions were performed on a CFX96 Touch Real-Time PCR Detection System (BioRad) using iQ Custom SYBR Green Supermix (BioRad). We preformed RNA expression PCR (RT–qPCR (quantitative PCR)) and ChIP-qPCR for H3K4me3 and H3K9ac on promoters of *odc1*, *eef1a1o*, *rnf146*, *tor1a*, *zic1*, *cdc14b*, *eomes*, *xrcc1*, *drosha*, *gdf3*, *t*, *tbx2*, *fastkd3*, *gs17* (seeSupplementary Methods for primer sequences). ChIP-qPCR enrichment over background was calculated using the average of 5 negative loci.

Detection of Enriched Regions

We used MACS2 (ref. 59) with standard settings and a *q*-value of 0.05. Fragment size was determined using phantompeakqualtools[60]. Broad settings (--BROAD) were used for H3K4me1, H3K36me3, H3K27me3, H3K9me2, H3K9me3, H4K20me3 and RNAPII. Broad and narrow peaks were merged for H3K4me3. For H3K9ac narrow peaks were used. For p300 broad peaks were used in the ChomHMM analysis, narrow p300 peaks were used for super-enhancer and MaD versus ZyD analyses. All peaks were called relative to an input control track. Peaks that showed at least 75% overlap with 1 kb regions that have more than 65 input reads, and peaks that have a ChIP-seq RPKM higher than the 95 percentile of random background regions are excluded from further analysis. Only scaffolds 1–10 (the chromosome-sized scaffolds) were included in the analysis. Relative RPKM was calculated by dividing the ChIP-seq RPKM of the peaks by the ChIP-seq RPKM of the 95 percentile of random background regions.

We used MAnorm[61] to determine differentially enriched regions in α-amanitin and control embryos. We used merged peak sets of replicate 1, replicate 2 and stage 10.5 to avoid bias caused by peak calling. Lost, gained and unchanged peaks per biological replicate were determined using the following parameters: lost peaks have *M*-values >1 and a −log base 10(*P*value) >5 (for H3K27me3) or 1.3 (for H3K4me3 and p300) and have a relative RPKM (background corrected) >1 in stage 11 control (no cut-off was used for st.11 control of H3K27me3 rep.1), stage 10.5 (H3K4me3 and p300) or stage 12 (H3K27me3); *increased peaks* have M-values smaller than -1 and a -log base 10(*P* value) >5 (H3K27me3) or 1.3 (H3K4me3 and p300) and have a rel. RPKM greater than 1 in stage 11 α-amanitin, stage 10.5 (H3K4me3 and p300) or stage 12 (H3K27me3); unchanged peaks are neither gained nor lost and have a rel. RPKM >1 in stage 11 control (no cut-off was used for st.11 control of H3K27me3 rep.1), stage 11 α-amanitin, stage 10.5 (H3K4me3 and p300) or stage 12 (H3K27me3). Maintained peaks are peaks that are not lost and have a rel. RPKM >1 in stage 11 control (no cut-off was used for st.11 control of H3K27me3 rep.1), stage 11 α-amanitin, stage 10.5 (H3K4me3 and p300) or stage 12 (H3K27me3). Common lost, gained, unbiased and maintained peaks are present in both replicates. All other peaks are considered not defined (ND). Replicate-specific peaks were only used for Supplementary Fig. 4b, for all other figures the common peaks were used.

DNA methylation levels in Supplementary Fig. 4d was calculated using previously published Bio-CAP data[62]. Bio-CAP RPKM levels of stage 11–12 were calculated for H3K4me3, H3K27me3 and p300 peaks, and corrected for input values. For Fig. 4c genes were considered 'hypomethylated' if the Bio-CAP/Input ratio on the promoter (±1 kb from TSS) was >1.

RNA expression analysis was performed as previously published[35]. Embryonic transcripts were separated based on the clustering of maximum expression levels per stage in Fig. 3d of Paranjpe*et al.*[35] (cluster 1=blastula, cluster 5=gastrula, clusters 3 and 4=neurula, clusters 2 and 6=tailbud).

Enhancer clusters were called as previously described[43]. Enhancer Clusters are called per stage and merged to determine the total Enhancer Cluster region. Percentage of the EC region is calculated relative to the total Enhancer Cluster region.

MaD and ZyD Classification

MaD peaks emerge at or before stage 11 and are also acquired in α-amanitin treated embryos in both replicates. Zygotically defined (ZyD) peaks appear at or before stage 11 and are lost in α-amanitin treated embryos in both replicates, or emerge after stage

11. To classify MaD and ZyD H3K4me3 genes we ran MAnorm on promoters (±250 bp from TSS) only, using similar restrictions as described in 'Detection of enriched regions'. MaD H3K4me3 genes have a maintained promoter in both replicates, ZyD H3K4me3 genes have a lost promoter H3K4me3 peak in both α-amanitin replicates, or a peak that emerges after stage 11. MaD H3K27me3 genes have at least one MaD peak in the vicinity of their promoter (±2.5 kb from TSS). ZyD H3K27me3 genes have at least one ZyD peak in their promoter and lack a MaD peak. ND peaks or genes do meet the criteria for neither MaD nor ZyD. For p300 the total number of ZyD and MaD peaks was counted in GREAT[63] regions of genes.

ChomHMM Analysis

Chromatin states were discovered and characterized using ChromHMM v1.10 (ref. 26), an implementation of a hidden Markov model. As input we used the enriched regions from ten tracks (H3K27me3, H3K36me3, H3K4me1, H3K4me3, H3K9ac, H3K9me2, H3K9me3, H4K20me3, p300 and RNAPII) across five developmental stages. We trained and ran the model with a range of states, and determined the 19 emission states model as the optimal number of states that could sufficiently capture the biological variation in co-occurrence of chromatin marks. We subsequently classified the states into seven main groups based on the presence and absence of specific chromatin marks.

The segmentation files of the seven main groups per stage were binned in 200 base pairs intervals. An $m \times n$ matrix was created, where m corresponds to the 200 base pair intervals and $n$to the developmental stages (9–30). Each element $a(i,j)$ represents the chromatin state of interval $i$at stage j. For each chromatin group occurrences were counted per stage n. The changes between stage n and n+1 were plotted using Sankey diagrams (https://github.com/tamc/Sankey), a flow diagram closely related to alluvial diagrams.

Motif Analyses

For the prediction of motif contribution to p300 recruitment (Fig. 5a) we have implemented the ISMARA method developed by Balwierz *et al.*[64] This method uses motif activity response analysis to determine the transcription factors that drive the observed changes in chromatin state across samples. As input we used the number of known motifs found per p300 binding site and the RPKM of the p300 peaks per developmental stage. The model infers the unknown motif activities from the equation in which the changes in signal levels are explained with the number of binding sites and the unknown motif activities. Motifs that showed a z-score activity that was >13 are shown in Fig. 5a. Enriched motifs (Supplementary Fig. 7) were detected with gimme diff, a tool from the GimmeMotifs package[65]. The vertebrate motifs used in this script were obtained from CISBP (http://cisbp.ccbr.utoronto.ca/)[66] and clustered using gimme cluster from GimmeMotifs. The motifs are available at http://dx.doi.org/10.6084/m9.figshare.1555851 (ref. 67).

Generation of Plots and Heatmaps

All heatmaps were generated using fluff (http://simonvh.github.com/fluff)[13] or gplots (http://cran.r-project.org/web/packages/gplots/index.html). For all heatmap clustering, the Euclidean distance metric was used. Other plots were generated using ggplot2 (http://ggplot2.org/).

REFERENCES

1. Mikkelsen, T. S. et al. Genome-wide maps of chromatin state in pluripotent and lineage-committed cells. Nature 448, 553–560 (2007).
2. Strahl, B. D. & Allis, C. D. The language of covalent histone modifications. Nature 403, 41–45 (2000).
3. Lee, J. S., Smith, E. & Shilatifard, A. The language of histone crosstalk. Cell 142, 682–685 (2010).
4. Jenuwein, T. & Allis, C. D. Translating the histone code. Science 293, 1074–1080 (2001).
5. Gerstein, M. B. et al. Integrative analysis of the Caenorhabditis elegans genome by the modENCODE project. Science 330, 1775–1787 (2010).
6. Roy, S. et al. Identification of functional elements and regulatory circuits by Drosophila modENCODE. Science 330, 1787–1797 (2010).
7. Tadros, W. & Lipshitz, H. D. The maternal-to-zygotic transition: a play in two acts. Development 136, 3033–3042 (2009).
8. Kane, D. A. & Kimmel, C. B. The zebrafish midblastula transition. Development 119, 447–456 (1993).
9. O'Farrell, P. H., Stumpff, J. & Su, T. T. Embryonic cleavage cycles: how is a mouse like a fly? Curr. Biol. 14, R35–R45 (2004).
10. Paranjpe, S. S. & Veenstra, G. J. Establishing pluripotency in early development. Biochim. Biophys. Acta 1849, 626–636 (2015).
11. Newport, J. & Kirschner, M. A major developmental transition in early Xenopus embryos: I. characterization and timing of cellular changes at the midblastula stage. Cell 30, 675–686 (1982).
12. Newport, J. & Kirschner, M. A major developmental transition in early Xenopus embryos: II. Control of the onset of transcription. Cell 30, 687–696 (1982).
13. van Heeringen, S. J. et al. Principles of nucleation of H3K27 methylation during embryonic development. Genome Res. 24, 401–410 (2014).
14. Akkers, R. C. et al. A hierarchy of H3K4me3 and H3K27me3 acquisition in spatial gene regulation in Xenopus embryos. Dev. Cell 17, 425–434 (2009).
15. Vastenhouw, N. L. et al. Chromatin signature of embryonic pluripotency is established during genome activation. Nature 464, 922–926 (2010).
16. Lindeman, L. C. et al. Prepatterning of developmental gene expression by modified histones before zygotic genome activation. Dev. Cell 21, 993–1004

(2011).

17. Kim, T. K. et al. Widespread transcription at neuronal activity-regulated enhancers. Nature 465, 182–187 (2010).
18. Visel, A. et al. ChIP-seq accurately predicts tissue-specific activity of enhancers. Nature 457, 854–858 (2009).
19. Heintzman, N. D. et al. Histone modifications at human enhancers reflect global cell-type-specific gene expression. Nature 459, 108–112 (2009).
20. Heintzman, N. D. et al. Distinct and predictive chromatin signatures of transcriptional promoters and enhancers in the human genome. Nat. Genet. 39, 311–318 (2007).
21. Roh, T. Y., Wei, G., Farrell, C. M. & Zhao, K. Genome-wide prediction of conserved and nonconserved enhancers by histone acetylation patterns. Genome Res. 17, 74–81 (2007).
22. Bernstein, B. E. et al. Genomic maps and comparative analysis of histone modifications in human and mouse. Cell 120, 169–181 (2005).
23. Santos-Rosa, H. et al. Active genes are tri-methylated at K4 of histone H3. Nature 419, 407–411 (2002).
24. Barski, A. et al. High-resolution profiling of histone methylations in the human genome. Cell 129, 823–837 (2007).
25. Schotta, G. et al. A silencing pathway to induce H3-K9 and H4-K20 trimethylation at constitutive heterochromatin. Genes Dev. 18, 1251–1262 (2004).
26. Ernst, J. & Kellis, M. ChromHMM: automating chromatin-state discovery and characterization. Nat. Methods 9, 215–216 (2012).
27. Chafin, D. R., Guo, H. & Price, D. H. Action of alpha-amanitin during pyrophosphorolysis and elongation by RNA polymerase II. J. Biol. Chem. 270, 19114–19119 (1995).
28. Sible, J. C., Anderson, J. A., Lewellyn, A. L. & Maller, J. L. Zygotic transcription is required to block a maternal program of apoptosis in Xenopus embryos. Dev. Biol. 189, 335–346 (1997).
29. Skirkanich, J., Luxardi, G., Yang, J., Kodjabachian, L. & Klein, P. S. An essential role for transcription before the MBT in Xenopus laevis. Dev. Biol. 357, 478–491 (2011).
30. Clouaire, T. et al. Cfp1 integrates both CpG content and gene activity for accurate H3K4me3 deposition in embryonic stem cells. Genes Dev. 26, 1714–1728 (2012).
31. Clouaire, T., Webb, S. & Bird, A. Cfp1 is required for gene expressiondependent H3K4 trimethylation and H3K9 acetylation in embryonic stem cells. Genome Biol. 15, 451 (2014).
32. Thomson, J. P. et al. CpG islands influence chromatin structure via the CpG-binding protein Cfp1. Nature 464, 1082–1086 (2010). 33. van Kruijsbergen, I., Hontelez, S. & Veenstra, G. J. Recruiting polycomb to chromatin. Int. J. Biochem. Cell Biol. 67, 177–187 (2015).

33. Ng, H. H., Robert, F., Young, R. A. & Struhl, K. Targeted recruitment of Set1 histone methylase by elongating Pol II provides a localized mark and memory of recent transcriptional activity. Mol. Cell 11, 709–719 (2003).

34. Paranjpe, S. S., Jacobi, U. G., van Heeringen, S. J. & Veenstra, G. J. A genomewide survey of maternal and embryonic transcripts during Xenopus tropicalis development. BMC Genomics 14, 762 (2013).

35. Lauberth, S. M. et al. H3K4me3 interactions with TAF3 regulate preinitiation complex assembly and selective gene activation. Cell 152, 1021–1036 (2013).

36. Vermeulen, M. et al. Selective anchoring of TFIID to nucleosomes by trimethylation of histone H3 lysine 4. Cell 131, 58–69 (2007).

37. Sims, 3rd R. J. et al. Recognition of trimethylated histone H3 lysine 4 facilitates the recruitment of transcription postinitiation factors and pre-mRNA splicing. Mol. Cell 28, 665–676 (2007).

38. Gentsch, G. E. et al. In vivo T-box transcription factor profiling reveals joint regulation of embryonic neuromesodermal bipotency. Cell Rep. 4, 1185–1196 (2013).

39. Yasuoka, Y. et al. Occupancy of tissue-specific cis-regulatory modules by Otx2 and TLE/Groucho for embryonic head specification. Nat. Commun. 5, 4322 (2014).

40. Chiu, W. T. et al. Genome-wide view of TGFbeta/Foxh1 regulation of the early mesendoderm program. Development 141, 4537–4547 (2014).

41. Pott, S. & Lieb, J. D. What are super-enhancers? Nat. Genet. 47, 8–12 (2015).

42. Whyte, W. A. et al. Master transcription factors and mediator establish super-enhancers at key cell identity genes. Cell 153, 307–319 (2013).

43. Hnisz, D. et al. Super-enhancers in the control of cell identity and disease. Cell 155, 934–947 (2013).

44. Blythe, S. A., Cha, S. W., Tadjuidje, E., Heasman, J. & Klein, P. S. beta-Catenin primes organizer gene expression by recruiting a histone H3 arginine 8 methyltransferase, Prmt2. Dev. Cell 19, 220–231 (2010).

45. Bogdanovic, O. et al. Temporal uncoupling of the DNA methylome and transcriptional repression during embryogenesis. Genome Res. 21, 1313–1327 (2011).

46. Potok, M. E., Nix, D. A., Parnell, T. J. & Cairns, B. R. Reprogramming the maternal zebrafish genome after fertilization to match the paternal methylation pattern. Cell 153, 759–772 (2013).

47. Jiang, L. et al. Sperm, but not oocyte, DNA methylome is inherited by zebrafish early embryos. Cell 153, 773–784 (2013).

48. Lawrence, M. S. et al. Discovery and saturation analysis of cancer genes across 21 tumour types. Nature 505, 495–501 (2014).

49. Huether, R. et al. The landscape of somatic mutations in epigenetic regulators across 1,000 paediatric cancer genomes. Nat. Commun. 5, 3630 (2014).

50. Wu, G. et al. The genomic landscape of diffuse intrinsic pontine glioma and pediatric non-brainstem high-grade glioma. Nat. Genet. 46, 444–450 (2014).

51. Simmer, F. et al. Comparative genome-wide DNA methylation analysis of colorectal tumor and matched normal tissues. Epigenetics 7, 1355–1367 (2012).

52. Brinkman, A. B. et al. Sequential ChIP-bisulfite sequencing enables direct genome-scale investigation of chromatin and DNA methylation cross-talk. Genome Res. 22, 1128–1138 (2012).

53. Jallow, Z., Jacobi, U. G., Weeks, D. L., Dawid, I. B. & Veenstra, G. J. Specialized and redundant roles of TBP and a vertebrate-specific TBP paralog in embryonic gene regulation in Xenopus. Proc. Natl Acad. Sci. USA 101, 13525–13530 (2004).

54. Akkers, R. C., Jacobi, U. G. & Veenstra, G. J. Chromatin immunoprecipitation analysis of Xenopus embryos. Methods Mol. Biol. 917, 279–292 (2012).

55. Bogdanovic, O., Fernandez-Minan, A., Tena, J. J., de la Calle-Mustienes, E. & Gomez-Skarmeta, J. L. The developmental epigenomics toolbox: ChIP-seq and MethylCap-seq profiling of early zebrafish embryos. Methods 62, 207–215 (2013).

56. Lister, R. et al. Global epigenomic reconfiguration during mammalian brain development. Science 341, 1237905 (2013).

57. Lister, R. et al. Hotspots of aberrant epigenomic reprogramming in human induced pluripotent stem cells. Nature 471, 68–73 (2011).

58. Zhang, Y. et al. Model-based analysis of ChIP-Seq (MACS). Genome Biol. 9, R137 (2008).

59. Kharchenko, P. V., Tolstorukov, M. Y. & Park, P. J. Design and analysis of ChIP-seq experiments for DNA-binding proteins. Nat. Biotechnol. 26, 1351–1359 (2008).

60. Shao, Z., Zhang, Y., Yuan, G. C., Orkin, S. H. & Waxman, D. J. MAnorm: a robust model for quantitative comparison of ChIP-Seq data sets. Genome Biol. 13, R16 (2012).

61. Long, H. K. et al. Epigenetic conservation at gene regulatory elements revealed by non-methylated DNA profiling in seven vertebrates. eLife 2, e00348 (2013).

62. McLean, C. Y. et al. GREAT improves functional interpretation of cis-regulatory regions. Nat. Biotechnol. 28, 495–501 (2010).

63. Balwierz, P. J. et al. ISMARA: automated modeling of genomic signals as a democracy of regulatory motifs. Genome Res. 24, 869–884 (2014).

64. van Heeringen, S. J. & Veenstra, G. J. GimmeMotifs: a de novo motif prediction pipeline for ChIP-sequencing experiments. Bioinformatics 27, 270–271 (2011).

65. Weirauch, M. T. et al. Determination and inference of eukaryotic transcription factor sequence specificity. Cell 158, 1431–1443 (2014).

66. van Heeringen, S. J. et al.Vertebrate motif clusters v3.0. figshare http:// dx.doi.org/10.6084/m9.figshare.1555851 (2015).

9

CHAPTER

THE EVX1/EVX1AS GENE LOCUS REGULATES ANTERIOR-POSTERIOR PATTERNING DURING GASTRULATION

Charles C. Bell[1], Paulo P.Amara[l2], Anton Kalsbeek[2,3], GrahamW. Magor[1], Kevin R.Gillinder[1], PierreTangermann[2], Lorena di Lisio1, SethW. Cheetham[1,4], FranziskaGruhl[1,4], Jessica Frith[2,5], Michael R.Tallack[1,2], Ke-Lin Ru[2,5], JoannaCrawford[2], John S. Mattick[3,6], Marcel E. Dinger[3,4,6] & AndrewC. Perkins[1,2,7]

[1]Mater Research, Translational Research Institute, University of Queensland, Brisbane, Queensland, 4101, Australia

[2]The Institute for Molecular Bioscience, University of Queensland, Brisbane, Queensland, Australia

[3] Garvan Institute of Medical Research, Sydney, Australia

[4] Diamantina Institute; Translational Research Institute, University of Queensland, Brisbane, Queensland, 4102, Australia

[5] The Australian Institute for Bioengineering and Nanotechnology, University of Queensland, Brisbane, Queensland, 4102, Australia

[6] St Vincents Clinical School, Faculty of Medicine, UNSW Australia, Sydney, Australia

[7] The Princess Alexandra Hospital, Brisbane, Queensland, 4102, Australia

ABSTRACT

Thousands of sense-antisense mRNA-lncRNA gene pairs occur in the mammalian genome. While there is usually little doubt about the function of the coding transcript, the function of the lncRNA partner is mostly untested. Here we examine the function of the homeotic *Evx1-Evx1as* gene locus. Expression is tightly co-regulated in posterior mesoderm of mouse embryos and in embryoid bodies. Expression of both genes is enhanced by BMP4 and WNT3A, and reduced by Activin. We generated a suite of deletions in the locus by CRISPR-Cas9 editing. We show EVX1 is a critical downstream effector of BMP4 and WNT3A with respect to patterning of posterior mesoderm. The lncRNA, *Evx1as*arises from alternative

Citation: Charles C. Bell, Paulo P. Amaral, Anton Kalsbeek, Graham W. Magor, Kevin R. Gillinder, Pierre Tangermann, Lorena di Lisio, Seth W. Cheetham, Franziska Gruhl, Jessica Frith, Michael R. Tallack, Ke-Lin Ru, Joanna Crawford, John S. Mattick, Marcel E. Dinger & Andrew C. Perkins, The Evx1/Evx1as gene locus regulates anterior-posterior patterning during gastrulation, doi:10.1038/srep26657.

promoters and is difficult to fully abrogate by gene editing or siRNA approaches. Nevertheless, we were able to generate a large 2.6 kb deletion encompassing the shared promoter with *Evx1* and multiple additional exons of *Evx1as*. This led to an identical dorsal-ventral patterning defect to that generated by micro-deletion in the DNA-binding domain of EVX1. Thus, *Evx1as* has no function independent of EVX1, and is therefore unlikely to act *in trans*. We predict many antisense lncRNAs have no specific *trans* function, possibly only regulating the linked coding genes in *cis*.

INTRODUCTION

Gastrulation is a critical early developmental process in which mesoderm and definitive endoderm are generated, and then specified. The three germ layers (ectoderm, mesoderm and definitive endoderm) eventually form all of the tissues of the adult organism. In the mouse, a proximal-distal (P-D) axis is established first from ~E5.0 in response to nodal signalling, then an anterior-posterior (A-P) axis is established during gastrulation in response to BMP, FGF and WNT morphogen gradients1·2. Specification of epiblast cells into mesendodermal cell fates occurs as they migrate through the primitive streak and activate transcriptional programs in response to these gradients3. The cells which migrate first are specified to a posterior fate in response to high local concentrations of BMPs, WNTs and FGFs. These cells form blood, the vasculature and heart. Cells which migrate last through the streak take on an anterior fate in response to high local concentrations of activin/nodal morphogens and inhibitors of BMPs and WNTs such as Chordin, Noggin, Dkk1, Cerberus-like 1 (Cer1) and Lefty 1 and 22. They form notochord, somites and definitive endoderm. Cells from the mid-streak region form intermediate mesoderm derivatives such as kidney.

Hox genes have an established role in patterning during bilaterian embryonic development4. *Evx1* is located 50 kb downstream of the HoxA cluster5. Individual members of this cluster have graded anterior boundaries of expression and control rostral-caudal neural fates4. *Evx1*is expressed in the posterior primitive streak from ~E6.56, which is earlier in development than any members of the HoxA cluster7. Therefore, it is considered not to be co-regulated with the HoxA cluster during gastrulation. In Xenopus and zebrafish, the homologs of EVX1 (*xhox3, eve1*) play roles in 'posteriorization' of nascent mesoderm8·9. In humans, EVX1 is thought to function in a regulatory network with GOOSECOID (GSC) and BRACHYURY (T) to control anterior-posterior cell fates10. Recent evidence also points to it being a direct target of the WNT signalling pathway, as β-catenin binds to the *EVX1* promoter in differentiating hESCs11. Other related members of 'non-clustered' Hox gene families play master regulatory roles in A-P and dorsal-ventral (D-V) patterning in *Xenopus laevis*, zebrafish and mammals. For example,*Mixl1* is essential for posterior fate specification in mice12, and many Mix/Bix family members are required for ventral specification in response to BMPs in frogs13·14·15·16·17. *Mezzo*18, *dharma*/*bozozok*19 and*Mtx1*20 are required for formation or patterning of mesendoerm in zebrafish, and goosecoid or related anteriorly-expressed genes are required for anterior specification in most vertebrate species21.

Initially, EVX1 was thought to be critical for murine development, as EVX1$^{-/-}$ embryos were reported as failing to implant in the uterine wall22. However, other groups subsequently demonstrated that EVX1 is not required for viable embryonic development and suggested the initial EVX1$^{-/-}$ phenotype was due to disruption of another closely linked gene23 (Gail Martin, personal communication). In fact, EVX1$^{-/-}$mice display no overt signs of gastrulation defects after birth, other than minor tail kinks in some offspring and genetic backgrounds (Gail Martin, personal communication). Therefore, the importance of EVX1 in the regulation of gastrulation remains controversial.

Evx1 is expressed from a complex locus, which also expresses a long non-coding RNA (lncRNA), known as *Evx1as*5. There are other well studied lncRNAs associated with the HoxA cluster, including*Haunt/Halr1* which is located ~40 kb upstream of *Hoxa1*24, and *HOTTIP*which is immediately downstream of *Hoxa13*, the most 3′ member of the cluster25 (Fig. 1). Both have complex functions in global regulation or fine tuning of expression of the HoxA cluster. LncRNAs are an emerging class of regulatory molecules which have been implicated in a number of developmental processes26·27·28·29. They have been proposed to function via regulation of transcription, splicing, and chromatin dynamics. Antisense/bidirectional transcription of protein coding genes and lncRNAs is a common feature in the mammalian genome, in particular at highly expressed and developmentally regulated genes30. We have previously reported *Evx1* and *Evx1as* are dynamically and concomitantly expressed during embryoid body (EB) differentiation5, a commonly used model of early *in vivo* embryonic development and the function of lncRNAs24·31. *Evx1as* is abundant and has stability similar to*Evx1* transcripts, suggesting function5.

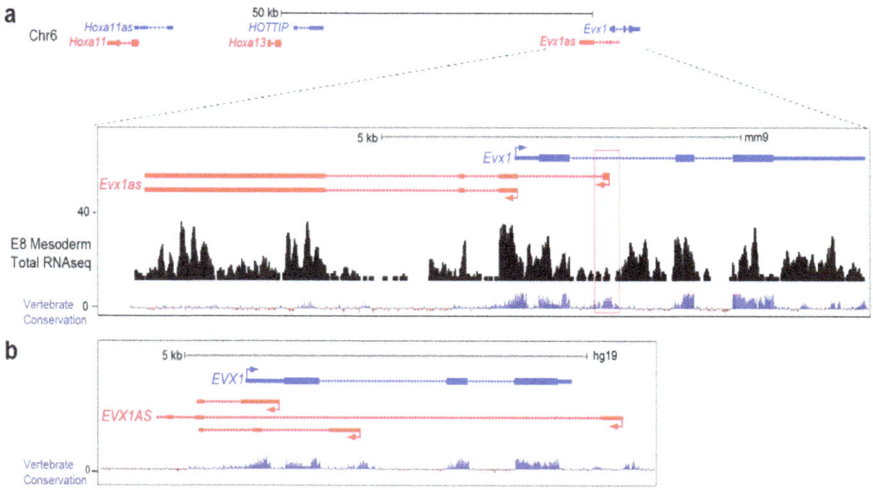

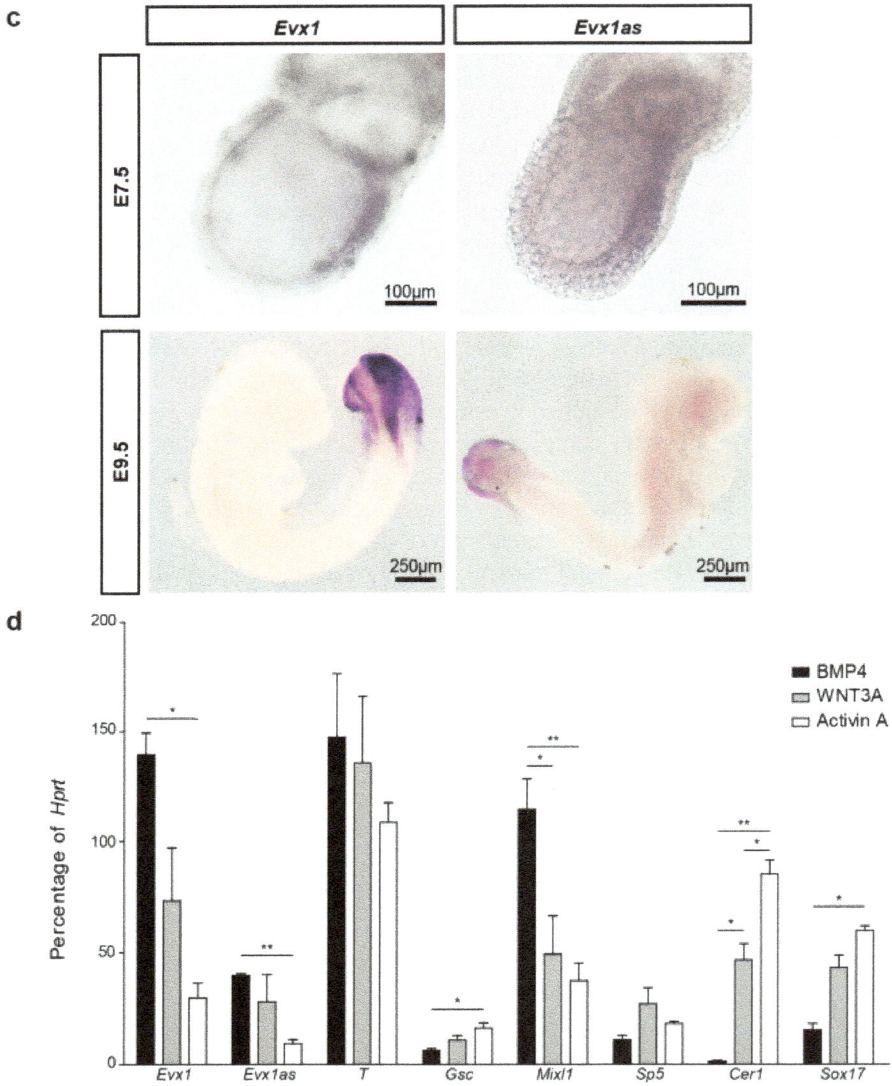

Figure 1: ***Evx1*** **and** ***Evx1as*** **are co-expressed in the primitive streak during gastrulation.**

(**a**) Schematic of the *Evx1/Evx1as* locus, and its proximity to the HoxA cluster modified from the UCSC genome browser. Wiggle track of Total RNAseq from E8 mouse Pre-Somatic Mesoderm (PSM) and Vertebrate Conservation tracks are also shown. Conservation of the*Evx1as* P1 region is boxed in red. (**b**) UCSC browser shot of the human *EVX1* locus and the syntenic *EVX1AS* transcripts. (**c**) WISH of E7.5 and E9.5 mouse embryos using probes against*Evx1* and *Evx1as*. Purple indicates the presence of the transcript. (**d**) Expression profiling of mESCs differentiated to day 4 EBs under serum-free conditions with the addition of BMP4 (10 ng/ml), WNT3A (20 ng/ml) or Activin A (10 ng/ml). Three replicates were performed for each condition. *Indicates p-value < 0.05, **indicates p-value < 0.01. All error bars indicate SEM.

We sought to dissect the *Evx1/Evx1as* locus in order to elucidate the potential role for EVX1 and/or *Evx1as* during gastrulation, and to gain insights into the broader functional significance of antisense/bidirectional lncRNAs. To investigate the function of EVX1, we generated bi-allelic small frameshift deletions in the homeodomain-encoding region using CRISPR/Cas9. We generated stable murine ES cell clones and performed RNAseq at day 4 of embryoid body (EB) differentiation in direct comparison with parental non-edited ES cells. We found that disruption of EVX1 results in upregulation of anterior visceral endoderm (AVE) and definitive endoderm genes including *Cer1*and *Sox17*, at the expense of posterior genes, such as *Mixl1* and *Kdr* (Flk-1). We also show EVX1 is likely to function as a downstream effector of BMP4 and WNT signalling pathways, to regulate posterior tissue patterning.

To test whether *Evx1as* has a function independent of EVX1, we generated a suite of CRISPR-Cas9 mediated deletions using a similar approach to that recently reported for *Haunt*24. With all of these DNA manipulations, we were not able to disrupt expression of*Evx1as* without also disrupting expression of *Evx1*. We performed RNAseq to convincingly show loss of EVX1 and *Evx1as* produced identical aberrations in A-P gene expression patterns to those which we observed in the EVX1 loss-of-function cell lines. Together, our results strongly suggest there is no independent function for *Evx1as* beyond that of EVX1. However, we cannot rule out a function for *Evx1as* in the regulation of *Evx1 in cis*.

RESULTS

Organization, Conservation and Expression of the *Evx1/Evx1as* Locus

The *Evx1* locus is located 50 kb downstream of the HoxA gene cluster on chromosome 6 (Fig. 1a). *Evx1* and *Evx1as* are developmentally regulated, displaying peak and concordant expression during gastrulation5. They are both also highly expressed *in vivo* in the pre-somitic mesoderm (PSM)32. Whole mount *in situ* hybridization (WISH) of E7.5 and E9.5 embryos demonstrates the *Evx1* and *Evx1as* are co-expressed in the primitive streak during gastrulation (Fig. 1c). At E7.5, both *Evx1* and*Evx1as* are expressed at the posterior-proximal side of the embryo, which is the location of the primitive streak. At E9.5, both transcripts localize to the tail bud, which contains the embryological remnants of the primitive streak. Thus, *Evx1* and *Evx1as* are co-expressed during gastrulation.

Like many lncRNA-coding gene pairs30, *Evx1* and *Evx1as* are expressed from opposite DNA strands in a sense-antisense configuration (Fig. 1a). Interestingly, there are two other lncRNAs within the distal end of the HoxA cluster, *Hox11as* and *HOTTIP*25, which are expressed in an antisense direction with respect to *Hox11* and *Hox13* coding genes, respectively (Fig. 1a). *Evx1as* is expressed as at least two different isoforms according to EST and depostied cDNA data; the second exon overlaps with the start of *Evx1* by ~70 bp and the first exon resides in the first intron of *Evx1* (Fig. 1a). We confirmed only 1% of the *Evx1as*transcript contains the first exon in D4 EBs (Supplementary Figure 1A). EVX1 is highly conserved throughout the protein coding sequence, whereas *Evx1as* displays notable conservation at the promoter (Fig. 1a), like many lncRNAs33. The human *EVX1-EVX1AS* locus is syntenic with the murine locus. *EVX1AS* has two conserved promoters (intronic or bidirectional), but is also expressed from one additional promoter, located 3′ of EVX1 (Fig. 1b).

EVX1 Regulates Anterior-Posterior Patterning During Gastrulation

In order to study the function of EVX1 and *Evx1as*, we used EB differentiation as a model for early *in vivo* mouse embryonic development. Day 4 EBs grown in serum or serum free media (SFM) with recombinant BMP4 display a very similar transcriptional program to embryos undergoing gastrulation7. *Evx1* and *Evx1as* were most highly expressed when differentiated under BMP4 and WNT3A conditions (Fig. 1d), which suggests that they might be either downstream targets of the BMP4 and WNT3A signalling pathways or that BMP4 and WNT3A enhance the generation of posterior mesoderm cell types, wherein the*Evx1* locus is expressed. *Evx1* and *Evx1as* were expressed at a much lower level in EBs grown in SFM with Activin A, which is an anterior/dorsal mesoderm inducing factor34. Our results are consistent with expression of human EVX1 in these same growth factors10. *Evx1*and *Evx1as* display a similar expression pattern to *Mixl1* (Fig. 1d), a related homeodomain transcription factor which is also expressed in nascent posterior mesoderm12. Expression of pan-mesoderm genes such as *Brachyury* (T) and *Sp5* was similar in BMP4, WNT3A and Activin A, while expression of anterior markers *Cer1* and *Sox17* was significantly higher in WNT3A and Activin A (Fig. 1d).

We generated EVX1 'knockout' cell lines using CRISPR-Cas9 gene micro-editing of the homeodomain coding region in exon 2, thus leaving *Evx1as* unperturbed (Fig. 3a). Two independent compound heterozygous *Evx1* gene edited clones were selected based on our previously described NGS screening strategy35. Both clones have small frameshift mutations in both alleles (Fig. 2a), which are predicted to result in a premature stop codons within 20 bps of the deletion site. Thus, no functional DNA-binding domain of EVX1 could be generated by either of these clones. We refer to these CRISPR-edited lines as 'EVX1-Δfs' from here on. EVX1-Δfs lines and W9.5 controls were differentiated in serum to D4 under suspension culture conditions to identify a potential regulatory role for EVX1 during gastrulation. EVX1-Δfs D4 EBs appeared phenotypically normal, and indistinguishable from the wildtype controls (Supplementary Figure 1B). However, qPCR analysis revealed that both EVX1-Δfs clones display differences in expression of key anterior and posterior patterning markers, including *Mixl1* and *Cer1*(Supplementary Figure 1C). There were no significant differences between the two clones, therefore only one was used for subsequent analyses (Clone 14) (Supplementary Figure 1C).

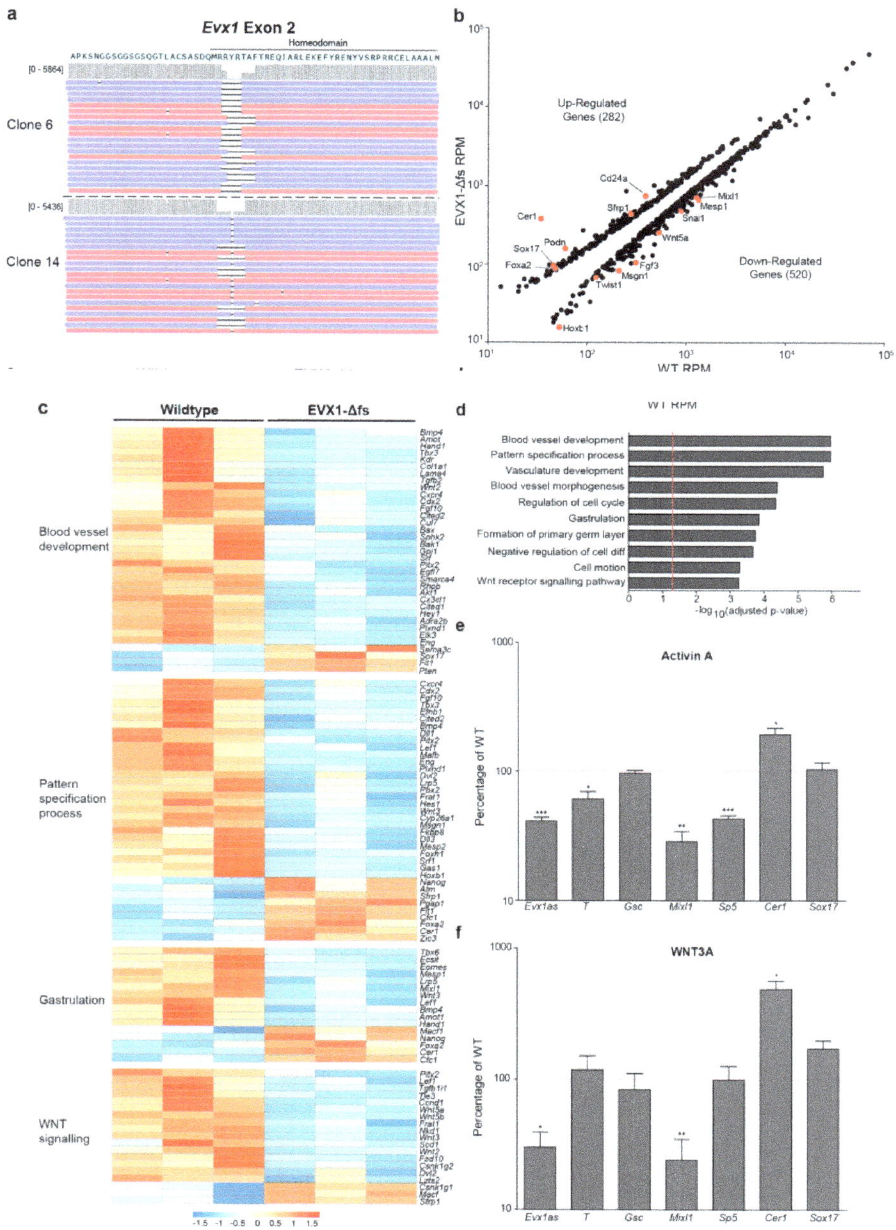

Figure 2: EVX1 is required to regulate anterior-posterior patterning during gastrulation.

(**a**) Visualization of CRISPR-induced mutations in two independent mESC clones (clone 6 and 14) using Integrated Genome Viewer (IGV). The peptide sequence of EVX1 is shown. Red and blue indicate different read strands. (**b**) Scatterplot of average counts of DEGs from mRNAseq comparing 3 replicates of WT and EVX1-Δfs Day 4 EBs. Known markers and regulators of tissue specification are shown in red. RPM = Reads per million reads.(**c**) Heatmap of counts for differentially expressed genes corresponding to particular Gene Ontology (GO) terms in (**d**). Each category is hierarchically clustered. Each value is normalized to its mean expression across all samples. Red indicates higher than average expression, blue indicates lower than average expression. (**d**) GO analysis (DAVID) of DEGs shows the biological processes disrupted in the EVX1-Δfs D4 EBs. Red line indicates an adjusted p-value of 0.05. (**e,f**) Expression profiling of EVX1-Δfs in SFM supplemented with (**e**) Activin A (10 ng/ul) or (**f**) WNT3A (20 ng/ul). Expression of each gene in each sample was first normalized to *Hprt* then normalized to WT expression. 4 biological replicates were performed. *Indicates a p-value < 0.05, **indicates a p-value < 0.01, ***indicates a p-value < 0.001 when compared to WT. All error bars indicate SEM.

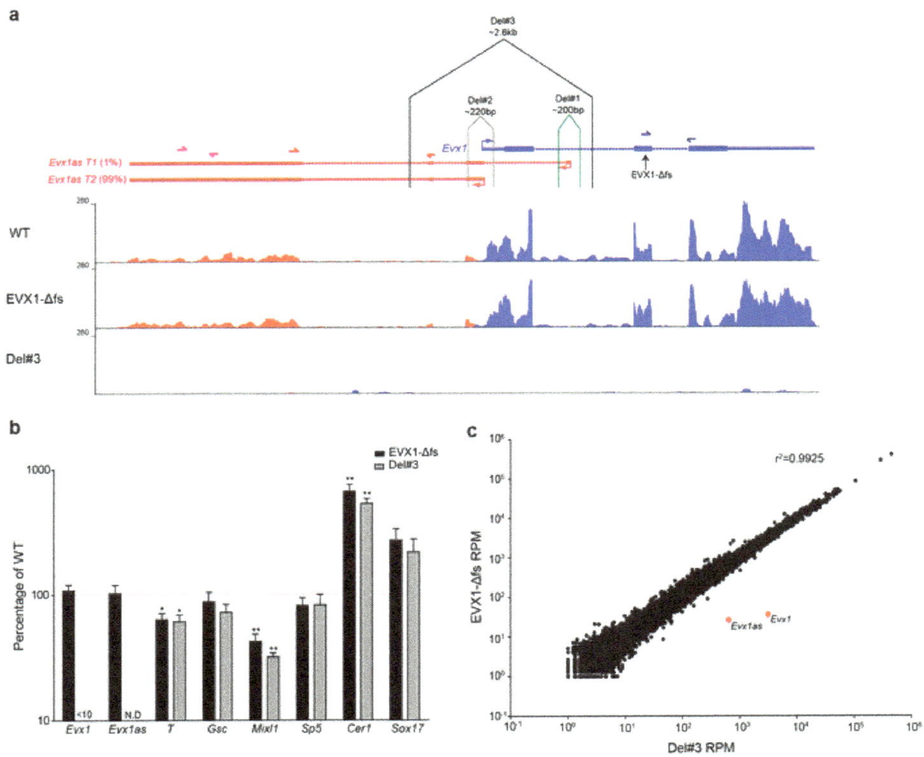

Figure 3: ***Evx1as*** **does not have a function independent of EVX1.**

(**a**) Schematic of the strategy for dissecting the *Evx1*/*Evx1as* locus modified from USCS Genome Browser. The three different deletions performed are shown on the diagram. Wiggle tracks of WT, EVX1-Δfs and Del#3 mRNAseq from D4 EBs are shown. Corresponding qPCR primers pairs are shown in the same colour. The location of the *Evx1* deletion is indicated by the arrow. Del = deletion, T1 = transcript 1,

T2 = transcript 2. (**b**) Expression profiling of EVX1-Δfs and Del#3. Expression of each gene in each sample was first normalized to *Hprt* then normalized to WT expression. 4 biological replicates were performed. Error bars show SEM. *Indicates a p-value < 0.05, **indicates a p-value < 0.01, when compared to WT. No significant differences were found when comparing EVX1-Δfs and Del#3. N.D = not detected. <10 = less than 10% of WT. (**c**) Scatter plot of average mRNAseq counts from three replicates comparing EVX1-Δfs and Del #3 D4 EBs. *Evx1* and *Evx1as* are shown in red. A value of 1 was added to all RPM values to improve visualization. R^2 was obtained from Pearson Correlation. RPM = Reads per million reads.

To determine whether EVX1 has a global role in anterior-posterior patterning, mRNA-seq was performed on three biological replicates of W9.5 and EVX1-Δfs EBs differentiated for 4 days in serum without LIF (see Methods). Using DESeq2, we identified 802 differentially expressed genes (DEGs). In the absence of functional EVX1, 282 genes were upregulated and 520 genes were downregulated (Fig. 2b). The upregulated genes included many which are preferentially expressed in anterior visceral endoderm and/or anterior mesendoderm (e.g. *Sox17, Cer1*, and *Foxa2*). Many of the down regulated genes are normally expressed in posterior mesoderm (e.g. *Mixl1, Mesp1, Wnt5a* and *Fgf3*) (Fig. 2b). This suggests there is an imbalance in cellular composition of EVX1-Δfs EBs with loss of posterior cell types and relative expansion of anterior cell types (both mesoderm and endoderm).

Using Gene Ontology (GO) we found the most highly significantly enriched cell biological process terms among the DEGs were: blood vessel development (35) and morphogenesis (28), pattern specification (38), regulation of the cell cycle (29), gastrulation (16) or formation of primary germ layer (12), cell differentiation (25), cell motion (37) and WNT signalling (20) (Fig. 2d). We undertook hierarchical clustering of the genes within each of the GO Ontology categories (Fig. 2c). Each category contains genes down-regulated (blue) or up-regulated (red) in EVX1-Δfs cells. Within the blood vessel development, patterning and gastrulation categories, there is down-regulation of many ligands, receptors, signalling molecules and transcription factors which are implicated in posterior patterning of mesoderm. These include *Bmp4, Wnt2, Fgf10, Kdr (Flk1*), *Cxcr4, Cited1, Cited2, Cdx2, Hand1, Mixl1, Hes1, Tbx3, Tbx6, Mesp1, Eomes, Snai1* and others. There is also upregulation of a smaller set of key patterning genes which are normally expressed in the AVE, anterior mesendoderm or epiblast. These include *Sox17, Nanog, Atm, Zic3*, and *Foxa2*. We also found many genes in the WNT signalling pathway are downregulated in EVX1-Δfs EBs, including *Wnt3, Wnt2, Wnt5a, Wnt5b, Lef1, Pitx2, Dvl2, Lrp5* and others. On the other hand, WNT pathway antagonists such as *Cer1, Cfc1/cripto* and *Sfrp1* are upregulated. Interestingly, *Evx1* is downregulated in WNT3A KO embryos36, demonstrating that there is a mutual dependence (direct or indirect) between WNT signalling and EVX1. There were no significant changes in expression of any members of the HoxA cluster under these differentiation conditions. Full lists of differentially expressed genes are provided in Supplementary Table 2.

Previously EVX1 has been shown to directly repress GSC in human ESCs under Activin A growth conditions10. *Gsc* was not upregulated in either EVX1-Δfs clone by qPCR or RNAseq under serum differentiation conditions (Supplemental Table 2, Supplementary Figure 1C). To ensure that the lack of *Gsc* upregulation was not due to differences in the differentiation conditions used, we differentiated EVX1-Δfs and W9.5 EBs in SFM with Activin A. We did not find significant upregulation of*Gsc* (Fig. 2e). Importantly, the A-P gene expression defect in EVX1-Δfs persisted in Activin A. Defective expression of *Mixl1* and *T* remained, and additional significant reduction in expression of the pan mesoderm marker *Sp5* was also present in EVX1-Δfs EBs (Fig. 2e). The upregulation of AVE and endoderm markers *Cer1* and *Sox17* was less dramatic in Activin A, presumably because selective expansion of AVE and anterior tissues dilutes the differences in anterior gene expression between wildtype and EVX1-Δfs EBs (Fig. 2f).

From this RNAseq data, it was unclear whether EVX1 exerts its A-P patterning function by directly regulating BMP4 and/or WNT signalling pathways or whether changes in *Bmp4/Wnt* expression are an indirect result of changes in cellular composition within the EVX1-Δfs EBs. BMP4 and WNT direct EB differentiation towards posterior cell types37, so the aberrant transcriptome in EVX1-Δfs EBs may be primarily due to downregulation of these factors. To address this possibility, we attempted to rescue EVX1-Δfs EBs by culture in SFM with exogenous recombinant WNT3A (20 ng/ml) or BMP4 (10 ng/ml). Under WNT3A conditions, we observed no rescue of *Mixl1* gene expression; i.e. it remained significantly down regulated (Fig. 2f). Likewise there was no reversal of the up-regulation of *Cer1*, suggesting the A-P patterning defect was not rescued by WNT3A (Fig. 2f). Similar results were found for EBs grown in BMP4 (Supplementary Figure 1D). Thus, downregulation of *Bmp4* and *Wnt* genes (*Wnt3, Wnt5a* and *Wnt5b*) is likely to be an indirect result of changes in cellular composition of EVX1-Δfs EBs and not due to direct regulation of these factors by EVX1. This data also shows EVX1 is required for appropriate transcriptional responses and differentiation downstream of BMP4 and WNT signalling pathways. This function appears to be conserved in humans, as β-catenin, the major effector of WNT signalling, has been shown to bind to regulatory regions in the *EVX1* promoter in differentiated hESCs11. Together, these results confirm EVX1 plays a key role in A-P patterning of nascent mesoderm and endoderm and is essential for regulating the posteriorizing effects of WNTs and BMP4.

CRISPR/Cas9 Mediated Removal of the *Evx1/Evx1as* Locus results in an Identical D-V Pattering Defect as EVX1-Δfs

Bidirectional/antisense transcription from highly expressed, developmentally regulated genes is a common phenomenon in the mammalian genome30. However the functional significance and possible mechanisms of action of such antisense lncRNAs remain topics of debate38,39. The *Evx1* locus contains a stable, highly expressed antisense lncRNA associated with *Evx1*, so it is an ideal locus to study. Visual inspection of RNAseq data from D4 EBs suggests that the majority of the *Evx1as* transcript is expressed as *Evx1as* T2 (Fig. 3a). We further validated this by qPCR using primers to distinguish between the different isoforms of *Evx1as*. Only 1% of the total transcript derived from the overlapping intronic promoter (Supplementary Figure 1A).

We initially used RNAi, delivered via shRNA containing lentiviral vectors40 to attempt stable knockdown of *Evx1as* without disturbing the genomic locus (Supplementary Figure 2A) (see Methods). Such an approach has been successful for some lncRNAs but challenging for others41·42. In some cases RNAi has obtained different and more severe phenotypes than constitutive deletions, which may be due to off target or toxicity effects in some cases or due to compensatory events in others43·44. We generated stable clones targeting *Evx1as* using three independent shRNAs (Supplementary Figure 2A). Unfortunately, none of the clones from any of the three siRNAs displayed knockdown of *Evx1as*to <50% of WT levels, so we could not make concrete conclusions about the potential function of *Evx1as* from these experiments. There were no significant changes in expression of key mesoderm genes such as*Brachyury, Mixl1*, or *Evx1* itself (data not shown).

Genetic deletion or mutation remains the gold standard to demonstrate the requirement of a gene's function43. We therefore performed 3 different CRISPR-Cas9 mediated manipulations of the *Evx1/Evx1as*locus (Fig. 3a). The first manipulation 'Del#1' removed the entire first intronic promoter of *Evx1as*, which is specific to isoform *Evx1as* T1 (1% of total transcript) and conserved (Fig. 1a). Removal of this promoter, which generates a small proportion of *Evx1as* RNA, did not affect expression of *Evx1as* or *Evx1* (data not shown). 'Del#2' removed the shared promoter region between *Evx1* and *Evx1as*. This deletion removed 220 bps centred around the beginning of *Evx1* and *Evx1as*transcription, based on ESTs and RNAseq data. This deletion resulted in a dramatic loss of *Evx1* reducing the transcript to ~16% of WT levels (Supplementary Figure 2B). However this deletion only decreased*Evx1as* to ~46% of WT levels. This result may be explained by a more malleable transcriptional start site and/or weaker constraints on transcriptional regulation at the lncRNA promoter. As expected, 'Del#2' displayed reduction of *Mixl1* and significant upregulation of *Cer1*, supporting the results from the EVX1-Δfs RNAseq data (Supplemental Figure 2B).

Since *Evx1as* was still expressed at 46% of WT in 'Del#2', we could not draw any definitive conclusions about its potential function. Therefore, we generated a large ~2.6 kb deletion ('Del#3') that removed the promoters from both versions of the transcript (to prevent the possibility of isoform switching between the two alternative promoters), as well as the second/third exon (Fig. 3a). Successful deletion was validated by PCR and later by RNAseq (Fig. 3a,Supplementary Figure 2C). This deletion resulted in almost complete ablation of *Evx1* by qRT-PCR (Fig. 3b). Since the *Evx1as* qPCR primers sit inside the deletion site (Fig. 3a), primers were designed for the last exon of the transcript to detect any residual expression. Approximately 15% of WT levels of expression of the last exon of *Evx1as* remained in the 'Del#3' D4 EBs (Supplementary Figure 2D).

To determine whether *Evx1as* has any function independent of EVX1, we performed a direct comparison between the transcriptome of EVX1-Δfs and the *Evx1as/ Evx1* double KO ('Del#3'). We reasoned that if *Evx1as*has an independent function from EVX1, then differences should be observed between the two transcriptomes, regardless of whether it functions in transcriptional or post-transcriptional regulation. Transcriptional differences, if present, could be either due to direct transcriptional regulation by *Evx1as* or an indirect consequence of*Evx1as* regulating post-transcriptional pathways.

No significant differences were observed by qPCR between the two lines for a number of gastrulation and A-P patterning genes (Fig. 3b). Both the EVX1-Δfs and *Evx1/Evx1as* Δ2.6 kb ('Del#3') EBs displayed significantly different expression of *Mixl1, T* and *Cer1* relative to WT EBs (Fig. 3b). To determine whether there was any *trans* function for *Evx1as*RNA independent of the function of EVX1, we searched for differences genome-wide by mRNAseq. Remarkably, differential gene expression analysis identified only 3 DEGs, two of which were *Evx1* and *Evx1as*, themselves. In the *Evx1/Evx1as* double KO ('Del#3'), *Evx1* and *Evx1as* are dramatically reduced to 1% and 4% of EVX1-Δfs (in which *Evx1* and*Evx1as* are expressed the same as WT under serum conditions). The other DEG (*Grb10*) has no reported role during gastrulation and is unlikely to be of any biological significance. The lack of differences is reflected in the very high correlation observed between the two samples (R^2=0.9925, Pearson) (Fig. 3c). We therefore conclude that*Evx1as* does not have function independent of EVX1, and therefore is unlikely to function *in trans*. This does not rule out a potential function for *Evx1as* in the regulation of *Evx1 in cis* (see Discussion).

DISCUSSION

Anterior-posterior patterning determines cell fates and tissue specification during gastrulation. Although, EVX1$^{-/-}$ mice are viable and display no dramatic gastrulation phenotype after birth, we show EVX1 is an important downstream effector of BMP4 and WNT signalling in EBs, and regulates A-P patterning. Genes which are normally highly expressed on the posterior side of the embryo are markedly reduced in EVX1-Δfs EBs. Thus, many genes associated with blood, blood vessel and cardiac outcomes (e.g. *Kdr (Flk1), Hand1, Cited2, Tbx3*)45,46,47,48,49 are markedly downregulated in the absence of EVX1. On the other hand, genes expressed throughout the mesoderm such as *Brachyury (T)* and*Sp5*50 are only mildly affected. Genes expressed on the anterior-dorsal side of the embryo such as *Cer1* and *Sox17*51 are increased. This is consistent with the functions of EVX1 in *Xenopus laevis (xhox3)*, zebrafish (*eve1*) and humans (EVX1); i.e. they all play essential roles in posteriorization of nascent mesoderm8,9,10.

Members of the homeobox gene superfamily can act as either transcriptional activators and/or repressors depending on cofactors or context. EVX1 may function as a transcriptional activator of posterior mesoderm genes, such as *Mixl1*, and/or as a transcriptional repressor of anterior/dorsal genes such as *Sox17* and *Cer1*. Thus, it could function to maintain the integrity of the posterior region of the embryo by preventing ectopic expression of anterior genes. Indeed, such a role has been proposed for human EVX110. Both scenarios are possible and not mutually exclusive. For example, EVX1 might co-operate with transcriptional activators or repressors, or recruit co-activators and/or co-repressors in different contexts. ChIPseq for EVX1 would be able to distinguish these possibilities; however, no specific mouse EVX1 antibodies are available and we found human antibodies did not work in the mouse.

As is the case for EVX1, disruption of some other key regulators of gastrulation does not always result in strong gastrulation phenotypes in the intact mouse. For example, the GSC, SP5 and CER1 knockout mice display no overt signs of gastrulation defects, despite each of these genes having well established roles in regulating patterning in ES cells or model organisms36·52·53·54·55. The lack of an overt phenotype may be explained by genetic redundancy and/or compensation by other genes and pathways. This is the case for *Cer1*, which has an overlapping expression profile and redundant function with *Lefty1*56. Unlike either of the single knockout mice, the CER1$^{-/-}$/LEFTY1$^{-/-}$ mouse has an expanded primitive streak or multiple primitive streaks56. Whether redundancy explains the lack of phenotype in the EVX1$^{-/-}$ mice is unclear. *Evx1* has a paralog, *Evx2*, which is thought to have arisen from vertebrate specific duplication of *Evx1*57. It is similar to *Evx1* (82.5% at the nucleotide level), and only differs at one amino acid in the homeodomain58. However, *Evx2* is not expressed during gastrulation *in vivo*57 or in EBs, so it is unlikely to act redundantly with *Evx1*.

Compensation within key signalling pathways has been shown to 'buffer' mis-regulation of some genes during development and dampen phenotypic consequences of gene knockouts59·60·61·62. Since BMP, WNT, TGFβ and FGF signalling pathways act on a large number of genes, it is possible that changes in expression or activity of components of these pathways could compensate for the loss of EVX1 in the context of an entire embryo1·59·61. This does not appear to occur in EVX1-Δfs EBs, since BMP4 and WNT3A cannot rescue the posterior mesoderm gene expression program. Perhaps only in the context of an intact embryo, can compensation be fully active. Alternatively, compensation may occur at the level of alternated expression of transcription factors (such as *Mixl1*) *in vivo*, but not in EBs. In short, it remains unclear why EVX1$^{-/-}$ mice display minimal defects, despite a clear function for EVX1 in regulating A-P patterning in EBs.

LncRNAs are an important new class of genetic material with a variety of biochemical activities. Some have proven key regulatory roles in development and differentiation. The abundance of sense-antisense coding-lncRNA gene pairs, such as *Evx1/Evx1as*, is widely appreciated but the functional significance of the associated lncRNA transcript remains mostly untested. This is partly because functional studies have focused primarily on intergenic lncRNAs to avoid disrupting overlapping coding transcripts63. Certainly, some lncRNAs have well established functions *in trans*28·63. However, despite being stable, spliced, abundant and developmentally regulated, *Evx1as* does not function to regulate gene expression independently of EVX1; it is therefore unlikely to function *in trans*. This does not rule out a possible *cis* function for*Evx1as* in the regulation of *Evx1*. Indeed, other examples of *cis* acting lncRNAs have been reported5·25·64·65. Further coding-lncRNA pairs need be carefully dissected using precise genetic tools to develop general principles which accurately describe their biological functions. We suggest that it is likely that some lncRNAs may function *in trans* to exert broad effects on gene expression, some function in *cis* to regulate their coding partner, and some may simply be non-functional bi-products of intense transcriptional activity at the partner coding gene's promoter or enhancer.

METHODS

ESC CULTURE and Embryoid Body Differentiation

W9.5 mESCs were maintained on gelatanized plates in mESC media with 1000 U/ml LIF at 37 °C, 5% CO_2, as described previously7. To generate EBs, mESCs were plated at a density of 2×10^4 cells/ml on STAR™ low adherence plates and allowed to differentiate for 4 days in suspension. Serum free differentiation was conducted in ESGRO complete basal media (SF002, Millipore) as described previously66. Serum-free media was supplemented with recombinant BMP4 at 10 ng/ml, recombinant Activin A at 10 ng/ml and recombinant WNT3A at 20 ng/ml. BMP4, Activin A and WNT3A were purchased from R and D Biosystems. Differentiations performed for side by side comparisons between conditions were set up on the same day to minimize technical and biological variation. At least 3 differentiations (biological replicates) were performed for each comparison with each replicate set up on a different day.

RNA Extraction and qRT-PCR

RNA was extracted from cells harvested at D4 using Trizol™ (Thermo-Fischer) according to manufacturer's instructions. cDNA was synthesized using Superscript III™ (Thermo-Fischer) according to manufacturer's instructions using random hexamer primers. qRT-PCR was performed using SYBR-Green (Thermo-Fischer) on the AB7900 (Applied Biosystems). Delta Ct was calculated by comparing to *Hprt* as a house keeping gene. Delta Delta Ct was calculated by normalizing each condition to WT expression levels. Statistical significance was determined by t-tests. Each condition was tested for significantly different expression to WT. At least 3 biological replicates were performed for each condition in each experiment. For testing the relative abundance of each *Evx1as* isoform, standard curves were performed with known quantities of DNA to calculate primer efficiencies. These primers efficiencies were taken into account in the Delta Ct calculations. All graphs and statistical tests were performed in Prism 5 (GraphPad). Primer sequences are available in Supplementary Table 1.

CRISPR-Cas9 Clone Generation and Screening

All sgRNAs were designed using the CRISPR design tool (crispr.mit.edu) and cloned into the pSpCas9(BB)-2A-GFP plasmid (#48138, Addgene) as described previously67. pSpCas9(BB)-2A-GFP (PX458) was a gift from Feng Zhang. For generating the EVX1-Δfs mESCs, a single guide was targeted to the second exon of *Evx1* in the homeodomain region. mESCs were transfected with the pCas9-GFP plasmid also expressing the sgRNA targeting *Evx1*. Sorting and screening was conducted as described previously, with minor modifications35. Specifically, the top 5% of GFP+cells were sorted into a 10 cm^2 dish and allowed to grow for two days before picking individual clones. This was done to prevent high numbers of mixed clones, which results from retention of the plasmid through cell division. Large deletions ('Del#1', 'Del#2' and 'Del#3') were generated using the dual sgRNA strategy67. mESCs were transfected with two plasmids, each containing pCas9-GFP and the appropriate sgRNA. The top 5% of

GFP+cells were sorted into a 10 cm^2dish and allowed to grow for two days before picking individual clones. Individual clones were transferred to a 96 well plate and allowed to grow until confluent. DNA was extracted and PCR was performed using primers outside and inside the deleted region to identify clones containing the intended deletions. Clones with the appropriate deletion were expanded and used for subsequent analyses. sgRNA and screening primer sequences are available in Supplementary Table 1.

Whole Mount *In Situ* Hybridization (WISH)

WISH was conducted as described previously, with minor modifications to reduce Proteinase K digestion of E7.5 embryos5. Primers used to amplify and clone probes for *Evx1* and *Evx1as* are available inSupplementary Table 1.

RNA Sequencing and Data Analysis

Total RNA was analysed on the Bioanalyser (Agilent) to confirm high RNA quality (Rinn value>8). mRNA selection of total RNA was performed using Ambion Dynabeads© mRNA DIRECT™ Micro Purification Kit (#61021, Life Technologies). Barcoded RNAseq libraries were generated from the mRNA using the Ion Total RNAseq Kit v3 (#4475936) and Ion Xpress™ RNAseq Barcode 1–16 Kit (#4475485, Life Technologies). RNAseq libraries were sequenced on the Ion Proton sequencing instrument using the Ion P1 Hi-Q Sequencing 200 Kit with the Ion P1 v2 chip. Each sample was sequenced to a depth of least 15 million reads. RNAseq data was mapped to the Mus Musculus reference genome (mm9) using TopHat2 and then TMAP68. Tophat2 is required to map spliced junctions, while TMAP is designed for IonTorrent sequencing data. Transcript counts were obtained by using HTSeq on the mapped data using the standard settings69. Differential gene expression analysis was performed on the HTSeq output using DESeq2 using the standard settings70. Hierarchically clustered heatmaps were generated in R using pHeatmap (Bioconductor). Scatterplots were generated in Prism 5 (GraphPad). Gene Ontology analysis was performed with all DEGs on DAVID71. Raw data is available via the GEO repository (http://www.ncbi.nlm.nih.gov/geo/query/acc.cgi?acc=http://www.ncbi.nlm.nih.gov/geo/) under the accession numbers: GSE75735, GSE75737.

Lentiviral Transduction of shRNAs

We employed the iRNAi v2.1 software to design siRNAs targeting the*Evx1as* gene. We tested for potential off targeting in the genome by performing BLAST searches to exclude candidates which bound to additional genomic sites. We selected three siRNAs for cloning (Supplementary Table 1). We purchased ssDNA oligos designed for annealing to leave 5′ and 3′ overhands for cloning into the *Cla1* and *Mlu1*restriction sites of pLVTHM40. We annealed 2 ul of sense and antisense oligos (1 ug/ul) in 1× DNA annealing solution (100 mM potassium acetate, 30 mM Hepes, 2 mM magnesium acetate) in a 50 final volume by heating to 94 °C for 5 mins and slowly cooling at 37 °C. dsDNA was phosphorylated with PNK and then ligated to pLVTHM cut with *Mlu1*and *Cla1* and dephosphorylated with CIAP. Ligations were performed by standard techniques. Plasmid preps were checked for inserts and concatamer exclusion by digestion with *EcoR1* and *Xba1* and Sanger sequencing. To generate lentiviruses, 293T cells were transduced Opti-MEM with Lipofectamine-2000 and shRNA-pLVTHM together with pCMV-dR8.2 and pMD2G-VSV-G, as described40. Viral stocks were harvested from media at 48–72 hours, filtered (0.45 μ) and snap frozen. For ES cell transductions, adherent ES cells were cultured for 48 hours with viral supernatant (20% in ES cell media) and polybrene (4 μg/ml). Transduced cells were sorted as eGFP+ by FACS and seeded onto MEFs. GFP+ clones were then selected for molecular and functional analyses.

Additional Information

How to cite this article: Bell, C. C. *et al.* The *Evx1/Evx1as* gene locus regulates anterior-posterior patterning during gastrulation. *Sci. Rep.* **6**, 26657; doi: 10.1038/srep26657 (2016).

REFERENCES

1. Tam, P. P. & Loebel, D. A. Gene function in mouse embryogenesis: get set for gastrulation. Nat Rev Genet 8, 368–381 (2007).
2. Arnold, S. J. & Robertson, E. J. Making a commitment: cell lineage allocation and axis patterning in the early mouse embryo. Nature reviews. Molecular cell biology 10, 91–103, doi: 10.1038/nrm2618 (2009).
3. Kinder, S. J. et al. The orderly allocation of mesodermal cells to the extraembryonic structures and the anteroposterior axis during gastrulation of the mouse embryo. Development 126, 4691–4701 (1999).
4. Pearson, J. C., Lemons, D. & McGinnis, W. Modulating Hox gene functions during animal body patterning. Nat Rev Genet 6, 893–904, doi: 10.1038/nrg1726 (2005).
5. Dinger, M. E. et al. Long noncoding RNAs in mouse embryonic stem cell pluripotency and differentiation. Genome Res 18, 1433–1445, doi: 10.1101/gr.078378.108 (2008).
6. Dush, M. K. & Martin, G. R. Analysis of mouse Evx genes: Evx-1 displays graded expression in the primitive streak. Dev Biol 151, 273–287 (1992).
7. Bruce, S. J. et al. Dynamic transcription programs during ES cell differentiation towards mesoderm in serum versus serumfreeBMP4 culture. BMC Genomics 8, 365, doi: 10.1186/1471-2164-8-365 (2007).
8. Ruiz i Altaba, A. & Melton, D. A. Involvement of the Xenopus homeobox gene Xhox3 in pattern formation along the anteriorposterior axis. Cell 57, 317–326 (1989).
9. Seebald, J. L. & Szeto, D. P. Zebrafish eve1 regulates the lateral and ventral fates of mesodermal progenitor cells at the onset of gastrulation. Dev Biol 349, 78–89, doi: 10.1016/j.ydbio.2010.10.005 (2011).

10. Kalisz, M., Winzi, M., Bisgaard, H. C. & Serup, P. EVEN-SKIPPED HOMEOBOX 1 controls human ES cell differentiation by directly repressing GOOSECOID expression. Dev Biol 362, 94–103, doi: 10.1016/j.ydbio.2011.11.017 (2012).
11. Funa, N. S. et al. beta-Catenin Regulates Primitive Streak Induction through Collaborative Interactions with SMAD2/SMAD3 and OCT4. Cell Stem Cell 16, 639–652, doi: 10.1016/j.stem.2015.03.008 (2015).
12. Hart, A. H. et al. Mixl1 is required for axial mesendoderm morphogenesis and patterning in the murine embryo. Development 129, 3597–3608 (2002).
13. Mead, P. E., Brivanlou, I. H., Kelley, C. M. & Zon, L. I. BMP-4-responsive regulation of dorsal-ventral patterning by the homeobox protein Mix.1. Nature 382, 357–360 (1996).
14. Rosa, F. M. Mix.1, a homeobox mRNA inducible by mesoderm inducers, is expressed mostly in the presumptive endodermal cells of Xenopus embryos. Cell 57, 965–974 (1989).
15. Tada, M., Casey, E. S., Fairclough, L. & Smith, J. C. Bix1, a direct target of Xenopus T-box genes, causes formation of ventral mesoderm and endoderm. Development 125, 3997–4006 (1998).
16. Lemaire, P., Darras, S., Caillol, D. & Kodjabachian, L. A role for the vegetally expressed Xenopus gene Mix.1 in endoderm formation and in the restriction of mesoderm to the marginal zone. Development 125, 2371–2380 (1998).
17. Henry, G. L. & Melton, D. A. Mixer, a homeobox gene required for endoderm development. Science 281, 91–96 (1998).
18. Poulain, M. & Lepage, T. Mezzo, a paired-like homeobox protein is an immediate target of Nodal signalling and regulates endoderm specification in zebrafish. Development 129, 4901–4914 (2002).
19. Fekany, K. et al. The zebrafish bozozok locus encodes Dharma, a homeodomain protein essential for induction of gastrula organizer and dorsoanterior embryonic structures. Development 126, 1427–1438 (1999).
20. Kikuchi, Y. et al. The zebrafish bonnie and clyde gene encodes a Mix family homeodomain protein that regulates the generation of endodermal precursors. Genes Dev 14, 1279–1289 (2000).
21. Cho, K. W., Blumberg, B., Steinbeisser, H. & De Robertis, E. M. Molecular nature of Spemann's organizer: the role of the Xenopus homeobox gene goosecoid. Cell 67, 1111–1120 (1991).
22. Spyropoulos, D. D. & Capecchi, M. R. Targeted disruption of the even-skipped gene, evx1, causes early postimplantation lethality of the mouse conceptus. Genes Dev 8, 1949–1961 (1994).
23. Moran-Rivard, L. et al. Evx1 is a postmitotic determinant of v0 interneuron identity in the spinal cord. Neuron 29, 385–399 (2001).
24. Yin, Y. et al. Opposing Roles for the lncRNA Haunt and Its Genomic Locus in Regulating HOXA Gene Activation during Embryonic Stem Cell Differentiation. Cell Stem Cell 16, 504–516, doi: 10.1016/j.stem.2015.03.007 (2015).

25. Wang, K. C. et al. A long noncoding RNA maintains active chromatin to coordinate homeotic gene expression. Nature 472, 120–124, doi: 10.1038/nature09819 (2011).
26. Mercer, T. R., Dinger, M. E. & Mattick, J. S. Long non-coding RNAs: insights into functions. Nat Rev Genet 10, 155–159, doi: 10.1038/nrg2521 (2009).
27. Mattick, J. S., Amaral, P. P., Dinger, M. E., Mercer, T. R. & Mehler, M. F. RNA regulation of epigenetic processes. Bioessays 31, 51–59, doi: 10.1002/bies.080099 (2009).
28. Rinn, J. L. & Chang, H. Y. Genome regulation by long noncoding RNAs. Annual review of biochemistry 81, 145–166, doi: 10.1146/ annurev-biochem-051410-092902 (2012).
29. Fatica, A. & Bozzoni, I. Long non-coding RNAs: new players in cell differentiation and development. Nat Rev Genet 15, 7–21, doi: 10.1038/nrg3606 (2014).
30. Sigova, A. A. et al. Divergent transcription of long noncoding RNA/mRNA gene pairs in embryonic stem cells. Proc Natl Acad Sci USA 110, 2876–2881, doi: 10.1073/pnas.1221904110 (2013).
31. Bertani, S., Sauer, S., Bolotin, E. & Sauer, F. The noncoding RNA Mistral activates Hoxa6 and Hoxa7 expression and stem cell differentiation by recruiting MLL1 to chromatin. Mol Cell 43, 1040–1046, doi: 10.1016/j.molcel.2011.08.019 (2011).
32. Werber, M., Wittler, L., Timmermann, B., Grote, P. & Herrmann, B. G. The tissue-specific transcriptomic landscape of the midgestational mouse embryo. Development 141, 2325–2330, doi: 10.1242/dev.105858 (2014).
33. Derrien, T. et al. The GENCODE v7 catalog of human long noncoding RNAs: analysis of their gene structure, evolution, and expression. Genome Res 22, 1775–1789, doi: 10.1101/gr.132159.111 (2012).
34. Green, J. B. & Smith, J. C. Graded changes in dose of a Xenopus activin A homologue elicit stepwise transitions in embryonic cell fate. Nature 347, 391–394, doi: 10.1038/347391a0 (1990).
35. Bell, C. C., Magor, G. W., Gillinder, K. R. & Perkins, A. C. A high-throughput screening strategy for detecting CRISPR-Cas9 induced mutations using next-generation sequencing. BMC Genomics 15, 1002, doi: 10.1186/1471-2164-15-1002 (2014).
36. Dunty, W. C., Jr., Kennedy, M. W., Chalamalasetty, R. B., Campbell, K. & Yamaguchi, T. P. Transcriptional profiling of Wnt3a mutants identifies Sp transcription factors as essential effectors of the Wnt/beta-catenin pathway in neuromesodermal stem cells. PLoS One 9, e87018, doi: 10.1371/journal.pone.0087018 (2014).
37. Nostro, M. C., Cheng, X., Keller, G. M. & Gadue, P. Wnt, activin, and BMP signaling regulate distinct stages in the developmental pathway from embryonic stem cells to blood. Cell Stem Cell 2, 60–71 (2008).
38. Ponjavic, J., Ponting, C. P. & Lunter, G. Functionality or transcriptional noise? Evidence for selection within long noncoding RNAs. Genome Res (2007).

39. Wei, W., Pelechano, V., Jarvelin, A. I. & Steinmetz, L. M. Functional consequences of bidirectional promoters. Trends Genet 27, 267–276, doi: 10.1016/j.tig.2011.04.002 (2011).
40. Wiznerowicz, M. & Trono, D. Conditional suppression of cellular genes: lentivirus vector-mediated drug-inducible RNA interference. J Virol 77, 8957–8961 (2003).
41. Paralkar, V. R. et al. Lineage and species-specific long noncoding RNAs during erythro-megakaryocytic development. Blood 123, 1927–1937, doi: 10.1182/blood-2013-12-544494 (2014).
42. Lin, N. et al. An evolutionarily conserved long noncoding RNA TUNA controls pluripotency and neural lineage commitment. Mol Cell 53, 1005–1019, doi: 10.1016/j.molcel.2014.01.021 (2014).
43. Kornienko, A. E., Guenzl, P. M., Barlow, D. P. & Pauler, F. M. Gene regulation by the act of long non-coding RNA transcription. BMC biology 11, 59, doi: 10.1186/1741-7007-11-59 (2013).
44. Rossi, A. et al. Genetic compensation induced by deleterious mutations but not gene knockdowns. Nature 524, 230–233, doi: 10.1038/nature14580 (2015).
45. Risebro, C. A. et al. Hand1 regulates cardiomyocyte proliferation versus differentiation in the developing heart. Development 133, 4595–4606, doi: 10.1242/dev.02625 (2006).
46. Li, Q., Pan, H., Guan, L., Su, D. & Ma, X. CITED2 mutation links congenital heart defects to dysregulation of the cardiac gene VEGF and PITX2C expression. Biochem Biophys Res Commun 423, 895–899, doi: 10.1016/j.bbrc.2012.06.099 (2012).
47. Hoogaars, W. M. et al. The transcriptional repressor Tbx3 delineates the developing central conduction system of the heart. Cardiovascular research 62, 489–499, doi: 10.1016/j.cardiores.2004.01.030 (2004).
48. Choi, K., Kennedy, M., Kazarov, A., Papadimitriou, J. C. & Keller, G. A common precursor for hematopoietic and endothelial cells. Development 125, 725–732 (1998).
49. Huber, T. L., Kouskoff, V., Fehling, H. J., Palis, J. & Keller, G. Haemangioblast commitment is initiated in the primitive streak of the mouse embryo. Nature 432, 625–630 (2004).
50. Treichel, D., Becker, M. B. & Gruss, P. The novel transcription factor gene Sp5 exhibits a dynamic and highly restricted expression pattern during mouse embryogenesis. Mech Dev 101, 175–179 (2001).
51. Arkell, R. M. & Tam, P. P. Initiating head development in mouse embryos: integrating signalling and transcriptional activity. Open biology 2, 120030, doi: 10.1098/rsob.120030 (2012).
52. Yamada, G. et al. Targeted mutation of the murine goosecoid gene results in craniofacial defects and neonatal death. Development 121, 2917–2922 (1995).
53. Meno, C. et al. lefty-1 is required for left-right determination as a regulator of lefty-2 and nodal. Cell 94, 287–297 (1998).

54. Rivera-Perez, J. A., Mallo, M., Gendron-Maguire, M., Gridley, T. & Behringer, R. R. Goosecoid is not an essential component of the mouse gastrula organizer but is required for craniofacial and rib development. Development 121, 3005–3012 (1995).

55. Simpson, E. H. et al. The mouse Cer1 (Cerberus related or homologue) gene is not required for anterior pattern formation. Dev Biol 213, 202–206, doi: 10.1006/dbio.1999.9372 (1999).

56. Perea-Gomez, A. et al. Nodal antagonists in the anterior visceral endoderm prevent the formation of multiple primitive streaks. Dev Cell 3, 745–756 (2002).

57. Dolle, P., Fraulob, V. & Duboule, D. Developmental expression of the mouse Evx-2 gene: relationship with the evolution of the HOM/Hox complex. Dev Suppl, 143–153 (1994).

58. Bastian, H. & Gruss, P. A murine even-skipped homologue, Evx 1, is expressed during early embryogenesis and neurogenesis in a biphasic manner. Embo J 9, 1839–1852 (1990).

59. Eldar, A. et al. Robustness of the BMP morphogen gradient in Drosophila embryonic patterning. Nature 419, 304–308, doi: 10.1038/ nature01061 (2002).

60. Zhu, L. et al. Cerberus regulates left-right asymmetry of the embryonic head and heart. Curr Biol 9, 931–938 (1999).

61. Goentoro, L. & Kirschner, M. W. Evidence that fold-change, and not absolute level, of beta-catenin dictates Wnt signaling. Mol Cell 36, 872–884, doi: 10.1016/j.molcel.2009.11.017 (2009).

62. Halstead, A. M. & Wright, C. V. Disrupting Foxh1-Groucho interaction reveals robustness of nodal-based embryonic patterning. Mech Dev 136, 155–165, doi: 10.1016/j.mod.2014.12.002 (2015).

63. Ulitsky, I. & Bartel, D. P. lincRNAs: genomics, evolution, and mechanisms. Cell 154, 26–46, doi: 10.1016/j.cell.2013.06.020 (2013).

64. Jiang, W., Liu, Y., Liu, R., Zhang, K. & Zhang, Y. The lncRNA DEANR1 facilitates human endoderm differentiation by activating FOXA2 expression. Cell reports 11, 137–148, doi: 10.1016/j.celrep.2015.03.008 (2015).

65. Zhang, X., Weissman, S. M. & Newburger, P. E. Long intergenic non-coding RNA HOTAIRM1 regulates cell cycle progression during myeloid maturation in NB4 human promyelocytic leukemia cells. RNA biology 11, 777–787 (2014).

66. ten Berge, D. et al. Wnt signaling mediates self-organization and axis formation in embryoid bodies. Cell Stem Cell 3, 508–518, doi: 10.1016/j.stem.2008.09.013 (2008).

67. Ran, F. A. et al. Genome engineering using the CRISPR-Cas9 system. Nature protocols 8, 2281–2308, doi: 10.1038/nprot.2013.143 (2013).

68. Kim, D. et al. TopHat2: accurate alignment of transcriptomes in the presence of insertions, deletions and gene fusions. Genome Biol 14, R36, doi: 10.1186/gb-2013-14-4-r36 (2013).

69. Anders, S., Pyl, P. T. & Huber, W. HTSeq–a Python framework to work with high-throughput sequencing data. Bioinformatics 31, 166–169, doi: 10.1093/bioinformatics/btu638 (2015).
70. Love, M. I., Huber, W. & Anders, S. Moderated estimation of fold change and dispersion for RNA-seq data with DESeq2. Genome Biol 15, 550, doi: 10.1186/s13059-014-0550-8 (2014).
71. Huang, D. W. et al. The DAVID Gene Functional Classification Tool: a novel biological module-centric algorithm to functionally analyze large gene lists. Genome Biol 8, R183, doi: 10.1186/gb-2007-8-9-r183 (2007).

SECTION II. STEM CELLS: RESEARCH AND APPLICATIONS

10

CHAPTER

SINGLE CELL LINEAGE ANALYSIS OF MOUSE EMBRYONIC STEM CELLS AT THE EXIT FROM PLURIPOTENCY

Jamie Trott[1,2], and Alfonso Martinez Arias[1]

[1] Department of Genetics, University of Cambridge, Cambridge CB2 3EH, UK

[2] Wellcome Trust Centre for Stem Cell Research, University of Cambridge, Cambridge CB2 1QR, UK

SUMMARY

Understanding how interactions between extracellular signalling pathways and transcription factor networks influence cellular decision making will be crucial for understanding mammalian embryogenesis and for generating specialised cell types in vitro. To this end, pluripotent mouse Embryonic Stem (mES) cells have proven to be a useful model system. However, understanding how transcription factors and signalling pathways affect decisions made by individual cells is confounded by the fact that measurements are generally made on groups of cells, whilst individual mES cells differentiate at different rates and towards different lineages, even in conditions that favour a particular lineage. Here we have used single-cell measurements of transcription factor expression and Wnt/β-catenin signalling activity to investigate their effects on lineage commitment decisions made by individual cells. We find that pluripotent mES cells exhibit differing degrees of heterogeneity in their expression of important regulators from pluripotency, depending on the signalling environment to which they are exposed. As mES cells differentiate, downregulation of *Nanog* and *Oct4* primes cells for neural commitment, whilst loss of *Sox2* expression primes cells for primitive streak commitment. Furthermore, we find that Wnt signalling acts through *Nanog* to direct cells towards a primitive streak fate, but that transcriptionally active β-catenin is associated with both neural and primitive streak commitment. These observations confirm and extend previous suggestions that pluripotency genes influence lineage

Citation: Jamie Trott, Alfonso Martinez Arias, Single cell lineage analysis of mouse embryonic stem cells at the exit from pluripotency, doi: 10.1242/bio.20135934.

commitment and demonstrate how their dynamic expression affects the direction of lineage commitment, whilst illustrating two ways in which the Wnt signalling pathway acts on this network during cell fate assignment.

INTRODUCTION

The mammalian embryo is derived from a small group of cells, the epiblast, which is first segregated from extraembryonic tissue just before implantation, at the blastocyst stage (Tam and Loebel, 2007;Rossant and Tam, 2009). This group of cells forms an epithelium (the postimplantation epiblast) that undergoes a proliferative expansion before giving rise to different lineages through the spatio-temporal segregation of patterns of gene expression. The first symmetry-breaking event in the embryonic epiblast occurs around the time of implantation (E4.5), with the definition of an anteroposterior axis, mediated by signalling between the extraembryonic endoderm and the epiblast (Takaoka et al., 2006). This event positions the primordium for the brain (ectoderm) anteriorly, and the primitive streak, that will give rise to the endoderm and the mesoderm, posteriorly (Kimura et al., 2000; Perea-Gomez et al., 2002; Yamamoto et al., 2004; Rivera-Pérez and Magnuson, 2005; Arkell and Tam, 2012). The transition from the blastocyst to the patterned epiblast is a continuous process that preludes the diversification and spatial organization of the embryo and the signalling pathways involved have been studied in detail (Tam and Loebel, 2007). However, the molecular events that initially specify a cell as anterior or posterior in the epiblast, have not been identified, as they coincide with implantation of the embryo. Embryonic Stem (ES) cells provide an alternative and complementary system to study these events.

Mouse Embryonic Stem (mES) cells are clonal populations of cultured cells derived from the blastocyst of the embryo that can give rise to all of the cell types that constitute the adult organism, i.e. they are pluripotent (Bradley et al., 1984; Smith, 2001). Several studies have revealed that pluripotent cells in vitro fluctuate between a ground state of pluripotency and one in which cells are primed for lineage commitment (Chambers et al., 2007; Hayashi et al., 2008; Toyooka et al., 2008; Kalmar et al., 2009;Canham et al., 2010; Martinez Arias and Brickman, 2011). Manipulation of different extracellular signalling pathways can shift the population as a whole towards ground state pluripotency or cause germ layer differentiation, depending on the combination of signals the cells are exposed to (Ying et al., 2008; Wray et al., 2010). Leukemia Inhibitory Factor (LIF) and Bone Morphogenetic Protein (BMP) are sufficient for self-renewal of pluripotency of an mES cell population in a chemically defined medium like N2B27 (Ying et al., 2003a). However, culturing cells in the presence of inhibitors of Extracellular-signal-Regulated Kinases (ERK) and Glycogen Synthase Kinase 3 (GSK3), hereafter referred to as “2i” conditions, shifts cells to ground state pluripotency (Ying et al., 2008). Conversely, in N2B27 basal medium alone, which contains the Retinoic Acid precursor Vitamin A and insulin, pluripotent cells adopt a neural fate, whilst activation of Wnt or Nodal signalling elicits a primitive streak like population that serves as a precursor for the endoderm and the mesoderm (Ying et al., 2003b; Gadue et al., 2006; Bakre et al., 2007; Nostro et al., 2008; Hansson et al., 2009). On the assumption that mES cells reflect an embryonic population, they can be used as a model to study cell fate assignment and maintenance as well as the early stages of mammalian development.

Here we have analyzed the controlled exit of mES cells from pluripotency under established neural differentiation conditions (N2B27) by measuring gene expression in single cells. This technique has previously been used to study mouse blastocysts (Guo et al., 2010), pluripotent mES cells (Hayashi et al., 2008; Tang et al., 2010; MacArthur et al., 2012; Trott et al., 2012) and haematopoietic stem cells (Pina et al., 2012; Moignard et al., 2013), but our study represents the first analysis of the early stages of differentiation of mES cells.

We find that initial differentiation of mES cells in N2B27 is not restricted to neural lineages, but that during the first four days of differentiation, individual cells express markers for a variety of lineages. Furthermore, our single cell analysis suggests that the expression of individual components of the core pluripotency network (*Nanog*, *Oct4* and *Sox2*) restricts differentiation to specific lineages as cells exit pluripotency. Our findings lead us to suggest that during the first four days of differentiation in culture, cells undergo a transition that mimics events in vivo between the blastocyst and the postimplantation epiblast, by the end of which cells are biased towards commitment to different lineages. We also find that, during this period, ES cells display a widespread activation of canonical Wnt/β-catenin signalling which is not affiliated with specific lineages and discuss the possible implications of this observation.

RESULTS

Pluripotency at the Single Cell Level

Mouse ES cells self-renew under a variety of culture conditions and attempts to reveal key molecules supporting pluripotency revealed that LIF is central in the promotion of "naïve pluripotency" (Smith et al., 1988; Williams et al., 1988). Subsequently it was found that mES cells placed in synthetic medium (N2B27) in the presence of LIF and BMP4 maintained their pluripotency, which demonstrated that BMP is the component in Serum responsible for inhibition of differentiation (Ying et al., 2003a). A number of experiments have since established that heterogeneity in the expression of *Rex1*, *Nanog*, *Klf4* and*Stella* is a hallmark of naïve pluripotent mES cells, with cells expressing high levels being pluripotent and cells expressing low levels being primed for differentiation, as reflected by their increased propensity to exit pluripotency permanently (Chambers et al., 2007; Hayashi et al., 2008; Toyooka et al., 2008; Kalmar et al., 2009). Furthermore, in Serum and LIF or LIF and BMP, cells can transit between different expression states. Culturing mES cells in 2i conditions eliminates differentiation-primed cells from the culture and leads to a robust state of pluripotency that has been termed "ground state pluripotency" and that can be propagated in these growth conditions (Ying et al., 2008; Wray et al., 2010). Accordingly, culturing cells in 2i increases colony formation and chimaera contribution rates.

To determine whether the enhanced pluripotency of mES cells cultured in 2i has a transcriptional basis at the level of single cells, expression of several key markers of pluripotency and lineage commitment was measured directly in individual cells by qPCR (Fig. 1A). In the presence of LIF and BMP4, it is possible to observe some cells expressing differentiation markers: *Sox1* (neural) and *T/Bra* (Primitive Streak, the precursor of the mesoderm and the endoderm) (Herrmann et al., 1990; Wood and Episkopou, 1999;Smith, 2004). These cells express low levels of pluripotency markers and are likely to represent the differentiating population known to be present in these conditions. When cells are placed in 2i (in this case supplemented with LIF), the heterogeneities disappear and ~80% of the cells express high and homogeneous levels of *Oct4*, *Sox2* and *Nanog* alongside negligible levels of differentiation markers (Fig. 1B). This demonstrates that the enhanced pluripotency of cells cultured in 2i is associated with more stable expression of key pluripotency regulators at the level of single cells and consequently a lack of differentiation in such cultures.

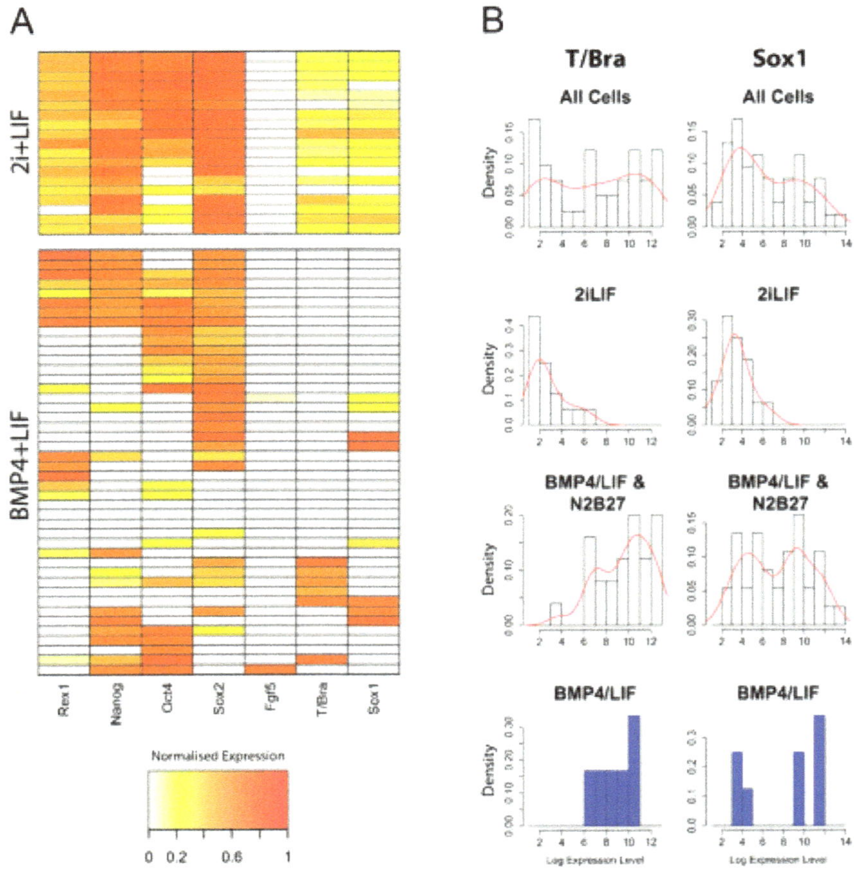

Figure. 1: Gene expression in individual wild-type (E14Tg2A) pluripotent mES cells in N2B27 supplemented with BMP4+LIF or 2i+LIF.

(A) Heat maps of pluripotency (*Rex1*, *Nanog*, *Oct4* and *Sox2*) and differentiation (*Fgf5, T/Bra and Sox1*) marker gene expression in mES cells maintained in 2i+LIF or BMP4+LIF. Plotted values are natural logarithms of *Gapdh*-normalised expression levels, scaled such that cells with no expression of a particular gene have a value of 0 whilst cells that express the maximum detected level of a gene have a value of 1. (B) Histograms of primitive streak (*T/Bra*) and neural (*Sox1*) marker gene expression in different groups of mES cells. Natural logarithms of *Gapdh*-normalised expression levels are plotted as histograms (black outlined bars) with the corresponding probability density functions overlaid (red lines). The top row of histograms show data for all cells cultured in 2i+LIF, BMP4+LIF and after one to four days BMP4+LIF withdrawal (seeMaterials and Methods), whilst the lower three rows show data for subsets of the above. Only cells with detectable expression of the gene in question are included. The histograms show bimodal expression level distributions for both *Sox1*and *T/Bra*, suggesting there are discrete groups of cells expressing high or low levels of these genes. These differentiation markers are only ever expressed at low levels by cells cultured in 2i+LIF, and are often coexpressed in individual cells (Fig. 1A). Conversely, in BMP4+LIF and after BMP4+LIF withdrawal (N2B27), these genes can be expressed at high or low levels, but tend not to be coexpressed (**Fig. 1A**). These observations suggest that cells expressing high levels of these genes are differentiating, whilst low-level expression of these genes is not indicative of commitment to a particular lineage. The fact that some cells in BMP4+LIF express high levels of *Sox1* or *T/Bra* suggests these populations contain a proportion of differentiating cells.

The Exit From the Pluripotent State

Removal of LIF and BMP4 from an N2B27 based culture medium is often used to trigger loss of pluripotency and the differentiation of the culture towards neural lineages (Ying et al., 2003b); in these conditions, addition of Retinoic Acid (RA) is used to enhance this effect and we have observed that it accelerates differentiation, though it increases cell death (J.T. and A.M.A., unpublished observations). As mES cells begin to differentiate, they undergo two sequential transitions: firstly from a multi-layered to a monolayer epithelium – which presumably reflects the transition to an EpiSC state – and secondly, after five or six days, through an Epithelial Mesenchymal Transition (EMT), to a phenotypically heterogeneous population. While some cells form rosette like structures typical of neural cells, others have morphologies typical of mesenchymal or epithelial tissues (Fig. 2A). Prolonged culture in N2B27 appears to eliminate the latter and allows the expansion of a population with neural characteristics, such that by six days cells express high levels of *Sox1* (Ying et al., 2003b; Abranches et al., 2009; Engberg et al., 2010; Stavridis et al., 2010). These observations raised the possibility that N2B27 might not provide an environment to direct differentiation strictly into neural fates during the exit from pluripotency and led us to monitor gene expression across different lineages as cells exit pluripotency in N2B27.

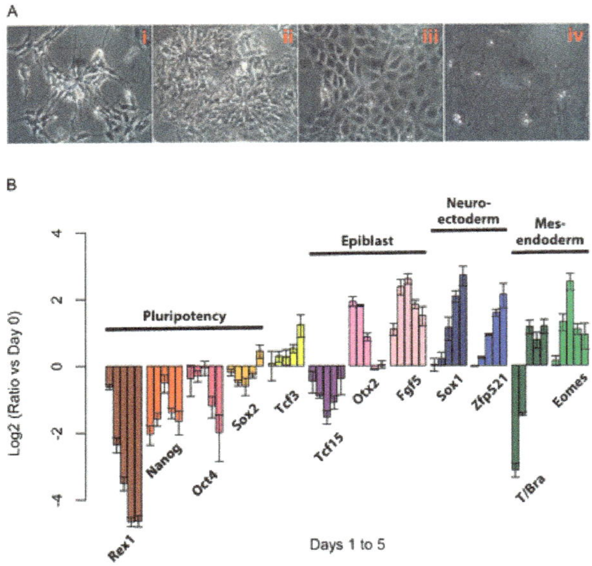

Figure. 2: Changes in morphology and gene expression as mES cells are differentiated over five days in N2B27 medium following BMP4+LIF withdrawal (see Materials and Methods).

(A) Examples of cellular morphologies observed after five days in N2B27: neural rosettes (i,ii), epithelial (iii) and mesenchymal (iv). (B) Population level changes in gene expression over the course of five days of differentiation. Expression of markers of pluripotency (*Rex1*, *Nanog*, *Oct4* and *Sox2*), epiblast (*Tcf15*, *Otx2* and *Fgf5*), neuroectoderm (*Sox1* and*Zfp521*) and mesendoderm (*T/Bra* and *Eomes*) was measured. For each gene on each of the days from day 1 (left most bar) to day 5 (right most bar), the value plotted is the $\log_2$ of the ratio on that day compared to the level in BMP4+LIF (both values normalised to the levels of the gene *Ppia*). Error bars above and below each bar represent the standard deviation of the mean of 3 or more qPCR replicates.

Cells that had been grown in the presence of LIF and BMP4 were placed in N2B27 and the expression of several genes, some associated with pluripotency and some with differentiation, was monitored daily over time (Fig. 2B). We observed the expected rapid decrease in *Rex1*, *Nanog* and *Oct4* expression as cells exit pluripotency, as well as transient expression of the epiblast affiliated genes *Fgf5* and *Otx2*. There is also an expected, steady increase in the expression of genes associated with neural fates (*Sox1*and *Zfp521*) and an early decrease in the expression of *Sox2* as cells exit pluripotency, followed by an increase as cells adopt neural fates. However, we also observed expression of *T/Bra* and *Eomes*, which are genes associated with Primitive Streak and endodermal fates (the sharp decline in *T/Bra* levels on day 1 most likely represents the failure of cells expressing this gene in BMP4+LIF conditions to re-adhere when reseeded). These profiles additionally reveal a small, but significant and reproducible, rise in the levels of *Nanog* and *Oct4* on day 3 of differentiation, which can be observed, but is rarely remarked on, in many reports. Dynamic expression of *Nanog* and *Oct4* probably reflects reactivation of these genes as cells pass through the epiblast state, where both are expressed (Brons et al., 2007). These results demonstrate that the early stages of differentiation in N2B27 involve expression of fates other than neural commitment.

The Exit from Pluripotency from the Perspective of the Pluripotency Network

Population analyses, like the one described above, provide a useful guide to the behaviour of an ensemble of cells, but are less helpful when trying to determine how individual cells choose between different lineages during differentiation. Therefore, we used single cell qPCR to identify changes in gene expression that occur as cells exit pluripotency. Individual cells were selected at random from mES cell populations over four days of differentiation in N2B27 and processed for measuring gene expression. We were able to obtain 19 cells grown in 2i+LIF, 45 cells grown in LIF and BMP4, and 44, 42, 36 and 34 cells for each of one, two, three and four days following withdrawal of LIF and BMP4 (the cells were taken from the same populations for which bulk gene expression was presented in Fig. 2B). Expression of genes representative of pluripotency (*Rex1*, *Nanog*, *Oct4* and *Sox2*), epiblast differentiation (*Fgf5*), neural commitment (*Sox2* and *Sox1*) and primitive streak commitment (*T/Bra*) was measured to determine the differentiation state of individual cells over the course of the experiment (for details of the procedure, see Materials and Methods).

Principal Components Analysis (PCA) of the gene expression profiles of individual cells shows that in 2i conditions cells have very similar expression profiles, which gradually diverge as they differentiate (Fig. 3A). In 2i, cells cluster at the bottom right of the PCA plot alongside cells from other conditions that also express *Rex1* (Fig. 3B), indicating that cells in this region of the plot are pluripotent. After four days differentiation, most cells are to be found towards the top- or bottom-left of the plot where cells that express the primitive streak marker *T/Bra* or the neural marker *Sox1* are found. On days 1 and 2 following BMP4+LIF withdrawal, cells are dispersed between these two extremes, which suggests that, in vitro, individual cells proceed towards lineage commitment at different rates. Furthermore, coexpression of significant levels of *Brachyury* and *Sox1* is observed very rarely suggesting multi-lineage priming does not occur frequently in this system.

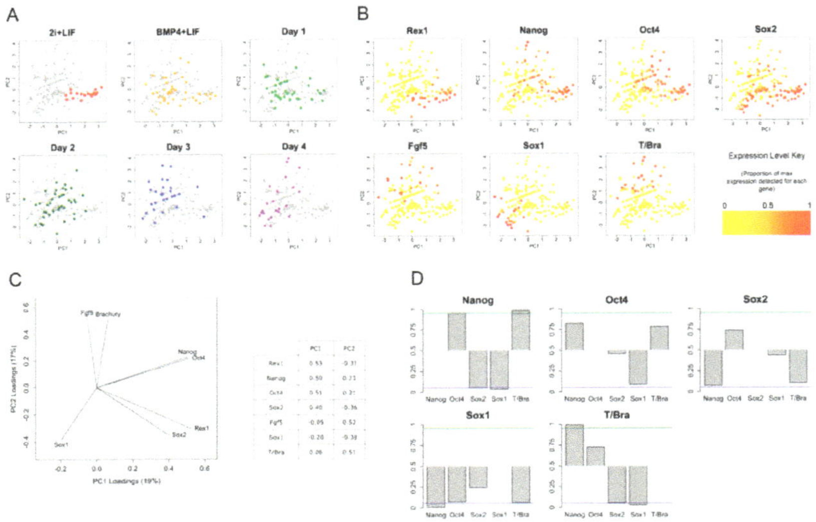

Figure. 3: Relationships between the expression levels of markers of pluripotency and lineage commitment in mES cells exiting pluripotency.

The expression of markers of pluripotency (*Rex1*, *Nanog*, *Oct4* and *Sox2*) and lineage commitment (*Fgf5*, *Sox1* and *T/Bra*) was measured in mES cells maintained in 2i+LIF, BMP4+LIF or differentiated over four days in N2B27 following withdrawal of BMP4 and LIF. Natural logarithms of *Gapdh*-normalised expression levels were used for all calculations relating to this figure and for plotting. (A,B,C) Principal Components Analysis (PCA) of gene expression in individual mES cells. Each point in the plots in panels A and B represents a single cell, the position of which is a function of the expression of seven genes (C) in that cell. The positions of individual cells are identical in each of the plots shown. The cells analysed here, with the exception of those cultured in 2i+LIF, were taken from the same population as those used to produce the population level gene expression data in Fig. 2. Individual cells are coloured according to treatment/differentiation day (A) or their expression of a particular gene (B). In panel B the natural logarithms of the *Gapdh*-normalised expression levels (including cells in which expression was not detected) are scaled such that they range between 0 and 1. Colours are then mapped to these values as shown by the colour key. The principal component loading values used to calculate the positions of individual cells and the percentage of the variability in the dataset that is explained by each axis are shown in panel C. (D) Relationships between the expression of pluripotency and differentiation genes after three to four days differentiation in N2B27. Cells were segregated (in silico) according to whether or not they express a first gene (plot titles). The mean level of a second gene (horizontal labels below each bar) in cells that express the first gene was then calculated. Bootstrapping was used to determine the probability that the mean level of the second gene was higher in cells that express the first gene than would be expected by chance (*P* value). These *P* values were plotted as grey bars. Bootstrapping involves shuffling the levels of the second gene across the population and repeating the calculation described above 100,000 times; the *P* value is the proportion of the bootstrap means that are lower than the real mean. A high *P* value indicates that the levels of two genes are positively correlated, whilst a low *P* values indicates they are negatively correlated. The green and blue lines at *P*values of 0.95 and 0.05 denote statistically significant relationships.

Patterns of pluripotency gene expression across the plot suggest that individual pluripotency genes influence the direction of lineage commitment (Fig. 3B,C). Pluripotent cells at the bottom right of the plot coexpress *Nanog*, *Oct4* and *Sox2* homogeneously. However, differentiating cells express different combinations of these genes, suggesting that the sequence in which the expression of pluripotency genes is downregulated differs between differentiating cells. Interestingly, cells that retain high levels of*Sox2* expression often coexpress *Sox1*, but very rarely express *T/Bra*, whilst the opposite is true of cells that retain high levels of *Nanog* and/or *Oct4* expression. These observations suggest that retention of*Sox2* expression predisposes cells towards neural commitment, whilst retention of *Nanog* or *Oct4*expression biases cells towards primitive streak commitment.

To test these relationships more formally we looked at probabilities of coexpression after four days differentiation (Fig. 3D). This analysis confirmed the strong anti-correlation between expression of*Nanog* and *T/Bra* on the one hand, and *Sox2* and *Sox1* on the other, in differentiating cells. The relationships between expression of *Oct4* and the other genes were similar to those of *Nanog*, but generally much weaker. These findings confirm earlier suggestions that different pluripotency genes specifically inhibit commitment to individual lineages, and additionally demonstrate how these interactions are relieved over time in order to guide cells towards different lineages.

β-catenin Biases Lineage Commitment through the Regulation of *Nanog*

Nanog acts as a gatekeeper for pluripotency but, as we have shown here and others have observed previously, it is also associated with the emergence of a primitive streak fate (Hart et al., 2004;Chambers et al., 2007; Osorno et al., 2012). Similarly, β-catenin stimulation has been shown to promote both pluripotency and primitive streak commitment (Liu et al., 1999; Huelsken et al., 2000; Wray et al., 2011; Yi et al., 2011). Since β-catenin has been shown to enhance *Nanog* expression (Pereira et al., 2006; Yi et al., 2008), we tested whether β-catenin regulates pluripotency and lineage commitment via upregulation of *Nanog* expression by assessing the effects of activating Wnt/β-catenin signalling in the absence of *Nanog* (Fig. 4). *Nanog* mutant mES cells exhibit an unstable pluripotency as shown by their high rates of differentiation. However, Chiron, which activates β-catenin signalling, is able to block upregulation of *Fgf5* expression in the absence of *Nanog*, demonstrating that Chiron maintains pluripotency independently of *Nanog*.

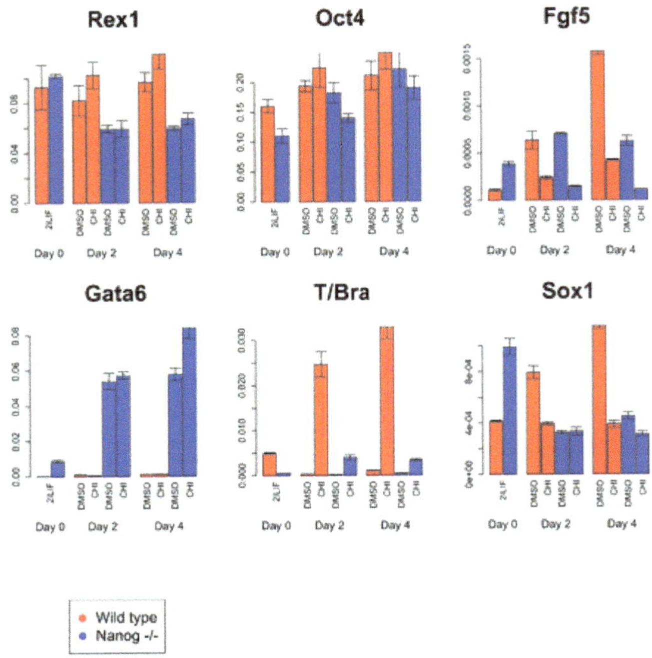

Figure. 4: The effects of β-catenin stimulation on pluripotency and lineage marker gene expression in wild-type (E14Tg2A) and*Nanog*−/− **(44Cre6) cells.**

Wild-type (red bars) and *Nanog*−/− (blue bars) mES cells maintained in 2i+LIF on gelatin-coated tissue culture plastic were transferred to serum+LIF to allow a proportion of cells to exit pluripotency (a percentage of mES cells differentiate spontaneously in serum+LIF), in the presence or absence of the β-catenin agonist Chiron. The effects of Chiron stimulation of β-catenin on expression of markers of pluripotency (*Rex1*, *Nanog* and *Oct4*), epiblast (*Fgf5*), neuroectoderm (*Sox1* and *Zfp521*), mesendoderm (*T/Bra*) and primitive endoderm (*Gata6*) were then assessed by qPCR after two and four days. Plotted are the levels of individual genes in each sample relative to the levels of the reference gene *Gapdh*. Error bars above and below each bar represent the standard deviation of the mean of at least 3 qPCR replicates. DMSO is used a control in this experiment as Chiron is dissolved in DMSO.

We notice that *Nanog* mutant cells have a strong tendency to upregulate expression of the endoderm marker *Gata6*, which most likely reflects increased primitive endoderm commitment, consistent with the established role of *Nanog* in blocking primitive endoderm commitment (Mitsui et al., 2003). Nonetheless, *Nanog* mutant cells are able to commit to both neural and primitive streak fates (Chambers et al., 2007). However, we observe that in the absence of *Nanog*, β-catenin is unable to block*Sox1* expression or stimulate *T/Bra* expression significantly. These observations indicate that the ability of β-catenin to predispose cells towards primitive streak commitment is dependent on *Nanog*.

Wnt Signalling as a Signature for the exit from Pluripotency

When LIF is substituted for Retinoic Acid (RA) in Serum, mES cells undergo preferential neural differentiation and we have observed that they activate a reporter of Wnt signalling (Faunes et al., 2013). This is surprising as Wnt signalling inhibits the neural fate (Aubert et al., 2002; Haegele et al., 2003; Engberg et al., 2010). Analysis of the reporter at the single cell level reveals that not every cell activates Wnt signalling, and therefore raises the possibility that only those cells that are negative for Wnt signalling will develop into neural precursors. If this were the case, there should be a negative correlation between reporter expression and neural differentiation. To test this, we made use of the fact that, in a population of cells self-renewing in Serum and LIF, some cells exhibit low levels of Wnt signalling. Using a fluorescent Wnt reporter, we selected cells in the top 10% of the distribution for analysis of gene expression (Faunes et al., 2013). These cells amount to ~1% of the population (**Fig. 5A**) and *Axin2* is only detectable in these cells, in agreement with their high levels of Wnt reporter activity.

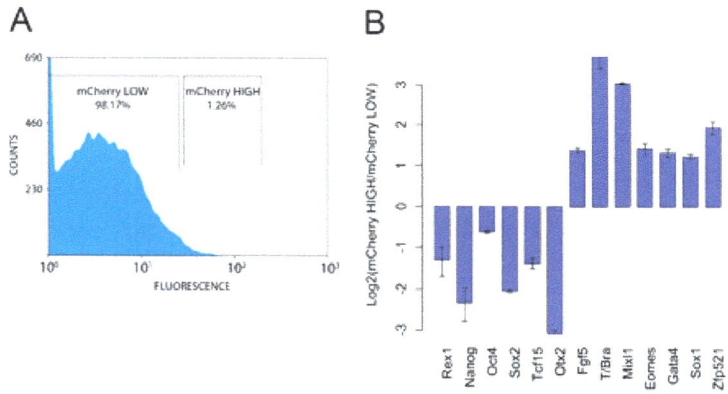

Figure. 5: Wnt signalling marks the exit from pluripotency.

(A) Fluorescence profile of mES cells cultured in serum+LIF that express an mCherry reporter of transcriptionally active β-catenin (Ferrer-Vaquer et al., 2010). The proportion of cells in each of the gates is shown. (B) Gene expression in cells sorted for expression of mCherry using the gates shown in panel A. The two populations were assessed for their expression of markers of pluripotency (*Rex1*, *Nanog*, *Oct4* and *Sox2*), epiblast differentiation (*Fgf5*), primitive streak and its derivatives (*T/Bra*, *Mixl1*, *Eomes* and *Gata4*) and neural differentiation (*Sox1* and *Zfp521*) using qPCR. Each bar shows the $\log_2$ value of the ratio of gene expression in mCherry high cells versus mCherry low cells. All values were first normalised to*Ppia* expression. Error bars above and below each bar represent the standard deviation of the mean of 3 or more qPCR replicates. *Axin2*, a direct transcriptional target of β-catenin signalling, was also measured in both populations and, as expected, was only expressed in the mCherry high population. Increased levels of β-catenin-mediated transcription are associated with loss of pluripotency gene expression and upregulation of markers of multiple opposing lineages.

Cells that express the Wnt reporter exhibit hallmarks of differentiation (Fig. 5B): *Fgf5* expression and low levels of the pluripotency genes *Rex1*, *Nanog*, *Oct4*, *Sox2*, *Tcf15* and *Otx2* relative to the rest of the population. These cells also express lineage associated genes, but without a bias for particular fates. Thus we observe expression of markers of primitive streak and its derivatives (*T/Bra*, *Mixl1*, *Eomes* and*Gata4*), as expected, and also genes associated with neural development (*Sox1* and *Zfp521*). These results suggest that, as cells exit pluripotency in culture, they activate Wnt signalling and that this activation is not associated with specific lineages, but rather appears to be a feature of the early stages of differentiation. It is known that during self-renewal there is little expression of Wnt genes and that as cells differentiate they express many members of the Wnt family and increase expression of those expressed during self-renewal (Nordin et al., 2008). The increase in reporter expression is likely to be associated with this. The generalized expression of differentiation genes in this population is, at this moment, more difficult to explain.

DISCUSSION

Here we have analysed gene expression in individual mES cells during self-renewal and the early stages of differentiation. Our results confirm suggestions from population studies that naïve pluripotency (mES cells grown in N2B27 supplemented with BMP4 and LIF) is associated with variable but generally high levels of expression of pluripotency genes (*Nanog*, *Oct4*, *Sox2* and *Rex1*), as well as differentiation markers in a proportion of cells. In ground state pluripotency (2i conditions), these genes are expressed more homogeneously and the expression of differentiation markers is reduced, though we observe a small degree of heterogeneity, which is in agreement with previous observations (Canham et al., 2010) (C. L. Lim, Investigating the dynamics of Nanog heterogeneity in mouse embryonic stem cells, PhD thesis, University of Cambridge, 2011).

When cells exit pluripotency, the expression of the pluripotency markers, with the exception of *Rex1*, does not disappear completely but individual pluripotency genes become restricted to subsets of cells alongside genes affiliated with different lineages: *Oct4* and *Nanog* become associated with the primitive streak gene *T/Bra*, whilst *Sox2* becomes associated with *Sox*1, which marks the neural fate (Wood and Episkopou, 1999; Rivera-Pérez and Magnuson, 2005). Others have reported that individual pluripotency factors are associated with differentiation towards specific lineages (Avilion et al., 2003;Graham et al., 2003; Hart et al., 2004; Osorno et al., 2012). Furthermore, it was recently demonstrated that *Oct4* and *Sox2* specifically inhibit neural and primitive streak differentiation, respectively (Thomson et al., 2011). Our experiments extend these analyses to reveal how dynamic expression of the pluripotency genes during differentiation is used to guide individual cells towards particular lineages. In support of our model, we find that very few cells maintain high-level coexpression of *Nanog*, *Oct4* and*Sox2* during the differentiation process. In this context our experiments reveal that the ability of Wnt/β-catenin signalling to promote Primitive Streak differentiation is dependent on *Nanog* and that this interaction is independent of the roles of Nanog and β-catenin in the maintenance of pluripotency (Silva et al., 2009; Wray et al., 2011; Yi et al., 2011).

In the course of our experiments we were surprised to observe cells primed for endoderm and mesoderm in N2B27, which is commonly used as a base medium to generate neural tissue (Ying et al., 2003b; Engberg et al., 2010; Stavridis et al., 2010). Furthermore, after five days in culture most of the cells can be coaxed towards neural fates if treated with high levels of RA, whilst T/Bra protein is rarely detected under these differentiation conditions (A.M.A., unpublished observations). This suggests that the initial phases of differentiation involve a general priming of multiple lineages at the level of single cells, preceding a process of selection that, in N2B27, results in most cells eventually expressing *Sox1*and undergoing neural commitment and differentiation (Abranches et al., 2009). We also notice a very low proportion of single cells coexpressing markers of different lineages, although the number of cells analyzed in our experiment is not very large. This would suggest that multilineage priming, in the manner that has been shown to occur in hematopoietic precursors (Cross and Enver, 1997; Hu et al., 1997) is a minor event during loss of pluripotency. Instead it appears that even the earliest stages of differentiation of mES cells are mediated by changes in gene expression that are specific to individual lineages.

Our experiments confirm previous observations that differentiation is associated with an increase in Wnt/β-catenin signalling (Faunes et al., 2013). Surprisingly, we observe that this activity is not associated with a particular fate and that cells with high levels of Wnt/β-catenin signalling express genes associated with neural as well as primitive streak fates. It may be that this is part of the 'priming' process and that in the same manner that cells are primed for particular fates by lineage specific transcription factors, they are also primed for signalling, i.e. signalling is activated indiscriminately and only later becomes restricted to the cells dependent on it, i.e. those fated to the become the Primitive Streak. Further studies should shed light on the meaning of this observation. We have shown that single cell analysis of the early stages of differentiation of mES cells in culture can reveal features of this process that are masked by population averaging. It will be important to extend these studies to include more genes and to carry out a comparative study of single cells from embryos at the equivalent stages.

MATERIALS AND METHODS

Cell Culture and Differentiation

E14Tg2A wild-type (129/Ola), *Nanog−/−* (44Cre6) (Chambers et al., 2007) and Wnt reporter (Faunes et al., 2013) mES cells were routinely cultured on gelatin-coated tissue culture plastic in serum-containing medium (Glasgow Minimal Essential Medium (Sigma, G5154), Foetal Bovine Serum (10%), non-essential amino acids 0.1 mM (Life Technologies, 11140-050), GlutaMAX 2 mM (Life Technologies, 35050-038), sodium pyruvate 1 mM (Sigma, S8636), 2-mercaptoethanol 100 nM (Life Technologies, 31350-010)) supplemented with 100 U/mL LIF (Leukaemia Inhibitory Factor – Recombinant human, Department of Biochemistry, University of Cambridge). Prior to initiating experiments, cells were transferred to NDiff N2B27 (StemCells Inc, SCS-SF-NB-02) supplemented with 100 U/mL LIF and 10 ng/mL BMP4 (Department of Biochemistry, University of Cambridge) or 3 μM Chiron (CHIR99021, Division of signal transduction, University of Dundee) and 2 μM PD03 (PD0325901, Division of signal transduction, University of Dundee) and cultured for two passages. Cells were propagated by trypsinisation every two days with a split ratio of approximately 1 in 5. For differentiations, cells were trypsinised and reseeded in N2B27 alone at a density of 10,000 cells/cm^2.

Quantitative Polymerase Chain Reaction (qPCR)

RNA was isolated from ~5×10^5 trypsinised and pelleted mES cells using the RNeasy Mini kit (Qiagen, 74104) according to the manufacturer›s instructions, and resuspended in 30 μL distilled water. RNA was reverse transcribed as follows. RNA samples (1 μg in 38 μL nuclease free water) were combined with 2 μL Oligo-dT anchored primers (Life Technologies, 12577-011) and incubated at 80°C for 2 minutes before transferring immediately to ice for 2 minutes. PCR master mix was then added to each sample: 1.5 μl dNTPs (Life Technologies, 18427-013), 12 μl 5× First Strand buffer (Life Technologies, 18080-400), 3 μl 0.1 M DTT (Life Technologies, D-1532), 1.5 μl RNaseOUT (Life Technologies, 10777-019) and 2 μl Superscript III Reverse Transcriptase (Life Technologies, 18080-400). Thermal cycling was carried out as follows: 25°C 10 minutes, 50°C 30 minutes, 70°C 15 minutes. To remove RNA from the resultant cDNA samples 1 μL RNaseH (Life Technologies, 18021071) was added and samples were incubated at 37°C for 20 minutes and then at 80°C for 20 minutes. Samples were diluted 1 in 10 before being analysed for gene expression using a Rotor-gene Q instrument. Reactions were set up using a Qiagility liquid-handling robot (12.5 μL 2× QuantiFast SYBR Green Master Mix (Qiagen, 204052), 0.5 μL primer mix (50 μM each), 9 μL PCR grade water and 3 μL diluted cDNA) and analysed using the 2-step protocol described in the Quantifast manual (95°C for 5 minutes, then 40 cycles of 95°C for 10 s + 60°C for 30 s, followed by melt curve analysis). Fluorescence curves were quantified using the MAK3 algorithm found in the R package qpcR. Primer sequences are listed in supplementary material Table S1.

Single Cell qPCR

FACS was used to distribute individual mES cells into the wells of a 96-well plate containing lysis buffer (see below). The transcriptomes of isolated cells were then amplified according to a previously published protocol (up to step 30) (Tang et al., 2010). The resultant cDNA samples were purified using a PCR purification kit (Qiagen, 28104) and diluted 1 in 10 in PCR-grade water. Gene expression was measured as described above.

Flow Cytometry/FACS

Cells were trypsinised and dissociated, spun down in a microcentrifuge, resuspended in serum-containing medium, then analysed and/or sorted on a Beckman–Coulter Cytomation MoFlo High Performance Cell Sorter.

ACKNOWLEDGEMENTS

J.T. was funded by the Wellcome Trust PhD programme. A.M.A. was funded by an ERC Senior Investigator Grant. The *Nanog−/−* (44Cre6) cells described were a gift from Prof. Austin Smith. We want to thank members of the Martinez Arias group for discussions throughout the course of this work.

REFERENCES

1. Abranches, E., Silva, M., Pradier, L., Schulz, H., Hummel, O., Henrique, D. and Bekman, E. (2009). Neural differentiation of embryonic stem cells in vitro: a road map to neurogenesis in the embryo. PLoS ONE 4, e6286.
2. Arkell, R. M. and Tam, P. P. (2012). Initiating head development in mouse embryos: integrating signalling and transcriptional activity. Open Biol 2, 120030.
3. Aubert, J., Dunstan, H., Chambers, I. and Smith, A. (2002). Functional gene screening in embryonic stem cells implicates Wnt antagonism in neural differentiation. Nat. Biotechnol. 20, 1240-1245.
4. Avilion, A. A., Nicolis, S. K., Pevny, L. H., Perez, L., Vivian, N. and Lovell-Badge, R. (2003). Multipotent cell lineages in early mouse development depend on SOX2 function. Genes Dev. 17, 126-140.
5. Bakre, M. M., Hoi, A., Mong, J. C., Koh, Y. Y., Wong, K. Y. and Stanton, L. W. (2007). Gneration of multipotential mesendodermal progenitors from mouse embryonic stem cells via sustained Wnt pathway activation. J. Biol. Chem. 282, 31703-31712.
6. Bradley, A., Evans, M., Kaufman, M. H. and Robertson, E. (1984). Formation of germ-line chimaeras from embryo-derived teratocarcinoma cell lines. Nature 309, 255-256.
7. Brons, I. G., Smithers, L. E., Trotter, M. W., Rugg-Gunn, P., Sun, B., Chuva de Sousa Lopes, S. M., Howlett, S. K., Clarkson, A., Ahrlund-Richter, L., Pedersen, R. A. et al. (2007). Derivation of pluripotent epiblast stem cells from mammalian embryos. Nature 448, 191-195.
8. Canham, M. A., Sharov, A. A., Ko, M. S. and Brickman, J. M. (2010). Functional heterogeneity of embryonic stem cells revealed through translational amplification of an early endodermal transcript. PLoS Biol. 8, e1000379.
9. Chambers, I., Silva, J., Colby, D., Nichols, J., Nijmeijer, B., Robertson, M., Vrana, J., Jones, K., Grotewold, L. and Smith, A. (2007). Nanog safeguards pluripotency and mediates germline development. Nature 450, 1230-1234.
10. Cross, M. A. and Enver, T. (1997). The lineage commitment of haemopoietic progenitor cells. Curr. Opin. Genet. Dev. 7, 609-613.
11. Engberg, N., Kahn, M., Petersen, D. R., Hansson, M. and Serup, P. (2010). Retinoic acid synthesis promotes development of neural progenitors from mouse embryonic stem cells by suppressing endogenous, Wnt-dependent nodal signaling. Stem Cells 28, 1498-1509.

12. Faunes, F., Hayward, P., Descalzo, S. M., Chatterjee, S. S., Balayo, T., Trott, J., Christoforou, A., Ferrer-Vaquer, A., Hadjantonakis, A. K., Dasgupta, R. et al. (2013). A membrane-associated b-catenin/Oct4 complex correlates with ground-state pluripotency in mouse embryonic stem cells. Development 140, 1171-1183.

13. Ferrer-Vaquer, A., Piliszek, A., Tian, G., Aho, R. J., Dufort, D. and Hadjantonakis, A. K. (2010). A sensitive and bright single-cell resolution live imaging reporter of Wnt/ß-catenin signaling in the mouse. BMC Dev. Biol. 10, 121.

14. Gadue, P., Huber, T. L., Paddison, P. J. and Keller, G. M. (2006). Wnt and TGF-beta signaling are required for the induction of an in vitro model of primitive streak formation using embryonic stem cells. Proc. Natl. Acad. Sci. USA 103, 16806-16811.

15. Graham, V., Khudyakov, J., Ellis, P. and Pevny, L. (2003). SOX2 functions to maintain neural progenitor identity. Neuron 39, 749-765.

16. Guo, G., Huss, M., Tong, G. Q., Wang, C., Li Sun, L., Clarke, N. D. and Robson, P. (2010). Resolution of cell fate decisions revealed by single-cell gene expression analysis from zygote to blastocyst. Dev. Cell 18, 675-685.

17. Haegele, L., Ingold, B., Naumann, H., Tabatabai, G., Ledermann, B. and Brandner, S. (2003). Wnt signalling inhibits neural differentiation of embryonic stem cells by controlling bone morphogenetic protein expression. Mol. Cell. Neurosci. 24, 696-708.

18. Hansson, M., Olesen, D. R., Peterslund, J. M., Engberg, N., Kahn, M., Winzi, M., Klein, T., Maddox-Hyttel, P. and Serup, P. (2009). A late requirement for Wnt and FGF signaling during activin-induced formation of foregut endoderm from mouse embryonic stem cells. Dev. Biol. 330, 286-304.

19. Hart, A. H., Hartley, L., Ibrahim, M. and Robb, L. (2004). Identification, cloning and expression analysis of the pluripotency promoting Nanog genes in mouse and human. Dev. Dyn. 230, 187-198.

20. Hayashi, K., Lopes, S. M., Tang, F. and Surani, M. A. (2008). Dynamic equilibrium and heterogeneity of mouse pluripotent stem cells with distinct functional and epigenetic states. Cell Stem Cell 3, 391-401.

21. Herrmann, B. G., Labeit, S., Poustka, A., King, T. R. and Lehrach, H. (1990). Cloning of the T gene required in mesoderm formation in the mouse. Nature 343, 617-622.

22. Hu, M., Krause, D., Greaves, M., Sharkis, S., Dexter, M., Heyworth, C. and Enver, T. (1997). Multilineage gene expression precedes commitment in the hemopoietic system. Genes Dev. 11, 774-785.

23. Huelsken, J., Vogel, R., Brinkmann, V., Erdmann, B., Birchmeier, C. and Birchmeier, W. (2000). Requirement for beta-catenin in anterior-posterior axis formation in mice. J. Cell Biol. 148, 567-578.

24. Kalmar, T., Lim, C., Hayward, P., Mun˜ oz-Descalzo, S., Nichols, J., Garcia-Ojalvo, J. and Martinez Arias, A. (2009). Regulated fluctuations in nanog expression mediate cell fate decisions in embryonic stem cells. PLoS Biol. 7, e1000149.

25. Kimura, C., Yoshinaga, K., Tian, E., Suzuki, M., Aizawa, S. and Matsuo, I. (2000). Visceral endoderm mediates forebrain development by suppressing posteriorizing signals. Dev. Biol. 225, 304-321.

26. Liu, P., Wakamiya, M., Shea, M. J., Albrecht, U., Behringer, R. R. and Bradley, A. (1999). Requirement for Wnt3 in vertebrate axis formation. Nat. Genet. 22, 361- 365.

27. MacArthur, B. D., Sevilla, A., Lenz, M., Mu¨ller, F. J., Schuldt, B. M., Schuppert, A. A., Ridden, S. J., Stumpf, P. S., Fidalgo, M., Ma'ayan, A. et al. (2012). Nanogdependent feedback loops regulate murine embryonic stem cell heterogeneity. Nat. Cell Biol. 14, 1139-1147.

28. Martinez Arias, A. and Brickman, J. M. (2011). Gene expression heterogeneities in embryonic stem cell populations: origin and function. Curr. Opin. Cell Biol. 23, 650- 656.

29. Mitsui, K., Tokuzawa, Y., Itoh, H., Segawa, K., Murakami, M., Takahashi, K., Maruyama, M., Maeda, M. and Yamanaka, S. (2003). The homeoprotein Nanog is required for maintenance of pluripotency in mouse epiblast and ES cells. Cell 113, 631-642.

30. Moignard, V., Macaulay, I. C., Swiers, G., Buettner, F., Schu¨tte, J., Calero-Nieto, F. J., Kinston, S., Joshi, A., Hannah, R., Theis, F. J. et al. (2013). Characterization of transcriptional networks in blood stem and progenitor cells using high-throughput single-cell gene expression analysis. Nat. Cell Biol. 15, 363-372.

31. Nordin, N., Li, M. and Mason, J. O. (2008). Expression profiles of Wnt genes during neural differentiation of mouse embryonic stem cells. Cloning Stem Cells 10, 37-48.

32. Nostro, M. C., Cheng, X., Keller, G. M. and Gadue, P. (2008). Wnt, activin, and BMP signaling regulate distinct stages in the developmental pathway from embryonic stem cells to blood. Cell Stem Cell 2, 60-71.

33. Osorno, R., Tsakiridis, A., Wong, F., Cambray, N., Economou, C., Wilkie, R., Blin, G., Scotting, P. J., Chambers, I. and Wilson, V. (2012). The developmental dismantling of pluripotency is reversed by ectopic Oct4 expression. Development 139, 2288-2298.

34. Perea-Gomez, A., Vella, F. D., Shawlot, W., Oulad-Abdelghani, M., Chazaud, C., Meno, C., Pfister, V., Chen, L., Robertson, E., Hamada, H. et al. (2002). Nodal antagonists in the anterior visceral endoderm prevent the formation of multiple primitive streaks. Dev. Cell 3, 745-756.

35. Pereira, L., Yi, F. and Merrill, B. J. (2006). Repression of Nanog gene transcription by Tcf3 limits embryonic stem cell self-renewal. Mol. Cell. Biol. 26, 7479-7491.

36. Pina, C., Fugazza, C., Tipping, A. J., Brown, J., Soneji, S., Teles, J., Peterson, C. and Enver, T. (2012). Inferring rules of lineage commitment in haematopoiesis. Nat. Cell Biol. 14, 287-294.

37. Rivera-Pe´rez, J. A. and Magnuson, T. (2005). Primitive streak formation in mice is preceded by localized activation of Brachyury and Wnt3. Dev. Biol. 288, 363-371.

38. Rossant, J. and Tam, P. P. (2009). Blastocyst lineage formation, early embryonic asymmetries and axis patterning in the mouse. Development 136, 701-713.

39. Silva, J., Nichols, J., Theunissen, T. W., Guo, G., van Oosten, A. L., Barrandon, O., Wray, J., Yamanaka, S., Chambers, I. and Smith, A. (2009). Nanog is the gateway to the pluripotent ground state. Cell 138, 722-737.

40. Smith, A. G. (2001). Embryo-derived stem cells: of mice and men. Annu. Rev. Cell Dev. Biol. 17, 435-462.

41. Smith, J. C. (2004). Role of T-box genes during gastrulation. In Gastrulation: From Cells To Embryo (ed. C. D. Stern), p. 731. Cold Spring Harbor, NY: Cold Spring Harbor Laboratory Press.

42. Smith, A. G., Heath, J. K., Donaldson, D. D., Wong, G. G., Moreau, J., Stahl, M. and Rogers, D. (1988). Inhibition of pluripotential embryonic stem cell differentiation by purified polypeptides. Nature 336, 688-690.

43. Stavridis, M. P., Collins, B. J. and Storey, K. G. (2010). Retinoic acid orchestrates fibroblast growth factor signalling to drive embryonic stem cell differentiation. Development 137, 881-890.

44. Takaoka, K., Yamamoto, M., Shiratori, H., Meno, C., Rossant, J., Saijoh, Y. and Hamada, H. (2006). The mouse embryo autonomously acquires anterior-posterior polarity at implantation. Dev. Cell 10, 451-459.

45. Tam, P. P. and Loebel, D. A. (2007). Gene function in mouse embryogenesis: get set for gastrulation. Nat. Rev. Genet. 8, 368-381.

46. Tang, F., Barbacioru, C., Nordman, E., Li, B., Xu, N., Bashkirov, V. I., Lao, K. and Surani, M. A. (2010). RNA-Seq analysis to capture the transcriptome landscape of a single cell. Nat. Protoc. 5, 516-535.

47. Thomson, M., Liu, S. J., Zou, L. N., Smith, Z., Meissner, A. and Ramanathan, S. (2011). Pluripotency factors in embryonic stem cells regulate differentiation into germ layers. Cell 145, 875-889.

48. Toyooka, Y., Shimosato, D., Murakami, K., Takahashi, K. and Niwa, H. (2008). Identification and characterization of subpopulations in undifferentiated ES cell culture. Development 135, 909-918.

49. Trott, J., Hayashi, K., Surani, A., Babu, M. M. and Martinez-Arias, A. (2012). Dissecting ensemble networks in ES cell populations reveals micro-heterogeneity underlying pluripotency. Mol. Biosyst. 8, 744-752.

50. Williams, R. L., Hilton, D. J., Pease, S., Willson, T. A., Stewart, C. L., Gearing, D. P., Wagner, E. F., Metcalf, D., Nicola, N. A. and Gough, N. M. (1988). Myeloid leukaemia inhibitory factor maintains the developmental potential of embryonic stem cells. Nature 336, 684-687.

51. Wood, H. B. and Episkopou, V. (1999). Comparative expression of the mouse Sox1, Sox2 and Sox3 genes from pre-gastrulation to early somite stages. Mech. Dev. 86, 197-201.

52. Wray, J., Kalkan, T. and Smith, A. G. (2010). The ground state of pluripotency. Biochem. Soc. Trans. 38, 1027-1032.

53. Wray, J., Kalkan, T., Gomez-Lopez, S., Eckardt, D., Cook, A., Kemler, R. and Smith, A. (2011). Inhibition of glycogen synthase kinase-3 alleviates Tcf3 repression of the pluripotency network and increases embryonic stem cell resistance to differentiation. Nat. Cell Biol. 13, 838-845.

54. Yamamoto, M., Saijoh, Y., Perea-Gomez, A., Shawlot, W., Behringer, R. R., Ang, S. L., Hamada, H. and Meno, C. (2004). Nodal antagonists regulate formation of the anteroposterior axis of the mouse embryo. Nature 428, 387-392.

55. Yi, F., Pereira, L. and Merrill, B. J. (2008). Tcf3 functions as a steady-state limiter of transcriptional programs of mouse embryonic stem cell self-renewal. Stem Cells 26, 1951-1960.

56. Yi, F., Pereira, L., Hoffman, J. A., Shy, B. R., Yuen, C. M., Liu, D. R. and Merrill, B. J. (2011). Opposing effects of Tcf3 and Tcf1 control Wnt stimulation of embryonic stem cell self-renewal. Nat. Cell Biol. 13, 762-770.

57. Ying, Q. L., Nichols, J., Chambers, I. and Smith, A. (2003a). BMP induction of Id proteins suppresses differentiation and sustains embryonic stem cell self-renewal in collaboration with STAT3. Cell 115, 281-292.

58. Ying, Q. L., Stavridis, M., Griffiths, D., Li, M. and Smith, A. (2003b). Conversion of embryonic stem cells into neuroectodermal precursors in adherent monoculture. Nat. Biotechnol. 21, 183-186.

59. Ying, Q. L., Wray, J., Nichols, J., Batlle-Morera, L., Doble, B., Woodgett, J., Cohen, P. and Smith, A. (2008). The ground state of embryonic stem cell self-renewal. Nature 453, 519-523.

11
CHAPTER

GENERATION OF FAMILIAL AMYLOIDOTIC POLYNEUROPATHY-SPECIFIC INDUCED PLURIPOTENT STEM CELLS

Kaori Isono[a,b], Hirofumi Jono[c,d], Yuki Ohya [b], Nobuaki Shiraki [e], Taiji Yamazoe[e] , Ayaka Sugasaki [a], Takumi Era [f], Noemi Fusaki [g,h,i], Masayoshi Tasaki [a,c], Mitsuharu Ueda[c], Satoru Shinriki [c], Yukihiro Inomata [b], Shoen Kume [e], Yukio Ando [a,c]

[a] Department of Neurology, Graduate School of Medical Science, Kumamoto University, Kumamoto, Japan

[b] Department of Transplantation and Pediatric Surgery, Graduate School of Medical Science, Kumamoto University, Kumamoto, Japan

[c] Department of Diagnostic Medicine, Graduate School of Medical Science, Kumamoto University, Kumamoto, Japan

[d] Department of Pharmacy, Kumamoto University Hospital, Kumamoto University, Kumamoto, Japan

[e] Department of Stem Cell Biology, Institute of Molecular Embryology and Genetics, Kumamoto University, Kumamoto, Japan

[f] Department of Cell Modulation, Institute of Molecular Embryology and Genetics, Kumamoto University, Kumamoto, Japan

[g] DNAVEC Corporation, Ibaraki, Japan

[h] Japan Science and Technology Agency (JST), PRESTO, Saitama, Japan

[i] Ophthalmology, School of Medicine, Keio University, Japan

ABSTRACT

Familial amyloidotic polyneuropathy (FAP) is a hereditary amyloidosis induced by amyloidogenic transthyretin (ATTR). Because most transthyretin (TTR) in serum

---

Citation: Kaori Isono, Hirofumi Jono, Yuki Ohya, Nobuaki Shiraki, Taiji Yamazoe, Ayaka Sugasaki, Takumi Era, Noemi Fusaki, Masayoshi Tasaki, Mitsuharu Ueda, Satoru Shinriki, Yukihiro Inomata, Shoen Kume, Yukio Ando, Generation of familial amyloidotic polyneuropathy-specific induced pluripotent stem cells, Stem Cell Research, Volume 12, Issue 2, March 2014, Pages 574-583, ISSN 1873-5061, http://dx.doi.org/10.1016/j.scr.2014.01.004.

is synthesized by the liver, liver transplantation (LT) is today the only treatment available to halt the progression of FAP, even though LT is associated with several problems. Despite the urgent need to develop alternatives to LT, the detailed pathogenesis of FAP is still unknown; also, no model fully represents the relevant processes in patients with FAP. The induction of induced pluripotent stem (iPS) cells has allowed development of pluripotent cells specific for patients and has led to useful models of human diseases. Because of the need for a tool to elucidate the molecular pathogenesis of FAP, in this study we sought to establish heterozygous ATTR mutant iPS cells, and were successful, by using a Sendai virus vector mixture containing four transcription factors (Oct3/4, Sox2, Klf4, and c-Myc) to reprogram dermal fibroblasts derived from FAP patients. Moreover, FAP-specific iPS cells had the potential to differentiate into hepatocyte-like cells and indeed expressed ATTR. FAP-specific iPS cells demonstrated the possibility of serving as a pathological tool that will contribute to understanding the pathogenesis of FAP and development of FAP treatments.

INTRODUCTION

Transthyretin (TTR) is a β-sheet-rich protein that is mainly synthesized by the liver (Buxbaum and Reixach, 2009). TTR normally serves as a plasma transport protein for thyroid hormone and retinol-binding protein with vitamin A (Kanai et al., 1968). Mutant forms of TTR, however, cause familial amyloidotic polyneuropathy (FAP), which is the most common type of autosomal-dominant hereditary systemic amyloidosis (Saraiva et al., 1983 and Ando et al., 2005). As of today, more than 100 different points of mutation and a deletion in the TTR gene have been reported (Westermark et al., 2002, Ando and Ueda, 2012 and Benson and Kincaid, 2007), with the Val30Met mutation being the most common. Systemic amyloid depositions in FAP cause various symptoms, including cardiac and renal dysfunctions, gastrointestinal disorders, glandular and autonomic dysfunctions, and peripheral neuropathy (Ando and Suhr, 1998, Ando et al., 1993, Ando et al., 1997 and Araki, 1984). Because the liver synthesizes most of the TTR in the serum, liver transplantation (LT) is the only treatment available to halt the progression of FAP (Ando et al., 1995a and Suhr et al., 1995). Although LT is widely accepted as the only lifesaving treatment option for FAP patients (Ando et al., 1995b), LT involves several problems, such as a shortage of liver donors, the effects of immunosuppressants, and the progression of ocular disorders caused by a continuing production of amyloidogenic TTR (ATTR) by the retina (Ong et al., 1994 and Kawaji et al., 2005).

Despite an urgent need to develop alternatives to LT, details about the mechanism of amyloid formation in FAP are still unknown. Although attempts were made to establish experimental models of FAP, a suitable tool is not yet available (Buxbaum et al., 2003,Pokrzywa et al., 2007 and Berg et al., 2009). Because FAP is an autosomal-dominant inherited disease, TTR secreted into plasma is the heterotetrameric mixture of wild-type TTR and variant TTR. However, all TTRs used in experiments have been homotetramers, and an artificial Val30Met-overexpressed cell system does not fully represent the relevant processes in patients with FAP (Sousa et al., 2000, Cardoso et al.,

2008 and Sato et al., 2007). Therefore, an urgent need exists to establish an experimental model such as heterozygous ATTR mutant cells derived from FAP patients.

Induced pluripotent stem (iPS) cells have an unlimited replicative ability and the potential to differentiate into most cell types in organisms (Takahashi and Yamanaka, 2006,Takahashi et al., 2007 and Yu et al., 2007). The creation of iPS cells has permitted the development of patient-specific pluripotent cells and has led to useful models of human diseases (Saha and Jaenisch, 2009). Recent studies reported success in generating patient-specific iPS cells for various diseases including neurologic (Dimos et al., 2008,Ebert et al., 2009 and Soldner et al., 2009), hematologic (Raya et al., 2009), and metabolic disorders (Maehr et al., 2009). A report on spinal muscular atrophy-specific iPS cells suggested applications to disease modeling and drug screening by showing the disease-specific changes in cell survival and function (Ebert et al., 2009). A report on Fanconi anemia-specific iPS cells also indicated a potential value for cell therapy by correcting the genetic defect before iPS cell derivation (Raya et al., 2009). Thus, although disease-specific iPS cells may help therapeutic research, iPS cells from patients with FAP have not yet been generated.

In this study, we first report the generation of iPS cells from patients with FAP ATTR Val30Met, which we achieved by reprogramming their fibroblasts with a mixture of Sendai virus (SeV) vector, which does not integrate into the host genome (Li et al., 2000) and has a low risk of tumorigenicity, encoding four transcription factors: octamer 3/4 (Oct3/4), sex-determining region Y box 2 (Sox2), Kruppel-like factor 4 (Klf4), and c-Myc (Fusaki et al., 2009 and Ban et al., 2011). FAP-specific iPS cells indeed differentiated into hepatocyte-like cells (Shiraki et al., 2008 and Shiraki et al., 2011) and expressed Val30Met ATTR. FAP-specific iPS cells may thus provide valuable experimental tools to elucidate the molecular pathogenesis of FAP with a potential value for cell therapy applications.

MATERIALS AND METHODS

Reagents

Reagents were purchased and used at the designated concentrations as follows: recombinant human Activin A (HumanZyme, Chicago, IL), 100 ng/ml; recombinant human basic fibroblast growth factor (ReproCELL, Yokohama, Japan); recombinant human Bone morphogenetic protein 4 (R&D systems, Minneapolis, MN), 10 ng/ml; recombinant human Fibroblast growth factor 10 (PeproTech, London, UK), 10 ng/ml; recombinant human hepatocyte growth factor (PeproTech), 10 ng/ml; dexamethasone (Dex; Sigma, St. Louis, MO), 1 μM; Y-27632 (Rho-associated kinase inhibitor; Wako Chemical, Osaka, Japan), 10 μM; Dulbecco›s Modified Eagle Medium (Gibco, Funakoshi, Tokyo, Japan); RPMI-1640 (Invitrogen, Glasgow, UK); B27 supplement (Invitrogen), nonessential amino acids (Gibco); L-glutamine (Gibco); knockout serum replacement (Gibco); polyclonal rabbit anti-human TTR antibody (Dako, Glostrup, Denmark); Mouse anti-Oct3/4 antibody (Santa Cruz Biotechnology, Texas, USA), Rabbit anti-Alphafeto protein antibody (Dako), Goat anti-human Albumin antibody

(Bethyl, Texas, USA), Goat anti-Sox17 antibody (R&D systems); and Rabbit anti-HNF-4α antibody (Santa Cruz Biotechnology).

Patients

Skin biopsy samples were obtained from three Japanese female patients with FAP ATTR Val30Met in Kumamoto University Hospital. All FAP patients in this study had a definitive diagnosis of FAP on the basis of genetic investigations and clinical manifestations of FAP (Table 1). The research followed the guidelines of the Kumamoto University Ethical Committee.

Table 1: Characterization of established clones.

Clone no.	Fibroblasts	Disease	P	Age/sex	SeV vector	Average induction efficiency (%)	Established SeV (–) clones
1	BJ	Normal	8	Neonatal/ M	Kit	0.46	NT
2	Patient A002	FAP	11	51/F	Kit	0.054	30
3	Patient A003	FAP	9	33/F	Kit	0.41	36
4	Patient A004	FAP	12	44/F	Kit	0.12	33

Generation of iPS Cells

Skin fibroblasts from patients with FAP ATTR Val30Met were maintained in Dulbecco's Modified Eagle Medium (Gibco) supplemented with 10% fetal bovine serum. Induction of human iPS cells was performed as described previously (Fusaki et al., 2009 and Ban et al., 2011). In brief, 1×10^6 human fibroblasts BJ from neonatal foreskin (American Type Culture Collection, Manassas, VA) or FAP patients' fibroblasts were infected with conventional SeV vectors carrying Oct3/4, Sox2, Klf 4, and a temperature-sensitive vector, which has c-Myc at the HNL position at a multiplicity of infection of 3. One week after infection, cells were collected and replated on mitomycin C-treated mouse embryonic fibroblast feeder cells. The next day, medium was changed to a primate embryonic stem cell medium supplemented with 10 ng/ml basic fibroblast growth factor. Embryonic stem cell-like colonies were picked up 28 days after infection. At passage 3, many colonies were negative for the SeV vector; if the vector was present, SeV-negative clones were obtained by incubating cells at nonpermissive temperature (38 °C for 3 days). The reprogramming efficiency was calculated as the number of alkaline phosphatase (ALP)-positive embryonic stem cell-like colonies formed per number of infected cells seeded.

Differentiation into Hepatocyte-Like Cells

FAP-specific iPS cells were differentiated into hepatocyte-like cells using feeder free method modified from reported protocols (Shiraki et al., 2008 and Shiraki et al., 2011), with the following modifications: FAP-specific iPS cells were pretreated overnight with 10 μM Y-27632 (Wako Chemical) and were then dissociated by using 0.25% trypsin–EDTA and were plated at 100,000 cells per well in 96-well plates that had been previously coated with fibronectin. The cells were cultured for 5 days in DMEM supplemented with 100 ng/ml Activin A (HumanZyme), 2% B27 supplement (Invitrogen), nonessential amino acids (Gibco), L-glutamine (Gibco), penicillin and streptomycin, and β-mercaptoethanol. For differentiation from D5 to D7, RPMI-1640 (Invitrogen) supplemented with 10 ng/ml BMP4, 10 ng/ml Fgf10 and 2% B27 supplement, nonessential amino acids (Gibco), L-glutamine (Gibco), penicillin and streptomycin, and β-mercaptoethanol. The medium was then changed to hepatic differentiation medium-DMEM supplemented with 10% knockout serum replacement (Gibco), 10 ng/ml hepatocyte growth factor (PeproTech), 1 μM Dex (Sigma), nonessential amino acids, L-glutamine, penicillin and streptomycin, and β-mercaptoethanol—for up to 20 days. Medium was replaced every day (Day 0–5) or every 2 days (Day 5–20) with fresh differentiation medium supplemented with growth factor. Human iPS cell lines (201B7), the first iPS cells established from human dermal fibroblasts by Dr. Shinya Yamanaka, and HepG2 cells, a human hepatocellular carcinoma cell line, were used as control (Takahashi et al., 2007). For albumin and TTR secretion assay, definitive endoderm cells (Day 5) were dissociated with 0.25% trypsin–EDTA (Invitrogen) and then plated at 200,000 cells/well on a synthemax (corning) pre-coated 96-well plate, and culture up to Day 20 using the abovementioned protocol.

ALP Staining

ALP staining was performed with ALP substrate (1-Step NBT/BCIP; Pierce, IL, USA) after fixation with 10% neutral buffered formalin solution (Wako Chemical), as previously described (Fusaki et al., 2009 and Ban et al., 2011).

Reverse-Transcription Polymerase Chain Reaction (RT-PCR) Analysis

RNA extraction, cDNA synthesis, and RT-PCR were performed as described previously (Fusaki et al., 2009 and Shiraki et al., 2008). The following primers were used: NANOG homeobox (NANOG): forward 5′-TACCTCAGCCTCCAGCAGAT-3′, reverse 5′-TGCGTCACACCATTGCTATT-3′; telomerase reverse transcriptase (TERT): forward 5′-TGCCCGGACC TCCATCAGAGCCAG-3′, reverse 5′-TCAGTCCAGGATGGTC TTGAAGTCTG-3′; SeV: forward 5′-TGGCTAAGAACATCGGAAGG-3′, reverse 5′-GTTTTGCAACCAAGCACTCA-3′; and β-actin: forward 5′-CAACCGCGAGAAGA TGAC-3′, reverse 5′-AGGAAGGCTGGAAGAGTG-3′.

Real-Time PCR Analysis

The real-time PCR conditions were as follows: denaturation at 95 °C for 15 s and annealing and extension at 60 °C for 60 s, for up to 40 cycles. Target messenger RNA (mRNA) levels were expressed as arbitrary units and were determined by using the standard curve method. The following primers were used: Oct3/4: forward 5′-AGGTGTGGGGGATTCCCCCAT-3′, reverse 5′-GCGATGTGGCTGATCTGCTGC-3′; Sox17: forward 5′-ACTGCAACTA

TCCTGACGTG-3′, reverse 5′-AGGA AATG GAGGAAGC TGTT-3′; alpha-fetoprotein (AFP): forward 5′-TGCCAACTC AGTGAGGA CAA-3′, reverse 5′-TCCAACAGGCCTGA GAAATC-3′; albumin (ALB): forward 5′-GATGTCTTCC TGGGCATGTT-3′, reverse 5′-ACATTTGCTGCCCACTTTTC-3′; TTR: forward 5′-CATTCTTGGCAGGA TGGCTTC-3′, reverse 5′-CT CCCAGGTGTC ATCAGCAG-3′; and GAPDH: forward 5′-CGAGATCCCTCC AAAATCAA-3′, reverse 5′-CATGAGTCCTTCCACGATACCAA-3′.

Immunocytochemistry

For whole-mount immunocytochemical analysis, iPS cell cultures were fixed in 4% paraformaldehyde in phosphate-buffered saline (PBS) for 30 min, followed by permeabilization with 0.1% Triton-X (Nakalai Tesque) in PBS for 10 min at room temperature, rinsed several times with PBS then incubated with diluted antibody in 20% Blocking One (Nakalai Tesque) in PBST (0.1% Tween-20 in PBS) in a humidified chamber overnight at 4 °C. Cells were washed in PBST, and incubated with secondary antibody in 20% Blocking One for 2 h at room temperature in the dark. After washing off the secondary antibody in PBST, cells were counterstained with 6-diamidino-2-phenylindole (DAPI) (Roche Diagnostics, Indianapolis, IN). The following antibodies were used as first antibodies: rabbit anti-AFP (Dako), Goat anti-human Albumin antibody (Bethyl), anti-Oct3/4 antibody (Santa Cruz Biotechnology), goat anti-Sox17 antibody (R&D systems), rabbit anti-HNF-4α antibody (Santa Cruz Biotechnology), and anti-HN monoclonal antibody IL4.1 (Fusaki et al., 2009). Secondary antibodies used were Alexa 568-conjugated and Alexa 488-conjugated antibodies (Invitrogen). To assess the efficiency of differentiation, Sox17-, AFP-, and ALB-positive cells versus total cells (DAPI-positive cells) were quantified using ImageXpress Micro cellular imaging system (Molecular Devices Sunnyvale, CA).

PAS Analysis

The cultured cells were fixed in 3.3% formalin for 10 min, and intracellular glycogen was stained using a PAS staining solution (Muto Pure Chemicals, Tokyo, Japan), according to the manufacturer's instructions.

Albumin Secretion Assay

The differentiation medium was changed to fresh medium 48 h before the assay. Albumin secretion was measured by the central clinical laboratory at Kumamoto University, Kumamoto, Japan.

PCR-Restriction Fragment Length Polymorphism Analysis (PCR-RFLP)

Extraction of RNA from HepG2 cells and the frozen samples of livers from FAP patients A002 and cDNA synthesis were performed as previously described (Fusaki et al., 2009). RNA was extracted from FAP-specific iPS cells with the RNeasy Micro Kit (Qiagen, Hilden, Germany) according to the manufacturer's protocol, and cDNA synthesis was performed as described elsewhere (Sueyoshi et al., 2012). PCR primers, which were the same as those used in the real-time PCR reaction, were designed to amplify the Val30Met mutation in exon 2 of the TTR gene. The PCR conditions were 5 min at 98 °C; 35 cycles at 94 °C for 30 s, 55 °C for 30 s, and 72 °C for 1 min; followed by 7 min at 72 °C. The PCR product of 199 bp was digested at 37 °C for 2 h with 5 U *Nsi*1 restriction enzyme (New England Biolabs, Ipswich, MA, USA), which recognized the mutation site. PCR products were evaluated via a microchip electrophoresis system (Cosmo-I SV1210; Hitachi Electronics Engineering, Tokyo, Japan). Measurements were obtained according to the manufacturer's manual.

Surface-Enhanced Laser Desorption/Ionization time-of-Flight Mass Spectrometry (SELDI-TOF MS)

Serum specimens and culture supernatants were evaluated with the PCS 4000 SELDI-TOF MS instrument (Bio-Rad Laboratories, Hercules, CA, USA) by using the following protocol (Ueda et al., 2009), unless otherwise specified: ion focus mass, 13,800 *m*/*z*; laser energy, 2000 nJ; matrix attenuation, 500 *m*/*z*; sample rate, 800; shots/pixel, 5; partition, 1 of 4; and acquired mass range from 0 to 100,000 *m*/*z*. Baseline smoothing: smoothing before fitting baseline, window 25 points. Baseline width: automatic. These baseline settings were the default settings. External calibration of the instrument was performed by using the All-in-One protein molecular mass standard (Bio-Rad Laboratories). Conditions: Q10 ProteinChip in 50 mmol/l phosphate buffer, pH 7.0.

Liquid Chromatography–Tandem Mass Spectrometry (LC–MS/MS)

The Amicon Ultra centrifugal filter 10 K (Millipore, Billerica, MA, USA) was used to make 7–10-fold concentrated culture supernatants. The concentration was incubated overnight at 4 °C with 5 µg of polyclonal rabbit anti-human TTR antibody (Dako). PureProteome Protein G Magnetic Beads (Millipore) were added to the reaction to capture the immune complexes and were agitated for 2 h at 4 °C. After the immune complexes were washed three times with PBST, 25 µl of sample buffer (Bio-Rad Laboratories) was added and incubation proceeded for 5 min at 95 °C. Sample buffers (25 µl), which eluted TTR protein from the beads, were fractionated via sodium dodecyl sulfate-polyacrylamide gel electrophoresis. Silver staining of gels was performed with ProteoSilver™ Plus Silver Stain Kit (Sigma-Aldrich) according to the manufacturer›s protocol. The bands of TTR were excised from the gel, destained according to the manufacturer›s protocol, digested with sequence-grade modified trypsin (Promega, Madison, WI). The peptide mixtures were dried and redissolved with 40 µl of MS-grade water containing 0.1% trifluoroacetic acid and 2% acetonitrile, and were used for nano-flow reversed-phase LC–MS/MS (LTQ Velos Pro; Thermo Fisher Scientific, San Jose, CA). A capillary reversed-phase LC–MS/MS system composed of an Advance Splitless Nano-Capillary LC dual solvent delivery system (Bruker-Michrom, Auburn, CA), an HTS-xt PAL autosampler (CTC Analytics, Zwingen, Switzerland), and LTQ Velos Pro equipped with an XYZ nanoelectrospray ionization source (AMR, Tokyo, Japan) was used. The samples were injected into a peptide L-trap column (Chemical Evaluation Research Institute, Tokyo, Japan). The peptides were separated by using a capillary reversed-phase C18 column (Chemical Evaluation Research Institute) with gradient elution and an ion spray into the mass spectrometer at a spray voltage of 2.3 kV. The peptide and fragment mass tolerances were 2.0 Da and 0.8 Da, respectively. To calculate the ratios of wild-type TTR and ATTR to total TTR in culture supernatants, we measured the areas of the peptide mass peaks derived from wild-type TTR and ATTR, respectively. Peptides (22–34 peptides; GSPAINVAVHVFR) derived from wild-type TTR had a molecular mass of 684 *m*/*z*. Peptides (22–34 peptides; GSPAINVAMHVFR) derived from ATTR had a molecular mass of 700 *m*/*z*.

Western Blotting

As described in the LC–MS/MS section, the immune complexes, made from 20-fold concentrated culture supernatants, were washed three times with PBS, 20 μl of sample buffer (Bio-Rad Laboratories) was added and incubation proceeded for 5 min at 95 °C. Sample buffers (10 μl), which eluted TTR protein from the beads, were fractionated via sodium dodecyl sulfate-polyacrylamide gel electrophoresis and transferred to nitrocellulose membranes (GE Healthcare, Buckinghamshire, UK). Membranes were blocked with 5% nonfat dried milk and PBST and were then incubated overnight at 4 °C with antibodies against TTR in 5% bovine serum albumin (Sigma) and PBST. After the membranes were washed, they were incubated in biotinylated secondary antibodies for 1 h and then in horseradish peroxidase-conjugated streptavidin for 1 h. After this process, specific protein bands were detected with an enhanced chemiluminescence system (Amersham Pharmacia Biotech, Buckinghamshire, UK).

Enzyme-Linked ImmunoSorbent Assay (ELISA)

The concentration of TTR protein in the media of differentiated FAP-specific iPS cells was measured by ELISA assay. The differentiation medium was changed to fresh medium 48 h before the assay. The wells of Nunc-Immuno™ plate II (Thermo Fisher Scientific) were coated with 7000-fold dilution of TTR sheep anti-human polyclonal antibody (LifeSpan BioSciences, Seattle, USA) in carbonate/bicarbonate buffer and incubated overnight at 4 °C. The coating buffer was removed and the wells were washed three times with 200 μl PBST. 250 μl blocking buffer and 0.5% gelatin in the coating buffer were added per well and incubated for 1 h at room temperature. The blocking buffer was removed and the wells were washed as described above. 100 μl of appropriately diluted samples and serum as the standard were added to each well and incubated for 1 h at room temperature. The samples and standard were removed and the wells were washed as described above. 100 μl of 10,000-fold diluted polyclonal rabbit anti-human TTR antibody (Dako) in 0.5% gelatin in the PBST was added to each well and incubated for 1 h at room temperature. The detection antibody was removed and wells were washed as described above. 100 μl of 5000-fold diluted polyclonal goat anti-rabbit immunoglobulins/HRP (Dako) in 0.5% gelatin in the PBST was added to each well and incubated for 1 h at room temperature. The secondary antibody was removed and the wells washed as described above. 100 μl of TMB (KPL, Gaithersburg, USA) solution was added to each well and 1 and half min later, equal volume of stop solution (1 M HCL) was added and read the optical density at 450 nm.

RESULTS

Generation of iPS Cells from Patients with FAP

Dermal fibroblasts obtained from heterozygotic FAP Val30Met patients (Table 1) were cultured and infected with SeV vectors encoding the reprogramming factors Oct3/4, Sox2, Klf4, and c-Myc. Forms of TTR in the serum of one FAP patient (A002) were analyzed by using SELDI-TOF MS, which detected peaks of approximately 13,761 Da for wild-type TTR and 13,792 Da for ATTR Val30Met (Fig. 1). The reprogrammed cells were positive for ALP activity (Fig. 2A). RT-PCR confirmed the expression of pluripotency markers such as NANOG and TERT in FAP-specific iPS cells. SeV vectors were diluted during cell growth and removed after the temperature shift treatment (at 38 °C for 3 days) (Fig. 2B). Immunostaining with the antibody against SeV protein demonstrated the clones free of viral proteins (Fig. 2C). The reprogrammed cells exhibited the ES-like morphology and the expression of pluripotency marker Oct3/4 was confirmed by immunostaining (Fig. 2D). The induction efficiency of FAP-specific iPS cells ranged from 0.054% to 0.41% for each individual (Table 1).

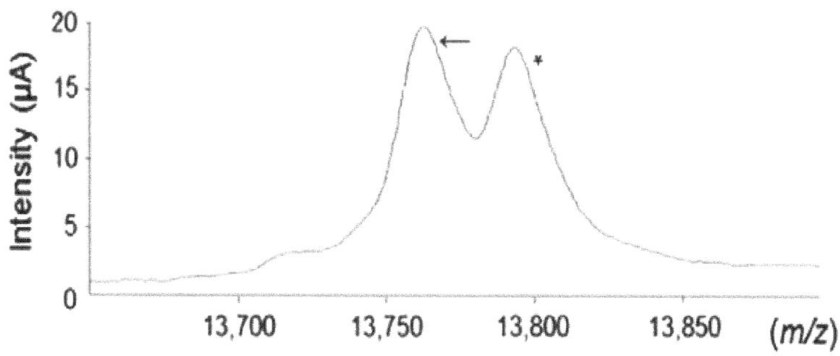

Figure 1: SELDI-TOF MS analysis of TTR forms in the serum of an FAP patient from whom FAP-specific iPS cells were generated. Peaks at 13,761 and 13,792 Da were free forms of wild-type TTR and ATTR Val30Met, respectively. Arrows indicate the wild-type TTR peaks. Asterisks indicate the ATTR Val30Met peaks.

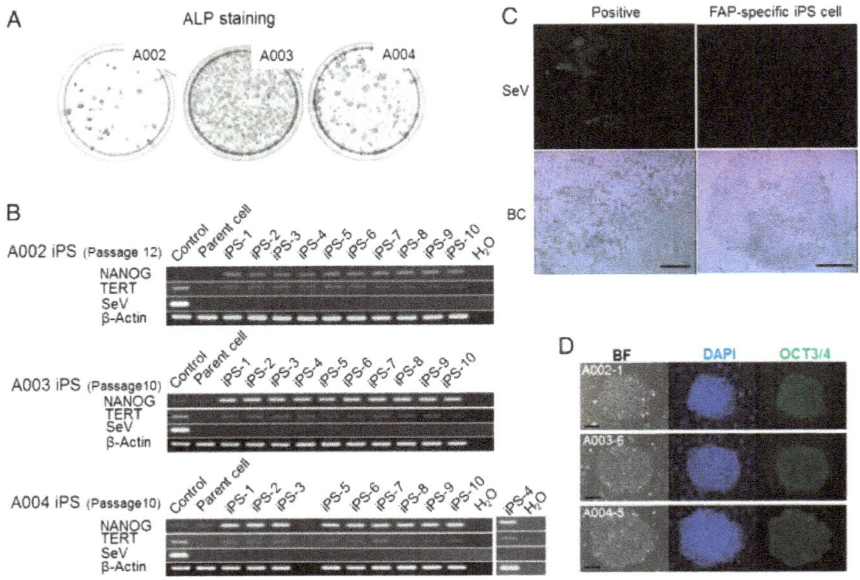

Figure 2: Efficient generation of FAP-specific iPS cells by SeV vectors. (A) ALP staining of cells growing on 100-mm dishes. (B) RT-PCR analysis of expression of NANOG, TERT, SeV, and β-actin. iPS-1 to iPS-10 are established clones. (C) Typical staining of FAP-specific iPS cell colonies with anti-SeV antibodies. The colony was partially positive (left panel) and negative (right panel) for SeV. BC, bright contrast. Scale bar, 50 μm and 100 μm. (D) Immunostaining of established clones with pluripotency marker Oct3/4 (green). Nuclei were stained with DAPI (blue). BF, bright field. Scale bar, 100 μm.

Differentiation of FAP-Specific iPS Cells into Hepatocyte-like Cells

For cell differentiation, FAP-specific iPS cells were cultured with serial changes of media as shown in Fig. 3A. To test whether FAP-specific iPS cells can differentiate into hepatocyte-like cells, we analyzed several markers via real-time PCR analysis on Day 5 (D5), D13 and D20 differentiated FAP-specific iPS cells (Fig. 3B). A decrease in expression of the pluripotency marker Oct3/4 was accompanied by differentiation of FAP-specific iPS cells. Expression of the endoderm marker Sox17 was observed on D5 differentiation and decreased gradually after the medium was changed to hepatic differentiation medium on D7 (Fig. 3B). The hepatic progenitor marker AFP and the mature hepatocyte marker ALB were obviously expressed on D13 and D20. In addition, immunocytochemical analyses showed Sox17 expression on D5, both HNF-4α and AFP expression on D13, and ALB cytoplasmic staining on D20 (Fig. 3C). Quantitative imaging analysis revealed that approximately 78 ± 0.6% of cells were Sox17-positive on D5 and approximately 88 ± 1.1% of cells were AFP-positive on D13 and approximately 29 ± 0.9% of cells were ALB-positive on D20 (Fig. 3D). The ALB secretion in the media of differentiated FAP-specific iPS cells on D20 was approximately 20 μg/ml (Fig. 3E). Moreover, these D20 differentiated FAP-specific iPS cells were also periodic acid-Schiff (PAS)-positive, indicating cytoplasmic glycogen storage (Fig. 3F). These results clearly indicated that FAP-specific iPS cells had the potential to differentiate into hepatocyte-like cells.

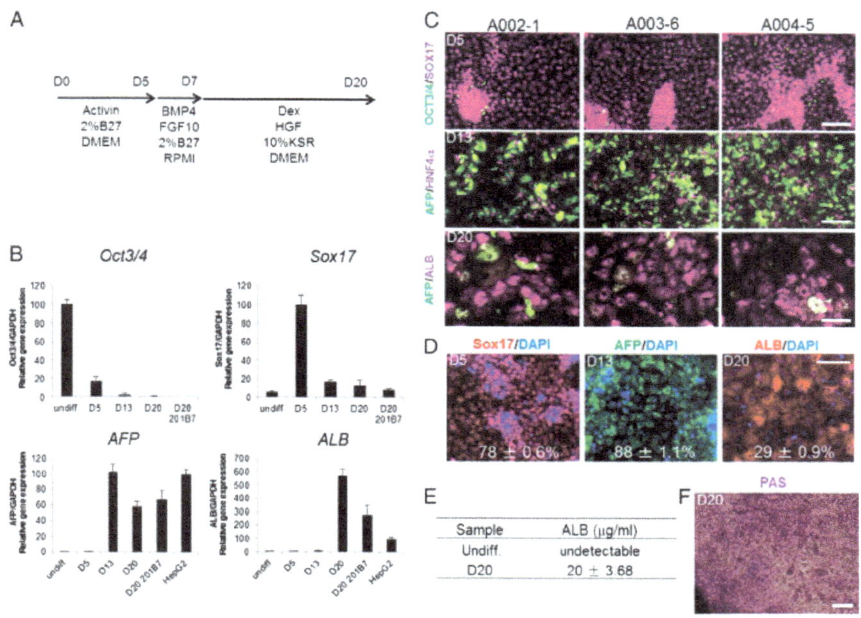

Figure 3: Differentiation of FAP-specific iPS cells. (A) Schematic of the experimental procedure. FAP-specific iPS cells were differentiated on Day (D) 0 to Day 5 in DMEM supplemented with 100 ng/ml Activin and 2% B27 supplement. From D5 to D7, RPMI-1640 supplemented with 10 ng/ml BMP4, 10 ng/ml Fgf10, and 2% B27 supplement were used. At D7, the supplements were changed to 1 μM dexamethasone (Dex) and 10 ng/ml HGF in KSR with DMEM. (B) Quantification via real-time PCR analyses of the relative expression levels of the pluripotency marker Oct3/4, the definitive endoderm marker Sox17, and the hepatic markers AFP and ALB in differentiated FAP-specific iPS cells at D5, D13, and D20 and in undifferentiated iPS cells (Undiff) ($n = 3$). The expression levels were normalized to that of GAPDH. RNA from 201B7 which were differentiated into hepatocyte-like cells the same as FAP-specific iPS cells and the hepatic cell line HepG2 were examined as a positive control ($n = 3$). (C) Differentiated FAP-specific iPS cells on D5 were stained for Oct3/4 (green) and Sox 17 (purple), D13 were stained for AFP (green) and HNF-4α (purple), D20 were stained for AFP (green) and Albumin (purple). Scale bar, 100 μm. (D) FAP-specific iPS cells differentiated on D5, D13 and D20 were stained for Sox 17 (red), AFP (green) and ALB (red), and counterstained with DAPI (blue). Efficiency of differentiation assessed by dividing the number of positive cells for Sox 17, AFP and ALB by the number of total cells ($n = 3$). Scale bar, 100 μm. (E) Albumin secretion by differentiated FAP-specific iPS cells on D20 and undifferentiated iPS cells was measured ($n = 3$). The differentiation medium was changed to fresh medium 48 h before the assay. (F) PAS staining of differentiated FAP-specific iPS cells on D20 indicated numerous hepatocyte-like cells within the colonies with cytoplasmic glycogen storage. Scale bar, 100 μm.

Production of Wild-Type TTR and ATTR Val30Met by Differentiated Hepatocyte-like Cells

We next sought to determine whether hepatocyte-like cells differentiated from FAP-specific iPS cells would indeed express TTR. Expression of TTR mRNA was detectable from D13 (Fig. 4A). To confirm the presence of TTR protein, Western blotting of culture supernatant was performed with an anti-human TTR antibody. As Fig. 4B shows, expression of TTR protein levels was confirmed in the concentrated D20 culture supernatant. The approximate concentration of TTR protein in the media of differentiated FAP-specific iPS cells was approximately 2.28 μg/ml (Fig. 4C). In addition, the PCR-RFLP method showed that hepatocyte-like cells differentiated from FAP-specific iPS cells indeed expressed ATTR Val30Met and wild-type TTR mRNA, similar to expression in the liver of the FAP patient from whom FAP-specific iPS cells were generated (Fig. 5A). Moreover, LC–MS/MS analysis clearly showed that hepatocyte-like cells differentiated from FAP-specific iPS cells indeed expressed both ATTR Val30Met and wild-type TTR at protein level and the ratio of TTR protein to ATTR protein is approximately 1 to 1 in the D20 culture supernatant (Fig. 5B). Other FAP-specific iPS cell lines also expressed both ATTR Val30Met and wild-type TTR at protein level in the culture supernatant (data not shown).

A

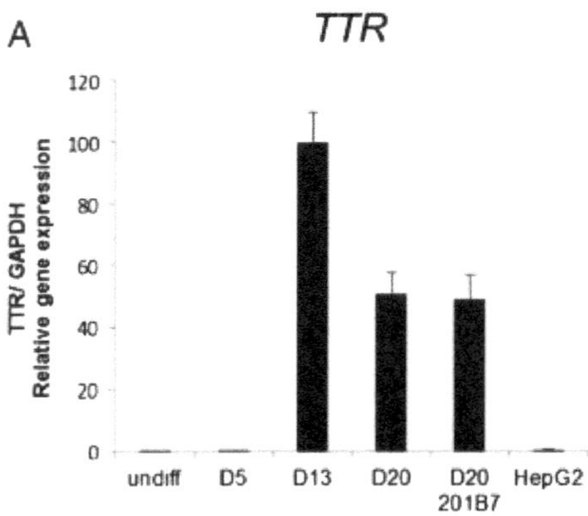

B

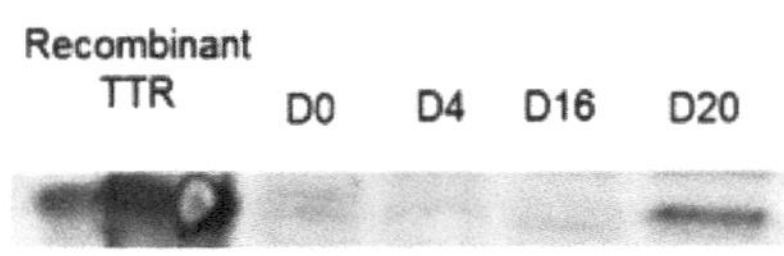

C

Sample	TTR (μg/ml)
Undiff.	undetectable
D20	2.28 ± 0.92

Figure 4: TTR expression in differentiated FAP-specific iPS cells. (A) Quantification via real-time PCR analyses of the relative expression levels of TTR mRNA in differentiated FAP-specific iPS cells on D5, D13, D20, and 201B7 on D20 and in undifferentiated iPS cells (Undiff) ($n = 3$). The expression levels were normalized to that of GAPDH. (B) TTR protein expression determined in the D20 supernatant by Western blotting with anti-TTR polyclonal antibody. (C) TTR secretion by differentiated FAP-specific iPS cells on D20 and undifferentiated iPS cells was measured by ELISA assay ($n = 3$).

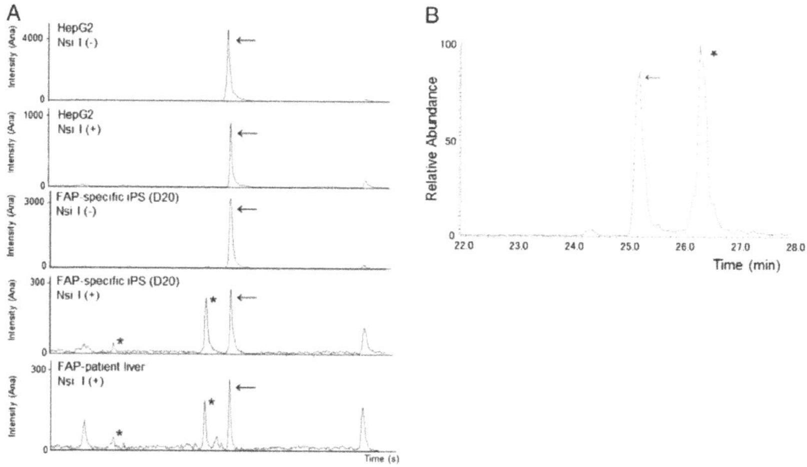

Figure 5: TTR and ATTR expression in differentiated FAP-specific iPS cells. (A) PCR-RFLP analysis of TTR and ATTR mRNA expression of differentiated hepatocyte-like cells on D20. Arrows indicate the wild-type TTR peaks. Asterisks indicate the ATTR Val30Met peaks after digestion with the *Nsi*1 restriction enzyme. The liver of the FAP patient (A002) and HepG2 cells was used as a control. (B) LC–MS/MS analysis of the D20 culture supernatant showed both TTR and ATTR at protein levels. Arrows indicate the wild-type TTR peaks. Asterisks indicate the ATTR Val30Met peaks.

DISCUSSION

In the present study, we generated iPS cells from patients with FAP ATTR Val30Met by introducing four reprogramming factors (Oct3/4, Sox2, Klf 4, and c-Myc) into dermal fibroblasts via SeV vector infection. FAP-specific iPS cells had the potential to differentiate into hepatocyte-like cells, a major TTR-producing cell, and indeed expressed ATTR Val30Met and wild-type TTR protein. FAP-specific iPS cells demonstrated the possibility of serving as a pathological tool, which may contribute to elucidating the molecular pathogenesis of FAP and developing novel therapeutic strategies for FAP.

Attempts to establish the experimental models of FAP (Buxbaum et al., 2003, Pokrzywa et al., 2007 and Berg et al., 2009) have not been as successful as researchers would wish. It has been reported that transgenic mice overexpressing mutated TTR did not display signs of neuropathology (Buxbaum et al., 2003). It is still controversial whether the pathogenic mechanism of FAP in *Drosophila* model is in the same manner as human (Pokrzywa et al., 2007 and Berg et al., 2009). In addition, a review of previous studies about the molecular pathogenesis of FAP indicates that most of the studies investigated the pathologic effect of ATTR using non-mutant cells or an artificial high expression system (Sousa et al., 2000, Cardoso et al., 2008 and Sato et al., 2007). Indeed, the neurodegeneration induced by ATTR or endoplasmic reticulum quality control of ATTR was studied with non-mutant mammalian cells transfected with the receptor for advanced glycation end products (Sousa et al., 2000) or wild-type TTR and ATTR (Sato et al., 2007). Because FAP is an autosomal-dominant hereditary disease and those systems did not utilize heterozygous ATTR mutant cell models, those systems do not fully represent the relevant processes in patients with FAP and are unsuitable for evaluation of therapy. The ratio of expression of TTR protein to ATTR protein varies among cells. As with iPS cells designed to be specific for other diseases, iPS cells generated from patients who were heterozygous for FAP are more useful for elucidating the pathogenesis of FAP: TTR and ATTR in heterozygous FAP form heterotetramers, which produce amyloid, and recombinant TTR forms a homotetramer, which is a totally different form of TTR in plasma and tissues. Generation of various FAP-specific iPS cells may help clarify the mechanism of the phenotypic variations among individuals with the same genotype (Kato-Motozaki et al., 2008) and the cellular and molecular pathogenesis. In this study, hepatocyte-like cells differentiated from FAP-specific iPS cells indeed secreted TTR and ATTR protein in the ratio of almost 1 to 1 as heterozygous ATTR mutant cells (Fig. 5B). The amount of TTR protein was approximately 2.28 μg/ml (Fig. 4C), which is comparable with the previous result (0.2–2 μg/ml) in functional hepatocyte-like cells differentiated from normal iPS cells (Sullivan et al., 2010), and exhibits 50 times higher than the previous result (0.034–0.057 μg/ml) in human hepatocytes in primary culture (Wigmore et al., 1997). Moreover, differentiation of the same FAP-specific iPS cell line into various cells such as neurons and cardiomyocytes may aid understanding of the tissue selectivity of amyloidosis, which differs according to variants (Hammarstrom et al., 2003, Sekijima et al., 2003 and Yamashita et al., 2005). Studies of other FAP-specific iPS cell lines generated from TTR variants other than the Val30Met mutation are currently in progress. FAP-specific iPS cells may also be used for screening novel drug candidates such as inhibitors of amyloid formation. Additional investigations are needed to evaluate the phenotype of FAP-specific iPS cells in greater detail and confirm the value of these cells as a novel experimental tool of FAP.

In this study, we took full advantage of the special characteristics of this SeV vector to establish FAP-specific iPS cells. Because SeV vectors replicate only in the cytoplasm of infected cells and do not integrate into the host genome (Li et al., 2000), no risk of modifying the host genome exists. Using retrovirus or lentivirus vectors results in integration of viral transgenes into the host genome, which includes a risk of tumorigenicity (Hussein et al., 2011 and Gore et al., 2011). To solve this problem, plasmids (Okita et al., 2008), a Cre/loxP system (Soldner et al., 2009), adenoviruses (Stadtfeld et al., 2008), piggyback (Woltjen et al., 2009), a minicircle vector (Jia et al., 2010), and proteins (Zhou et al., 2009) have been developed. The risk of integration into the genome still remains, however, for DNA-type vectors (Harui et al., 1999), and those methods also demonstrated low induction efficiency. Thus, the SeV vector that we used is believed to have a significant advantage compared with available methods because of its safety, efficiency, and convenience. SeV vectors are slowly diluted and disappear during iPS cell division, and SeV vector-positive cells can be removed by means of an anti-SeV-HN antibody (Fusaki et al., 2009). Moreover, with the temperature-sensitive SeV vector used in this study, vectors, even if present, could easily be removed after the temperature shift treatment and would not be reactivated in iPS cells (Ban et al., 2011). FAP-specific iPS cells would thus be safe and may be a source for cell replacement therapy.

CONCLUSIONS

We successfully generated, for the first time, FAP-specific iPS cells. Such FAP-specific iPS cells may serve as valuable experimental tools to clarify the molecular pathogenesis of FAP, with potential value in cell therapy applications.

ACKNOWLEDGMENTS

We thank Dr. Konen Obayashi for technical assistance and Mrs. Hiroko Katsura for her technical support during histopathologic investigations.

REFERENCES

1. Ando, Y., Suhr, O.B., 1998. Autonomic dysfunction in familial myloidotic polyneuropathy (FAP). Amyloid 5, 288–300.
2. Ando, Y., Ueda, M., 2012. Diagnosis and therapeutic approaches to transthyretin amyloidosis. Curr. Med. Chem. 19, 2312–2323. Ando, Y., Araki, S., Ando, M., 1993. Transthyretin and familial
3. amyloidotic polyneuropathy. Intern. Med. 32, 920–922.
4. Ando, Y., Tanaka, Y., Nakazato, M., et al., 1995a. Change in variant transthyretin levels in patients with familial amyloidotic polyneuropathy type I following liver transplantation. Biochem. Biophys. Res. Commun. 211, 354–358.
5. Ando, Y., Tanaka, Y., Ando, E., et al., 1995b. Effect of liver transplantation on autonomic dysfunction in familial amyloidotic polyneuropathy type I. Lancet 345, 195–196.
6. Ando, E., Ando, Y., Okamura, R., et al., 1997. Ocular manifestations of familial amyloidotic polyneuropathy type I: long-term follow up. Br. J. Ophthalmol. 81, 295–298.
7. Ando, Y., Nakamura, M., Araki, S., 2005. Transthyretin-related familial amyloidotic polyneuropathy. Arch. Neurol. 62, 1057–1062.
8. Araki, S., 1984. Type I, familial amyloidotic polyneuropathy (Japanese type). Brain Dev. 6, 128–133.
9. Ban, H., Nishishita, N., Fusaki, N., et al., 2011. Efficient generation of transgene-free human induced pluripotent stem cells (iPSCs) by temperature-sensitive Sendai virus vectors. Proc. Natl. Acad. Sci. U. S. A. 108, 14234–14239.
10. Benson, M.D., Kincaid, J.C., 2007. The molecular biology and clinical features of amyloid neuropathy. Muscle Nerve 36, 411–423.
11. Berg, I., Thor, S., Hammarstrom, P., 2009. Modeling familial amyloidotic polyneuropathy (transthyretin V30M) in Drosophila melanogaster. Neurodegener. Dis. 6, 127–138.
12. Buxbaum, J.N., Reixach, N., 2009. Transthyretin: the servant of many masters. Cell. Mol. Life Sci. 66, 3095–3101.
13. Buxbaum, J., Tagoe, C., Gallo, G., et al., 2003. The pathogenesis of transthyretin tissue deposition: lessons from transgenic mice. Amyloid 10 (Suppl. 1), 2–6.

14. Cardoso, I., Brito, M., Saraiva, M.J., 2008. Extracellular matrix markers for disease progression and follow-up of therapies in familial amyloid polyneuropathy V30M TTR-related. Dis. Markers 25, 37–47.
15. Dimos, J.T., Rodolfa, K.T., Niakan, K.K., et al., 2008. Induced pluripotent stem cells generated from patients with ALS can be differentiated into motor neurons. Science 321, 1218–1221.
16. Ebert, A.D., Yu, J., Rose Jr., F.F., et al., 2009. Induced pluripotent stem cells from a spinal muscular atrophy patient. Nature 457, 277–280.
17. Fusaki, N., Ban, H., Nishiyama, A., et al., 2009. Efficient induction of transgene-free human pluripotent stem cells using a vector based on Sendai virus, an RNA virus that does not integrate into the host genome. Proc. Jpn. Acad. Ser. B 85, 348–362.
18. Gore, A., Li, Z., Fung, H.L., et al., 2011. Somatic coding mutations in human induced pluripotent stem cells. Nature 471, 63–67.
19. Hammarstrom, P., Sekijima, Y., White, J.T., et al., 2003. D18G transthyretin is monomeric, aggregation prone, and not detectable in plasma and cerebrospinal fluid: a prescription for central nervous system amyloidosis? Biochemistry 42, 6656–6663.
20. Harui, A., Suzuki, S., Kochanek, S., et al., 1999. Frequency and stability of chromosomal integration of adenovirus vectors. J. Virol. 73, 6141–6146.
21. Hussein, S.M., Batada, N.N., Vuoristo, S., et al., 2011. Copy number variation and selection during reprogramming to pluripotency. Nature 471, 58–62.
22. Jia, F., Wilson, K.D., Sun, N., et al., 2010. A nonviral minicircle vector for deriving human iPS cells. Nat. Methods 7, 197–199.
23. Kanai, M., Raz, A., Goodman, D.S., 1968. Retinol-binding protein: the transport protein for vitamin A in human plasma. J. Clin. Invest. 47, 2025–2044.
24. Kato-Motozaki, Y., Ono, K., Shima, K., et al., 2008. Epidemiology of familial amyloid polyneuropathy in Japan: identification of a novel endemic focus. J. Neurol. Sci. 270, 133–140.
25. Kawaji, T., Ando, Y., Nakamura, M., et al., 2005. Transthyretin synthesis in rabbit ciliary pigment epithelium. Exp. Eye Res. 81, 306–312.
26. Li, H.O., Zhu, Y.F., Asakawa, M., et al., 2000. A cytoplasmic RNA vector derived from nontransmissible Sendai virus with efficient gene transfer and expression. J. Virol. 74, 6564–6569.
27. Maehr, R., Chen, S., Snitow, M., et al., 2009. Generation of pluripotent stem cells from patients with type 1 diabetes. Proc. Natl. Acad. Sci. U. S. A. 106, 15768–15773.
28. Okita, K., Nakagawa, M., Hyenjong, H., et al., 2008. Generation of mouse induced pluripotent stem cells without viral vectors. Science 322, 949–953.
29. Ong, D.E., Davis, J.T., O'Day, W.T., et al., 1994. Synthesis and secretion of retinol-binding protein and transthyretin by cultured retinal pigment epithelium. Biochemistry 33, 1835–1842.

30. Pokrzywa, M., Dacklin, I., Hultmark, D., et al., 2007. Misfolded transthyretin causes behavioral changes in a Drosophila model for transthyretin-associated amyloidosis. Eur. J. Neurosci. 26, 913–924.
31. Raya, A., Rodriguez-Piza, I., Guenechea, G., et al., 2009. Diseasecorrected haematopoietic progenitors from Fanconi anaemia induced pluripotent stem cells. Nature 460, 53–59.
32. Saha, K., Jaenisch, R., 2009. Technical challenges in using human induced pluripotent stem cells to model disease. Cell Stem Cell 5, 584–595.
33. Saraiva, M.J., Costa, P.P., Birken, S., et al., 1983. Presence of an abnormal transthyretin (prealbumin) in Portuguese patients with familial amyloidotic polyneuropathy. Trans. Assoc. Am. Physicians 96, 261–270.
34. Sato, T., Susuki, S., Suico, M.A., et al., 2007. Endoplasmic reticulum quality control regulates the fate of transthyretin variants in the cell. EMBO J. 26, 2501–2512.
35. Sekijima, Y., Hammarstrom, P., Matsumura, M., et al., 2003. Energetic characteristics of the new transthyretin variant A25T may explain its atypical central nervous system pathology. Lab. Invest. 83, 409–417.
36. Shiraki, N., Umeda, K., Sakashita, N., et al., 2008. Differentiation of mouse and human embryonic stem cells into hepatic lineages. Genes Cells 13, 731–746.
37. Shiraki, N., Yamazoe, T., Qin, Z., et al., 2011. Efficient differentiation of embryonic stem cells into hepatic cells in vitro using a feeder-free basement membrane substratum. PLoS One 6, e24228.
38. Soldner, F., Hockemeyer, D., Beard, C., et al., 2009. Parkinson's disease patient-derived induced pluripotent stem cells free of viral reprogramming factors. Cell 136, 964–977.
39. Sousa, M.M., Yan, S.D., Stern, D., et al., 2000. Interaction of the receptor for advanced glycation end products (RAGE) with transthyretin triggers nuclear transcription factor κB (NF-κB) activation. Lab. Invest. 80, 1101–1110.
40. Stadtfeld, M., Nagaya, M., Utikal, J., et al., 2008. Induced pluripotent stem cells generated without viral integration.
41. Science 322, 945–949.
42. Sueyoshi, T., Jono, H., Shinriki, S., et al., 2012. Therapeutic approaches targeting midkine suppress tumor growth and lung metastasis in osteosarcoma. Cancer Lett. 316, 23–30.
43. Suhr, O.B., Holmgren, G., Steen, L., et al., 1995. Liver transplantation in familial amyloidotic polyneuropathy. Follow-up of the first 20 Swedish patients. Transplantation 60, 933–938.
44. Sullivan, G.J., Hay, D.C., Park, I.H., et al., 2010. Generation of functional human hepatic endoderm from human induced pluripotent stem cells. Hepatology 51 (1), 329–335.

45. Takahashi, K., Yamanaka, S., 2006. Induction of pluripotent stem cells from mouse embryonic and adult fibroblast cultures by defined factors. Cell 126, 663–676.
46. Takahashi, K., Tanabe, K., Ohnuki, M., et al., 2007. Induction of pluripotent stem cells from adult human fibroblasts by defined factors. Cell 131, 861–872.
47. Ueda, M., Misumi, Y., Mizuguchi, M., et al., 2009. SELDI-TOF mass spectrometry evaluation of variant transthyretins for diagnosis and pathogenesis of familial amyloidotic polyneuropathy. Clin. Chem. 55, 1223–1227.
48. Westermark, P., Benson, M.D., Buxbaum, J.N., et al., 2002. Amyloid fibril protein nomenclature — 2002. Amyloid 9, 197–200.
49. Wigmore, S.J., Fearon, K.C., Ross, J.A., 1997. Modulation of human hepatocyte acute phase protein production in vitro by n–3 and n–6 polyunsaturated fatty acids. Ann. Surg. 225 (1), 103–111.
50. Woltjen, K., Michael, I.P., Mohseni, P., et al., 2009. piggyback transposition reprograms fibroblasts to induced pluripotent stem cells. Nature 458, 766–770.
51. Yamashita, T., Ando, Y., Katsuragi, S., et al., 2005. Muscular amyloid angiopathy with amyloidogenic transthyretin Ser50Ile and Tyr114Cys. Muscle Nerve 31, 41–45.
52. Yu, J., Vodyanik, M.A., Smuga-Otto, K., et al., 2007. Induced pluripotent stem cell lines derived from human somatic cells. Science 318, 1917–1920.
53. Zhou, H., Wu, S., Joo, J.Y., et al., 2009. Generation of induced pluripotent stem cells using recombinant proteins. Cell Stem Cell 4, 381–384.

12
CHAPTER

DERIVATION OF DIVERSE HORMONE-RELEASING PITUITARY CELLS FROM HUMAN PLURIPOTENT STEM CELLS

Bastian Zimmer,[1,2] Jinghua Piao,[1,3] Kiran Ramnarine,[1,4] Mark J. Tomishima,[1,4] Viviane Tabar,[1,3] and Lorenz Studer[1,2,3]

[1]The Center for Stem Cell Biology

[2]Developmental Biology Program Sloan Kettering Institute for Cancer Research, 1275 York Avenue, New York, NY 10065, USA

[3]Department of Neurosurgery, Memorial Sloan Kettering Cancer Center, 1275 York Avenue, New York, NY 10065, USA

[4]SKI Stem Cell Research Facility, 1275 York Avenue, New York, NY 10065, USA

INTRODUCTION

Human pluripotent stem cells (hPSCs) provide a unique resource for basic as well as translational research. Both human embryonic stem cells (hESCs) and human induced pluripotent stem cells (hiPSCs) are widely used to study early human development (Zhu and Huangfu, 2013), assess the toxic effects of chemicals (Dreser et al., 2015, Zimmer et al., 2012), model human diseases or cancer (Bellin et al., 2012, Funato et al., 2014, Merkle and Eggan, 2013), and discover novel potential drugs (Lee et al., 2012). Furthermore, access to greatly improved protocols for lineage-specific differentiation has led to the first experimental applications of hPSC-derived lineages in regenerative medicine such as in patients with macular degeneration (Schwartz et al., 2015). Other hPSC-based applications that are being pursued intensely include the replacement of hormone-producing cells such as in type 1 diabetes (Pagliuca et al., 2014, Rezania et al., 2014). Replacing hormone-producing cells is a particularly attractive approach for cell therapy, especially if restoration of feedback mechanisms with subsequent dynamic release of hormones can be achieved by the grafted cells.

Citation: Bastian Zimmer, Jinghua Piao, Kiran Ramnarine, Mark J. Tomishima, Viviane Tabar, and Lorenz Studer, Derivation of Diverse Hormone-Releasing Pituitary Cells from Human Pluripotent Stem Cells, http://dx.doi.org/10.1016/j.stemcr.2016.05.005.

The pituitary gland is considered the master gland of hormone function. Hypopituitarism is a disease condition with insufficient or absent function of the pituitary gland. Pituitary tumors are the most common cause but many other triggers can induce pituitary dysfunction including inborn genetic defects, brain trauma, immune and infectious diseases, or radiation therapy. The prevalence of hypopituitarism has been estimated at 46 per 100,000 (Regal et al., 2001), but this is likely an underestimation. The consequences of pituitary dysfunction are particularly serious in children where they can lead to severe learning disabilities, growth and skeletal problems, as well as effects on puberty and sexual function (Chemaitilly and Sklar, 2010). Chronic hypopituitarism requires lifelong complex hormone replacement therapies that are very costly and compromise quality of life. Furthermore, static delivery of hormones can only poorly mimic the dynamic secretion of the intact pituitary gland, which reacts to feedback mechanisms such as the hypothalamic-pituitary-adrenal (HPA) axis or the circadian clock. Therefore, there is a considerable clinical need to direct current treatment paradigms toward a more physiological and complete hormone replacement therapy (Smith, 2004).

It is conceivable that replacing the damaged cells via cell transplantation can restore pituitary function and permanently cure chronic hypopituitarism. Previous work in mouse ESCs has shown that anterior pituitary cells, capable of hormone secretion, can be generated in 3D cultures by recapitulating some of the complex morphogenetic interaction between the developing hypothalamic and oral ectoderm tissues in vitro (Suga et al., 2011). Our laboratory has recently reported a first attempt at generating functional adenohypophyseal cells from human PSCs (Dincer et al., 2013), and very recently pituitary cells have been generated from hPSCs using a 3D organoid approach (Ozone et al., 2016). While these studies represent a promising proof of concept, current protocols remain inefficient, poorly defined, and unsuitable for developing current good manufacturing practice (cGMP)-compatible culture conditions that will be eventually required for human therapeutic use.

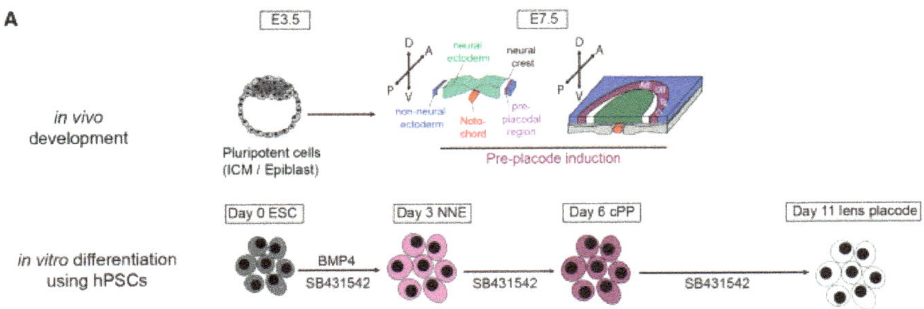

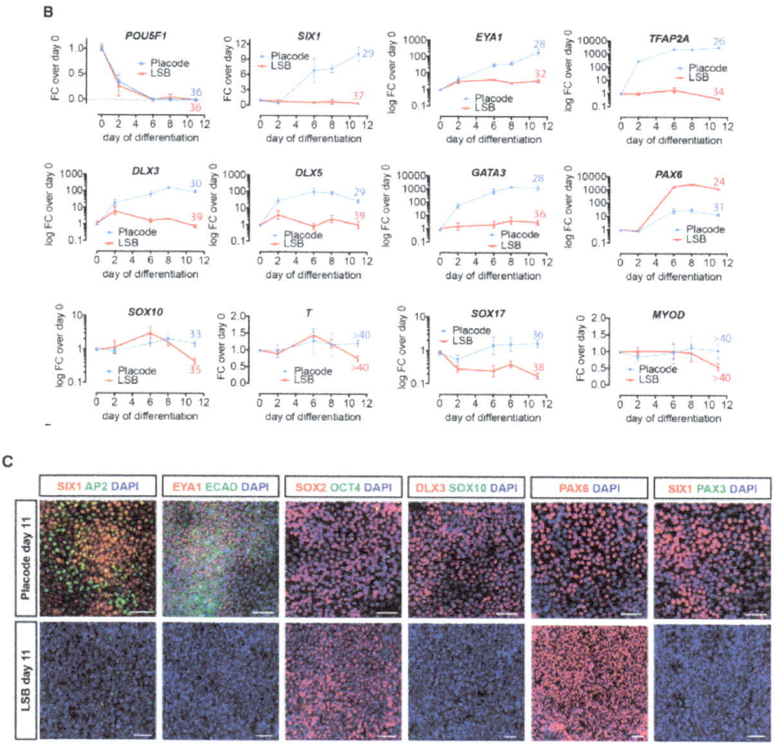

Figure 1: Differentiation of hPSCs into Cranial Placode Using Chemically Defined Conditions

(A) Schematic representation of cranial placode in vivo development and protocol for directed differentiation of hPSCs. ICM, inner cell mass. (B) Real-time PCR gene expression time course of key cranial placode (SIX1, EYA1) and non-neural ectoderm (TFAP2A, DLX3/5, GATA3) genes as well as genes probing for potential contaminates (SOX10, T, SOX17, MYOD). Values are normalized to GAPDH and expression on inefficient, poorly defined, and unsuitable for developing current good manufacturing practice (cGMP)-compatible culture conditions that will be eventually required for human therapeutic use.

Here, we report the efficient derivation of anterior pituitary cells from hPSCs in clinically compatible and scalable culture conditions. We further characterize the diversity of anterior pituitary subtypes achieved in vitro using single-cell mRNA expression analysis. The resulting hPSC-derived pituitary cells are functional in vitro, respond to appropriate stimuli, and are capable of secreting hormones in an animal model of hypopituitarism in vivo. Importantly, our data indicate that pituitary cell fate can be induced independent of mimicking the complex 3D organization of the developing gland. We demonstrate that by providing appropriate signals to purified placode precursor cells, pituitary identity can be specified at high efficiency, and that further manipulations of morphogen gradients enable controlled changes in the relative composition of hormonal cell types. In conclusion, we provide a robust differentiation platform to access diverse hormone-producing cell types suitable for further development toward a cell-based treatment of hypopituitarism.

RESULTS

Derivation of Cranial Placode from hPSCs under Fully Defined Conditions

The anterior pituitary gland is derived from cranial placode cells that form from the oral ectoderm. Therefore, the first step in establishing a defined protocol is the efficient induction of cranial placode cells competent in generating anterior pituitary lineages. The cranial placode induction protocol (PIP) presented here relies on serum-free monolayer-based induction conditions, uses fully defined cGMP-ready components, and eliminates ill-defined factors such as knockout serum replacement (KSR), Matrigel, or mouse embryonic fibroblast feeders that were part of previous protocols (Dincer et al., 2013). The specific factors used to trigger placode induction are based on signals shown to specify placode development in vivo (Figure 1A). We observed that exposure to moderate concentrations of bone morphogenetic protein 4 (BMP4) efficiently induced cranial placode cells that adopted by "default" a lens fate (**Figures S1**A–S1C). These results are in agreement with previous efforts of inducing placode fate from hPSCs and with data in the developing chick embryo reporting a lens default in the absence of fibroblast growth factor (FGF) signals (Bailey et al., 2006, Leung et al., 2013). Temporal expression analysis revealed the rapid loss of pluripotency markers and induction of key placode genes such as *SIX1*,*EYA1*, and *DLX3/5* after 6 days of differentiation at both mRNA (Figure 1B) and protein (Figure 1C) levels. Transcripts of contaminating cell types such as*SOX10* (neural crest), *SOX17* (endoderm), *T/ Brachyury* (mesoderm), or *MYOD*(myogenic lineages) were not induced under these conditions (Figure 1B). Our previous study reported PAX3$^+$ trigeminal placode as the "default" identity of hPSC-derived placode cells (Dincer et al., 2013). Our current method, using the defined placode induction conditions, shows PAX6 rather than PAX3 expression (Figure 1C). To further quantify the yield and selectivity of placode induction and PAX6 expression, we used hESC genetic reporter lines for *SIX1::H2B-GFP*,*PAX6::H2B-GFP*, *SOX10::GFP* (Chambers et al., 2012, Mica et al., 2013). Flow analysis confirmed robust induction of PAX6 and SIX1 without SOX10, consistent with anterior cranial placode in the absence of contaminating neural crest cells (**Figures S1**D and S1E).

Patterning of Anterior Pituitary from hPSC-Derived Cranial Placode

The complex morphogenetic development of the pituitary gland occurs during early embryonic stages (~embryonic day 10 [E10] in mouse and between embryonic weeks 4 and 8 in human). Both the anterior and intermediate lobes of the pituitary gland are derived from oral ectoderm, which corresponds to the pituitary placode while the posterior pituitary gland develops from the neural ectoderm (Figure 2A). Inductive tissue interactions as well as various defined signaling pathways including FGFs, BMPs, and Sonic hedgehog (SHH) are thought to be important for proper gland development and hormonal subtype specifications in vivo (Zhu et al., 2007) (Figure 2A, upper panel). Here we assessed the ability of these developmental pathways to induce pituitary identity (Figure 2A, lower panel). Our data show that timed exposure to SHH, FGF8, and FGF10 robustly induces gene expression associated with anterior pituitary development including *PITX1/2*, *LHX3/4*, *HESX1*, and *SIX6* (Figure 2B). Where possible, we confirmed expression at the protein level using antibodies against PITX1, LHX3, HESX1, and SIX6 (Figure 2C). We compared our cGMP-ready E8/E6-based induction protocol with our published KSR-based PIP (Dincer et al., 2013). To compensate for KSR lot-to-lot variability, which can dramatically affect differentiation efficiency, we performed PIP using two distinct concentrations of the BMP inhibitor LDN-193189. After 15 days of differentiation, the cells were analyzed using qRT-PCR probing for pan-placodal markers such as *SIX1* as well as the pan-pituitary markers *PITX1*, *PITX2*, *LHX3*, and *LHX4*. In addition, the neuroectoderm marker *PAX6* and the non-neural ectoderm transcription factor*TFAP2A* were included in the analysis (Figure S2A). Cell identity was further confirmed at the protein level using immunofluorescence staining for SIX1 and LHX3 (Figure S2B). The KSR lot used for these experiments failed to effectively induce pituitary or placode identity as shown by the low expression of *SIX1* and*TFAP2A* and high expression of *PAX6*. Lowering the LDN-193189 concentration was able to partially but not fully rescue that effect compared with our new E8/E6-based protocol. The cGMP-ready protocol presented here works reliably and with comparable efficiency across various hESC and hiPSC lines (Figures S3A–S3C). Furthermore, we confirmed that the recombinant protein SHH can be replaced by small-molecule smoothened agonists such as purmorphamine and SAG (Figures S3D and S3E). However, despite robust induction of anterior pituitary-lineage markers, we observed an increase in cell death when using the small-molecule-based induction conditions, which prompted us to use recombinant SHH for subsequent studies.

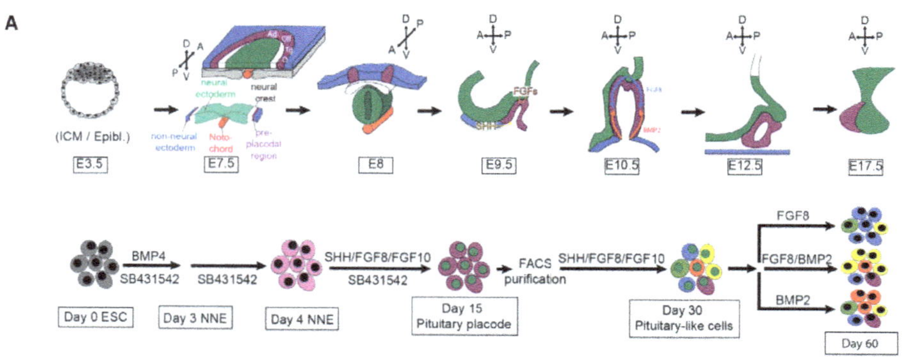

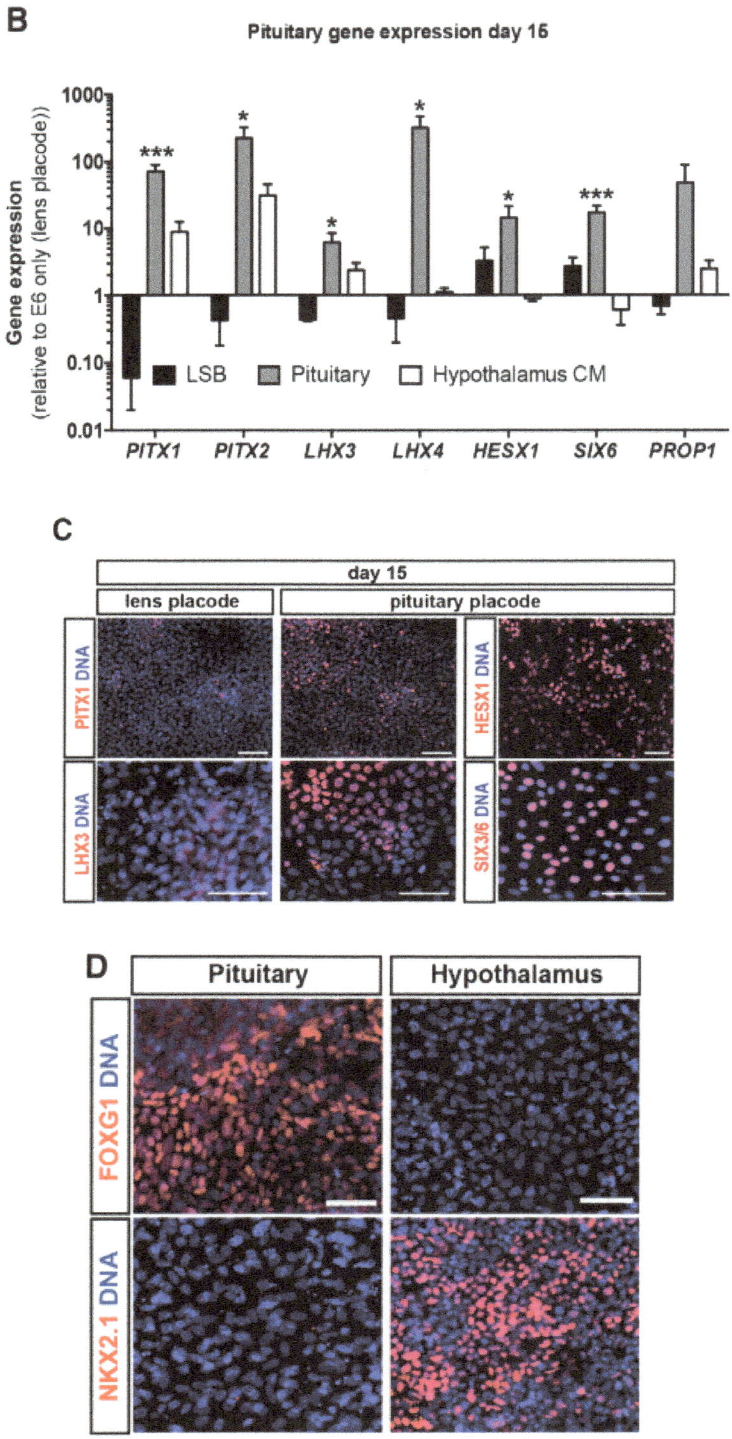
B
Pituitary gene expression day 15
Gene expression
(relative to E6 only (lens placode))
1000
100
10
1
0.10
0.01
***
*
*
*
*
***
LSB
Pituitary
Hypothalamus CM
PITX1
PITX2
LHX3
LHX4
HESX1
SIX6
PROP1
C
day 15
lens placode
pituitary placode
PITX1 DNA
HESX1 DNA
LHX3 DNA
SIX3/6 DNA
D
Pituitary
Hypothalamus
FOXG1 DNA
NKX2.1 DNA

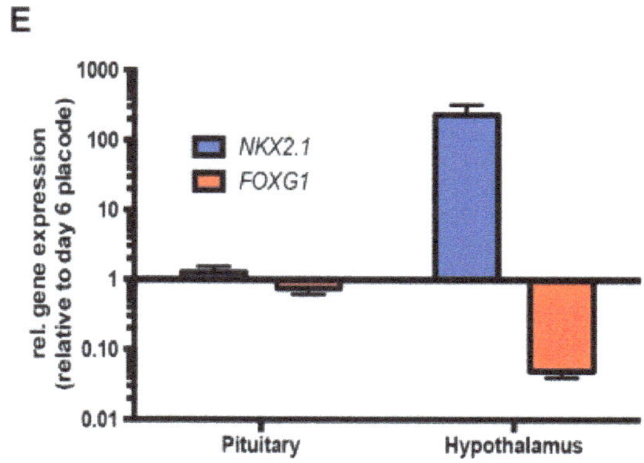

Figure 2: Pituitary Specification of Anterior Cranial Placode-Derived hPSCs

(A) Schematic representation of pituitary gland in vivo development and protocol for directed differentiation of hPSCs into anterior pituitary-like cells. ICM/Epibl., inner cell mass/epiblast; NNE, non-neural ectoderm. (B) Real-time PCR analysis comparing expression of key genes involved in pituitary development in LSB, pituitary condition, and medium conditioned by hypothalamic neuroectoderm (Hypothalamus CM) after 15 days of differentiation in the respective medium. Values are normalized to GAPDH and gene expression on day 15 of lens differentiation (E6 only) and plotted as mean ± SEM of at least four independent experiments. $*p < 0.05$, $***p < 0.001$ compared with E6 only condition on day 15. (C) Immunofluorescence analysis comparing expression of PITX1 and LHX3 after 15 days of differentiation under lens or pituitary conditions as well as expression of HESX1 and SIX3/6 on day 15 of pituitary differentiation. Scale bars, 50 mm. (D) Immunofluorescence comparison of cells differentiated for 15 days under either pituitary or hypothalamus condition. Cells were stained for either FOXG1 (Pituitary) or NKX2.1 (Hypothalamus). Scale bars, 50 mm. (E) qRT-PCR analysis of day 0-15 cells differentiated under pituitary or hypothalamic ectoderm condition probing for NKX2.1 and FOXG1. Values have been normalized to GAPDH and expression in day-6 placode cells and are plotted as means ± SEM of two to four independent experiments.

Interestingly, medium conditioned by hPSC-derived hypothalamic anlage (Maroof et al., 2013, Merkle et al., 2015) (Figures 2D and 2E) was not sufficient to robustly induce pituitary marker expression (such as *LHX4*, *HESX1*, and *SIX6*;Figure 2B). *LHX4* and *SIX6* have been implicated in pituitary progenitor expansion, which might explain the reduced cell yield using hypothalamic-conditioned medium. Despite compelling evidence to the contrary (Suga et al., 2011), our data suggest that defined extrinsic cues might be sufficient to induce pituitary placode identity. To directly assess the role of hypothalamic tissue during pituitary placode induction and differentiation, we made use of the*SIX1::H2B-GFP* reporter cell line. Early placode cells differentiated under default conditions were sorted at day 6 of differentiation for *SIX1::H2B-GFP* expression followed by further differentiation under either lens conditions (default), in the presence of SHH, FGF8, and FGF10 (pituitary conditions) or in the presence of medium conditioned by hPSC-derived hypothalamic neuroectoderm (Figure 3A). Gene-expression analysis revealed that even purified *SIX1::H2B-GFP*$^{+}$ cells, devoid of any hypothalamic lineage cells, are capable of expressing all key pituitary markers in response to defined cues. In contrast, hypothalamic conditioned medium failed to induce *LHX4* expression above levels observed under default lens conditions (Figure 3B). The levels of induction, especially*LHX4*, were slightly lower in the *SIX1::H2B-GFP* purified cells compared with our standard pituitary PIP. This is possibly due to the relatively late pituitary induction, which started on day 6 (versus day 4) to exclude patterning of the pre-placode tissue. Alternatively, the sorting process or the associated cell dissociation may have also contributed to the slight decrease in induction efficiency.

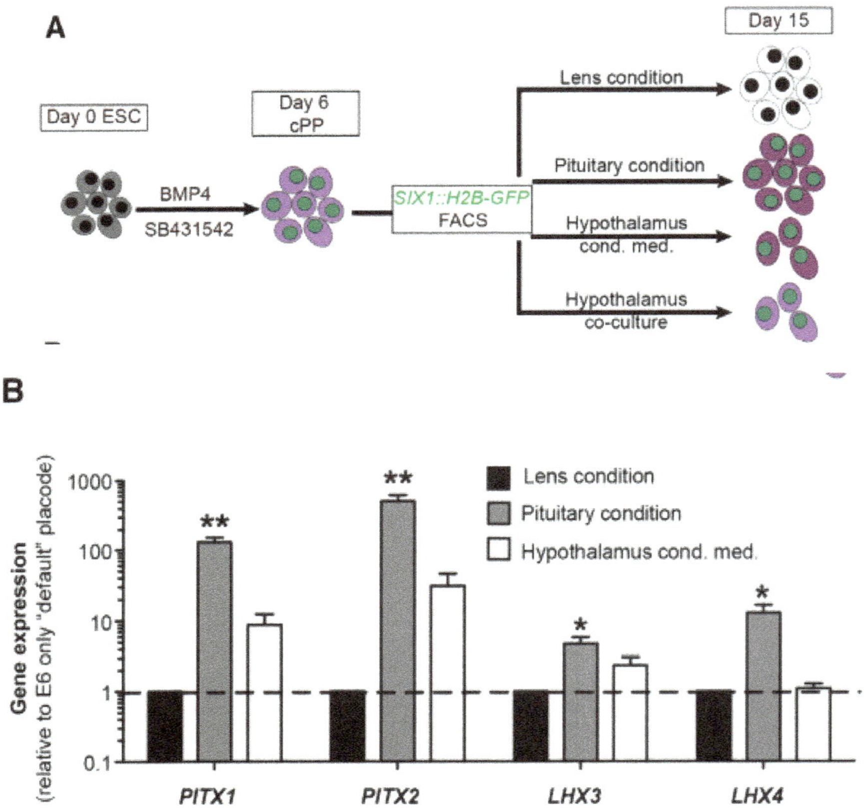

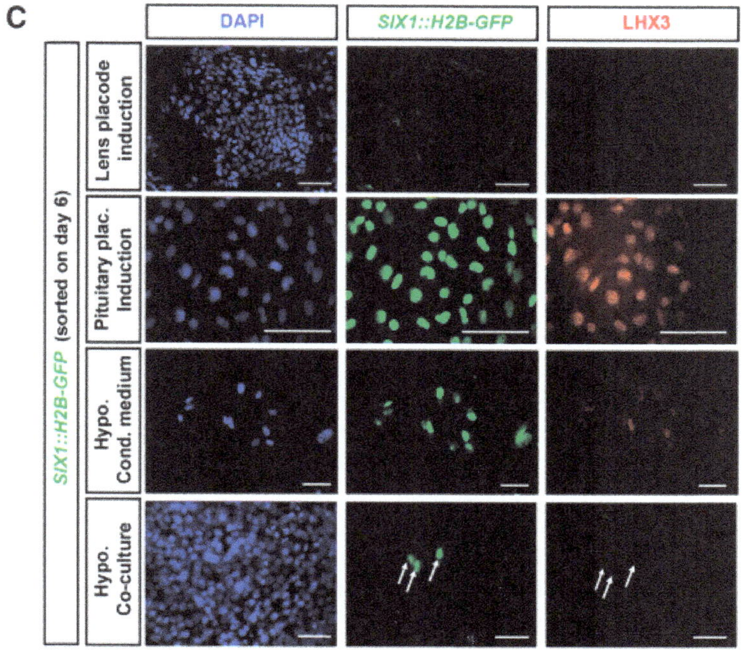

Figure 3: Pituitary Placode Induction from Purified SIX1:GFPH2B Cells

(A) Schematic representation of the experimental outline. hESC were differentiated under default conditions for 6 days. Unpatterned SIX1+ cells were FACS purified and cultured for an additional 9 days in various conditions. Cells were analyzed on day 15. (B) Gene-expression analysis of key pituitary genes in cells grown in three conditions described in (A). Values are normalized to GAPDH and gene expression on day 15 of lens differentiation (E6 only) and plotted as mean ± SEM of at least four independent experiments. $*p < 0.05$, $**p < 0.01$ compared with E6-only condition on day 15. (C) Immunofluorescence analysis of SIX1 sorted cells after 9 days of differentiation in respective medium condition. Arrows indicate absence of LHX3 expression in SIX1+ cells in co-culture condition. Scale bars, 50 mm.

In a separate set of experiments, we co-cultured day-6 sorted, *SIX1::H2B-GFP*$^+$cells in direct contact with hPSC-derived hypothalamic anlage and stained the cells for the pan-placodal marker SIX1 and the pituitary marker LHX3 after 9 days of additional co-culture (day 15). As additional controls, we included the "default" lens conditions as well as our standard pituitary conditions and hypothalamic-conditioned medium (Figure 3A). While the lens condition formed lentoid-like clusters and downregulated SIX1 expression, the pituitary condition resulted in SIX1/LHX3 double-positive cells. The conditioned medium maintained SIX1 while inducing only low levels of LHX3. In the co-culture, SIX1 expression was maintained but no LHX3 expression was detected (Figure 3C). These findings support the idea that exposure to defined extrinsic cues is sufficient to direct hPSC into pituitary lineages cells. While we cannot rule out that further optimization of co-culture conditions may yield pituitary-lineage cells, we demonstrate that defined signals present a robust alternative that during cell manufacturing should greatly reduce variability inherent to complex co-culture systems.

hPSC-Derived Adenohypophyseal Cells Are Functional

The main function of the anterior pituitary (adenohypophysis) is to secrete six specific hormones controlling key events in the human body including stress response (adrenocorticotropic hormone [ACTH]), skeletal growth (growth hormone [GH]), metabolism (thyroid-stimulating hormone [TSH]), and reproductive functions (prolactin [PRL], follicle-stimulating hormone [FSH], luteinizing hormone [LH]). We therefore assessed the presence of hormonal subtypes in our culture. We were able to detect ACTH, GH, and PRL as well as FSH- and LH-expressing cells in our culture at day 30 of differentiation (Figure 4A). ELISA measurements of cell-culture supernatant confirmed basal secretion rates for ACTH, GH, and FSH (Figure 4B). Hormone release in the anterior pituitary gland is tightly regulated by several feedback mechanisms from the various target organs as well as from upstream hypothalamic signals delivered via the hypophyseal portal venous plexus. Therefore, the functional response of pituitary cells must integrate multiple, distinct regulatory stimuli. hPSC-derived pituitary cells at day 30 of differentiation showed induction of ACTH release in response to stimulation with corticotropin-releasing factor (CRF), stressin, or urocortin. In contrast, exposure to inappropriate stimuli such as ghrelin or somatocrinin did not trigger ACTH release (Figure 4C). On the other hand, somatocrinin but not CRF exposure triggered a significant increase in GH release (Figure 4D). Finally, the cultures also induced FSH release upon exposure to nafarelin (Figure 4E).

Single-Cell Gene-Expression Analysis Reveals Diversity of Hormonal Lineages

Differentiation of the hormonal progenitor lineages within the adenohypophysis (Tabar, 2011) is a tightly regulated spatial and temporal process involving various patterning events (Figure 2A). To address the diversity of progenitor fates in our hPSC-based culture system, we performed single-cell qRT-PCR analyses. We probed for 34 genes spanning pituitary development from undifferentiated PSCs to mature hormone-producing cells. We also included primers to survey for potential mesodermal or endodermal contaminants such as *T/Brachyury*, *MYOD*, and *SOX17*. Principal component analysis (PCA) of cells at day 30 and day 60 of differentiation showed a clear time-dependent change in expression with only few cells moving ahead of schedule (i.e., day-30 cells showing a day-60 profile) or being delayed (i.e., day-60 cells retaining a day-30 signature) (Figure 5A). The scree plot (Figure 5B) defined the PCA components that explain most of the variability of the data. Hierarchical clustering confirmed the separation of cells largely along the time axis, resulting in two main clusters interspersed with several smaller subclusters (Figure 5C). In addition, heatmaps based on the raw ct values are provided in Figure S4. We further validated our single-cell data by immunofluorescence staining in day-30 cultures for the progenitor marker HESX1, and for NEUROD1, a more mature marker transiently expressed in corticotrophs. Immunofluorescence analysis at day 15 of differentiation served as negative control for NEUROD1 (Figure S5A). We confirmed co-labeling of HESX1 and NEUROD1 in the same cell at day 30 of differentiation. However, the levels of HESX1 expression were much lower at day 30 compared with day 15.

Our analysis revealed that day-30 cultures contain an already high percentage of pituitary-like cells with ~70% of cells co-expressing pituitary transcripts such as*PITX1* and *LHX3*. Only four cells (~5% of all cells analyzed) expressed *T*,*SOX17*, or *MYOD*, suggesting a low percentage of contaminating cells. Most cells expressed *TBX19* (*TPIT*) a transcription factor shown to be crucial for the development of the *POMC* lineage (Lamolet et al., 2001). Furthermore, most cells expressed the pan-placodal marker *SIX1* and co-expressed *PAX6*compatible with pituitary placode fate. However, we also observed expression of other placode fates including *PAX2* (epibranchial), *PAX3* (trigeminal), or *PAX8*(otic) that together were detected in about 20% of the *SIX1*$^+$ population.

The ultimate functional units of the anterior pituitary are cells that secrete specific hormones. Our single-cell analysis showed that at day 30 of differentiation approximately 50% of the cells expressed at least one hormonal mRNA species. This percentage increased to about 80% by day 60, indicating further in vitro maturation (Figure 5D). This increase in the percentage of hormone-positive cells was one of the main factors responsible for the overall difference observed between day-30 and day-60 cells. There have been reports suggesting that both the developing and adult rodent pituitary gland contain cells that express more than a single hormone (Nunez et al., 2003, Villalobos et al., 2004). Indeed, in our hPSC-derived cultures we could detect expression of more than one hormonal transcript (hereafter termed "multiple hormone transcripts") in 10% of the cells by day 30 of differentiation. By day 60 of differentiation, this percentage increased to ~30% of the total cell population (Figure 5D). We found that the majority of multiple hormone transcript-expressing cells by day 60 expressed both *POMC* and *GH* (~10%). Cells expressing more than two hormone transcript types were only detected by day 60 and always included*POMC* expression (Figure S5B).

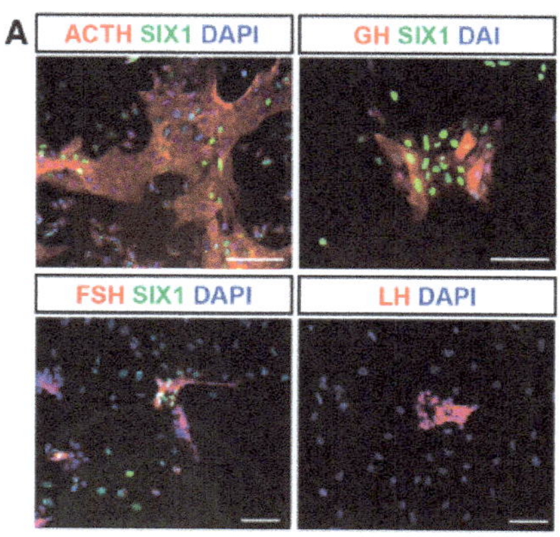

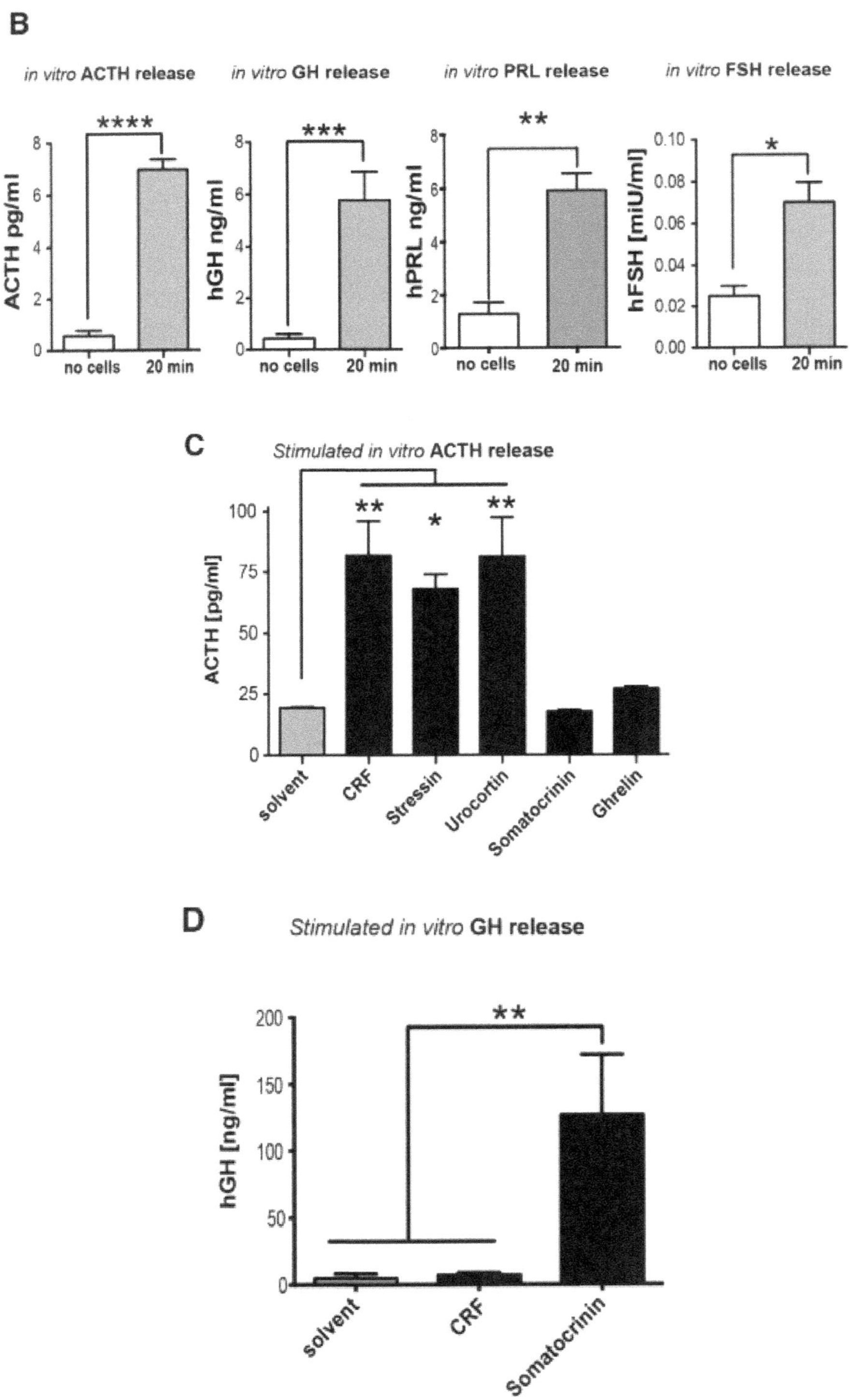
B
in vitro ACTH release
in vitro GH release
in vitro PRL release
in vitro FSH release
****
***
**
*
ACTH pg/ml
hGH ng/ml
hPRL ng/ml
hFSH [miU/ml]
no cells
20 min
C
Stimulated in vitro ACTH release
ACTH [pg/ml]
solvent
CRF
Stressin
Urocortin
Somatocrinin
Ghrelin
D
Stimulated in vitro GH release
hGH [ng/ml]
solvent
CRF
Somatocrinin

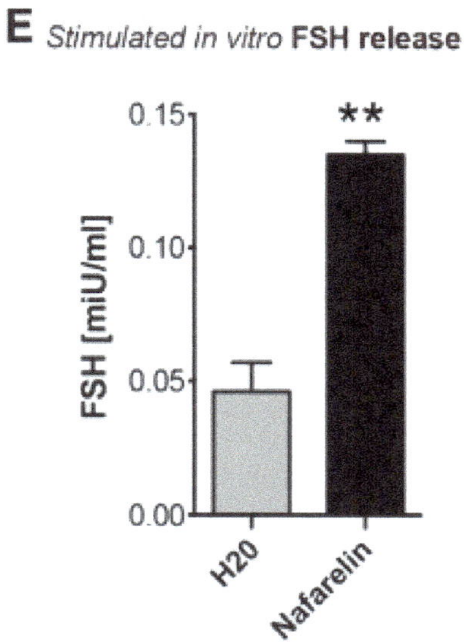

Figure 4: Functional Characterization of Anterior Pituitary Cells

(A) Immunofluorescence analysis of anterior pituitary cells after 30 days of differentiation. On day 30 the culture contains corticotrophs (ACTH), somatotrophs (GH), and gonadotrophs (FSH, LH). Scale bar, 50 mm. (B) In vitro basal hormone release on day 30 of differentiation as assessed by ELISA. Data are plotted as mean ± SEM of three independent experiments. *$p < 0.05$, **$p < 0.01$, ***$p < 0.001$, ****$p < 0.0001$ compared with no cells (differentiation medium only). (C–E) Quantification of hormone levels after 24 hr of in vitro stimulation using compounds triggering hormone release. ACTH release was specifically induced by CRF, stressin, or urocortin and not by somatocrinin or ghrelin (C), GH release was induced by somatocrinin but not CRF (D), and FSH release was induced by nafarelin (E). Data are plotted as mean ± SEM of three independent experiments. *$p < 0.05$, **$p < 0.01$ compared with the solvent control.

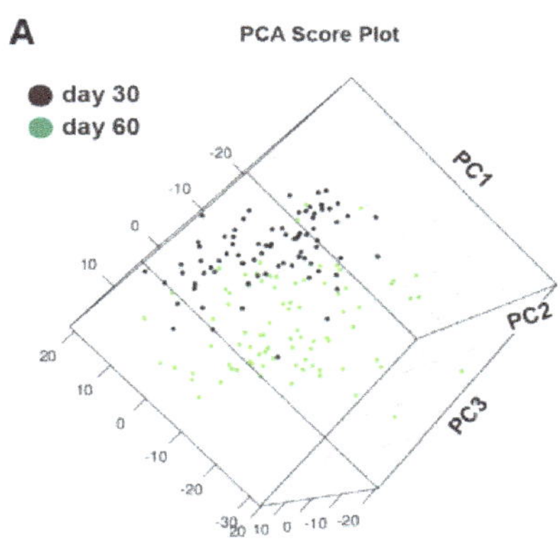

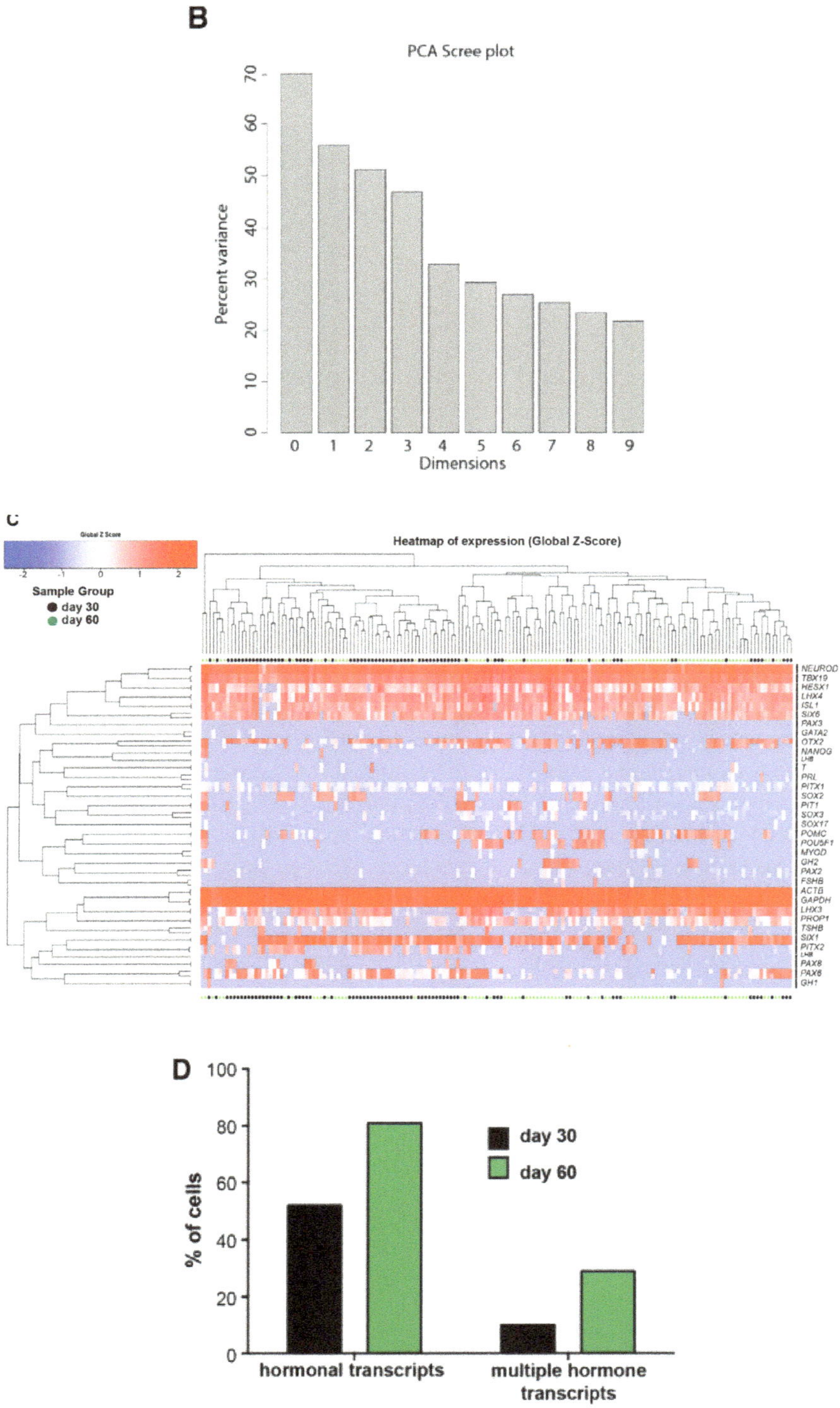
B
PCA Scree plot
Percent variance
0
10
20
30
40
50
60
70
0 1 2 3 4 5 6 7 8 9
Dimensions
C
Global Z Score
-2 -1 0 1 2
Sample Group
day 30
day 60
Heatmap of expression (Global Z-Score)
NEUROD
TBX19
HESX1
LHX4
ISL1
SIX6
PAX3
GATA2
OTX2
NANOG
LHB
T
PRL
PITX1
SOX2
PIT1
SOX3
SOX17
POMC
POU5F1
MYOD
GH2
PAX2
FSHB
ACTB
GAPDH
LHX3
PROP1
TSHB
SIX1
PITX2
PAX8
PAX6
GH1
D
% of cells
0
20
40
60
80
100
day 30
day 60
hormonal transcripts
multiple hormone transcripts

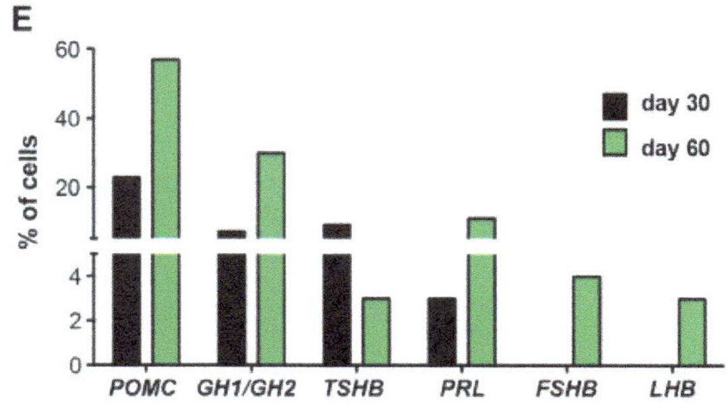

Figure 5: Temporal Single-Cell qRT-PCR Analysis of Anterior Pituitary Development In Vitro

(A and B) Principal component analysis of single cells on day 30 (black) and day 60 (green) of differentiation reveals two distinct populations of cells. (C) Unsupervised hierarchical clustering of day-30 and day-60 cells using 34 different primer pairs identifies two clusters of cells with very few leading cells (day-30 cells resembling day-60 cells) and cells lagging behind (day-60 cells still more closely resembling day-30 cells). (D) Quantification of hormone-expressing cells on day 30 and day 60 as well as percentage of cells expressing more than one hormonal transcript per cell. (E) Expression of individual hormonal transcripts per single cell on day 30 and day 60, respectively. See also Figures S4 and S5.

The most frequent hormonal transcript expressed in hPSC-derived pituitary cells at day 30 of differentiation was *POMC* (30% of total cells), thought to emerge from the dorsal pituitary anlage. The more ventral cell types such as *GH* or *TSH*made up about 20% of the total cell population by day 30 of differentiation. *PRL*was expressed in an even smaller subset of cells. Finally, *FSH* and *LH*, the two most ventral cell types, which appear only at later stages of development, were not detected by day 30 (Figure 5E). At day 60 of differentiation the number of*POMC*- and *GH*-expressing cells increased to 55% and 30%, respectively. Only few cells expressed *FSH* (~4%) and *LH* (~3%), even at day 60 of differentiation (Figure 5E). In addition to the single-cell PCR we characterized the cell-surface marker expression of the day-30 culture using the commercially available BD Lyoplate screening kit (Figure S6).

Dorsal-Ventral Patterning of Anterior Pituitary Cells In Vitro Using Patterning Factors

Hypopituitarism is a very diverse and complex disease. Depending on the cause of pituitary dysfunction, the type of hormones affected can vary. For example, GH deficits are commonly observed in patients with inborn genetic disease (van Gelderen and van der Hoog, 1981) but can also occur in patients following radiation treatment (Sklar and Constine, 1995). In contrast, lymphocytic hypophysitis, an autoimmune disease of the pituitary gland, affects primarily ACTH (Rivera, 2006). Therefore, for the broad application of hPSC-derived pituitary cells in the future, cell replacement therapy may need to be customized to the specific needs of a given patient population. Since our standard conditions mostly yield dorsal, ACTH+ cells, we asked whether additional signals can be used to enhance the production of more ventral cell types. It has been shown that FGF8 and BMP2 signaling gradients play an important role in dorsal-ventral patterning of the mouse pituitary gland (Rosenfeld et al., 2000) (Figure 2A). We therefore treated pituitary-lineage cells with high concentrations of either FGF8 (dorsalizing) or BMP2 (ventralizing), or with a mixture of the two patterning factors at intermediate concentrations to mimic morphogen gradients occurring in vivo. Gene-expression studies for key transcription factors of pituitary precursor lineage and hormonal subtypes confirmed the need for BMP2 to generate the most ventral cell types. *FSHB* and *LHB* were significantly upregulated in the presence of BMP2 while FGF8 exerted a negative effect on*FSHB* yield (Figure 6A).

We next performed single-cell qRT-PCR analysis to increase the resolution of our analysis. Unsupervised hierarchical clustering of single cells at day 60, treated with FGF8, BMP2, or a combination of both factors, revealed three larger clusters of cells (Figure 6B). In addition, heatmaps based on the raw ct values are provided in **Figure S4**. Cells corresponding to each of the three treatments were observed in every cluster. However, 49% of all the total cells in cluster 3 were from the BMP2-treated group while 56% of all the cells in cluster 2 were derived from the FGF8 group. Cluster 1 represented cells from all three treatments in roughly equal proportions. By analyzing single-cell expression for each of the anterior pituitary hormones, we confirmed findings predicted from mouse in vivo studies (Rosenfeld et al., 2000) in our hPSC-based culture system. High concentrations of FGF8 led to increased numbers of *POMC*+ cells compared with high concentrations of BMP2 (75% versus 40%, respectively) (Figure 6C). Furthermore, intermediate concentrations of both signaling molecules resulted in an increase of cells expressing *GH* and *TSHB* compared with either FGF8 or BMP2 alone (Figure 6C). Finally, high concentrations of BMP2 decreased the number of *POMC*+ dorsal cell types and increased the number of ventral *FSHB* and *LHB* cell types.

Immunofluorescence confirmed the single-cell qRT-PCR data for four hormones under the various dorsal-ventral patterning conditions (FGF8, FGF8/BMP2, BMP2; Figure 6D). Quantification showed a bias toward dorsal ACTH (*POMC*)-expressing cells in the FGF8 treated culture. PRL and GH were the most abundant hormones observed upon patterning with FGF8/BMP2 while FSH was the most abundant cell type in BMP2-treated cultures (Figure 6E). POMC is the precursor polypeptide of ACTH and 44 amino acids are removed during translation, which ultimately gives rise to the hormone ACTH.

Grafted Human ESC-Derived Anterior Pituitary Cells Function in a Model of Hypopituitarism

To assess the ability of hPSC-derived pituitary cells to survive and function in vivo, we transplanted day-30 cells into hypophysectomized rats. After surgical removal of the pituitary gland (Figure 7A) using a parapharyngeal approach, hypopituitarism in the rats was confirmed by measuring ACTH levels in the plasma via ELISA. Rats that were successfully hypophysectomized were divided into two experimental groups, the sham group (n = 4) receiving Matrigel-only injections and the grafting group (n = 7) receiving Matrigel containing hESC-derived pituitary cells (day 30, standard conditions without BMP2 treatment). Following transplantation of 2×10^6 cells subcutaneously, both treatment and control groups were followed for 7 weeks post-transplantation and pituitary hormone levels in the plasma were monitored (Figures 7B–7D). At 3 weeks after transplantation, hormone levels in the grafted group increased compared with the 1-week time point while the levels in the sham group remained largely unchanged. ACTH levels remained at higher levels in the grafted versus control group for the 7-week period of the experiment (Figure 7B). Increases in the level of two other hormones, namely GH and LH, were more variable and did not reach significance at all the time points tested. However, significant increases in GH levels were detected at 3 and 7 weeks after transplantation (Figure 7C). No significant increase in LH levels was observed (Figure 7D). Hormone levels in grafted animals were compared with levels in intact age-matched rats and found to be ~40% for ACTH, ~28% for GH, and ~20% for PRL (Figure S7). In a final step to evaluate the function of the transplanted cells, we performed measurements of the target hormones affected by pituitary hormone secretion. The normal HPA-axis response involves an increase in glucocorticoids secreted by the adrenal glands upon release of ACTH (Webster and Sternberg, 2004). In humans, ACTH triggers the release of cortisol, whereas the main glucocorticoid in rodents is corticosterone (Wand, 2008). We therefore measured corticosterone levels in both experimental groups (Figure 7E). The grafted group showed consistently higher levels of corticosterone than sham-grafted animals resulting in a statistically significant difference by 7 weeks after transplantation. These data indicate an appropriate response by the host adrenal glands to human ACTH released by grafted hPSC-derived cells. At 7 weeks after transplantation, the animals were euthanized and the graft was analyzed histologically. We were able to detect cells expressing each of the six anterior pituitary hormones within the grafts (Figures 7F and 7G), confirming cell survival and suggesting, at least in part, further in vivo differentiation and maturation of the cells. Stereological quantification of the grafts showed an average of $3.08 \times 10^6 \pm 0.42 \times 10^6$(average ± SD; n = 3) human cells per graft, with the majority (>95%) having a placode identity, as determined by co-expression of SIX1 and human nuclear antigen (hNA) by immunohistochemistry. The proportion of Ki67-positive proliferating cells in the graft was 9.6% ± 0.6% (average ± SD; n = 3) of the hNA^+ population. The entire graft stained negative for the pluripotency-associated surface markers SSEA-4 and Tra1-60. No signs of tumors were detected up to 7 weeks after transplantation. Stereological analyses showed an average of 18,212 ± 2,969 $ACTH^+$ cells per animal (optical fractionator method), and a graft volume of 81 ± 10 mm^3 (Cavalieri estimator; average ± SD, n = 3).

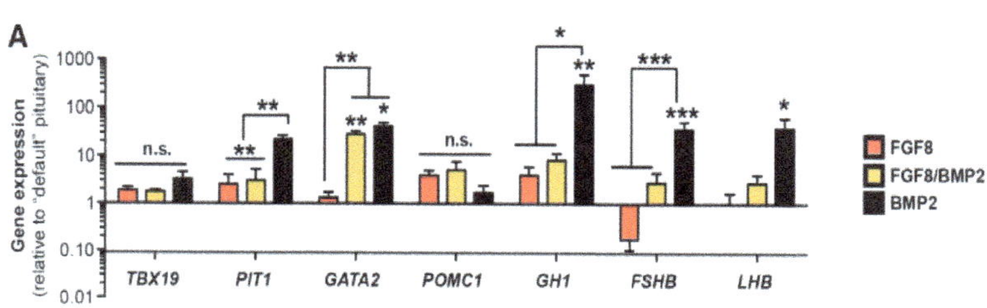

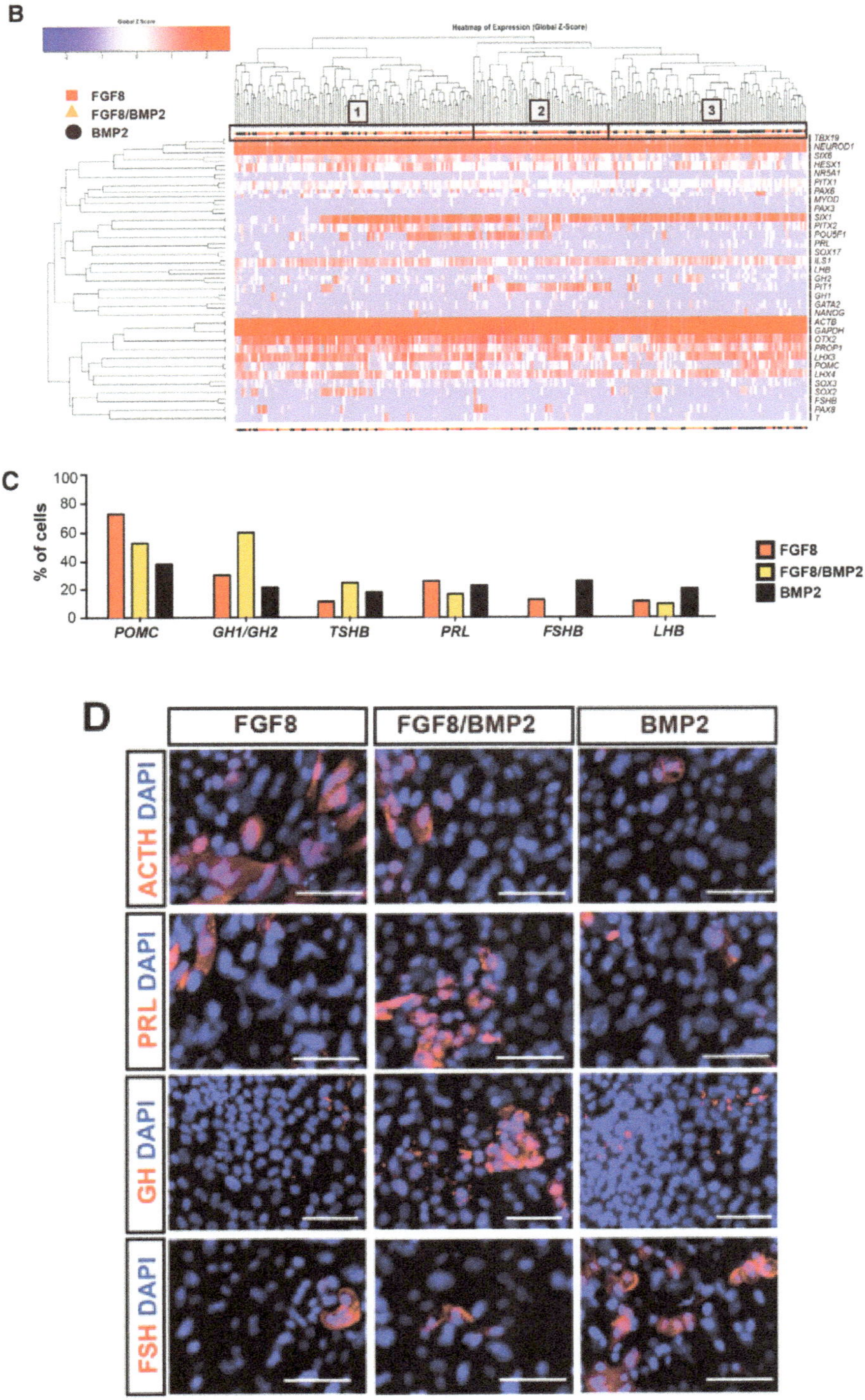
B
FGF8
FGF8/BMP2
BMP2
1
2
3
TBX19
NEUROD1
SIX6
HESX1
NR5A1
PITX1
PAX6
MYOD
PAX3
SIX1
PITX2
POU5F1
PRL
SOX17
ILS1
LHB
GH2
PIT1
GH1
GATA2
NANOG
ACTB
GAPDH
OTX2
PROP1
LHX3
POMC
LHX4
SOX3
SOX2
FSHB
PAX8
T
C
% of cells
100
80
60
40
20
0
POMC
GH1/GH2
TSHB
PRL
FSHB
LHB
FGF8
FGF8/BMP2
BMP2
D
FGF8
FGF8/BMP2
BMP2
ACTH DAPI
PRL DAPI
GH DAPI
FSH DAPI

E

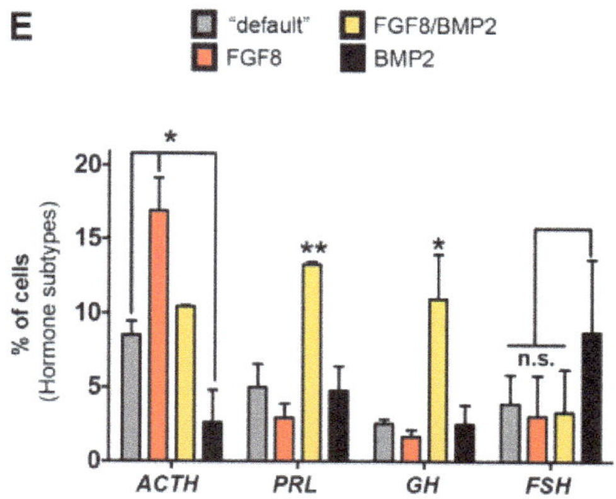

Figure 6: Specification of Hormonal Cells of the Pituitary In Vitro

(A) Bulk qRT-PCR analysis of day-60 cells patterned with FGF8, FGF8/BMP2, or BMP2 for 30 days. Patterning with BMP2 induced a more ventral cell identity (PIT1, GATA2, GH1, FSHB, and LHB) while FGF8 suppressed ventral cell types (FSHB). Data are plotted as mean ± SEM of two to four independent experiments. *p < 0.05, **p < 0.01, ***p < 0.001 (n.s., not significant) compared with the "default" pituitary differentiation on day 60. (B) Unsupervised hierarchical clustering of FGF8, FGF8/BMP2, and BMP2 patterned cells using 34 primer pairs identified three larger clusters of cells, with cluster 2 mainly comprising cells patterned by FGF8 (or FGF8/BMP2) and cluster 3 mainly comprising cells patterned by BMP2 (or FGF8/BMP2). (C) Quantification of hormonal transcripts per cell in different patterning conditions. Data are plotted as percentage of cells expressing the respective transcript (ct < 35 cycles in combination with a proper melting curve). (D) Immunofluorescence analysis (representative images) of hormone expression in cells patterned with FGF8, FGF8/BMP2, or BMP2 on day 60 of differentiation. Scale bars, 50 mm. (E) Quantification of hormone-expressing cells (per subtype) in different patterning conditions on day 60 of differentiation. High levels of FGF8 induced dorsal fate (ACTH) while intermediate levels of FGF8 and BMP2 induced dorsal/ventral fates (PRL and GH) compared with the default condition (E6 only). Data are plotted as mean ± SEM of two independent experiments. *p < 0.05, **p < 0.01 (n.s., not significant) compared with the "default" (E6 only) pituitary differentiation on day 60. See also Figure S4.

A

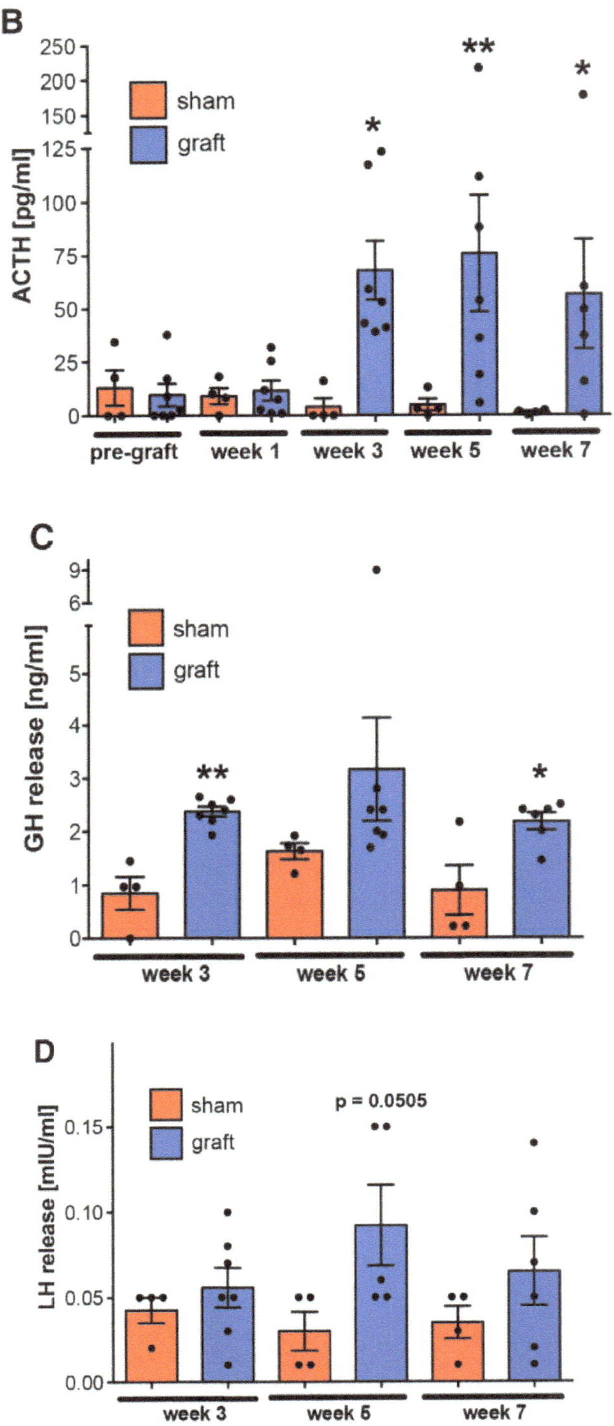
B
ACTH [pg/ml]
250
200
150
125
100
75
50
25
0
sham
graft
*
**
*
pre-graft
week 1
week 3
week 5
week 7
C
GH release [ng/ml]
9
6
5
4
3
2
1
0
sham
graft
**
*
week 3
week 5
week 7
D
LH release [mIU/ml]
0.15
0.10
0.05
0.00
sham
graft
p = 0.0505
week 3
week 5
week 7

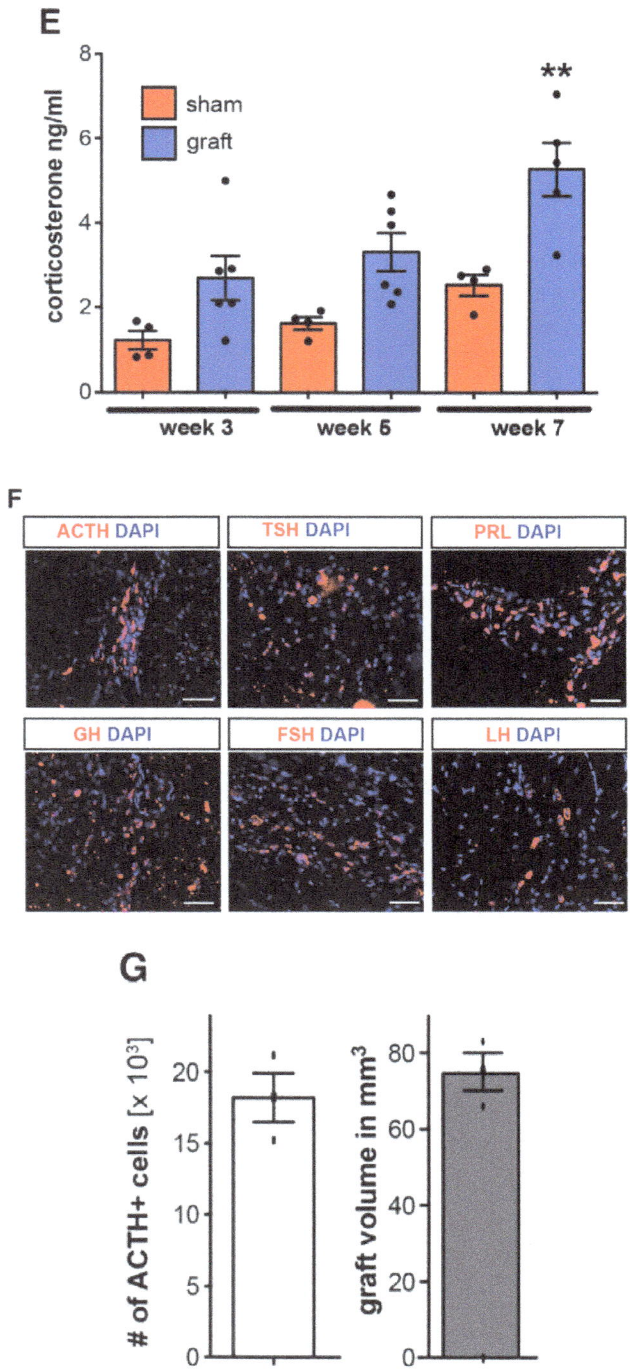

Figure 7: In Vivo Survival and Function of hPSC-Derived Anterior Pituitary Cells

(A) Schematic representation of experimental layout. After surgical removal of the pituitary gland and confirmation of hypopituitarism, cells embedded in Matrigel were transplanted subcutaneously. (B–E) ACTH (B), GH (C), LH (D), and corticosterone (E) levels were quantified in plasma for up to 7 weeks after transplantation of the cells using ELISA. Data are plotted as mean ± SEM with each dot representing an individual animal. $*p < 0.05$, $**p < 0.01$ compared with the corresponding sham control. (F) Immunohistological analysis of grafts 7 weeks after transplantation. Cells of each of the six hormonal lineages of the anterior pituitary gland were detectable within the graft. Scale bars, 50 mm. (G) Quantification of cells expressing ACTH and the corresponding graft volume 7 weeks after transplantation. Data are plotted as mean ± SEM with each dot representing an individual animal (three animals in total). See also Figure S7.

DISCUSSION

A goal of our study was the development of a defined and highly efficient protocol to generate anterior pituitary lineages that obviates the need for co-culture or complex media formulations and that should be suitable for clinical-grade cell manufacturing. Using fluorescence-activated cell sorting (FACS)-purified *SIX1::H2B-GFP*$^+$ placode cells, we demonstrate that pituitary fate can be induced at the expense of the default lens fate upon exposure to SHH, FGF8, and FGF10. Under these conditions, induction occurs in the absence of any hypothalamic cells previously thought to be critical for pituitary specification. While the signals essential for pituitary cell-fate induction from hPSCs do not require the presence of hypothalamic lineage cells, the interaction with the hypothalamic anlage is likely important for the complex tissue movements characteristic of pituitary gland development such as the formation of the Rathke›s pouch. In future studies it may be interesting to combine directed differentiation and organoid culture techniques to study developmental tissue interactions and to retain 3D cytoarchitecture. However, our current strategy offers obvious advantages for cell manufacturing of highly defined, therapeutically relevant cell types at scale. The cGMP-ready PIP presented here has several major advantages over the previously published protocol (Dincer et al., 2013). It is well established that KSR can exhibit considerable lot-to-lot variability (Rao, 2008, Schwartz et al., 2011). The protocol presented here overcomes these challenges and yields improved overall differentiation efficiencies under conditions of minimal lot-to-lot variability based on testing >10 different lots of Essential8 for hiPSC maintenance and >5 different lots of Essential6 for differentiation.

Our study made use of single-cell transcriptional profiling to define the heterogeneity of hormone-producing cells derived from hPSCs. The technology is a powerful tool that allowed us to determine the percentage of various hormone-producing cells and shifts in this distribution upon treatment with patterning factors or comparing different time points in vitro. However, one of the challenges of single-cell profiling is the largely binary readout of the expression data (McDavid et al., 2013). The HESX1/NEUROD1 validation experiment exemplifies this drawback, as the single-cell RT-PCR was not able to capture the decrease in HESX1 expression over time as observed by immunocytochemistry. Therefore, our binary approach of identifying expressed genes leads to the possibility that lowly expressed genes may be read as positive despite their expression levels not being high enough to achieve meaningful translation into protein. For example, we observed some instances of "inappropriate" expression patterns such as the co-expression of *SIX1* with *PAX6* and *PAX3* (though at different expression levels), which would indicate a mixed regional cranial placode identity. Similarly, the presence of putative plurihormonal cells was based purely on the expression of multiple hormonal transcripts. The presence of plurihormonal cells in the pituitary gland has been reported previously (Nunez et al., 2003, Villalobos et al., 2004) and is a well-known feature in pituitary adenomas (Scheithauer et al., 1986). However, it will be important to further assess in hPSC-derived pituitary cells whether transcripts for multiple hormones indeed result in the expression and secretion of multiple hormones from individual cells. To resolve some of the limitations of the single-cell gene-expression study, it will be important in the future to obtain appropriate positive and negative controls in order to define tissue-relevant thresholds of expression for each transcript. Unfortunately, developmentally matched human tissues are not readily available. In the present study we used immunofluorescence analysis as an independent method to quantify bona fide hormone-producing cells. Due to the lower sensitivity of this assay compared with qRT-PCR analysis, the percentages of hormone-positive cells were found to be lower but with the same relative proportions of the various hormone-producing cells.

We demonstrate that exposure to FGF8 and BMP2 can bias the dorsal-ventral composition of hormone-producing cells from $ACTH^+$ cells to FSH^+ or LH^+ cells. It should be noted that in vivo, cell types expressing different hormones are not necessarily located in specific areas but are scattered throughout the mature gland (Ericson et al., 1998, Olson et al., 2006). In addition, some reports argue for a less important role of extrinsic signaling compared with intrinsic factors in pituitary cell specification (Davis et al., 2011). Our human in vitro system provides a valuable tool to answer such questions. The current differentiation protocols only achieve partial enrichment of specific hormonal lineages. Therefore, lineage selection via cell sorting or the development of more sophisticated patterning strategies will be required to obtain purified populations such as GH^+ cells suitable for treating patients with selective GH deficiency (Bianchi et al., 2008). Our cell-surface epitope screen presents candidate markers that may be suitable for isolating specific hormone lineages. One cell type lacking in the current in vitro culture system are TSH^+ cells. Studies in sheep suggest that proper hypothalamic input is required for the development of thyrotrophs (Szarek et al., 2008). Therefore, it is conceivable that hypothalamic lineage cells may still be required for generating TSH^+ fates. Alternatively, TSH^+cells may simply require further optimization of our defined induction conditions or longer in vitro differentiation periods.

The in vivo studies show clear evidence of survival and hormone release up to 7 weeks after transplantation. Whether the lower percentage of hormone-producing cells in vivo is due to limited cell survival or increased proliferation of immature pituitary progenitors remains to be determined. Future studies also need to define the optimal transplantation time point for in vivo yield and hormone subtype and demonstrate the long-term survival of hPSC-derived pituitary cells. One key question is whether the functional integration of human cells will eventually require orthotopic placement of the graft into the pituitary gland or hypothalamic region. Preliminary experiments with orthotopic (median eminence) placement of the cells into lesioned animals resulted in low cell survival, suggesting the need for further optimization of such a transplantation paradigm. Ectopic placement in accessible locations does offer advantages from a translational perspective, namely the ability to access and remove the graft, should adverse side effects develop. However, hypothalamic release factors are the main triggers for hormonal release in the anterior pituitary gland (Smith and Vale, 2006). These release hormones have a short half-life and do not persist in the circulation. They are released into the hypophyseal portal venous plexus, which ensures immediate local delivery to the gland. Thus it is likely that the ectopically placed hPSC-derived grafts did not gain significant exposure to hypothalamic input. Our data therefore suggest a degree of autonomous hormonal release by the grafted cells, although a more comprehensive assessment of integration into homeostatic, endocrine mechanisms will be necessary. Based on existing literature on grafting primary cells (Maxwell et al., 1998, Naik et al., 1997), it is likely that proper regulation of at least some of the pituitary hormones will require orthotopic placement. Direct access to the skull base and pituitary gland in the mouse requires aggressive surgical procedures, such as transaural approaches or parapharyngeal neck dissection, complicating such studies. This is in marked contrast to human patients in whom surgical access to the pituitary gland can be safely performed through minimally invasive endoscopic transnasal routes (Tabar, 2011).

In conclusion, our study offers systematic access to human pituitary cell development and represents a powerful tool to obtain diverse hormone-producing cell types on demand and at scale. The platform also presents an important proof of concept in establishing functional and engraftable hormone-producing cells for regenerative medicine under conditions suitable for preclinical and clinical development.

ACKNOWLEDGMENTS

B.Z. was supported by an NYSTEM post-doctoral training fellowship, The SKI Stem Cell Research Facility is supported by NYSTEM grants C029153 and C024175 and The Starr Foundation. The work was further supported in part by NIH/NCI grants R21CA176700-02 and P30CA008748 . The authors thank M. Navare for technical support and A.F. Parlow and the National Hormone & Peptide program for the antisera against pituitary hormones, The PAX3 antibody developed by C.P. Ordahl, the TFAP2A antibody (3B5) developed by T.J. Williams, and the LHX3 antibody (Lim3, 67.4E12) developed by T.M. Jessell and S. Brenner-Morton were obtained from the Developmental Studies Hybridoma Bank, created by the NICHD of the NIH and maintained at the Department of Biology, University of Iowa, Iowa City.

REFERENCES

1. Bailey, A.P., Bhattacharyya, S., Bronner-Fraser, M., and Streit, A. (2006). Lens specification is the ground state of all sensory placodes, from which FGF promotes olfactory identity. Dev. Cell 11, 505–517.

2. Bellin, M., Marchetto, M.C., Gage, F.H., and Mummery, C.L. (2012). Induced pluripotent stem cells: the new patient? Nat. Rev. Mol. Cell Biol. 13, 713–726.
3. Bianchi, A., Giampietro, A., Pontecorvi, A., and De Marinis, L. (2008). Isolated growth hormone deficiency: clinical entity? J. Endocrinol. Invest. 31, 12–15.
4. Chambers, S.M., Qi, Y., Mica, Y., Lee, G., Zhang, X.J., Niu, L., Bilsland, J., Cao, L., Stevens, E., Whiting, P., et al. (2012). Combined small-molecule inhibition accelerates developmental timing and converts human pluripotent stem cells into nociceptors. Nat. Biotechnol. 30, 715–720.
5. Chemaitilly, W., and Sklar, C.A. (2010). Endocrine complications in long-term survivors of childhood cancers. Endocr. Relat. Cancer 17, R141–R159.
6. Davis, S.W., Mortensen, A.H., and Camper, S.A. (2011). Birthdating studies reshape models for pituitary gland cell specification. Dev. Biol. 352, 215–227.
7. Dincer, Z., Piao, J., Niu, L., Ganat, Y., Kriks, S., Zimmer, B., Shi, S.H., Tabar, V., and Studer, L. (2013). Specification of functional cranial placode derivatives from human pluripotent stem cells. Cell Rep. 5, 1387–1402.
8. Dreser, N., Zimmer, B., Dietz, C., Sugis, E., Pallocca, G., Nyffeler, J., Meisig, J., Bluthgen, N., Berthold, M.R., Waldmann, T., et al. (2015). Grouping of histone deacetylase inhibitors and other toxicants disturbing neural crest migration by transcriptional profiling. Neurotoxicology 50, 56–70.
9. Ericson, J., Norlin, S., Jessell, T.M., and Edlund, T. (1998). Integrated FGF and BMP signaling controls the progression of progenitor cell differentiation and the emergence of pattern in the embryonic anterior pituitary. Development 125, 1005–1015.
10. Funato, K., Major, T., Lewis, P.W., Allis, C.D., and Tabar, V. (2014). Use of human embryonic stem cells to model pediatric gliomas with H3.3K27M histone mutation. Science 346, 1529–1533.
11. Lamolet, B., Pulichino, A.M., Lamonerie, T., Gauthier, Y., Brue, T., Enjalbert, A., and Drouin, J. (2001). A pituitary cell-restricted T box factor, Tpit, activates POMC transcription in cooperation with Pitx homeoproteins. Cell 104, 849–859.
12. Lee, G., Ramirez, C.N., Kim, H., Zeltner, N., Liu, B., Radu, C., Bhinder, B., Kim, Y.J., Choi, I.Y., Mukherjee-Clavin, B., et al. (2012). Large-scale screening using familial dysautonomia induced pluripotent stem cells identifies compounds that rescue IKBKAP expression. Nat. Biotechnol. 30, 1244–1248.
13. Leung, A.W., Kent Morest, D., and Li, J.Y. (2013). Differential BMP signaling controls formation and differentiation of multipotent preplacodal ectoderm progenitors from human embryonic stem cells. Dev. Biol. 379, 208–220.
14. Maroof, A.M., Keros, S., Tyson, J.A., Ying, S.W., Ganat, Y.M., Merkle, F.T., Liu, B., Goulburn, A., Stanley, E.G., Elefanty, A.G., et al. (2013). Directed differentiation and functional maturation of ortical interneurons from human embryonic stem cells. Cell Stem Cell 12, 559–572.

15. Maxwell, M., Allegra, C., MacGillivray, J., Hsu, D.W., HedleyWhyte, E.T., Riskind, P., Madsen, J.R., and Black, P.M. (1998). Functional transplantation of the rat pituitary gland. Neurosurgery 43, 1157–1163.

16. McDavid, A., Finak, G., Chattopadyay, P.K., Dominguez, M., Lamoreaux, L., Ma, S.S., Roederer, M., and Gottardo, R. (2013).

17. Data exploration, quality control and testing in single-cell qPCRbased gene expression experiments. Bioinformatics 29, 461–467.

18. Merkle, F.T., and Eggan, K. (2013). Modeling human disease with pluripotent stem cells: from genome association to function. Cell Stem Cell 12, 656–668.

19. Merkle, F.T., Maroof, A., Wataya, T., Sasai, Y., Studer, L., Eggan, K., and Schier, A.F. (2015). Generation of neuropeptidergic hypothalamic neurons from human pluripotent stem cells. Development 142, 633–643.

20. Mica, Y., Lee, G., Chambers, S.M., Tomishima, M.J., and Studer, L. (2013). Modeling neural crest induction, melanocyte specification, and disease-related pigmentation defects in hESCs and patientspecific iPSCs. Cell Rep. 3, 1140–1152.

21. Naik, D.R., Das, S., Patnaik, L., and Samantaray, H. (1997). A novel and simple technique for ectopic transplantation of the pituitary gland. Gen. Comp. Endocrinol. 107, 304–310.

22. Nunez, L., Villalobos, C., Senovilla, L., and Garcia-Sancho, J. (2003). Multifunctional cells of mouse anterior pituitary reveal a striking sexual dimorphism. J. Physiol. 549, 835–843.

23. Olson, L.E., Tollkuhn, J., Scafoglio, C., Krones, A., Zhang, J., Ohgi, K.A., Wu, W., Taketo, M.M., Kemler, R., Grosschedl, R., et al. (2006). Homeodomain-mediated beta-catenin-dependent switching events dictate cell-lineage determination. Cell 125, 593–605.

24. Ozone, C., Suga, H., Eiraku, M., Kadoshima, T., Yonemura, S., Takata, N., Oiso, Y., Tsuji, T., and Sasai, Y. (2016). Functional anterior pituitary generated in self-organizing culture of human embryonic stem cells. Nat. Commun. 7, 10351.

25. Pagliuca, F.W., Millman, J.R., Gurtler, M., Segel, M., Van Dervort, A., Ryu, J.H., Peterson, Q.P., Greiner, D., and Melton, D.A. (2014). Generation of functional human pancreatic beta cells in vitro. Cell 159, 428–439.

26. Rao, M. (2008). Scalable human ES culture for therapeutic use: propagation, differentiation, genetic modification and regulatory issues. Gene Ther. 15, 82–88.

27. Regal, M., Paramo, C., Sierra, S.M., and Garcia-Mayor, R.V. (2001). Prevalence and incidence of hypopituitarism in an adult Caucasian population in northwestern Spain. Clin. Endocrinol. 55,735–740.

28. Rezania, A., Bruin, J.E., Arora, P., Rubin, A., Batushansky, I., Asadi, A., O'Dwyer, S., Quiskamp, N., Mojibian, M., Albrecht, T., et al.

29. (2014). Reversal of diabetes with insulin-producing cells derived in vitro from human pluripotent stem cells. Nat. Biotechnol. 32, 1121–1133.

30. Rivera, J.A. (2006). Lymphocytic hypophysitis: disease spectrum and approach to diagnosis and therapy. Pituitary 9, 35–45.

31. Rosenfeld, M.G., Briata, P., Dasen, J., Gleiberman, A.S., Kioussi, C., Lin, C., O'Connell, S.M., Ryan, A., Szeto, D.P., and Treier, M. (2000). Multistep signaling and transcriptional requirements for pituitary organogenesis in vivo. Recent Prog. Horm. Res. 55, 1–13, [discussion: 13–4].

32. Scheithauer, B.W., Horvath, E., Kovacs, K., Laws, E.R., Jr., Randall, R.V., and Ryan, N. (1986). Plurihormonal pituitary adenomas. Semin. Diagn. Pathol. 3, 69–82.

33. Schwartz, P.H., Brick, D.J., Nethercott, H.E., and Stover, A.E. (2011). Traditional human embryonic stem cell culture. Methods Mol. Biol. 767, 107–123.

34. Schwartz, S.D., Regillo, C.D., Lam, B.L., Eliott, D., Rosenfeld, P.J., Gregori, N.Z., Hubschman, J.P., Davis, J.L., Heilwell, G., Spirn, M., et al. (2015). Human embryonic stem cell-derived retinal pigment epithelium in patients with age-related macular degeneration and Stargardt's macular dystrophy: follow-up of two openlabel phase 1/2 studies. Lancet 385, 509–516.

35. Sklar, C.A., and Constine, L.S. (1995). Chronic neuroendocrinological sequelae of radiation therapy. Int. J. Radiat. Oncol. Biol. Phys. 31, 1113–1121.

36. Smith, J.C. (2004). Hormone replacement therapy in hypopituitarism. Expert Opin. Pharmacother. 5, 1023–1031.

37. Smith, S.M., and Vale, W.W. (2006). The role of the hypothalamicpituitary-adrenal axis in neuroendocrine responses to stress. Dialogues Clin. Neurosci. 8, 383–395.

38. Suga, H., Kadoshima, T., Minaguchi, M., Ohgushi, M., Soen, M., Nakano, T., Takata, N., Wataya, T., Muguruma, K., Miyoshi, H., et al. (2011). Self-formation of functional adenohypophysis in three-dimensional culture. Nature 480, 57–62.

39. Szarek, E., Farrand, K., McMillen, I.C., Young, I.R., Houghton, D., and Schwartz, J. (2008). Hypothalamic input is required for development of normal numbers of thyrotrophs and gonadotrophs, but not other anterior pituitary cells in late gestation sheep. J. Physiol. 586, 1185–1194.

40. Tabar, V. (2011). Making a pituitary gland in a dish. Cell Stem Cell 9, 490–491.

41. van Gelderen, H.H., and van der Hoog, C.E. (1981). Familial isolated growth hormone deficiency. Clin. Genet. 20, 173–175.

42. Villalobos, C., Nunez, L., and Garcia-Sancho, J. (2004). Phenotypic characterization of multi-functional somatotropes, mammotropes and gonadotropes of the mouse anterior pituitary. Pflugers Archiv. 449, 257–264.

43. Wand, G. (2008). The influence of stress on the transition from drug use to addiction. Alcohol Res. Health 31, 119–136.

44. Webster, J.I., and Sternberg, E.M. (2004). Role of the hypothalamicpituitary-adrenal axis, glucocorticoids and glucocorticoid receptors in toxic sequelae of exposure to bacterial and viral products. J. Endocrinol. 181, 207–221.

45. Zhu, Z., and Huangfu, D. (2013). Human pluripotent stem cells: an emerging model in developmental biology. Development 140, 705–717.

46. Zhu, X., Gleiberman, A.S., and Rosenfeld, M.G. (2007). Molecular physiology of pituitary development: signaling and transcriptional networks. Physiol. Rev. 87, 933–963.
47. Zimmer, B., Lee, G., Balmer, N.V., Meganathan, K., Sachinidis, A., Studer, L., and Leist, M. (2012). Evaluation of developmental toxicants and signaling pathways in a functional test based on the migration of human neural crest cells. Environ. Health Perspect. 120, 1116–1122.

SECTION IV. DEVELOPMENTAL DEFECTS: GENETIC AND ENVIRONMENTAL FACTORS

13

CHAPTER

PRICKLE1 MUTATION CAUSES PLANAR CELL POLARITY AND DIRECTIONAL CELL MIGRATION DEFECTS ASSOCIATED WITH CARDIAC OUTFLOW TRACT ANOMALIES AND OTHER STRUCTURAL BIRTH DEFECTS

Brian C. Gibbs[1], Rama Rao Damerla[1], Eszter K. Vladar[2], Bishwanath Chatterjee[1], Yong Wan[3], Xiaoqin Liu[1], Cheng Cui[1], George C. Gabriel[1], Maliha Zahid[1], Hisato Yagi[1], Heather L. Szabo-Rogers[3], Kaye L. Suyama[2], Jeffrey D. Axelrod[2] and Cecilia W. Lo[1]

[1] Department of Developmental Biology, University of Pittsburgh School of Medicine, Pittsburgh, PA 15201, USA

[2] Department of Pathology, Stanford University School of Medicine, Stanford, CA 94305, USA

[3] Department of Oral Biology, University of Pittsburgh School of Dental Medicine, Pittsburgh, PA 15261, USA

ABSTRACT

Planar cell polarity (PCP) is controlled by a conserved pathway that regulates directional cell behavior. Here, we show that mutant mice harboring a newly described mutation termed Beetlejuice (Bj) inPrickle1 (Pk1), a PCP component, exhibit developmental phenotypes involving cell polarity defects, including skeletal, cochlear and congenital cardiac anomalies. Bj mutants die neonatally with cardiac outflow tract (OFT) malalignment. This is associated with OFT shortening due to loss of polarized cell orientation and failure of second heart field cell

Citation: Brian C. Gibbs, Rama Rao Damerla, Eszter K. Vladar, Bishwanath Chatterjee, Yong Wan, Xiaoqin Liu, Cheng Cui, George C. Gabriel, Maliha Zahid, Hisato Yagi, Heather L. Szabo-Rogers, Kaye L. Suyama, Jeffrey D. Axelrod, Cecilia W. Lo, Prickle1 mutation causes planar cell polarity and directional cell migration defects associated with cardiac outflow tract anomalies and other structural birth defects, doi: 10.1242/bio.015750.

intercalation mediating OFT lengthening. OFT myocardialization was disrupted with cardiomyocytes failing to align with the direction of cell invasion into the outflow cushions. The expression of genes mediating Wnt signaling was altered. Also noted were shortened but widened bile ducts and disruption in canonical Wnt signaling. Using an in vitrowound closure assay, we showed Bj mutant fibroblasts cannot establish polarized cell morphology or engage in directional cell migration, and their actin cytoskeleton failed to align with the direction of wound closure. Unexpectedly, Pk1 mutants exhibited primary and motile cilia defects. Given Bj mutant phenotypes are reminiscent of ciliopathies, these findings suggest Pk1 may also regulate ciliogenesis. Together these findings show Pk1 plays an essential role in regulating cell polarity and directional cell migration during development.

INTRODUCTION

Planar cell polarity (PCP) is an evolutionarily conserved pathway that plays an important role in development. PCP has been examined in the context of developmental patterning in the *Drosophila*wing imaginal disk and compound eye, and refers to the polarization of cells within an epithelial sheet, orthogonal to the apical-basal polarity axis. Genetic analysis in *Drosophila* identified a group of interacting core PCP components that includes Van Gogh/Strabismus, Prickle, Frizzled, Dishevelled, Diego, and Flamingo (Devenport, 2014; Lawrence et al., 2007; Vladar et al., 2009). These proteins accumulate in asymmetrically localized complexes at proximal and distal apical cell junctions where they establish molecular cell polarity along the forming tissue axes via cell-cell communication (Axelrod, 2009). These proteins are conserved in vertebrates, and mutations in them cause a wide spectrum of developmental anomalies (Wansleeben and Meijlink, 2011), including the misalignment of hair cells in the cochlea, neural tube closure, brain and skeletal defects, and congenital heart disease (Cui et al., 2013).

In addition to epithelial planar polarization, PCP has been shown to regulate convergent-extension movements required for tissue morphogenesis. In *Xenopus* and mouse embryos, PCP-driven convergent-extension controls tissue elongation during gastrulation and neural tube closure (Cui et al., 2011; Dady et al., 2014; Juriloff and Harris, 2012; Wallingford and Harland, 2002). PCP also has been suggested to regulate directional migration of neural crest cells. This may contribute to the cardiac and craniofacial developmental anomalies in animal models with mutations in PCP core components (Montcouquiol et al., 2003; Simons et al., 2005; Simons and Mlodzik, 2008; Tada and Smith, 2000). However, the role of PCP in neural crest cell migration has been questioned (Pryor et al., 2014). Mice with mutations in the core PCP genes *Vangl2*, *Scrib* (Phillips et al., 2007), and *Dvl 1*, *2*, and *3* (Etheridge et al., 2008; Hamblet et al., 2002; Sinha et al., 2012) exhibit a similar spectrum of cardiac phenotypes involving outflow tract malalignment and septation defects (Boczonadi et al., 2014; Henderson et al., 2006). These cardiac defects likely involve not only perturbation of cardiac neural crest (CNC) cells, which are required for outflow septation, but also the second heart field (SHF). SHF cells migrate into the developing heart tube, forming most of the outflow tract, a structure that is often impacted by PCP mutations (Cohen et al., 2007; Schlessinger et al., 2009; Verzi et al., 2005).

Prickle1 (*Pk1*) is a PCP core component that has been shown to cause progressive monoclonic epilepsy-ataxis syndrome in human clinical studies (Tao et al., 2011). Studies in mice suggest *Pk1* also plays an important role in development. This is indicated by the finding of pregastrulation lethality of *Pk1*knockout mouse embryos (Tao et al., 2009). The recovery of a hypomorphic $Pk1^{C251X}$ mouse allele allowed survival to mid-gestation, making it possible to observe neural tube defects, cleft palate, and kidney defects, while heterozygote animals were used to model epilepsy (Liu et al., 2013; Sowers et al., 2014; Yang et al., 2014; Yates et al., 2010). Similarly, using a *Pk1* conditional allele, a spectrum of defects is observed that is described to phenocopy human Robinow syndrome with multiple organ system defects (Liu et al., 2014a).

In this study, we report findings from a novel *Pk1* missense allele, named *Beetlejuice* (*Bj*). Unlike other*Pk1* mutants, the *Bj* mutant survives to term, exhibiting a wide spectrum of developmental anomalies that include congenital heart defect, skeletal and craniofacial anomalies, and cochlea defects. We also show for the first time, *Pk1* mutation can cause biliary ductal hypoplasia in the spectrum of defects seen with biliary atresia. We provide evidence of a common mechanism involving disturbance of cell polarity and polarized cell migration contributing to the broad spectrum of developmental anomalies in the *Bj* mutant. This is associated with the disruption of canonical and noncanonical Wnt signaling.

RESULTS

We recovered a novel mutant line, *Beetlejuice* (*Bj*), from a large ethylnitrosourea (ENU) mouse mutagenesis screen in C57BL/6J mice designed to interrogate the genetic etiology of human congenital heart disease (Liu et al., 2014b). Cardiovascular phenotyping was conducted using noninvasive fetal echocardiography to identify fetuses with structural heart defects. Color flow imaging by ultrasound biomicroscopy (UBM) showed in a wild-type fetus, two blood-flow streams that criss-cross (Fig. 1A), delineating the normal position of the two outflow tracts (Fig. 1B-D). In contrast, in a *Bj*mutant fetus, we observed regurgitant flow (Fig. 1G), indicating the presence of outflow tract malalignment in conjunction with abnormal blood flow showing the presence of a ventricular septum defect (VSD) between the two ventricular chambers (Fig. 1G). Together, this would suggest a congenital heart defect known as an overriding aorta (OA) or double outlet right ventricle (DORV) (Fig. 1H-J). In DORV, the aorta position is shifted rightwards to lie more than 50% over the right ventricle (RV), becoming aligned parallel to the pulmonary artery. When the shift is less than 50% over the RV, it is referred to as OA. Follow up necropsy examination confirmed parallel positioning of the aorta and pulmonary artery in the *Bj* mutants (Fig. 1H). Further histopathology by episcopic confocal microscopy (ECM) confirmed in *Bj* mutants, the diagnosis of a DORV with a perimembranous VSD (Fig. 1I vs wild type in Fig. 1C). Further examination of the cardiac valves showed the atrioventricular and semilunar valves (see arrowheads) in the *Bj* mutant heart (Fig. 1J-L) were indistinguishable from that seen in wild-type littermate control (Fig. 1D-F), indicating the OFT malalignment defect in the *Bj* mutant model is not related to defects in valvular morphogenesis. Systematic analysis of *Bj* mutants (*n*=29) showed 63% have CHD comprising a DORV, while 17% have an overriding aorta. In 20.6% (*n*=6), only a simple perimembranous VSD was observed with normally positioned great arteries (Table S1). In the course of this analysis, no inflow tract anomalies nor any type of laterality defects were ever observed.

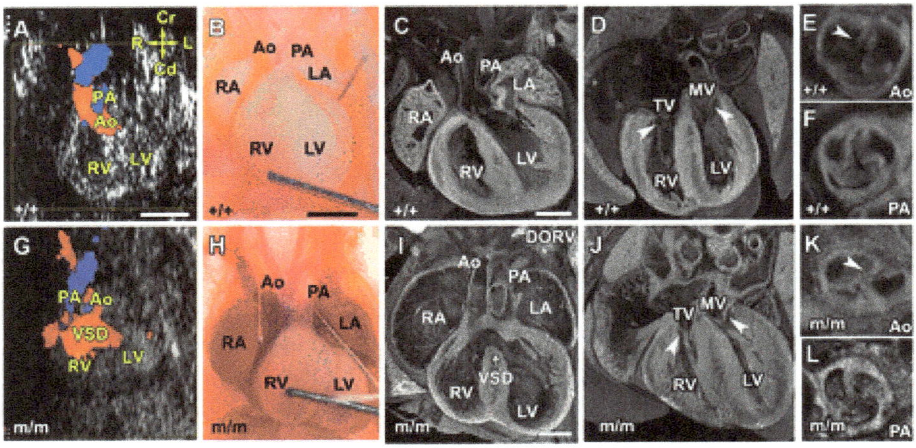

Figure. 1: ***Bj* mutant exhibits outflow tract** malalignment defects. (A,G) Echocardiography using color flow Doppler imaging showed in a normal E14.5 fetus (A), anterior positioning of blood flow from the aorta (Ao) emerging from the left ventricle (LV) and posterior positioning of blood flow from the pulmonary artery (PA) arising from the right ventricle (RV). In contrast, in the *Bj*mutant (G) fetus, blood flow streams from the outflow tract show parallel positioning, and this is associated with the mixing blood between the right ventricle and left ventricle, indicating aorta overriding the septum with a ventricular septal defect (VSD). Compass denotes orientation indicated by arrows; L, left; R, right; Cr, cranial; Cd, caudal. (B,H) Necropsy examination showed the aorta and pulmonary artery are abnormally positioned side by side in the mutant heart (H) compared to wild-type (B). RA, right atrium, LA, left atrium. (C,I) Further histopathology examination showed the *Bj* mutant aorta and pulmonary arteries are both connected to the right ventricle (RV), indicating it is a double outlet right ventricle (DORV) with a VSD (I). (D-F,J-L) The atrioventricular and semilunar valves are unaffected (J-L as compared to D-F, respectively; see arrowheads). TV, tricuspid valve; MV, mitral valve. Scale bars: 1 mm in A-H); 0.5 mm in C-L.

To recover the disease causing mutation in *Bj* mutants, we mated the *Bj* mouse line to wild-type mice of a different inbred strain background (C57BL10) and conducted a genome scan in the hybrid mutant offspring using polymorphic (B6/B10) markers to map the CHD causing mutation. This analysis showed the mutation is situated in a 19 Mb interval on chromosome 15, between position 83,618,701 to the telomeric end of the chromosome (Table S2). To identify the pathogenic mutation, we carried out whole mouse exome sequencing analysis at 50× coverage using DNA from one mutant animal. Comparison of the sequencing data obtained from the *Bj* mutant vs the C57BL/6J reference genome identified three homozygous coding mutations

(Tables S3, S4). Genotyping analysis of over 20 *Bj* mutants identified a missense mutation in *Pk1* (c.G482T:p.C161F) as the pathogenic mutation, as this mutation is the only one that is consistently homozygous in all of the Bj mutants. This missense mutation is in a highly conserved amino acid residue in the LIM1 domain, suggesting it is likely to be functionally important (Fig. S1). Consistent with this, analysis carried out using Polyphen-2, Provean, and Panther all predicted this mutation to be damaging or deleterious (Fig. S2) (Adzhubei et al., 2010; Brunham et al., 2005; Choi et al., 2012).

Second Heart Field and Cardiac Neural Crest Perturbations

Development of the looped heart tube requires recruitment of second heart field (SHF) cardiac precursors that contribute to lengthening of the heart tube, eventually giving rise to much of the right ventricle, atria, and outflow tract (Cai et al., 2003; Kelly et al., 2001; Waldo et al., 2001). In embryonic day (E)10.5 *Bj* mutant embryos, we observed a significant reduction in OFT length (Fig. 2C vs D,M; P=0.002), suggesting possible defect in the recruitment of the SHF cells. Antibody staining with Islet1, a SHF marker (Cai et al., 2003), showed the splanchnic mesoderm in the dorsal posterior wall (DPW) and transition zone (TZ), regions contiguous to the distal end of the OFT, are comprised of SHF cardiac precursors (Fig. 2I-L). This is observed similarly in both the mutant and wild-type embryos. However, while these Islet positive SHF cells exhibited a flat squamous epithelial morphology in control embryos, a distinct change to a cuboidal cell shape was observed in the *Bj* mutants (Fig. 2N,O). This suggested a possible convergent-extension defect that may play contribute to shortening of the OFT.

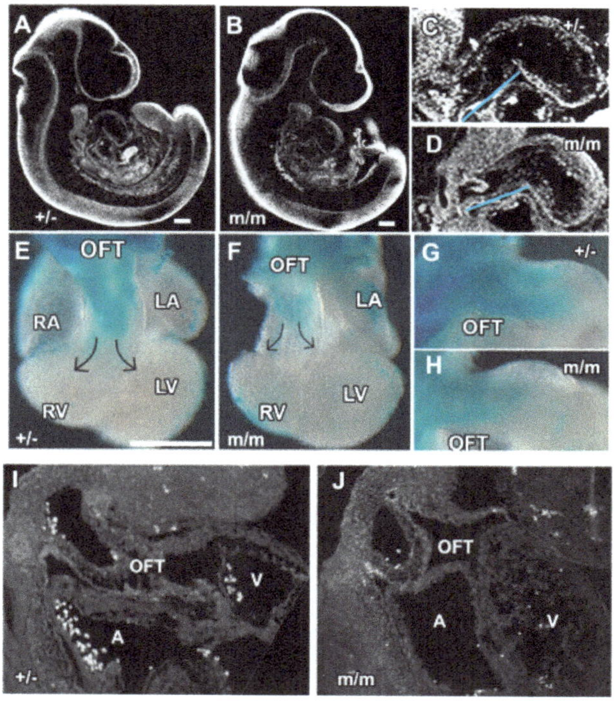

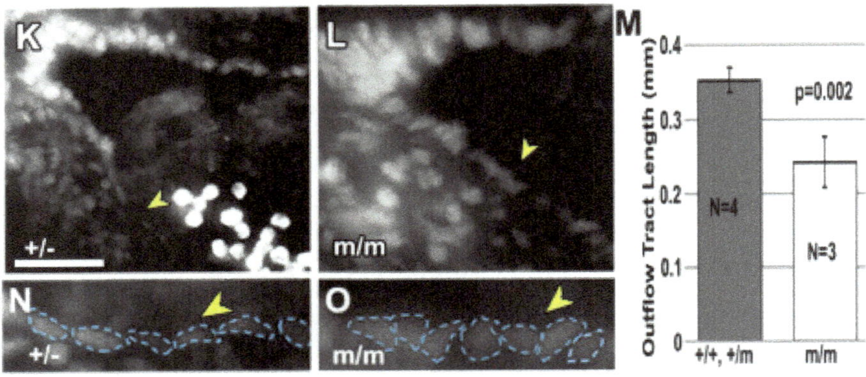

Figure. 2: ***Bj*** **mutants have shortened outflow tract and defects in neural crest and second heart field derivatives.** (A-D) Episcopic confocal histopathology in the sagittal plane of an E10.5 heterozygous (A) and *Bj* mutant (B) embryo is shown, with enlarged view of the outflow tract in (C) (n=4) and (D) (n=3) indicating reduction in length of the mutant OFT versus heterozygous (+/−) littermate control (see blue line in C,D). This was confirmed with further quantitative measurements (M). (E-H) Neural crest cells in the outflow tract of E10.5 embryos are visualized via a Cx43-*lacZ* transgene. This showed a reduction in neural crest cells in the homozygous *Bj* mutant heart (F,H) as compared to the heterozygous littermate control (E,G). (I-O) Sagittal sections of E10.5 embryo immunostained with anti-Islet1 (Isl1) antibody delineating second heart field cells in the dorsal pericardial wall (DPW) of the outflow tract of a wild-type (I,K) and *Bj* mutant embryo (J,L), with enlarged views shown in (K) and (L). Cells in the DPW wall (yellow arrowheads, denoted by blue outline in N,O) exhibit a flat squamous epithelial morphology in the wild-type embryo (N), but a distinct cuboidal morphology was observed in the homozygous *Bj* mutant embryo (O). A, atrium; V, ventricle. (M) Quantitative measurement using histopathology images showed a significant decrease in the length of the OFT in the homozygous *Bj* mutant hearts as compared to combined wild-type and heterozygous hearts. *P*-values were calculated with Student's *t*-test; error bars show standard deviation. Scale bars: 0.5 mm in A-D; 200 μm in E-H.

We also examined whether there may be changes in the distribution of cardiac neural crest cells (NCC), as CNC deficiency can also cause OFT defects (Kirby et al., 1983). For this analysis, we intercrossed into the *Bj* mutant line, a Cx43 promoter driven *lacZ* reporter transgene previously shown to label neural crest cells (Lo et al., 1997). X-gal staining of E10.5 heterozygous control embryos (Fig. 2E,G) and *Bj*mutant embryos (Fig. 2F,H) showed a reduction in CNC cells in the OFT of the *Bj* mutant embryos, indicating the *Pk1* mutation may have disrupted CNC migration into the OFT. The use of heterozygous embryos as controls is appropriate as heterozygous *Bj* mice are viable and indistinguishable from wild-type animals.

Disruption of Epithelial Cell Polarity in Second Heart Field Cardiac Precursors

To identify the basis for the altered cell morphology in the SHF cardiac precursors in the DPW and TZ, we examined the expression of various epithelia cell markers in the OFT of E10.5 embryos using confocal microscopy. This analysis showed no change in the cell surface localization of β-catenin in the DPW and TZ of the *Bj* mutant embryos (Fig. 3B,F vs A,E). This would suggest canonical Wnt signaling was not affected. Examination of laminin expression showed in the TZ, the presence of ectopic laminin localized apically, while laminin expression was lost in some cells basally (see Fig. 3C as compared to wild type in Fig. 3D). Laminin is normally not expressed in the DPW (Fig. 3E), but ectopic expression was observed in the *Bj* mutant embryo (Fig. 3F). Further examination for expression of PKC-ζ, a protein kinase required for oriented cell division (Whyte et al., 2010), showed disruption of the normal apical localization observed in the TZ and DPW (Fig. 3G,I). In the *Bj* mutant embryo, little or no PKC-ζ expression was observed apically, but low level PKC-ζ expression was observed basally (Fig. 3H,J).

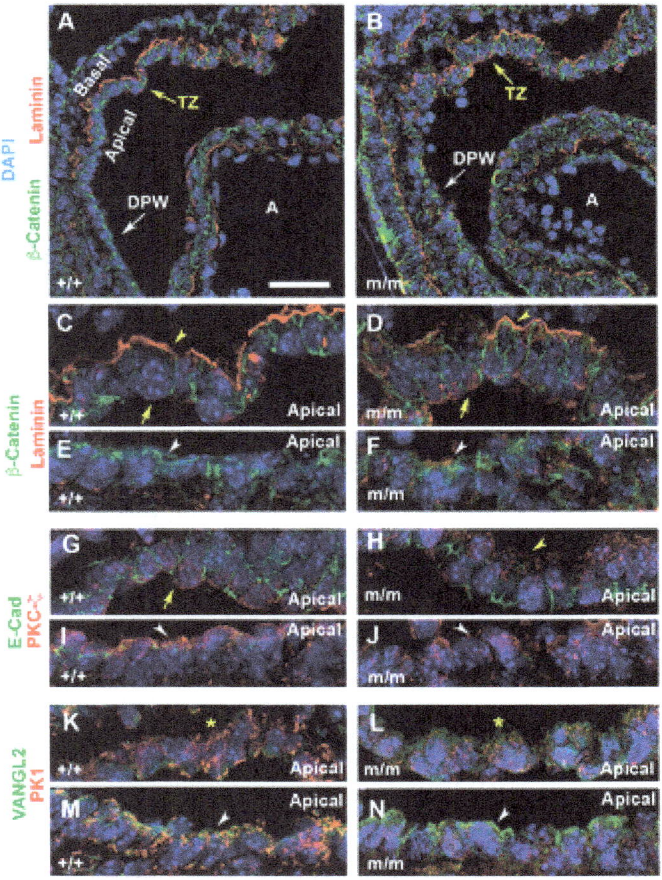

Figure. 3: Disruption of epithelial integrity in the developing outflow tract of the *Bj* **mutant embryos.** Immunostaining with antibodies against β-catenin (green) and laminin (red) (A-F), and E-cadherin (green) and PKCζ (red) (G-J), and Vangl2 (green) and Pk1 (red) (K-N) of wild-type (A,C,E,G,I,K,M) (*n*=3), and *Bj* mutant embryos (B,D,F,H,J,L,N) (*n*=3) showed marked disorganization of the epithelium in the transition zone (TZ) and dorsal pericardial wall (DPW) of the E10.5 *Bj* mutant embryo. Panels C,E and D,F are enlarged views of A and B, respectively. Panels

G-J and K-N are immunostaining of the same zones of A and B, respectively. Confocal imaging showed β-catenin is distributed along the cell surface in the control and *Bj* mutant embryos. Laminin is localized basally (arrowhead C,E) in the TZ of the control embryo, but in the mutant embryo, it is localized apically (arrow) and basally (arrow head D,F). While no laminin is detected in the DPW of the control embryo, it is ectopically expressed in the DPW of the *Bj* mutant embryo (compare E versus F). E-cadherin is widely expressed at regions of cell-cell contact, both in the TZ and in the DPW (G,I) of the control embryo. However, in *Bj* mutant embryos, E-cadherin expression is markedly reduced or extinguished in these regions (H,J). PKC-ζ expression is restricted to the apical (arrow) membrane of cells in the TZ and DPW epithelia (G,I). This polarized distribution is lost in the *Bj* mutant embryo, indicating the disruption of the epithelial organization of the DPW (H,J). Arrowheads in G-J indicate E-Cad and PKC-ζ distribution in the TZ and DPW. In the control embryo, Vangl2 and Pk1 are both expressed in the TZ and DPW of the outflow tract, with extensive co-localization observed (K,M). While Vangl2 expression is retained in the TZ and DPW of the *Bj* mutant, Pk1 expression is reduced (L,N; n=4; P=0.044). This is associated with an increase in overall thickness of the DPW. Arrowheads in M,N indicate Pk1 expression in the DPW, asterisks in K,L indicate Pk1 expression in TZ. Scale bar: 50μm in A-N. The *P*values were calculated with Student's *t*-test.

Together, these findings show the normal epithelial organization of the TZ and DPW of the OFT is disrupted, with the pseudostratified epithelial architecture of the DPW transformed into a thickened and disorganized multi-cell layered structure. To examine whether these changes may involve the perturbation of other PCP core components, we further examined the expression of Vangl2 and Scribble (Scrib). Both exhibited the same pattern of apical localization in the mutant and wild-type embryos (Fig. S3). We also examined the expression of Pk1 using an affinity purified anti-Pk1 antibody (see Fig. S4), and surprisingly, found Pk1 was markedly reduced in the *Bj* mutant embryo (Fig. 3L,N). This was confirmed with quantitative analysis of the Prickle1 immunofluorescence (P=0.044; mean intensity±standard deviation of 1.078±0.294 in wild-type vs 0.508±0.33 in homozygote mutant outflow tract;n=4). These findings suggest Pk1 is not required for specifying apical-basal cell polarity, but is required for the maintenance of epithelial tissue architecture in the TZ and DPW of the OFT.

Perturbation of Wnt Signaling in *Prickle1* Mutants

Given Wnt signaling is known to play a role in regulating OFT development (Cohen et al., 2012;Schleiffarth et al., 2007), we examined whether the OFT defects in the *Bj* mutant embryo may involve perturbation in Wnt signaling. For this analysis, we made use of the canonical Wnt *BAT-lacZ* reporter,crossing it into the *Bj* mutant mouse line and using X-gal staining to assay for canonical Wnt signaling. This analysis revealed a marked increase in *lacZ* expression at the base of the outflow tract in the *BJ*mutant heart, indicating increased canonical Wnt signaling (Fig. 4A,B). Quantitative real-time PCR analysis (Fig. 4C) was carried out using cDNA made from RNA obtained from the base of the OFT where*BAT-lacZ* expression was observed to be elevated. This analysis showed gene expression changes that suggested the perturbation of both canonical (*Ctnnb1*, *Apc*, *Tcf7*, *Wnt2b*) and noncanonical (*Wnt5a*,*Wnt11*, *RhoA*) Wnt signaling in the *Bj* mutant heart (Fig. 4C,D). These changes are surprising and perhaps reflect some type of feedback regulation.

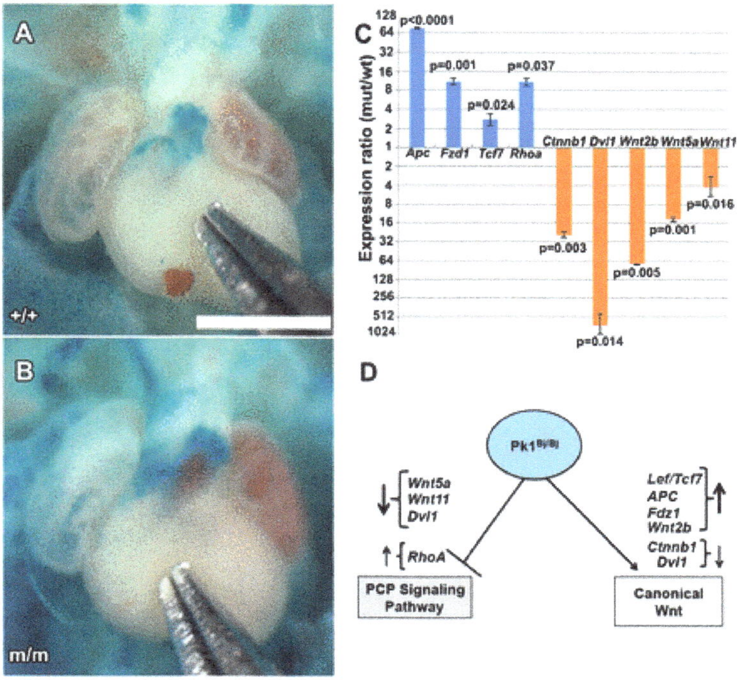

Figure. 4: ***Bj*** **mutants show disruption of canonical** Wnt signaling. (A,B) Analysis using the *BAT-lacZ* reporter showed increased *lacZ*reporter gene expression at the base of the outflow tract in the E13.5 *Bj* mutant heart (B) as compared to wild-type control (A). This suggests Wnt signaling is upregulated in the *Bj* mutant heart. Quantitative real-time PCR analysis (C) using RNA obtained from tissue isolated from the base of the OFT revealed changes in the expression of genes in both the canonical and noncanonical Wnt signaling pathway (D). Scale bar: 1 mm in A,B. The *P* values were calculated with Student's *t*-test; error bars show standard deviation.

Abnormal Myocardialization in the *Bj* Mutant

PCP has been shown to play a role in modulating cardiomyocyte migration to muscularize the outflow septum (Cui et al., 2013), a process referred to as myocardialization, that occurs after OFT septation (van den Hoff et al., 2001, 1999). Using MF20 immunostaining in the E13.5 wild-type mouse heart, the two prongs of invading myocardial cells can be visualized and are seen projecting into the outflow septum (see arrowheads in Fig. 5A). The direction of cardiomyocyte invasion is aligned with the orientation of actin filaments in the septal mesenchyme (visualized by phalloidin; Fig. 5A,C). Myofilaments in the invading cardiomyocytes are largely aligned with the direction of cell migration in the wild-type heart (asterisk Fig. 5E, inset in 5E). In contrast, in the E13.5 *Bj* mutant heart, the myocardial prongs are largely absent, and myofilaments in the cardiomyocytes are not well aligned with the direction of myocardialization (asterisk Fig. 5F, inset in 5F). Quantitative analysis showed a significant loss of oriented myofilaments alignment (Fig. 5G; P=0.0149). Phalloidin staining also revealed random orientation of actin filaments in the septal mesenchyme (Fig. 5D vs C). Together, these observations suggest that *Bj* mutants may fail to undergo normal myocardialization of the outflow septum due to a PCP defect that disrupts polarized alignment and migration of cardiomyocytes.

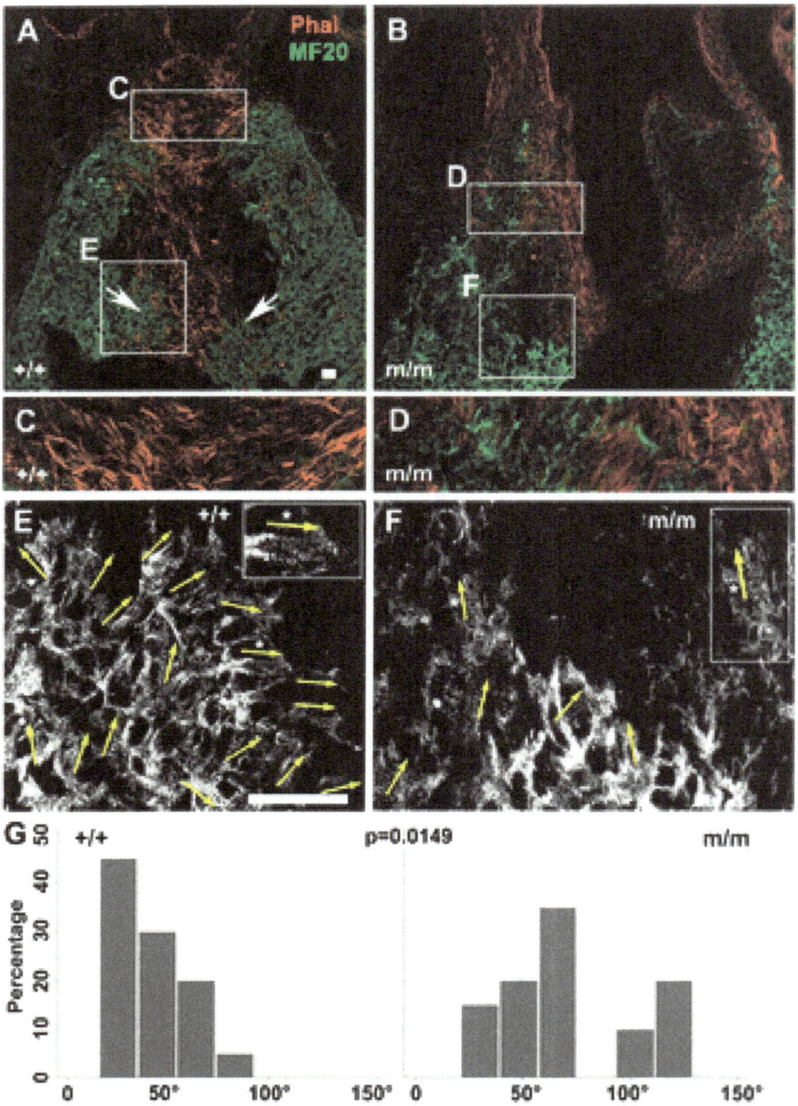

Figure. 5: Cell polarity and myocardialization defects in the OFT of *Bj* **mutants.** (A-F) Phalloidin (red) and MF20 (green) staining of paraffin sections of wild-type (A,C,E,) (n=3) and *Bj* mutant (B,D,F,) (n=3) heart. In the wild-type heart, phalloidin staining showed actin filament alignment with the direction of cell invasion into the OFT cushion (A,C), but in the *Bj* mutant, this pattern of actin alignment is not observed (B,D). Arrows in A indicate the two prongs of cardiomyocytes invading the cardiac septum. Examination of the striated banding pattern from the MF20 im-

munostain showed the developing myofilaments are closely aligned and oriented towards the direction of myocardialization in the wild-type embryo (arrows, E), but in the *Bj* mutant, the myofilaments are sparse and are largely oriented perpendicular to the direction of myocardialization and septum formation (arrows, F). (G) Quantitation of myofilaments showed wild-type (n=20) cardiomyocytes are polarized in the direction of cell invasion; in *Bj* mutant hearts (n=20), the cell orientation is not properly aligned (P=0.0149). Quantitative analysis showed cardiomyocytes (MF20 positive) in this conotruncal region of the heart is significantly reduced compared to controls (+/+, +/m 70%; m/m 22%: P=0.0025). Scale bar: 20 μm. The P values were calculated with a two-sample Wilcoxon rank-sum test.

Craniofacial Defects and Cranial Neural Crest Perturbation

Bj mutants exhibited various extracardiac defects that included skeletal anomalies and craniofacial defects (Fig. S5). The craniofacial defects included micrognathia, hypoplastic frontal bones, and shortened snouts. *Bj* mutants also had shortened limbs. This was confirmed by quantitative analysis of skeletal preparations, which showed reductions in the length of the long bones and metacarpals (Fig. S5). Histological analysis showed that chondrocytes in the growth plate failed to align along the proximal-distal axis, likely contributing to the reduction in skeletal outgrowth. This contrasts with the well-organized chondrocytes aligned in columns along the long axis of limb outgrowth in wild-type embryos (Fig. S6). We also examined the distribution of cranial neural crest cells in E10.5 *Bj* mutant versus wild-type embryos using the Cx43-*lacZ* transgene, as neural crest cells play essential roles in craniofacial development (Minoux and Rijli, 2010). This analysis showed a decrease in neural crest cells in the mandibular and maxillary prominences of the *Bj* mutant embryos, suggesting neural crest perturbation may contribute to the craniofacial defects observed in *Bj* mutants (Fig. S7).

PCP defects in the COCHLEA

We examined *Bj* mutants for cochlear defects using phalloidin staining to visualize the distribution of stereocilia and Vangl2 antibody labeling to assess molecular planar cell polarity. Normally, hair cells in the cochlea are arranged in repeating rows, with the actin-based stereociliary bundles exhibiting an identical polarized stereotypical chevron orientation, a patterning process that is PCP regulated (Fig. 6A,C) (Kelly and Chen, 2007). Analysis of three *Bj* mutant cochleae showed mild outer hair cell misalignment (Fig. 6B,D). While Vangl2 expression remained membrane localized in the supporting cells, the pattern of Vangl2 distribution was shifted, suggesting misalignment of the supporting cells adjacent to the inner hair cell (IHC) in the *Bj* mutant cochleae (Fig. 6D) compared to wild type (**Fig. 6**C). As this was observed in the basal, medial and apical regions of the cochlear duct, these changes may reflect global alternation in tissue architecture that could be elicited by a defect in convergent-extension cell movement.

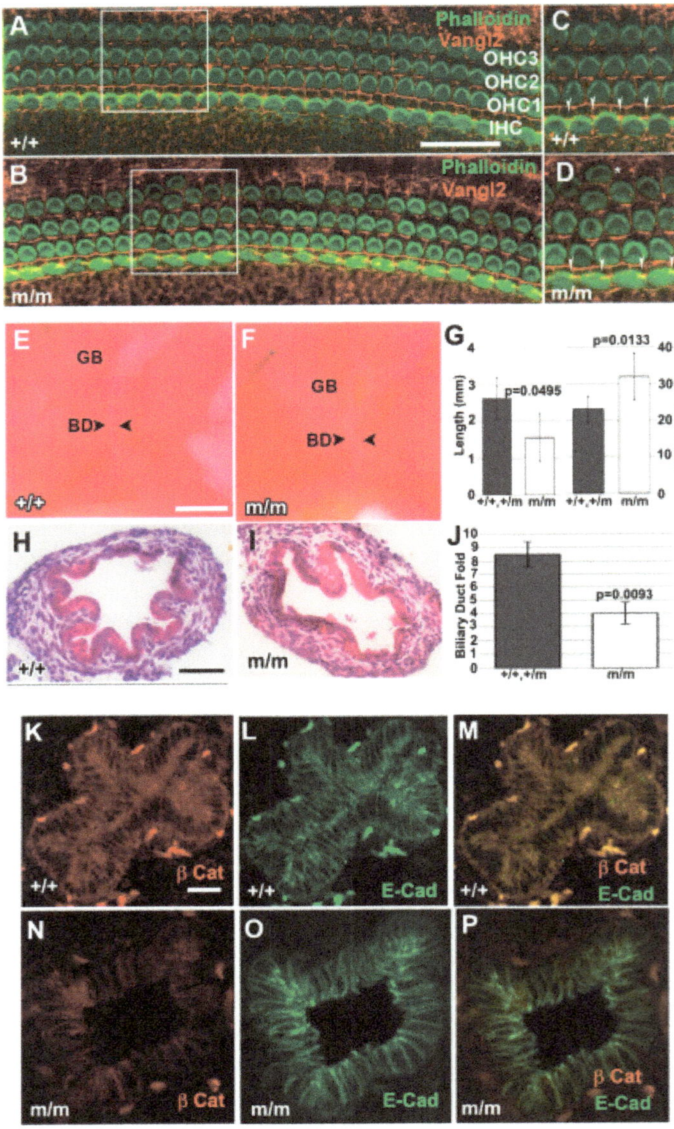

Figure. 6: *Bj* mutants show stereocilia patterning defects and biliary duct (BD) malformations. (A-D) Immunostaining with phalloidin (green) and Vangl2 (red) of the cochlea of near-term wild-type (A,C) and *Bj* mutant (B,D) embryo. The phalloidin staining delineated the normal chevron-shaped stereocilia in the outer hair cells (OHCs), showing malalignment in the *Bj* mutant (boxed B) compared to wild-type (boxed A). In addition, Vangl2 staining revealed a subtle misalignment of supporting cells connecting the inner hair cells (IHCs) in the *Bj* mutant (B), better seen in the enlarged views shown in panels C and D (see arrowheads). (E-J) Shown are bile ducts from a wild-type newborn mouse (E) (n=6) as compared to that of a *Bj* mutant (F) (n=6), which is noticeably shorter. Quantitative analysis showed a significant decrease in both the length and width of the *Bj*mutant biliary duct (G). Examination of hematoxylin and eosin

stained sections of newborn wild-type (H) (n=4) and *Bj*mutant (I) biliary ducts suggest a decrease in the mucosal folds of the mutant duct (I) (n=5). This was confirmed with quantitative analysis, which showed a significant decrease in mucosal folds in the mutant duct (J). Error bars show standard deviation. (K-P) Immunostaining with antibodies against β-catenin (red) and E-cadherin (green) showed β-catenin expression is markedly reduced in the *Bj* mutant biliary duct (N,P), but no change was observed for E-Cadherin (O,P,). Scale bars: 100 μm in A-E); 30 μm (K-P). The *P* values were calculated with a two-sample Wilcoxon rank-sum test. GB, gallbladder; BD, bile duct.

Biliary Ductal Abnormality

We examined the bile duct in the *Bj* mutant (Fig. 6F), and observed it was significantly shorter compared to wild type (Fig. 6E). Quantitative measurements revealed a decrease in the length, but an increase in width of the *Bj* mutant bile ducts (Fig. 6G). This was accompanied by reduction in the number of mucosal folds (Fig. 6H-J). Given several studies have suggested a role for β-catenin dependent canonical Wnt signaling in biliary fate determination (Tan et al., 2008), we conducted confocal immunohistology to examine the expression of both β-catenin and E-cadherin in the bile duct. In the wild-type bile duct, β-catenin and E-cadherin were highly expressed and show colocalization at the cell surface (Fig. 6K-M). In the *Bj* mutant, there was no change in the pattern of E-cadherin expression (Fig. 6O), but β-catenin expression was reduced (Fig. 6N,P). Further immunostaining was carried out to examine the distribution of laminin, Vangl2, PKC-ζ, and Pk1 in the mutant and wild-type biliary duct. This analysis showed the expression of laminin was unchanged while all three proteins were reduced compared to the wild-type bile duct, but the overall pattern of expression was unchanged (Fig. S8). These findings show *Pk1* is required for normal biliary duct morphogenesis, but this does not involve the modulation of apical-basal polarity.

Planar Cell Polarity and Polarized Cell Migration defect in *Bj* Mutant MEFs

Previous studies have shown that disruption of PCP components can perturb convergent-extension cell movements, and other types of directional cell migration. To examine the potential role of *Pk1* in modulating directional cell movement, we generated *Bj* mutant and wild-type MEFs for wound scratch assays. MEFS were grown to confluence, and a wound gap was created with a scratch in the monolayer. Wildtype MEFs quickly became aligned to the direction of wound closure, but the mutant MEFs remained randomly oriented (Fig. 7A,B). This was demonstrated quantitatively with analysis of Golgi orientation using a Golga2 antibody (Fig. 7C,D). In migrating wild-type MEFs, the Golgi is situated at the cell›s leading edge, aligned with the direction of cell migration. This is indicated by Golgi positioning that is mostly less than 60° relative to the direction of cell migration in the wound gap (Fig. 7C,E). In contrast, in *Bj* mutant MEFs, the distribution of Golgi orientation was broadened, indicating randomization (Fig. 7D,E; P<0.0001). These results indicate that *Bj* mutant MEFs are unable to establish the cell polarity required for efficient directional cell migration.

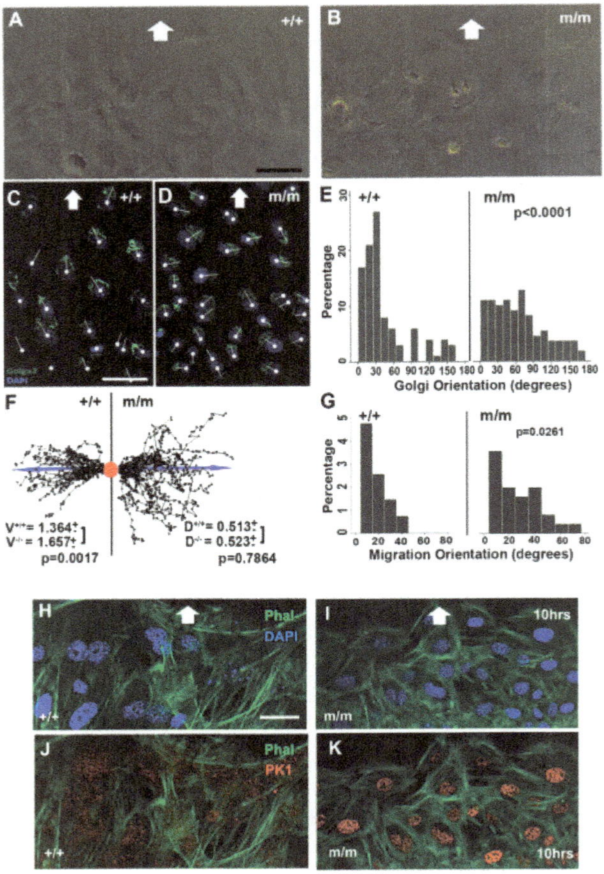

Figure. 7: Wound closure assay show defect in cell polarity and polarized cell migration in *Bj* **mutant MEFs.** (A-G). Analysis of cell polarity and directional cell migration using a MEF wound closure assay. Wildtype MEFs (A) are aligned with the direction of wound closure (white arrows), but *Bj* mutant MEFs (B) exhibit randomized pattern of cell migration. Cell orientation is delineated by examining the Golgi position after 24 h observed by Golga2 (green) immunostaining (C,D). Golgi orientation is indicated by a white line drawn from the cell nucleus (white circle) through the center of the Golgi. (E-G) Quantitative analysis of Golgi orientation relative to the direction of cell migration showed Golgi orientation is polarized along the direction of cell migration in wild-type (n=100), but not *Bj* mutant MEFs (n=150) (E). Tracings of the migratory path of MEFs after 8 h showed straight migration paths for wild-type MEFs (n=28), but tortuous migration paths with increased velocity for mutant MEFs (n=27) (F,G) V, speed of cell locomotion; D, directionality of cell movement. (H-K) Phalloidin (green) and Pk1 (red) antibody staining of wild-type (H,J) and *Bj* mutant (I,K) MEFs migrating into a wound gap after 10 h. This showed well formed actin stress fibers aligned with the direction of cell migration in the wild-type MEFs (H,J), but in the *Bj* mutants MEFs, the actin cytoskeleton exhibited a cortical basket formation (I,K). Interestingly, wild-type MEFs showed Pk1 localization in the cytoplasm (J), but in *Bj* mutant MEFs, Pk1 is localized to the nucleus (K). The P values were calculated with a two-sample Wilcoxon rank-sum test. Scale bars: 50 μm in A-D; 30 μm in H-K.

To examine further how *Pk1* may regulate directional cell migration, we conducted time-lapse videomicroscopy over 8 h to track the migratory behavior of individual cells during wound closure. By tracing the migratory paths of individual cells in the time-lapse videos, we observed a relatively straight migratory path in the wild-type MEFs (Fig. 7F,G), while a more tortuous migratory path was observed for the *Bj* mutant MEFs (Fig. 7F,G). This pattern of cell migration in the *Bj* mutant MEFS was correlated with a significant increase in the speed of cell locomotion (V in Fig. 7F; P=0.0017), but no net change in the directionality of cell movement (D in Fig. 7F; P=0.7864).

To examine for possible cytoskeletal changes that may account for the defect in polarized cell migration after 10 h, we carried out phalloidin staining to visualize the actin cytoskeleton. Wild-type MEFs showed well-formed actin stress fibers aligned with the direction of cell migration (Fig. 7H,J). In contrast, in *Bj*mutant MEFs, the actin stress fibers were organized in a basket configuration around the entire cell cortex, indicating no specific polarized orientation (Fig. 7I,K). We also carried out immunostaining with a*Pk1* antibody to determine if changes in Pk1 distribution may contribute to the defects observed in directional cell migration in the *Bj* mutant MEFs. Pk1 localization was observed in the cytoplasm in wild-type MEFs (Fig. 7J), but in the *Bj* mutant MEFs, it was largely localized to the nucleus (Fig. 7K). Pk1 was previously shown to bind REST, a transcriptional repressor, and sequester it in the cytoplasm to relieve transcription repression (Shimojo, 2006; Shimojo and Hersh, 2003). These findings are reminiscent of a previous study showing abnormal nuclear localization of a human *Pk1* mutant protein known to cause myoclonus epilepsy (Bassuk et al., 2008).

Primary and Motile Cilium defects in *Bj* Mutant Cells

Given previous studies indicating a role for cilia in constraining canonical versus noncanonical Wnt signaling (He, 2008) we further examined ciliogenesis in *Bj* mutants MEFs. Cilia were visualized by immunostaining with antibodies to acetylated α-tubulin and γ-tubulin to mark the axoneme and basal body, respectively, after serum starvation to stimulate ciliogenesis. *Bj* mutant MEFs (Fig. S9A,B) exhibited a reduction in ciliogenesis (Fig. S9D; P=0.0304) and the cilia formed were shorter compared to wild-type MEFs (Fig. S9C; P<0.0001). To examine for possible defects in motile cilia, which have been shown to be aligned by PCP signaling (Vladar et al., 2012), we obtained the tracheal epithelia of newborn*Bj* mutants and littermate controls and conducted videomicroscopy to examine ciliary motion. Surprisingly, the *Bj* mutant exhibited a significant reduction in cilia beat frequency (Fig. 8A). Transmission electron microscopy (TEM) showed normal orientation of the basal feet of cilia in the *Bj*mutant tracheal epithelia as observed in littermate controls (Fig. 8B,C). Scanning electron microscopy (SEM) (Assemat et al., 2004) showed abnormal large apical membrane bulges that incorporated axonemes in the *Bj* mutant tracheal epithelium (Fig. 8G,H). SEM revealed membrane blebs at the surface of ciliary axonemes (Fig. 8I′, insets), while cilia in wild-type tracheal were smooth (Fig. 8F′, insets). Also observed were abnormal axonemal microtubules (Fig. 8K,L), and compound axonemes (Fig. 8J). Enlarged views (Fig. 8M-P) show absent microtubule doublets in some ciliary axonemes (Fig. 8N). Taken together, these results suggest *Pk1* may play a role in the regulation of primary and motile cilia structure and function.

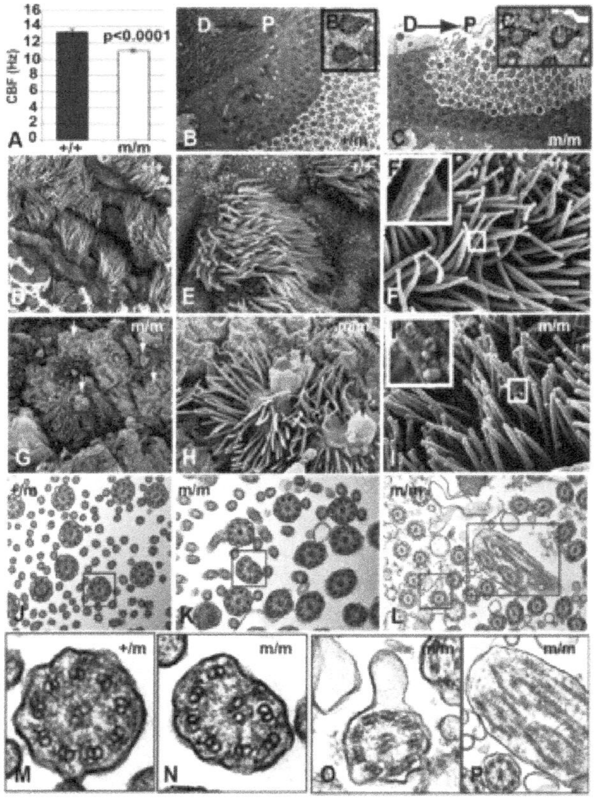

Figure. 8: Cilia defects in *Bj* **mutant tracheae.** (A) Quantitative analysis shows a reduction in cilia beat frequency of *Bj* mutant tracheal airway cilia. Error bars show standard deviation. (B,C) Transmission electron microscopy (TEM) of motile cilia in both mutant (C) and control (B) trachea show normal planar polarity based on the orientation of basal feet (see boxed insets) on individual cilia towards the proximal (oral) direction. D, dorsal; P, proximal. (D-I). Scanning electron microscopy of *Bj* mutant trachea (bottom G,H) shows that compared to control (D,E) nonciliated and multiciliated cells have large apical membrane bulges (arrows G,H), which in the multiciliated cells incorporate the axonemes. (I) Some mutant ciliary axonemes also have blebbed surface (boxed I′). (J-L) TEM of *Bj* mutant cilia show some abnormal axonemal microtubules (K,L), membrane blebs and compound axonemes compared to control (J). (M-P) Enlarged views of (J-L) show the absence of a microtubule doublet in the mutant axoneme (N), blebbed membrane (O), and a compound axoneme (P) compared to the control (M). Compound axonemes are likely contained within the membrane blebs observed by SEM (G,H).

DISCUSSION

We showed the *Bj* mutant exhibits a wide spectrum of developmental anomalies that included abnormalities involving the cardiac OFT, biliary duct, cochlea and skeletal and craniofacial anomalies. We showed these defects arise from a missense mutation in *Pk1*, a PCP core component. Consistent with this, phenotypes observed in the *Bj* mutant are similar to those seen with mutations in other PCP core component (Etheridge et al., 2008; Hamblet et al., 2002). While no change was found in apical-basal polarity, we observed defects in planar cell polarity and directional cell migration. In the OFT, SHF cells formed a disorganized multilayered aggregate, indicating a defect in convergent extension cell movement required for delamination of a cohesive epithelial sheet mediating OFT lengthening (Sinha et al., 2012). A myocardialization defect was also observed with cardiomyocytes in the conotruncal region of the heart failing to align with the direction of cell invasion into the outflow cushion. This cell migration defect was associated with perturbation of both canonical and noncanonical Wnt signaling. These findings suggest the DORV phenotype in the *Bj* mutant may arise from the combined disruption of OFT lengthening and a myocardialization defect in the OFT.

Similar OFT malalignment defects also have been observed with mutations in several other PCP components (Henderson et al., 2006; Phillips et al., 2007; Sinha et al., 2012), indicating this pathway plays an important role in the pathogenesis of DORV. In the *Bj* mutant, OFT malformations were observed in conjunction with craniofacial and skeletal anomalies that included micrognathia and shortened limbs, phenotypes reminiscent of those seen in Robinow syndrome (Soman and Lingappa, 2015; Webber et al., 1990). Robinow is associated with mutations in *ROR2*, a WNT signaling component (Afzal et al., 2000a,b; Butler and Wadlington, 1987). *Wnt5a* and *Ror2* have been shown to regulate proteasomal degradation of Pk1, suggesting Pk1 acts downstream of Wnt5a (Liu et al., 2014a; Yang et al., 2014). However, in the *Bj* mutant, we observed expression of *Wnt5a* is reduced in the conotruncal heart tissue. This may reflect feedback regulation, or possible tissue specific differences in the integration of Pk1 and Wnt signaling. The craniofacial/skeletal and OFT defects in *Bj* mutants also are reminiscent of phenotypes associated with velocardiofacial (VCF) or DiGeorge syndrome (Digilio et al., 2005). Many of the phenotypes observed in VCF/DiGeorge syndrome are known to involve neural crest perturbation (Bockman and Kirby, 1984; Bockman et al., 1987; Scambler, 2000), consistent with our observation of abnormal deployment of cardiac and cranial neural crest cells in the *Bj* mutant.

We demonstrated *Bj* mutant MEFs were unable to establish cell polarity mediating directional cell migration. Analysis of cell migratory behavior showed decreased directionality, but increased speed of cell locomotion. The mutant MEFs, unlike wild-type MEFs, failed to form actin stress fibers aligned with the direction of cell migration. Furthermore, we observed myofilaments in cardiomyocytes of the OFT were not well aligned with the direction of cell migration, contributing to the myocardialization defect in*Bj* mutant heart. As the actin cytoskeleton is also known to play important roles in ciliogenesis (Kim et al., 2010; Yan and Zhu, 2013), it is significant to note that both motile and primary cilia defects were observed in the *Bj* mutant. Thus a decrease in ciliation and primary cilia length was observed in *Bj*mutant MEFs, while ciliary beat frequency was reduced in the tracheal epithelia. Basal foot positioning was unaffected, but ultrastructural defects were observed indicating detachment of the axoneme from the ciliary membrane. The latter finding is reminiscent of those reported in mice harboring a mutation in*Tbc1d32* (Ko et al., 2010), where cilia ultrastructural defects were also linked to disruption of the actin cytoskeleton. These findings suggest *Pk1* may have a role in regulating ciliogenesis, an intriguing possibility given the similarity in the developmental phenotypes observed in the *Bj* mutant and that of ciliopathy mutant mouse models. This is supported by our previous study showing *Wdpcp*, a known PCP effector, is required for ciliogenesis and modulation of the actin cytoskeleton required for the specification of cell polarity and directional cell migration (Cui et al., 2013).

Overall, these studies showed the *Pk1* mutation can cause a wide spectrum of structural birth defects with overlap to those seen in various ciliopathies and in Robinow and velocardiofacial syndromes. This is associated with defects in cell polarity, directional cell migration, and PCP regulated convergent extension cell movements (Rao Damerla et al., 2014). Further study is needed to investigate whether the cilia defects observed in the *Bj* mutant may reflect the known link between actin and the regulation of ciliogenesis, and if this may underlie the ciliopathy related birth defect phenotypes seen in PCP related mutants.

MATERIALS AND METHODS

Institutional Approval for Animal Studies

All mouse experiments were conducted in accordance with animal protocols approved by the Institutional Animal Care and Use Committee at the University of Pittsburgh.

Necropsy and Histopathology Examination for Congenital Heart Disease

Newborn pups or E17.5 fetuses were collected, and fixed in 10% formalin. Necropsy examination was carried out to examine for external evidence of cardiac anomalies. To further evaluate for intracardiac defects, tissue was further analyzed by paraffin embedding and histopathology examination carried out using episcopic confocal microscopy (ECM). ECM imaging allows digital resectioning of the 2D serial image stack of the specimen and 3D reconstructions to uncover structural heart defects (Liu et al., 2014b).

Recovery of the *Pk1* Mutation and Mouse Breeding

Genomic DNA from line 019 (MGI: 5297388) mutant mice was sequence captured using Agilent SureSelect XT Mouse All Exon kit and sequenced with SOLiD 5500xl pair-end sequencing. Average 53.4× target coverage was achieved. Sequence reads were aligned to C57BL/6J mouse reference genome (mm9) and analyzed using LifeScope software (http://www.lifetechnologies.com). Sequence variants were annotated with ANNOVAR (www.openbioinformatics.org/annovar/) and custom scripts, and filtered against dbSNP and our in-house databases (Table S1). The three novel homozygous coding variants identified (Tables S2, S3) were then genotyped across all mutants from the same family. $Pk1^{c.G482T:p.C161F}$ was the only candidate mutation which was homozygous in all the mutants, thus indicating it is disease causing (Table S3).

Mouse Embryonic Fibroblast Isolation and Analysis of Primary Ciliogenesis

Primary mouse embryonic fibroblasts (MEFs) were isolated from E12.5-E13.5 mutant embryos and their wild-type littermates as previously described (Cui et al., 2013). Briefly, embryos were minced using a sterile razor blade, 500 ml of 0.25% trypsin-EDTA was added, and samples were incubated at 37°C for 20 min. After incubation for 20 min in 10 cm tissue culture dishes, the remaining fragments were pipetted multiple times to dissociate cells and incubated overnight in DMEM supplemented with 10% FBS, with penicillin and streptomycin. For primary cilia analysis, MEFs were serum starved in DMEM supplemented with 0.1% or 0.25% serum for 36 to 48 h to induce cilium growth, respectively. Cells were fixed 4% paraformaldehyde in PBS and immunostained with acetylated antibodies against α-tubulin (Sigma T7451, 1:1000), γ-tubulin (Sigma T6557, 1:1000) (St. Louis, MO, USA), and Ift88 (gift from Dr Gregory Pazour, University of Massachusetts Medical School, USA). Cells were then imaged using an Olympus Fluoview FV1000 or a Nikon Eclipse Ti-inverted laser scanning confocal microscopes.

Skeletal Preparation and Histo-Morphometric Analyses

Limbs of E15.5 and postnatal day (P)0 mice were fixed in 4% PFA at 4°C overnight and embedded in paraffin using standard protocols. For histomorphometry, 10-μm-thick longitudinal sections were cut through the forelimb and placed on SuperFrost Plus slides (Fisher Scientific 4951-001; Pittsburgh, PA, USA). For the wholemount skeleton staining to visualize bone and cartilage, E15.5 and P0 mice were fixed in 95% ethanol and stained with Alcian Blue/Alizarin Red. For visualization of the skull vault, neonates were euthanized and fixed in 4% paraformaldehyde overnight, cleared in 0.5% potassium hydroxide and stained in 0.1% Alizarin Red. Alcian Blue/Alizarin Red staining, and H&E, as well as Safranin O (shows maturation of cartilage in the growth plate) staining were performed as reported (Guo et al., 2004).

Histological Analysis and Confocal Microscopy

Embryo or tissue samples were fixed in 10% paraformaldehyde at 4°C overnight, and then dehydrated in a graded series of ethanol, followed by xylene infiltration and paraffin embedding. For cryoembedding, samples were fixed with 2% paraformaldehyde, followed by immersion in a 10-30% graded series of sucrose solution and then embedding in OCT. The paraffin blocks were sectioned at 9 μm using a Leica RM2235 rotary microtome. The cryoembedded specimens were sectioned at 12 μm using a Leica CM 1950.

For phalloidin staining, hearts from E13.5 embryos were cryoembedded, sectioned and stained with phalloidin and antibodies against MF20 (Hybridoma Bank, 1:40). Paraffin sections of the E10.5 embryos were incubated overnight at 4°C with the following antibodies: E-Cadherin (BD Transduction Laboratories, 1:100; San Jose, CA, USA), Laminin (Sigma-Aldrich, 1:200; St. Louis, MO, USA), Beta Catenin (BD Transduction Laboratories, 1:200), Protein Kinase C-zeta (Santa Cruz, 1:200;

Dallas, TX, USA), Scrib-11048 (Santa Cruz, 1:100), Vangl2 (R&D System, 1:200; Minneapolis, MN, USA), and Prickle1 (Bassuk et al., 2008). In addition, the biliary ducts were stained with Beta-Catenin and E-Cadherin (BD Transduction Laboratories). Secondary antibodies used included goat anti-mouse Alexa Fluor 488, donkey anti-sheep Alexa Fluor 488, donkey anti-rabbit Alexa Fluor A555, and donkey anti-mouse A555 (all 1:1000). The immunostained sections were imaged using either an Olympus Fluoview Fv1000 or Nikon Eclipse Ti inverted laser scanning confocal microscope.

Quantitative RT-PCR

Total RNA was isolated from MEFs and the base of the outflow tract of Prickle1 mutant embryos using the RNeasy Plus Mini Kit (Qiagen; Valencia, CA, USA). RNA was reverse transcribed into cDNA using High Capacity RNA-to-cDNA™ (Life Technologies; Grand Island, NY, USA) and qPCR was perform for several canonical and non-canonical WNT genes with β-Actin or GAPDH serving as internal control using the 7900HT Fast Real-Time PCR System (Life Technologies).

Whole-Mount β-gal Staining

For *in situ* whole-mount X-gal staining detection of *lacZ* expression via β-gal activity, embryos were fixed with 2% paraformaldehyde, 0.2% glutaraldehyde, and 0.02% Tween-20 in PBS for 1 h at 4°C. After several washes in PBS, whole embryos were incubated in a solution containing 40 mg/ml X-gal (5-bromo-4-chloro-3-indolyl-13-D-galactopyranoside), 20 mM potassium ferricyanide, 20 mM potassium ferrocyanide, 2 mM $MgCI_2$, and 10% Tween-20 in PBS. After overnight incubation at 37°C, the embryos were washed with PBT (PBS and 0.1% Tween-20) once for 10 min. Then the stained embryos were examined and imaged using a Leica stereomicroscope.

Scratch Wound Assay for Polarized Cell Migration

MEFs were cultured to confluence on 4 well chamber slides and a wound was generated using a 20 μl micropipette tip. Time lapse imaging was carried out using a 20× objective on an inverted microscope (Leica, DMIRE2; Buffalo Grove, IL, USA) with images captures every 10 min for 24 h using a Hamamatsu camera. To observe cell polarity, cells were fixed and immunostained with DAPI and Golgin subfamily A member 2 (GOLGA2; Sigma-Aldrich, 1:1000). GOLGA2 stains the Golgi apparatus in which is situated towards the direction of cell migration having Golgi situated in the front of the nucleus (forward facing) and aligned with migration direction. Polarity was scored as those cells with Golgi oriented within a 0°-180° sector along the direction of wound closure.

Statistical Methods

All continuous data was analyzed for normal distribution skewness and kurtosis normality test. Data that was normally distributed was compared using the Student's *t*-test or one-way analysis of variance for two or more than two groups, respectively. Non-normally distributed data was compared using the Wilcoxon rank-sum test. A two-tailed *P*-value of <0.05 was considered significant. All analyses were performed using Stata 12.0 (StataCorp, College Station, Texas).

Cilia analysis in the Mouse Tracheal Epithelia

Trachea was freshly harvested from newborn mice and opened longitudinally to expose the ciliated epithelia. The tissue was mounted for video microscopy to assess ciliary beat frequency and motility using previously described methods (Francis and Lo, 2013). For trachea cilia ultra structural analysis, transmission electron microscopy (TEM) and scanning electron microscopy (SEM) (Assemat et al., 2008) were carried out as reported (Vladar et al., 2012). Briefly, adult tracheas from mice of either gender were harvested and fixed in a solution containing 2% glutaraldehyde, 4% paraformaldehyde in 0.1 M Na Cacodylate buffer, pH 7.4 (Electron Microscopy Sciences) at 4°C overnight. Proximal airway direction was tracked throughout the procedure. Samples were osmicated, stained with uranyl acetate, then dehydrated with a graded ethanol series and infiltrated with EMbed-812 (Electron Microscopy Sciences; Hatfield, PA, USA). 80-100 nm sections were mounted onto copper grids and analyzed with a JEOL JEM-1400 microscope (Peabody, MA, USA) using a Gatan Orius Camera (Pleasanton, CA, UAS). For scanning electron microscopy (Assemat et al., 2008), longitudinally opened tracheal segments were fixed as for TEM, osmicated, dehydrated, dried with a Tousimis Autosamdri-815 critical point dryer, then tracheas were mounted luminal side up, sputter coated with 100 Å layer of Au/Pd and analyzed with a Hitachi S-3400N VP-SEM (New York, NY, USA) microscope operated at 10-15 kV, with a working distance of 7-10 mm and using secondary electron detection.

ACKNOWLEDGEMENTS

We would like to thank John Perrino and Lydia Joubert (Stanford University) for assistance with electron microscopy.

REFERENCES

1. Adzhubei, I. A., Schmidt, S., Peshkin, L., Ramensky, V. E., Gerasimova, A., Bork, P., Kondrashov, A. S. and Sunyaev, S. R. (2010). A method and server for predicting damaging missense mutations. Nat. Methods 7, 248-249.
2. Afzal, A. R., Rajab, A., Fenske, C., Crosby, A., Lahiri, N., Ternes-Pereira, E., Murday, V. A., Houlston, R., Patton, M. A. and Jeffery, S. (2000a). Linkage of recessive Robinow syndrome to a 4 cM interval on chromosome 9q22. Hum. Genet. 106, 351-354.
3. Afzal, A. R., Rajab, A., Fenske, C. D., Oldridge, M., Elanko, N., Ternes-Pereira, E., Tuysuz, B., Murday, V. A., Patton, M. A., Wilkie, A. O. et al. (2000b).
4. Recessive Robinow syndrome, allelic to dominant brachydactyly type B, is caused by mutation of ROR2. Nat. Genet. 25, 419-422.
5. Assemat, E., Bazellieres, E., Pallesi-Pocachard, E., Le Bivic, A. and MasseyHarroche, D. (2008). Polarity complex proteins. Biochim. Biophys. Acta 1778, 614-630.

6. Axelrod, J. D. (2009). Progress and challenges in understanding planar cell polarity signaling. Semin. Cell Dev. Biol. 20, 964-971.

7. Bassuk, A. G., Wallace, R. H., Buhr, A., Buller, A. R., Afawi, Z., Shimojo, M., Miyata, S., Chen, S., Gonzalez-Alegre, P., Griesbach, H. L. et al. (2008). A homozygous mutation in human PRICKLE1 causes an autosomal-recessive progressive myoclonus epilepsy-ataxia syndrome. Am. J. Hum. Genet. 83, 572-581.

8. Bockman, D. E. and Kirby, M. L. (1984). Dependence of thymus development on

9. derivatives of the neural crest. Science 223, 498-500.

10. Bockman, D. E., Redmond, M. E., Waldo, K., Davis, H. and Kirby, M. L. (1987). Effect of neural crest ablation on development of the heart and arch arteries in the chick. Am. J. Anat. 180, 332-341.

11. Boczonadi, V., Gillespie, R., Keenan, I., Ramsbottom, S. A., Donald-Wilson, C., Al Nazer, M., Humbert, P., Schwarz, R. J., Chaudhry, B. and Henderson, D. J. (2014). Scrib:Rac1 interactions are required for the morphogenesis of the

12. ventricular myocardium. Cardiovasc. Res. 104, 103-115.

13. Brunham, L. R., Singaraja, R. R., Pape, T. D., Kejariwal, A., Thomas, P. D. and Hyden, M. R. (2005). Accurate prediction of the functional significance of single nucleotide polymorphisms and mutations in the ABCA1 gene. PLoS Genet. 1, e83.

14. Butler, M. G. and Wadlington, W. B. (1987). Robinow syndrome: report of two patients and review of literature. Clin. Genet. 31, 77-85.

15. Cai, C.-L., Liang, X., Shi, Y., Chu, P.-H., Pfaff, S. L., Chen, J. and Evans, S. (2003). Isl1 identifies a cardiac progenitor population that proliferates prior to differentiation and contributes a majority of cells to the heart. Dev. Cell 5, 877-889.

16. Choi, Y., Sims, G. E., Murphy, S., Miller, J. R. and Chan, A. P. (2012). Predicting the functional effect of amino acid substitutions and indels. PLoS ONE 7, e46688.

17. Cohen, E. D., Wang, Z., Lepore, J. J., Lu, M. M., Taketo, M. M., Epstein, D. J. and Morrisey, E. E. (2007). Wnt/beta-catenin signaling promotes expansion of Isl-1- positive cardiac progenitor cells through regulation of FGF signaling. J. Clin. Invest. 117, 1794-1804.

18. Cohen, E. D., Miller, M. F., Wang, Z., Moon, R. T. and Morrisey, E. E. (2012). Wnt5a and Wnt11 are essential for second heart field progenitor development. Development 139, 1931-1940.

19. Cui, S., Capecci, L. M. and Matthews, R. P. (2011). Disruption of planar cell polarity activity leads to developmental biliary defects. Dev. Biol. 351, 229-241.

20. Cui, C., Chatterjee, B., Lozito, T. P., Zhang, Z., Francis, R. J., Yagi, H., Swanhart, L. M., Sanker, S., Francis, D., Yu, Q. et al. (2013). Wdpcp, a PCP protein required for ciliogenesis, regulates directional cell migration and cell polarity by direct modulation of the actin cytoskeleton. PLoS Biol. 11, e1001720.

21. Dady, A., Havis, E., Escriou, V., Catala, M. and Duband, J.-L. (2014). Junctional neurulation: a unique developmental program shaping a discrete region of the spinal cord highly susceptible to neural tube defects. J. Neurosci. 34, 13208-13221.
22. Devenport, D. (2014). Cell biology in development: the cell biology of planar cell polarity. J. Cell Biol. 207, 171-179.
23. Digilio, M., Marino, B., Capolino, R. and Dallapiccola, B. (2005). Clinical manifestations of Deletion 22q11.2 syndrome (DiGeorge/Velo-Cardio-Facial syndrome). Images Paediatr. Cardiol. 7, 23-34.
24. Etheridge, S. L., Ray, S., Li, S., Hamblet, N. S., Lijam, N., Tsang, M., Greer, J., Kardos, N., Wang, J., Sussman, D. J. et al. (2008). Murine dishevelled 3 functions in redundant pathways with dishevelled 1 and 2 in normal cardiac outflow tract, cochlea, and neural tube development. PLoS Genet. 4, e1000259.
25. Francis, R. and Lo, C. (2013). Ex vivo method for high resolution imaging of cilia motility in rodent airway epithelia. J. Vis. Exp. 78, e50343.
26. Guo, X., Day, T. F., Jiang, X., Garrett-Beal, L., Topol, L. and Yang, Y. (2004). Wnt/beta-catenin signaling is sufficient and necessary for synovial joint formation. Genes Dev. 18, 2404-2417.
27. Hamblet, N. S., Lijam, N., Ruiz-Lozano, P., Wang, J., Yang, Y., Luo, Z., Mei, L., Chien, K. R., Sussman, D. J. and Wynshaw-Boris, A. (2002). Dishevelled 2 is essential for cardiac outflow tract development, somite segmentation and neural tube closure. Development 129, 5827-5838.
28. He, X. (2008). Cilia put a brake on Wnt signalling. Nat. Cell Biol. 10, 11-13. Henderson, D. J., Phillips, H. M. and Chaudhry, B. (2006). Vang-like 2 and noncanonical Wnt signaling in outflow tract development. Trends Cardiovasc.
29. Med. 16, 38-45.
30. Juriloff, D. M. and Harris, M. J. (2012). A consideration of the evidence that genetic defects in planar cell polarity contribute to the etiology of human neural tube defects. Birth Defects Res. A Clin. Mol. Teratol. 94, 824-840.
31. Kelly, M. and Chen, P. (2007). Shaping the mammalian auditory sensory organ by the planar cell polarity pathway. Int. J. Dev. Biol. 51, 535-547.
32. Kelly, R. G., Brown, N. A. and Buckingham, M. E. (2001). The arterial pole of the mouse heart forms from Fgf10-expressing cells in pharyngeal mesoderm. Dev. Cell 1, 435-440.
33. Kim, J., Lee, J. E., Heynen-Genel, S., Suyama, E., Ono, K., Lee, K., Ideker, T., Aza-Blanc, P. and Gleeson, J. G. (2010). Functional genomic screen for modulators of ciliogenesis and cilium length. Nature 464, 1048-1051.
34. Kirby, M. L., Gale, T. F. and Stewart, D. E. (1983). Neural crest cells contribute to normal aorticopulmonary septation. Science 220, 1059-1061.
35. Ko, H. W., Norman, R. X., Tran, J., Fuller, K. P., Fukuda, M. and Eggenschwiler, J. T. (2010). Broad-minded links cell cycle-related kinase to cilia assembly and hedgehog signal transduction. Dev. Cell 18, 237-247.

36. Lawrence, P. A., Struhl, G. and Casal, J. (2007). Planar cell polarity: one or two pathways? Nat. Rev. Genet. 8, 555-563.

37. Liu, C., Lin, C., Whitaker, D. T., Bakeri, H., Bulgakov, O. V., Liu, P., Lei, J., Dong, L., Li, T. and Swaroop, A. (2013). Prickle1 is expressed in distinct cell populations of the central nervous system and contributes to neuronal morphogenesis. Hum. Mol. Genet. 22, 2234-2246.

38. Liu, C., Lin, C., Gao, C., May-Simera, H., Swaroop, A. and Li, T. (2014a). Null and hypomorph Prickle1 alleles in mice phenocopy human Robinow syndrome and disrupt signaling downstream of Wnt5a. Biol. Open 3, 861-870.

39. Liu, X., Francis, R., Kim, A. J., Ramirez, R., Chen, G., Subramanian, R., Anderton, S., Kim, Y., Wong, L., Morgan, J. et al. (2014b). Interrogating congenital heart defects with noninvasive fetal echocardiography in a mouse forward genetic screen. Circ. Cardiovasc. Imaging 7, 31-42.

40. Lo, C. W., Cohen, M. F., Huang, G.-Y., Lazatin, B. O., Patel, N., Sullivan, R., Pauken, C. and Park, S. M. J. (1997). Cx43 gap junction gene expression and gap junctional communication in mouse neural crest cells. Dev. Genet. 20, 119-132.

41. Minoux, M. and Rijli, F. M. (2010). Molecular mechanisms of cranial neural crest cell migration and patterning in craniofacial development. Development 137, 2605-2621.

42. Montcouquiol, M., Rachel, R. A., Lanford, P. J., Copeland, N. G., Jenkins, N. A. and Kelley, M. W. (2003). Identification of Vangl2 and Scrb1 as planar polarity genes in mammals. Nature 423, 173-177.

43. Phillips, H. M., Rhee, H. J., Murdoch, J. N., Hildreth, V., Peat, J. D., Anderson, R. H., Copp, A. J., Chaudhry, B. and Henderson, D. J. (2007). Disruption of planar cell polarity signaling results in congenital heart defects and cardiomyopathy attributable to early cardiomyocyte disorganization. Circ. Res. 101, 137-145.

44. Pryor, S. E., Massa, V., Savery, D., Andre, P., Yang, Y., Greene, N. D. E. and Copp, A. J. (2014). Vangl-dependent planar cell polarity signalling is not required for neural crest migration in mammals. Development 141, 3153-3158.

45. Rao Damerla, R., Gabriel, G. C., Li, Y., Klena, N. T., Liu, X., Chen, Y., Cui, C., Pazour, G. J. and Lo, C. W. (2014). Role of cilia in structural birth defects: insights from ciliopathy mutant mouse models. Birth Defects Res. C Embryo Today 102,

46. 115-125.

47. Scambler, P. J. (2000). The 22q11 deletion syndromes. Hum. Mol. Genet. 9, 2421-2426.

48. Schleiffarth, J. R., Person, A. D., Martinsen, B. J., Sukovich, D. J., Neumann, A., Baker, C. V. H., Lohr, J. L., Cornfield, D. N., Ekker, S. C. and Petryk, A. (2007). Wnt5a is required for cardiac outflow tract septation in mice. Pediatr. Res. 61, 386-391.

49. Schlessinger, K., Hall, A. and Tolwinski, N. (2009). Wnt signaling pathways meet Rho GTPases. Genes Dev. 23, 265-277.

50. Shimojo, M. (2006). Characterization of the nuclear targeting signal of REST/NRSF. Neurosci. Lett. 398, 161-166.

51. Shimojo, M. and Hersh, L. B. (2003). REST/NRSF-interacting LIM domain protein, a putative nuclear translocation receptor. Mol. Cell. Biol. 23, 9025-9031.

52. Simons, M. and Mlodzik, M. (2008). Planar cell polarity signaling: from fly development to human disease. Annu. Rev. Genet. 42, 517-540.

53. Simons, M., Gloy, J., Ganner, A., Bullerkotte, A., Bashkurov, M., Krönig, C., Schermer, B., Benzing, T., Cabello, O. A., Jenny, A. et al. (2005). Inversin, the gene product mutated in nephronophthisis type II, functions as a molecular switch between Wnt signaling pathways. Nat. Genet. 37, 537-543.

54. Sinha, T., Wang, B., Evans, S., Wynshaw-Boris, A. and Wang, J. (2012). Disheveled mediated planar cell polarity signaling is required in the second heart field lineage for outflow tract morphogenesis. Dev. Biol. 370, 135-144.

55. Soman, C. and Lingappa, A. (2015). Robinow syndrome: a rare case report and review of literature. Int. J. Clin. Pediatr. Dent. 8, 149-152.

56. Sowers, L. P., Yin, T., Mahajan, V. B. and Bassuk, A. G. (2014). Defective motile cilia in Prickle2-deficient mice. J. Neurogenet. 28, 146-152.

57. Tada, M. and Smith, J. C. (2000). Xwnt11 is a target of Xenopus Brachyury: regulation of gastrulation movements via Dishevelled, but not through the canonical Wnt pathway. Development 127, 2227-2238.

58. Tan, X., Yuan, Y., Zeng, G., Apte, U., Thompson, M. D., Cieply, B., Stolz, D. B., Michalopoulos, G. K., Kaestner, K. H. and Monga, S. P. (2008). Beta-catenin deletion in hepatoblasts disrupts hepatic morphogenesis and survival during mouse development. Hepatology 47, 1667-1679.

59. Tao, H., Suzuki, M., Kiyonari, H., Abe, T., Sasaoka, T. and Ueno, N. (2009). Mouse prickle1, the homolog of a PCP gene, is essential for epiblast apical-basal polarity. Proc. Natl. Acad. Sci. USA 106, 14426-14431.

60. Tao, H., Manak, J. R., Sowers, L., Mei, X., Kiyonari, H., Abe, T., Dahdaleh, N. S., Yang, T., Wu, S., Chen, S. et al. (2011). Mutations in prickle orthologs cause seizures in flies, mice, and humans. Am. J. Hum. Genet. 88, 138-149.

61. van den Hoff, M. J. B., Moorman, A. F. M., Ruijter, J. M., Lamers, W. H., Bennington, R. W., Markwald, R. R. and Wessels, A. (1999). Myocardialization of the cardiac outflow tract. Dev. Biol. 212, 477-490.

62. van den Hoff, M. J. B., Kruithof, B. P. T., Moorman, A. F. M., Markwald, R. R. and Wessels, A. (2001). Formation of myocardium after the initial development of the linear heart tube. Dev. Biol. 240, 61-76.

63. Verzi, M. P., McCulley, D. J., De Val, S., Dodou, E. and Black, B. L. (2005). The right ventricle, outflow tract, and ventricular septum comprise a restricted expression domain within the secondary/anterior heart field. Dev. Biol. 287, 134-145.

64. Vladar, E. K., Antic, D. and Axelrod, J. D. (2009). Planar cell polarity signaling: the developing cell's compass. Cold Spring Harb. Perspect. Biol. 1, a002964.

14

CHAPTER

ETHANOL EXPOSURE DISRUPTS EXTRAEMBRYONIC MICROTUBULE CYTOSKELETON AND EMBRYONIC BLASTOMERE CELL ADHESION, PRODUCING EPIBOLY AND GASTRULATION DEFECTS

Swapnalee Sarmah[1] , Pooja Muralidharan[1] , Courtney L. Curtis[1] , Jeanette N. McClintick[2] , Bryce B. Buente[1] , David J. Holdgrafer[1] , Osato Ogbeifun[1] , Opeyemi C. Olorungbounmi[1] , Liliana Patino[1] , Ryan Lucas[1] , Sonya Gilbert[1] , Evan S. Groninger[3] , Julia Arciero[4] , Howard J. Edenberg[2] and James A. Marrs[1]

[1] Department of Biology, Indiana University-Purdue University Indianapolis, 723 West Michigan Street, Indianapolis, IN 46202-5130, USA

[2] Department of Biochemistry and Molecular Biology, Indiana University School of Medicine, Indianapolis, IN 46202, USA

[3] Department of Mathematics, Wabash College, Crawfordsville, IN 47933, USA

[4] Department of Mathematics, Indiana University-Purdue University Indianapolis, Indianapolis, IN 46202, USA

SUMMARY

Fetal alcohol spectrum disorder (FASD) occurs when pregnant mothers consume alcohol, causing embryonic ethanol exposure and characteristic birth defects that include craniofacial, neural and cardiac defects. Gastrulation is a particularly sensitive developmental stage for teratogen exposure, and zebrafish is an outstanding model to study gastrulation and FASD. Epiboly (spreading blastomere cells over the yolk

Citation: Swapnalee Sarmah, Pooja Muralidharan, Courtney L. Curtis, Jeanette N. McClintick, Bryce B. Buente, David J. Holdgrafer, Osato Ogbeifun, Opeyemi C. Olorungbounmi, Liliana Patino, Ryan Lucas, Sonya Gilbert, Evan S. Groninger, Julia Arciero, Howard J. Edenberg, James A. Marrs, Ethanol exposure disrupts extraembryonic microtubule cytoskeleton and embryonic blastomere cell adhesion, producing epiboly and gastrulation defects, doi: 10.1242/bio.20135546.A cell), prechordal plate migration and convergence/extension cell movements are sensitive to early ethanol exposure. Here, experiments are presented that characterize mechanisms of ethanol toxicity on epiboly and gastrulation. Epiboly mechanisms include blastomere radial intercalation cell movements and yolk cell microtubule cytoskeleton pulling the embryo to the vegetal pole. Both of these processes were disrupted by ethanol exposure. Ethanol effects on cell migration also indicated that cell adhesion was affected, which was confirmed by cell aggregation assays. E-cadherin cell adhesion molecule expression was not affected by ethanol exposure, but E-cadherin distribution, which controls epiboly and gastrulation, was changed. E-cadherin was redistributed into cytoplasmic aggregates in blastomeres and dramatically redistributed in the extraembryonic yolk cell. Gene expression microarray analysis was used to identify potential causative factors for early development defects, and expression of the cell adhesion molecule protocadherin-18a (*pcdh18a*), which controls epiboly, was significantly reduced in ethanol exposed embryos. Injecting *pcdh18a* synthetic mRNA in ethanol treated embryos partially rescued epiboly cell movements, including enveloping layer cell shape changes. Together, data show that epiboly and gastrulation defects induced by ethanol are multifactorial, and include yolk cell (extraembryonic tissue) microtubule cytoskeleton disruption and blastomere adhesion defects, in part caused by reduced *pcdh18a* expression.

INTRODUCTION

Fetal alcohol spectrum disorder (FASD) is the most frequent preventable birth defect syndrome. Medical treatment, special education and lost productivity place a heavy toll on society (Popova et al., 2011). Developmental defects produced by ethanol exposure in animal models recapitulate defects seen in human FASD patients, including craniofacial, neural and cardiac defects (Ali et al., 2011;Haycock, 2009; Kelly et al., 2009; Sarmah and Marrs, 2013). Ethanol exposure could disrupt one or many developmental signaling mechanisms during early development (Muralidharan et al., 2013). Ethanol exposure is thought to affect embryonic gene expression, although mechanisms and specific changes are largely unknown (Haycock, 2009).

Gastrulation is a critical step in embryogenesis, producing the major embryonic axes (dorsal–ventral, anterior–posterior and left–right) (Solnica-Krezel and Sepich, 2012). Gastrulation is particularly sensitive to teratogen exposure (Gilbert-Barness, 2010), and ethanol exposure during gastrulation affects embryonic cell movements (Blader and Strähle, 1998; Sulik et al., 1981). Zebrafish (*Danio rerio*, Hamilton) is an exceptionally powerful model for studying gastrulation due to its rapid development, genetics, genomics, small size and experimental accessibility (Solnica-Krezel, 2006; Solnica-Krezel and Sepich, 2012). For example, injection of specific synthetic mRNA in early zebrafish embryos can rescue individual gene mutations or morpholino oligonucleotide gene knockdowns. Synthetic *shh* mRNA injection rescued effects of ethanol exposure in zebrafish (Loucks and Ahlgren, 2009).

Early embryogenesis includes cleavage and gastrulation stages. In zebrafish, epiboly cell movements occur during these stages, as blastomeres spread over the large yolk cell (Warga and Kimmel, 1990). Epiboly is coupled with gastrulation (Solnica-

Krezel and Driever, 1994; Strähle and Jesuthasan, 1993). Previously characterized epiboly mechanisms include: (i) yolk cell microtubule cytoskeleton pulling of the outer layer of cells, the enveloping layer, towards the vegetal pole (Solnica-Krezel and Driever, 1994;Strähle and Jesuthasan, 1993); and (ii) radial intercalation, where blastomere (deep) cells move and intercalate radially in the spherical embryo, causing blastomere layer thinning (Kane et al., 2005; Morita and Heisenberg, 2013; Solnica-Krezel, 2006; Song et al., 2013). Classical experiments showed that epiboly activities in the yolk cell are independent of the deep cell movements since yolk cell epiboly processes proceeded when the embryonic blastomeres were removed from the yolk cell (Betchaku and Trinkaus, 1978).

E-cadherin adhesion activity is required for epiboly and convergence/extension cell movements during gastrulation (Babb and Marrs, 2004; Kane et al., 2005; Morita and Heisenberg, 2013; Solnica-Krezel, 2006; Song et al., 2013). E-cadherin distribution and trafficking is regulated during gastrulation, particularly in the prechordal plate (the leading edge of the anterior mesendoderm) by critical early developmental signaling pathways, including non-canonical Wnt (Ulrich et al., 2005), heterotrimeric G-protein (Lin et al., 2009) and Pou5fl/Oct4 signaling pathways (Song et al., 2013).

Cell labeling, marker expression, protein distribution and live embryo imaging experiments were performed to dissect effects of ethanol on epiboly and gastrulation. Microtubule cytoskeleton in the yolk cell was disrupted by ethanol exposure, indicating that extraembryonic tissue effects contribute to early ethanol-sensitive developmental defects. Blastomere cell directional movements and cell adhesion activity was reduced by ethanol exposure, but ethanol effects on E-cadherin expression and distribution in blastomeres were minimal. Microarray analysis showed reduced *pcdh18a* gene expression in ethanol treated embryos, and *pcdh18a* mRNA injection partially rescued epiboly defects, showing that reduced protocadherin cell adhesion molecule expression is partially responsible for the complex effects of ethanol on zebrafish early embryo development.

RESULTS

Zebrafish embryos were treated with 100 mM ethanol in embryo medium beginning at 2 hpf; controls without ethanol were examined in parallel. This ethanol concentration produces highly reproducible phenotypes and is within levels attained in alcoholic patients. Epiboly progression was slowed (Fig. 1A–F; statistical comparison in Fig. 5F). Using *no tail (ntl) in situ* hybridization to mark the germ ring (70% epiboly stage; mesendodermal cells at the leading edge during epiboly progression) illustrates delayed epiboly progression (Fig. 1G,H). Dorsal forerunner cells (also marked by *ntl* staining) associate closely with the leading edge of the germ ring in control embryos, but these cells were dissociated from one another and advancing ahead of the germ ring in ethanol treated embryos at 6 hpf (Fig. 1G,H). Time lapse analysis of cell movements showed that the dorsal forerunner cells reaggregate by 10 hpf even in the continual presence of ethanol, and eventually form a single Kupffer›s vesicle at 12 hpf (S.S., C.L.C. and J.A.M., unpublished observations). Primary cilia assembly in the Kupffer›s vesicle, left–right asymmetry marker expression and heart looping were not disrupted in ethanol treated embryos (S.S. and J.A.M., unpublished observations; Sarmah and Marrs, 2013).

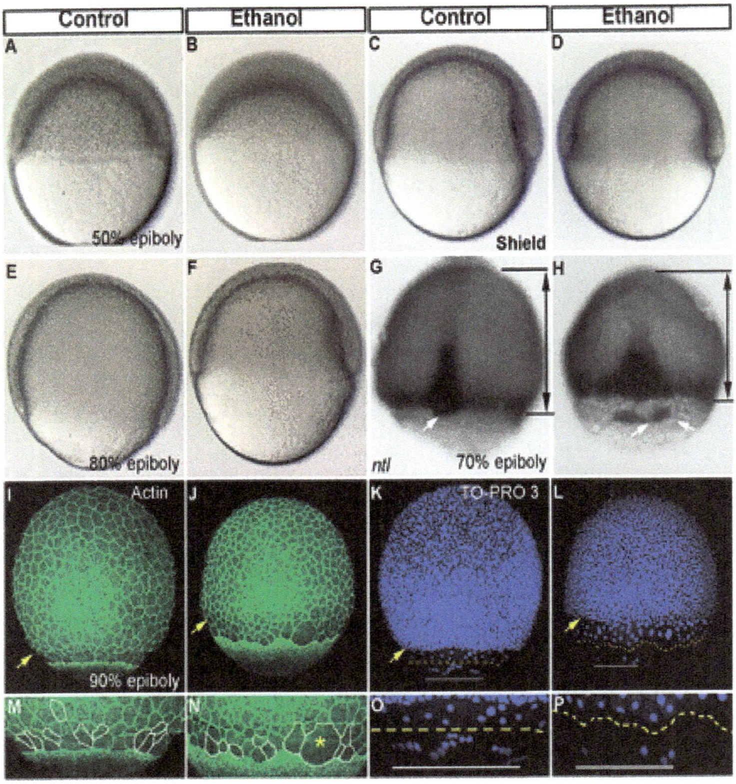

Figure. 1: Ethanol exposure reduces epiboly progression and dorsal forerunner cell aggregation.

(A–F) Live embryos at 50% epiboly (A,B), shield (C,D), and 80% epiboly stages (E,F) showed reduced epiboly progression in the ethanol treated embryos (B,D,F) compared to control (A,C,E). (G,H) *In situ* hybridization depicting *ntl* showed epiboly delay in the deep cells and obvious separation of the dorsal forerunner cells from the deep cell margin in the ethanol treated embryo (H). Black lines with arrows indicate the distance between the deep cell margin and the animal pole. White arrows: dorsal forerunner cells. (I,J) 3D renderings of confocal microscopy optical sections of phalloidin stained (F-actin) gastrulae. Yellow arrowhead: deep cell margin; yellow perforated line: EVL margin. (K,L) 3D renderings of confocal microscopy optical sections of TO-PRO-3 stained embryos showed deep cells nuclei, EVL cell nuclei and YSL nuclei. Yellow arrowhead: deep cell margin; yellow perforated line: EVL margin drawn from F-actin staining (I,J); white line: yolk syncytial nuclei margin. (M,N) High magnification images of control and ethanol treated embryos highlighting cell boundaries of a few EVL cells. Cells at the embryo margins in the control embryo showed elongated EVL cells, roughly perpendicularly aligned to the EVL margin (M).

Ethanol treated embryos showed rounder and not correctly aligned EVL cells (N). Yellow asterisk indicates big multinucleated cell. (O,P) High magnification images of the TO-PRO-3 stained control and ethanol treated embryos. Control embryos showed YSL nuclei proceeded beyond the EVL (O). Ethanol treated embryos showed fewer YSL nuclei proceeded beyond the EVL.

Microtubule Cytoskeleton Distribution was disrupted in Ethanol Treated Embryos

Epiboly can be controlled by purse string (actinomyosin) constriction and flow friction mechanisms, pulling the enveloping layer (which overlies and envelops the deep cell layer) toward the vegetal pole after 50% epiboly (Behrndt et al., 2012; Cheng et al., 2004; Köppen et al., 2006; Lin et al., 2009; Zalik et al., 1999). Actin and nuclear staining were used to mark the embryo location and evaluate cell shapes and cytoskeletal distribution. Actin cytoskeleton associated with the enveloping layer was present in ethanol treated embryos, similar to control embryos (90% epiboly stage; Fig. 1I,J,M,N). Yolk syncytial nuclei movements proceeded beyond the enveloping layer in control embryos, but these nuclei movements were reduced in ethanol treated embryos (Fig. 1K,L,O,P), suggesting that the epiboly movements of deep cells and the extraembryonic yolk cell nuclei were disrupted.

Tension created by epiboly movement produces long and thin cells in the enveloping layer at the embryo margin (Lin et al., 2009). Ethanol treated embryos had larger (sometimes multinucleate) enveloping layer cells at the embryo margins (90% epiboly stage; Fig. 1M,N) that had lower length-to-width ratios than control embryo cells (Fig. 5G); the ratio changed from 1.87 (s.d.=0.69, n=30) in control embryos to 1.00 (s.d.=0.47, n=33) in ethanol treated embryos (P-value <0.001), suggesting that there was less tension pulling the enveloping layer in ethanol treated embryos. Since the circumferential actin cytoskeleton was relatively normal in ethanol treated embryos, the microtubule cytoskeleton in the yolk cell, which pulls the enveloping layer toward the vegetal pole, was evaluated next.

Microtubule networks are associated with the enveloping layer, extending toward the vegetal pole of the yolk cell in 3 hpf embryos (Fig. 2A,B,A′,B′). These microtubules associate with the circumferential actin cytoskeleton at the enveloping layer margin. Disrupting these microtubules interferes with epiboly cell movements (Solnica-Krezel and Driever, 1994; Strähle and Jesuthasan, 1993). In ethanol treated embryos, the microtubule cytoskeleton displayed an abnormal distribution, having shorter microtubules, extending less distance toward vegetal regions of the yolk cell, indicating that ethanol disrupts microtubule arrays that drive epiboly cell movements.

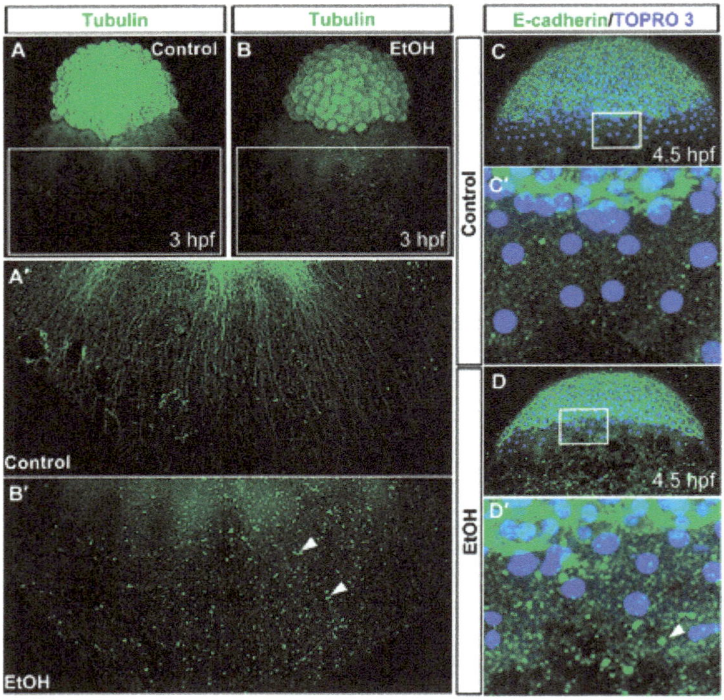

Figure. 2: Ethanol exposure disrupts yolk cell microtubule and E-cadherin distribution.

(A,B) 3D renderings of confocal microscopy optical sections of anti-α-tubulin antibody stained embryos showed microtubule organizations in the control (A) and ethanol treated (B) embryos at 3 hpf. (A′,B′) High magnification image of the boxed regions of embryos showed arrays of yolk cell microtubule extending toward the vegetal pole in the control (A′) and abnormal arrays of yolk cell microtubules in the ethanol treated embryo (B′). Arrowheads indicate aggregates of α-tubulin in the yolk cell. (C–D′) 3D images of confocal microscopy optical sections of anti-E-cadherin antibody stained embryos showed E-cadherin distribution in the control (C) and ethanol treated (D) embryos at 4.5 hpf. High magnification image of the boxed region of the control embryo showed relatively small aggregates of E-cadherin in the yolk cell (C′) than ethanol treated embryos, which had large aggregates of E-cadherin in the yolk cell (D′). Arrowhead indicates large E-cadherin aggregate.

The microtubule cytoskeleton in the yolk cell associates with E-cadherin, and loss of E-cadherin or microtubules disrupts this yolk cell membrane–cytoskeletal complex and interferes with epiboly cell movements (Babb and Marrs, 2004; Carvalho et al., 2009; Kane et al., 2005; Solnica-Krezel and Driever, 1994). In ethanol treated embryos (4.5 hpf), there were large aggregates of E-cadherin staining in the yolk cell near the enveloping layer cell border (Fig. 2C,D,C′,D′) in contrast to E-cadherin distribution in the enveloping and deep cells of the embryo proper.

Epiboly and Gastrulation Cell Movement defects in Ethanol Treated Embryos

A second mechanism that drives epiboly is radial intercalation cell movements, a process where cells at medial positions within the deep cell layer migrate radially and intercalate between cells within the outer and inner cell layers, producing a thinning and spreading of deep cells over the yolk cell (Morita and Heisenberg, 2013; Solnica-Krezel, 2006; Warga and Kimmel, 1990). Untreated control embryos showed more frequent radial intercalation events than ethanol treated embryos (Fig. 3; see also supplementary material Movies 1, 2). Migrating cells in control embryos directionally extended lamellipodia, leading to successful radial intercalation, but ethanol treated embryo cells extended lamellipodia in all directions and frequently failed to radially intercalate.

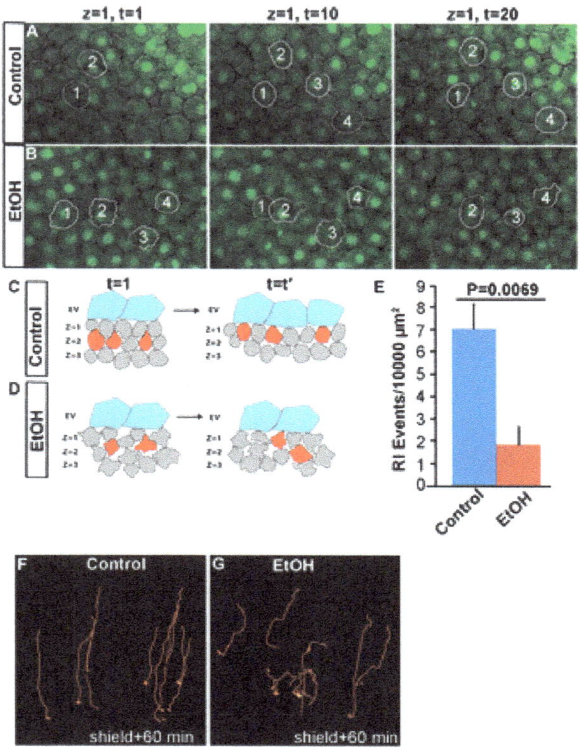

Figure. 3: Ethanol exposure affects radial intercalation and gastrulation cell movements.

(A,B) Confocal microscopy optical sections from the time-lapse image sequences at the most external epiblast layer of FITC-labeled histone-1 injected control (A) and ethanol (B) treated embryos at 4.5 hpf. Solid white line highlights representative cells in the upper layer, dotted line indicates cell in the lower layer. (C,D) Schematic diagrams illustrating radial intercalation (RI) events in the control (C) and ethanol treated (D) embryos. Unlike control, ethanol treated cells showed lamellipodia extension in all directions. (E) Histogram showing RI events in the control and ethanol treated embryos.

(F,G) The paths of ten deep cells in the mesendoderm of control and ethanol treated embryos. Cells in ethanol treated embryos showed abnormal trajectories. Animal pole, top; vegetal pole, bottom.

Cell movements in the shield, where mesendodermal cell involution first occurs during gastrulation, were examined from 6 to 7 hpf. Individual cell movements in control and ethanol treated embryos were directed toward the shield, and mesendodermal cells underwent involution (Fig. 3F,G). Deep cells in control and ethanol treated embryos moved at the same rate, having nearly identical average instantaneous velocities (Table 1). However, cells in ethanol treated embryos had abnormal paths, showing less directed movements toward the shield (Fig. 3F,G). These more random cell movements during gastrulation are reflected in a lower meandering index (Table 1; meandering index of 1.0 is a perfect straight line; the smaller the value, the greater the meandering of the track). The average meandering index was reduced to 0.770 in ethanol treated embryos from 0.959 in control embryos.

Table 1: Ethanol interferes with directional gastrulation cell movements, but not migration rate

	Control		Ethanol		
	Mean	s.d.	Mean	s.d.	*P*-value
Meandering index	0.959	0.006	0.770	0.072	0.01
Average velocity (μm/min)	1.621	0.148	1.710	0.180	0.54

s.d., standard deviation

Adhesion was disrupted in Ethanol Treated Embryos

To directly measure adhesion, dissociated cell aggregation assays using blastomeres from 4.5 hpf embryos were used to compare ethanol treated and control embryos. Cell aggregation was measured over a 3 hour time course and cell-to-object ratio was calculated. A ratio of 1.0 indicates no aggregation; greater the value, greater is the aggregation. Initially, the cell-to-object ratio was the same in control and ethanol treated cells (statistical comparison showed no significant difference at 1 hour), but at 2 and 3 hours, the cell-to-object ratio in control cells was significantly higher than ethanol treated cells (Fig. 4A).

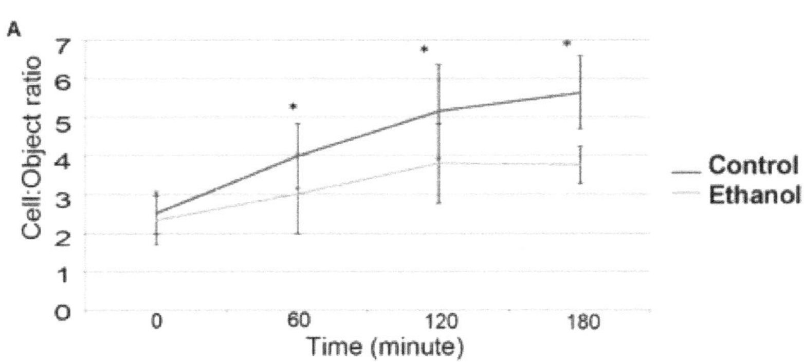

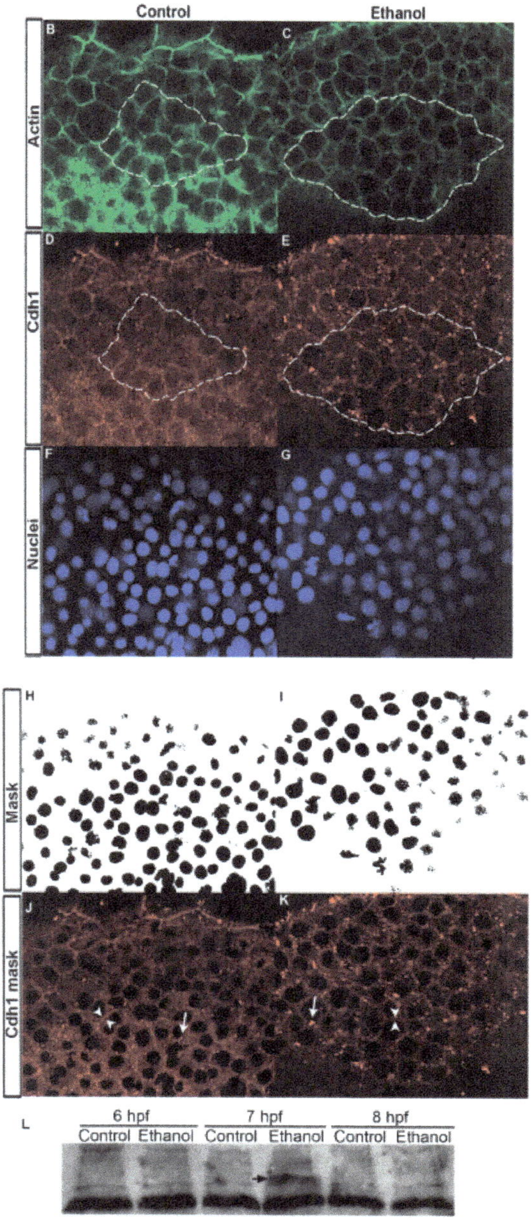

Figure. 4: Ethanol treatment reduces blastomere adhesion, but E-cadherin expression and distribution were only minimally affected.

(A) Cell adhesion assay using dissociated cells blastomeres. Graph showed cell-to-cell aggregation (object-to-cell ratios) of the dissociated control and ethanol treated blastomeres, which was reduced in ethanol treated cells. (B,C) Single confocal microscopy optical sections of the phalloidin stained prechordal plate cells in control and ethanol treated embryos at 8 hpf. (D,E) Single confocal microscopy optical sections showed E-cadherin distribution in the prechordal plate cells in the control and ethanol treated embryos at 8 hpf. White dotted lines indicate prechordal plate cells that were analyzed. (F,G) Co-staining with TO-PRO-3 labeled nuclei in the prechordal plate cells of control and ethanol treated embryos. (H,I) TO-PRO-3 was used as mask to exclude nuclei from E-cadherin intensity measurements. (J,K) E-cadherin distribution in the cytoplasm and cell surface after mask was applied. Arrowheads, cell surface; arrows, cytoplasmic aggregates. (L) Ethanol treatment did not change E-cadherin expression levels. Immunoblot analysis showed E-cadherin expression in the control and ethanol treated embryos at 6, 7 and 8 hpf. Arrow indicates an E-cadherin isoform that accumulated, which migrated more slowly.

Immunofluorescence using an E-cadherin specific antibody was performed on 8 hpf embryos that were treated with ethanol or untreated (Fig. 4D,E), and compared to cell surface and nuclei markers (Fig. 4B–K) in prechordal plate cells. E-cadherin distribution in the cytoplasm versus cell surface was not changed (Table 2), but cytoplasmic E-cadherin was found in larger cytoplasmic structures (see arrows in Fig. 4J,K). No difference in overall E-cadherin expression between control and ethanol treated embryos was detected at 6, 7 and 8 hpf (Fig. 4L). At 7 hpf, a band was consistently detected by the antibody that migrated more slowly in ethanol treated embryos. This may represent an accumulated E-cadherin isoform in ethanol exposed embryos (see arrow, Fig. 4L).

Table 2: Cdh1 distribution at the cell surface versus cytoplasm was unaffected after ethanol exposure in 8 hpf embryos

	Percent distribution				
	Cell surface	Cytoplasm	s.d.	s.e.m.	Number of embryos
Control	58.0	42.0	3.7	2.1	3
Ethanol	55.0	45.0	3.0	1.7	3

P-value is 0.24; s.d., standard deviation; s.e.m., standard error mean

Gene Expression Changes during Gastrulation Due to Ethanol Exposure

Gene expression changes were examined using Affymetrix GeneChip® microarray analysis. Many genes were identified that showed statistically significant changes in response to embryonic ethanol exposure between 2 and 8 hpf; a subset of those showing absolute changes ≥1.5-fold and P-values ≤0.001 are shown in Table 3 (the complete list is in supplementary material Table S1). Numerous genes were categorized by biological function based on published studies (see Materials and Methods), including cell adhesion (Table 3). Several gene expression changes were validated using quantitative PCR, indicating that the microarray analysis was sensitive and accurate (Table 4). There are functional categories, such as retinoid metabolism genes, that were expected based on previous studies. However, the dataset indicates that the transcriptional response to ethanol exposure in the early embryo is multifactorial.

Table 3: Ethanol responsive genes at 8 hpf. Expression level changes with fold change between ≤−1.50 or ≥1.5

Symbol	Gene name	Fold change
Cell proliferation, growth, and cell death		
odc1	ornithine decarboxylase 1	+1.81
vdac1	voltage-dependent anion channel 1	+1.66
anapc11	APC11 anaphase promoting complex subunit 11 homolog	+1.62
Epigenetic control		
ptp4a3	protein tyrosine phosphatase type IVA, member 3	+1.76
psat1	phosphoserine aminotransferase 1	+1.74
gstm	glutathione S-transferase M	+1.61
chmp4c	chromatin modifying protein 4C	−1.50
gatm	glycine amidinotransferase	−1.57
fbp1b	fructose-1,6-bisphosphatase 1b	−1.67
ckmt1	creatine kinase, mitochondrial 1	−1.87
bhmt	betaine-homocysteine methyltransferase	−1.94
gamt	guanidinoacetate N-methyltransferase	−2.05
Cell specification/differentiation and morphogenesis		
six3a	sine oculis homeobox homolog 3a	+1.79
hoxb1b	homeo box b1b	−1.72
a2ml	alpha-2 macroglobulin-like	−1.65
pcdh18a	protocadherin 18a	−1.77
hand2	heart and neural crest derivatives expressed transcript 2	−2.57
hoxc8a	homeo box C8a	−2.76
he1b	hatching enzyme 1b	−3.89
Neuronal function		
park7	parkinson disease (autosomal recessive, early onset) 7	+1.65
abat	4-aminobutyrate aminotransferase	+2.60
Retinol metabolism		
rbp4l	retinol binding protein 4 like	−1.98
Cholesterol metabolism		
dhcr7	7-dehydrocholesterol reductase	+1.62
Solute carrier protein		
slc7a3	solute carrier family 7 (cationic amino acid transporter, y+ system), member 3	+1.50
slc16a9a	solute carrier family 16 (mono-carboxylic acid transporters), member 9b	+2.56
slc1a4	solute carrier family 1 (glutamate/neutral amino acid transporter), member 4	

P≤0.01; False discovery rate for group ≤0.2; False discovery rate for treatment ≤0.1.

Table 4: Quantitative PCR confirms decrease in expression of selected cell-specification/differentiation and morphogenesis genes at 8 hpf after ethanol treatment

	PCR fold change mean ± s.d.	*P*-value (PCR)	Array fold change	*P*-value (array)
hand2	−2.84±0.23	<0.0001	−2.57	<0.001
pcdh18a	−1.79±0.11	<0.0001	−1.77	<0.001
he1b	−4.28±0.73	0.0002	−3.89	<0.001
hoxc8a	−3.70±1.18	0.002	−2.76	<0.001
hoxb1b	−2.14±0.6	0.0008	−1.53	<0.001

s.d., standard deviation

The gene encoding the adhesion molecules protocadherin-18a (*pcdh18a*) was identified, which showed reduced expression (1.77 fold by microarray analysis, and 1.79 fold by quantitative PCR analysis, Tables 3, 4). Previous studies showed that *pcdh18a* controls epiboly cell movements (Aamar and Dawid, 2008). Whole mount *in situ* hybridization showed that *pcdh18a* expression was generally reduced at all expression locations in ethanol treated 8 hpf embryos (Fig. 5A,B).

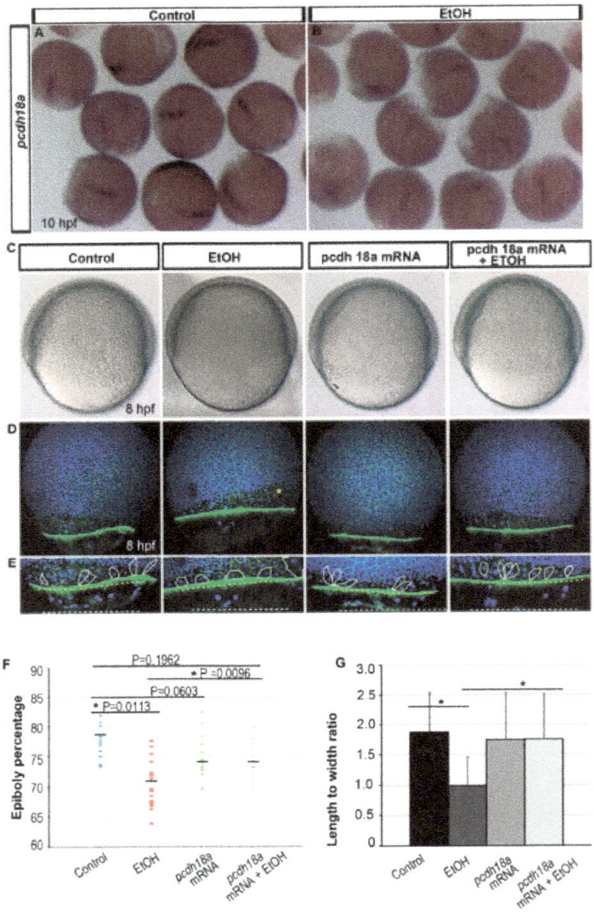

Figure. 5: Ethanol induced epiboly defect was rescued by *pcdh18a* **mRNA injection.**

(A,B) Whole mount *in situ* hybridization detecting *pcdh18a* mRNA showed reduced expression in the ethanol treated 10 hpf embryos as compared to control. (C) Bright field images focused at the EVL margin showed epiboly progression in the control, ethanol treated, *pcdh18a* mRNA injected, and *pcdh18a* mRNA injected plus ethanol treated embryos. (D) 3D renderings of confocal microscopy optical sections of phalloidin stained embryos co-labelled with TO-PRO-3 showed reduced epiboly in the ethanol treated embryos. Epiboly progression was similar in the control, *pcdh18a* mRNA injected, and *pcdh18a* mRNA injected plus ethanol treated embryos. Yellow asterisk indicates big multiucleated cells. (E) High magnification images of the embryos highlighting cell boundaries of a few EVL cells. Control, mRNA injected, and mRNA injected plus ethanol treated embryos showed elongated EVL cells; ethanol treated embryos showed rounder EVL cells that were not correctly aligned. Yellow perforated line: EVL margin; white line: yolk syncytial nuclei margin. Note: mRNA injected plus ethanol treated embryos showed more YSL nuclei beyond the EVL as compared to ethanol treated embryo. (F) Scatter plot representation shows reduced epiboly movement in ethanol treated embryos compared to control. These reduced epiboly movements were rescued by *pcdh18a* mRNA injection. (G) Histogram shows rescue by *pcdh18a* mRNA injection of the EVL cells length-to-width ratios at the embryo margins, which are reduced after ethanol treatment. Asterisks indicate statistical significance: $^*P<0.0001$.

To test whether reduced *pcdh18a* expression caused by ethanol exposure produces epiboly delay in zebrafish embryos, synthetic *pcdh18a* mRNA was injected into 2-to-4 cell stage embryos, and these embryos were compared with uninjected embryos treated with ethanol or untreated. Epiboly progression was evaluated (Fig. 5C,F). Epiboly delay induced by ethanol was restored to near control levels by injection of synthetic *pcdh18a* mRNA (EtOH+mRNA) (Fig. 5C,F). Injecting the same concentration of *pcdh18a* mRNA in the absence of ethanol did not change epiboly progression (Fig. 5C,F). In addition, synthetic *pcdh18a* mRNA injection rescued enveloping layer cell shape changes in ethanol treated embryos, showing greater length-to-width ratios as compared with ethanol treatment alone (Fig. 5D,E,G). Yolk syncytial nuclei movements were partially rescued in the *pcdh18a* mRNA injected, ethanol treated embryos.

DISCUSSION

Birth defects are induced by various environmental factors. Despite *in utero* ethanol exposure being the most frequent preventable birth defect, mechanisms of ethanol toxicity remain poorly understood. Better understanding of the genesis of birth defect syndromes can help inform clinical therapeutic interventions. Zebrafish is emerging a useful model to examine environmental toxins and birth defect syndromes (Ali et al., 2011). Ethanol exposure produces a diverse spectrum of birth defects, which can be recapitulated using zebrafish, including craniofacial, neural and cardiac defects, which can have their origin during gastrulation stages (Dlugos and Rabin, 2010; Muralidharan et al., 2013; Sarmah and Marrs, 2013). Gastrulation stages are particularly critical for animal development. In vertebrates, gastrulation developmental events establish the body axes (Solnica-Krezel and Sepich, 2012). Ethanol was previously shown to disrupt gastrulation cell movements (Blader and Strähle, 1998; Yelin et al., 2005;Zhang et al., 2010). Our experiments establish that abnormal epiboly and gastrulation cell movements are associated with yolk cell microtubule cytoskeleton, radial intercalation cell movement, shield blastomere cell movement, blastomere cell adhesion and, specifically, *pcdh18a* expression defects. Partial, but significant rescue in *pcdh18a* mRNA injection experiments indicate that *pcdh18a* expression defect is part of the mechanism of ethanol teratogenesis, but there are likely to be many other defects that contribute to the overall ethanol exposure phenotype. In zebrafish, ethanol treatment disrupts the E-cadherin-microtubule network in the yolk cell, which appears also to contribute to teratogenesis.

Adhesion defects in Gastrulation

E-cadherin cell adhesion and other cell adhesion mechanisms control epiboly and radial intercalation cell movements in the deep cells or blastomeres (Aamar and Dawid, 2008; Hammerschmidt and Wedlich, 2008; Morita and Heisenberg, 2013; Solnica-

Krezel, 2006; Ulrich et al., 2005). Gastrulation cell movements are also controlled by cell polarity mechanisms that respond to embryonic morphogen gradients, which often regulate differential cell adhesion in the early embryo (Roszko et al., 2009; Tada and Kai, 2012). Abnormal cell movements in the shield like those seen in ethanol treated embryos were also observed in embryos with cell adhesion defects, like E-cadherin loss-of-function embryos (Babb and Marrs, 2004; Kane et al., 2005). Our direct measurement in adhesion assays showed that ethanol affects cell adhesion activity. E-cadherin cell adhesion molecule expression and distribution in the cell surface versus cytoplasm was largely unaffected by ethanol exposure during early development. However, E-cadherin in the prechordal plate cells showed an abnormal distribution in larger cytoplasmic vesicles, and E-cadherin in the yolk cell was dramatically redistributed from the yolk cell surface associated with the embryo proper to large cytoplasmic vesicles. This redistribution coincides with microtubule redistribution in the yolk cell, and together, these effects could alter epiboly and gastrulation cell movements. Ethanol effects on *pcdh18a* gene expression may contribute to these cell adhesion and cell movement defects, perhaps playing an important role in ethanol induced zebrafish teratogenesis.

In zebrafish, E-cadherin adhesion during epiboly and gastrulation controls blastomere cohesion, and E-cadherin loss-of-function produces convergence/extension defects (Babb and Marrs, 2004; Kane et al., 2005). Mutations that cause epiboly defects were also identified in large scale screening, and *half-baked*mutations were found to be alleles of the E-cadherin gene, *cdh1* (Kane et al., 2005). Other gastrulation regulators, including Snail, Wnt11, heterotrimeric G-protein and Pou5f1 (Oct-4) signaling, regulate E-cadherin expression or intracellular trafficking (Ulrich et al., 2005; Esguerra et al., 2007; Lin et al., 2009;Speirs et al., 2010; Song et al., 2013).

Earlier studies showed that gastrulation stages are sensitive to ethanol exposure, producing persistent defects (Blader and Strähle, 1998; Zhang et al., 2010). Ethanol exposure during early development affects convergence/extension cell movements and prechordal plate migration. In zebrafish, these processes are coupled with epiboly movements. Our experiments confirm and support these previous findings (Blader and Strähle, 1998; Zhang et al., 2010), showing specific radial intercalation and shield cell migration defects. These cell migration defects were potentially caused by defects in cell adhesion mechanisms. Dissociated blastomere aggregation assays show that ethanol exposure reduced cell adhesion. Surprisingly, E-cadherin expression and distribution was largely unchanged. During early gastrulation, E-cadherin is needed to form the Kupffer›s vesicle and establish left–right asymmetry in the embryo (Essner et al., 2005; Kai et al., 2008; Matsui et al., 2011; Tay et al., 2013). Normal left–right asymmetry in ethanol exposed embryos (S.S. and J.A.M., unpublished observations; Sarmah and Marrs, 2013) supports our conclusion that there was not an overall E-cadherin deficiency.

Ethanol Induced gene Expression Changes: *pcdh18a*

To identify potential causative factors for the apparent adhesion defect in ethanol exposed embryos, gene microarray analysis was used to examine ethanol induced gene expression changes during gastrulation (8 hpf; mid-gastrulation). Hundreds of genes showed statistically significant expression changes, representing several functional categories including cell specification, differentiation, morphogenesis, proliferation, epigenetics, and retinol metabolism (Table 3). One category of genes that showed strong ethanol induced gene expression changes was solute transporter proteins (*slc*genes). It is interesting that *slc3a2* was shown to control yolk syncytial layer formation and microtubule distributions (Takesono et al., 2012). Analyzing the effects of solute transporter functional changes in response to ethanol exposure could be very instructive, particularly their effects on extraembryonic tissue development.

Previously, *shh* expression was shown to be reduced in ethanol treated zebrafish embryos (Loucks and Ahlgren, 2009), but our 8 hpf gene microarray did not detect differences in *shh* expression following ethanol treatment. A higher ethanol concentration (1–2.5%) was used in the previous study (Loucks and Ahlgren, 2009), as compared to 100 mM or 0.6% in our experiments.

Our study focused on cell adhesion functions. The cell adhesion molecule gene encoding Protocadherin-18a (*pcdh18a*) had reduced expression (Tables 3, 4). A previous study showed that Protocadherin-18a mediated adhesion regulates epiboly cell movements in the early embryo (Aamar and Dawid, 2008). The ability of synthetic *pcdh18a* mRNA injection experiments to partially rescue epiboly defects caused by ethanol illustrated that this gene plays a direct, mechanistic role in early development defects in the zebrafish FASD model. Protocadherin adhesion mechanisms are poorly understood. However, recent studies from Jontes and co-workers showed that protocadherin and cadherin proteins can physically interact, and the protocadherin/cadherin complex produces stronger adhesion than either adhesion molecule alone (Emond et al., 2011). Coordinated patterns of cell–cell adhesion control various cell movements during embryogenesis, including in the early gastrulation stage zebrafish embryo.

CONCLUSIONS

Our analysis of ethanol effects on early development showed that yolk cell E-cadherin/microtubule cytoskeleton was disrupted; blastomere radial intercalation was reduced; and shield cell migration was disorganized. Defects in cell adhesion induced by ethanol partially explain these epiboly and gastrulation defects. Synthetic *pcdh18a* mRNA injection rescues ethanol induced epiboly defects, showing that *pcdh18a* expression changes participate in the pathogenesis in this zebrafish model of FASD. Ethanol effects on yolk cell microtubule cytoskeleton and E-cadherin distribution suggests that extraembryonic tissues in the early embryo also contribute to the FASD phenotype.

In addition to *pcdh18a*, numerous other genes were identified whose expression changed when embryos were exposed to ethanol during early development (Table 3). The data do not point to a single effect that can explain a majority of these gene changes. Ethanol responsive genes fall into various pathways and functional categories, including retinoid metabolism, solute transport and neurogenic functions. Some genes identified in our gene microarray experiments (Table 3) were functionally characterized in previous studies (Aamar and Dawid, 2008; Inbal et al., 2007; Lagutin et al., 2003). The central importance of pluripotency regulators, Notch, Wnt and other pathways represented in the microarray experiments strongly indicate that the approach produced important data to test future hypotheses.

MATERIALS AND METHODS

Zebrafish Husbandry and Ethanol Treatment

Zebrafish (*Danio rerio*; Hamilton; TL strain) were raised and housed under standard laboratory conditions (Westerfield, 2000) in accordance with Indiana University Policy on Animal Care and Use. Embryos were exposed to ethanol by incubation in embryo medium containing 100 mM (0.6% vol./vol.) ethanol from 2 hours post fertilization (hpf) until the completion of the experiment, in Petri dishes wrapped with parafilm and maintained at 28.5°C.

Microscopy

Images of live embryos were collected using a Leica MZ12 microscope equipped with Leica DFC290 camera (Leica Microsystems Inc., Buffalo Grove, IL, USA). Confocal images were acquired using a Zeiss Observer Z1 LSM 700 confocal microscope (40× 1.1 NA W or 20× 0.8 NA objectives; Carl Zeiss Microscopy, Thornwood, NY, USA).

Cell labeling and time-Lapse Imaging

Embryos (1–4 cell stages) were injected with FITC-conjugated histone-1 protein. Embryos were dechorionated and mounted in low melting agarose at 4.3 hpf or 6 hpf. Images at several focal planes were captured every 2 minutes for 1–3 hours using Zeiss Observer Z1 LSM 700 confocal microscope (40× 1.1 NA W); images were focused at the shield region in 6 hpf embryos. All time-lapse movies were processed using Volocity software (Perkin Elmer, Waltham, MA, USA). Radial intercalation cell movements were evaluated manually. Average instantaneous velocities and meandering indexes (relative displacement) were calculated using Volocity software.

In situ Hybridization

Whole-mount *in situ* hybridization of zebrafish embryos was performed as described (Sarmah et al., 2010). Digoxigenin-labeled riboprobes for *ntl* and *pcdh18a* (generously provided by Drs C. Nusslein-Volhard and I. Dawid, respectively) were synthesized using DIG RNA Labeling Kit (Roche, Indianapolis, IN, USA) according to manufacturer's recommendations. Images were collected using a Leica MZ12 microscope equipped with Leica DFC290 camera.

Immunofluorescence, F-actin Staining and Image Analysis

Whole-mount immunostaining was performed as previously described (Clendenon et al., 2012) using primary antibodies against E-cadherin (Cdh1) and α-tubulin, at a dilution of 1:500 in blocking solution. Texas red or Alexa 488-conjugated anti-rabbit and Alexa 488-conjugated anti-mouse secondary antibodies were used at a 1:100 dilution (Molecular probes/Invitrogen). Alexa 488-conjugated phalloidin (Molecular

Probes/Life Technologies, Inc., Grand Island, NY, USA) was used at a 1:100 dilution. Nuclei were stained with TO-PRO-3 at a dilution of 1:1000.

Image analysis to measure cytoplasmic versus cell surface E-cadherin staining was performed as follows using Image J software. Embryos (8 hpf) were stained with E-cadherin antibody, phalloidin and TO-PRO-3 (Molecular Probes/Life Technologies, Inc.), and the prechordal plate region was imaged using confocal microscopy. Images containing the prechordal plate were identified in ethanol treated and control embryos. TO-PRO-3 images were used to make a mask that removed the nucleus area and staining from the E-cadherin. Actin staining was used to highlight the cell surface and cytoplasmic regions of the cells within the prechordal plate region. These highlighted regions were transferred to the corresponding E-cadherin image, and fluorescence intensity per unit area was calculated. Percentage of total staining at the cell surface versus cytoplasm were calculated in three independent experiments. Image J software was also used to measure length-to-width ratio of enveloping layer cells at the embryo margin.

Cell Adhesion Assay

Embryos at 4.5 hpf were dechorionated and incubated in 0.05% trypsin–EDTA for 10 minutes. Cells were disassociated using a glass Pasteur pipette, passing the cell suspension through the pipette 15–25 times to produce a primarily single cell suspension. Rinsing with 10% fetal bovine serum (in L-15 medium) was used to stop the trypsin reaction. Cells were harvested by centrifugation (1,800 rpm) for 3 minutes at 22°C. Cells were resuspended in L-15 medium with or without ethanol, and plated on 10 μg/ml fibronectin-coated, chambered coverslip slides. Images were collected at the same locations over a 3-hour time course using Zeiss Observer Z1 (20× 0.8 NA objective) equipped with a robotic stage and an Orca-AG CCD camera (Hamamatsu Photonics, K. K., Bridgewater, NJ, USA). Cells and cell aggregates (objects) were counted, and a cell-to-object ratio was calculated at each time point.

Immunoblotting

Immunoblotting was performed as previously described (Babb and Marrs, 2004). Cdh1 primary antibody was diluted 1:15,000, and anti-rabbit horseradish peroxidase (HRP)-conjugated secondary antibody (1:10,000 dilution; Amersham, Arlington Heights, IL, USA). Chemiluminescence (ECL kit, Amersham) according to manufacturer directions was used, and membranes were exposed to film (Kodak Bio-Max ML, Eastman Kodak, Rochester, NY, USA). Three independent experiments were compared.

RNA Isolation and Quantitative PCR Analysis

Total RNA was extracted from approximately 20 treated and untreated embryos at 8 hpf using TRIzol reagent (Sigma, St Louis, MO, USA). For quantitative PCR, 1 μg of total RNA was reverse transcribed to cDNA using M-MLV reverse transcriptase (Promega, Madison, WI, USA), and cDNA was diluted tenfold with RNase free water. Each 20 μl PCR reaction was performed with 1–4 μl of cDNA using Power SYBR Green PCR mix

(Applied Biosystems/Life Technologies, Inc.) and 0.5 μM of each primer. Primer sets used are listed in supplementary material Table S2. Three independent experiments in triplicate were performed using *rsp15* as internal control with either the 7300 Real Time PCR System (Applied Biosystems) or the LightCycler 480 (Roche).

Microarray Analysis

Embryos were exposed to ethanol from 2 to 8 hpf or left untreated in seven independent experiments. Total RNA was extracted from approximately 20 treated and untreated embryos for each treatment in each experiment, using TRIzol reagent (Sigma). The RNA samples were examined for quality using the Agilent Bioanalyzer RNA Nanochip (Agilent Technologies, Santa Clara, CA, USA). The RIN (RNA integrity number) for all samples was ≥9.0. The samples were labeled using the standard protocol for the Affymetrix 3'IVT Express kit (Affymetrix, Santa Clara, CA, USA) starting with 100 ng of total RNA. Individual labeled samples were hybridized to the Zebrafish Genome Array (Affymetrix) for 17 hours then washed, stained and scanned following the standard protocol. All 14 arrays were labeled, hybridized and scanned in one batch. Arrays were visually scanned for abnormalities or defects; none were found.

Affymetrix gene expression console software was used to generate MAS5 (MicroArray Suite 5.0) signals and detection calls; arrays were scaled to a target of 1000. To avoid analyzing genes that were not expressed in any condition, only those probe sets that had a fraction present ≥0.40 in at least one of the two treatments were analyzed (McClintick and Edenberg, 2006). MAS5 signals were imported into Partek Genomics Suite (Partek, Inc., St Louis, MO, USA) and $\log_2$ transformed. These $\log_2$ transformed signals were used for Principal Components Analysis (PCA), hierarchical clustering and signal histograms to determine if there were any outlier arrays. No outliers were detected. The PCA plot and hierarchical clustering indicated that there was a batch effect associated with the batch of embryos (perhaps because each experiment was performed using embryos derived from a single breeding pair). The $\log_2$ transformed signals were analyzed using a 2-way ANOVA with factors for treatment (alcohol vs control) and embryo batch (random effect). This analysis indicated that the embryo batch was indeed significant. The False Discovery Rate (FDR) was calculated using the Storey qvalue method (Storey and Tibshirani, 2003). Microarray data were deposited in the NCBI GEO database, accession number GSE48380. Performing PubMed searches identified protein functions using the gene names with absolute changes ≥1.5-fold and P-values ≤0.001. Functional categories shared by 2 or more genes were listed in Table 3.

mRNA Injection

For *pcdh18a* mRNA rescuing experiments, mRNA was synthesized from a pCS2+ *pcdh18a* vector (Aamar and Dawid, 2008) using a SP6 mMessage mMachine kit (Ambion, Austin, TX, USA). Synthetic mRNA (75 pg/embryo) was injected into the embryos at 1-cell stage. Injected and uninjected embryos were treated with and without 100 mM ethanol from 2 to 8 hpf. These embryos were fixed, dechorionated and imaged focusing on enveloping cell layer at the embryo margin. Percent epiboly progression was calculated using Image J software.

Statistical Analysis

In addition to microarray statistical analyses, analyses on cell migration, cell adhesion, cell shape, protein distribution, qPCR and epiboly were performed using unpaired two-tailed Student's *t*-test (GraphPad Software, La Jolla, CA, USA).

ACKNOWLEDGEMENTS

National Institutes of Health P50 AA07611 project and pilot funding to J.A.M. supported these studies. Microarray studies were carried out in the Center for Medical Genomics at Indiana University School of Medicine. We thank Drs C. Nusslein-Volhard (Max Planck Institute for Developmental Biology, Tubingen) and I. Dawid (National Institute of Child Health and Human Development, National Institutes of Health) for providing plasmid probes. We thank S. Armstrong, D. Caldwell, L. Fitzgerald, T. Moskowinski, R. Rehlander, D. Southern, D. Tonde and members of the Marrs lab for preliminary analysis and helpful discussion.

REFERENCES

1. Aamar, E. and Dawid, I. B. (2008). Protocadherin-18a has a role in cell adhesion behavior and migration in zebrafish development. Dev. Biol. 318, 335-346.
2. Ali, S., Champagne, D. L., Spaink, H. P. and Richardson, M. K. (2011). Zebrafish embryos and larvae: a new generation of disease models and drug screens. Birth Defects Res. C Embryo Today 93, 115-133.
3. Babb, S. G. and Marrs, J. A. (2004). E-cadherin regulates cell movements and tissue formation in early zebrafish embryos. Dev. Dyn. 230, 263-277.
4. Behrndt, M., Salbreux, G., Campinho, P., Hauschild, R., Oswald, F., Roensch, J., Grill, S. W. and Heisenberg, C. P. (2012). Forces driving epithelial spreading in zebrafish gastrulation. Science 338, 257-260.
5. Betchaku, T. and Trinkaus, J. P. (1978). Contact relations, surface activity, and cortical microfilaments of marginal cells of the enveloping layer and of the yolk syncytial and yolk cytoplasmic layers of fundulus before and during epiboly. J. Exp. Zool. 206, 381-426.
6. Blader, P. and Stra¨hle, U. (1998). Ethanol impairs migration of the prechordal plate in the zebrafish embryo. Dev. Biol. 201, 185-201.
7. Carvalho, L., Stu¨hmer, J., Bois, J. S., Kalaidzidis, Y., Lecaudey, V. and Heisenberg, C. P. (2009). Control of convergent yolk syncytial layer nuclear movement in zebrafish. Development 136, 1305-1315.
8. Cheng, J. C., Miller, A. L. and Webb, S. E. (2004). Organization and function of microfilaments during late epiboly in zebrafish embryos. Dev. Dyn. 231, 313-323.
9. Clendenon, S. G., Sarmah, S., Shah, B., Liu, Q. and Marrs, J. A. (2012). Zebrafish cadherin-11 participates in retinal differentiation and retinotectal axon projection during visual system development. Dev. Dyn. 241, 442-454.

10. Dlugos, C. A. and Rabin, R. A. (2010). Structural and functional effects of developmental exposure to ethanol on the zebrafish heart. Alcohol. Clin. Exp. Res. 34, 1013-1021.

11. Emond, M. R., Biswas, S., Blevins, C. J. and Jontes, J. D. (2011). A complex of Protocadherin-19 and N-cadherin mediates a novel mechanism of cell adhesion. J. Cell Biol. 195, 1115-1121.

12. Esguerra, C. V., Nelles, L., Vermeire, L., Ibrahimi, A., Crawford, A. D., Derua, R., Janssens, E., Waelkens, E., Carmeliet, P., Collen, D. et al. (2007). Ttrap is an essential modulator of Smad3-dependent Nodal signaling during zebrafish gastrulation and left-right axis determination. Development 134, 4381-4393.

13. Essner, J. J., Amack, J. D., Nyholm, M. K., Harris, E. B. and Yost, H. J. (2005). Kupffer's vesicle is a ciliated organ of asymmetry in the zebrafish embryo that initiates left-right development of the brain, heart and gut. Development 132, 1247-1260.

14. Gilbert-Barness, E. (2010). Teratogenic causes of malformations. Ann. Clin. Lab. Sci. 40, 99-114.

15. Hammerschmidt, M. and Wedlich, D. (2008). Regulated adhesion as a driving force of gastrulation movements. Development 135, 3625-3641. Haycock, P. C. (2009). Fetal alcohol spectrum disorders: the epigenetic perspective.

16. Biol. Reprod. 81, 607-617.

17. Inbal, A., Kim, S. H., Shin, J. and Solnica-Krezel, L. (2007). Six3 represses nodal activity to establish early brain asymmetry in zebrafish. Neuron 55, 407-415.

18. Kai, M., Heisenberg, C. P. and Tada, M. (2008). Sphingosine-1-phosphate receptors egulate individual cell behaviours underlying the directed migration of prechordal plate progenitor cells during zebrafish gastrulation. Development 135, 3043-3051.

19. Kane, D. A., McFarland, K. N. and Warga, R. M. (2005). Mutations in half baked/ Ecadherin block cell behaviors that are necessary for teleost epiboly. Development 132, 1105-1116.

20. Kelly, S. J., Goodlett, C. R. and Hannigan, J. H. (2009). Animal models of fetal alcohol spectrum disorders: impact of the social environment. Dev Disabil Res Rev 15, 200-208.

21. Ko¨ppen, M., Ferna´ndez, B. G., Carvalho, L., Jacinto, A. and Heisenberg, C. P. (2006). Coordinated cell-shape changes control epithelial movement in zebrafish and Drosophila. Development 133, 2671-2681.

22. Lagutin, O. V., Zhu, C. C., Kobayashi, D., Topczewski, J., Shimamura, K., Puelles, L., Russell, H. R., McKinnon, P. J., Solnica-Krezel, L. and Oliver, G. (2003). Six3 repression of Wnt signaling in the anterior neuroectoderm is essential for vertebrate forebrain development. Genes Dev. 17, 368-379.

23. Lin, F., Chen, S., Sepich, D. S., Panizzi, J. R., Clendenon, S. G., Marrs, J. A., Hamm, H. E. and Solnica-Krezel, L. (2009). Galpha12/13 regulate epiboly by inhibiting Ecadherin activity and modulating the actin cytoskeleton. J. Cell Biol. 184, 909-921.

24. Loucks, E. J. and Ahlgren, S. C. (2009). Deciphering the role of Shh signaling in axial defects produced by ethanol exposure. Birth Defects Res. A Clin. Mol. Teratol. 85, 556-567.
25. Matsui, T., Thitamadee, S., Murata, T., Kakinuma, H., Nabetani, T., Hirabayashi, Y., Hirate, Y., Okamoto, H. and Bessho, Y. (2011). Canopy1, a positive feedback regulator of FGF signaling, controls progenitor cell clustering during Kupffer's vesicle organogenesis. Proc. Natl. Acad. Sci. USA 108, 9881-9886.
26. McClintick, J. N. and Edenberg, H. J. (2006). Effects of filtering by Present call on analysis of microarray experiments. BMC Bioinformatics 7, 49.
27. Morita, H. and Heisenberg, C. P. (2013). Holding on and letting go: cadherin turnover in cell intercalation. Dev. Cell 24, 567-569.
28. Muralidharan, P., Sarmah, S., Zhou, F. C. and Marrs, J. A. (2013). Fetal Alcohol Spectrum Disorder (FASD) associated neural defects: complex mechanisms and potential therapeutic targets. Brain Sci. 3, 964-991.
29. Popova, S., Stade, B., Bekmuradov, D., Lange, S. and Rehm, J. (2011). What do we know about the economic impact of fetal alcohol spectrum disorder? A systematic literature review. Alcohol Alcohol. 46, 490-497.
30. Roszko, I., Sawada, A. and Solnica-Krezel, L. (2009). Regulation of convergence and extension movements during vertebrate gastrulation by the Wnt/PCP pathway. Semin. Cell Dev. Biol. 20, 986-997.
31. Sarmah, S. and Marrs, J. A. (2013). Complex cardiac defects after ethanol exposure during discrete cardiogenic events in zebrafish: Prevention with folic acid. Dev. Dyn. [Epub ahead of print] doi: 10.1002/dvdy.24015.
32. Sarmah, S., Barrallo-Gimeno, A., Melville, D. B., Topczewski, J., Solnica-Krezel, L. and Knapik, E. W. (2010). Sec24D-dependent transport of extracellular matrix proteins is required for zebrafish skeletal morphogenesis. PLoS ONE 5, e10367.
33. Solnica-Krezel, L. (2006). Gastrulation in zebrafish – all just about adhesion? Curr. Opin. Genet. Dev. 16, 433-441.
34. Solnica-Krezel, L. and Driever, W. (1994). Microtubule arrays of the zebrafish yolk cell: organization and function during epiboly. Development 120, 2443-2455.
35. Solnica-Krezel, L. and Sepich, D. S. (2012). Gastrulation: making and shaping germ layers. Annu. Rev. Cell Dev. Biol. 28, 687-717.
36. Song, S., Eckerle, S., Onichtchouk, D., Marrs, J. A., Nitschke, R. and Driever, W. (2013). Pou5f1-dependent EGF expression controls E-cadherin endocytosis, cell adhesion, and zebrafish epiboly movements. Dev. Cell 24, 486-501.
37. Speirs, C. K., Jernigan, K. K., Kim, S. H., Cha, Y. I., Lin, F., Sepich, D. S., DuBois, R. N., Lee, E. and Solnica-Krezel, L. (2010). Prostaglandin Gbetagamma signaling stimulates gastrulation movements by limiting cell adhesion through Snai1a stabilization. Development 137, 1327-1337.
38. Storey, J. D. and Tibshirani, R. (2003). Statistical significance for genomewide studies. Proc. Natl. Acad. Sci. USA 100, 9440-9445.

39. Stra¨hle, U. and Jesuthasan, S. (1993). Ultraviolet irradiation impairs epiboly in zebrafish embryos: evidence for a microtubule-dependent mechanism of epiboly. Development 119, 909-919.

40. Sulik, K. K., Johnston, M. C. and Webb, M. A. (1981). Fetal alcohol syndrome: embryogenesis in a mouse model. Science 214, 936-938.

41. Tada, M. and Kai, M. (2012). Planar cell polarity in coordinated and directed movements. Curr. Top. Dev. Biol. 101, 77-110.

42. Takesono, A., Moger, J., Farooq, S., Cartwright, E., Dawid, I. B., Wilson, S. W. and Kudoh, T. (2012). Solute carrier family 3 member 2 (Slc3a2) controls yolk syncytial layer (YSL) formation by regulating microtubule networks in the zebrafish embryo. Proc. Natl. Acad. Sci. USA 109, 3371-3376.

43. Tay, H. G., Schulze, S. K., Compagnon, J., Foley, F. C., Heisenberg, C. P., Yost, H. J., Abdelilah-Seyfried, S. and Amack, J. D. (2013). Lethal giant larvae 2 regulates development of the ciliated organ Kupffer's vesicle. Development 140, 1550-1559.

44. Ulrich, F., Krieg, M., Scho¨tz, E. M., Link, V., Castanon, I., Schnabel, V., Taubenberger, A., Mueller, D., Puech, P. H. and Heisenberg, C. P. (2005). Wnt11 functions in gastrulation by controlling cell cohesion through Rab5c and E-cadherin. Dev. Cell 9, 555-564.

45. Warga, R. M. and Kimmel, C. B. (1990). Cell movements during epiboly and gastrulation in zebrafish. Development 108, 569-580.

46. Westerfield, M. (2000). The Zebrafish Book. Eugene, OR: The University of Oregon Press. Yelin, R., Schyr, R. B., Kot, H., Zins, S., Frumkin, A., Pillemer, G. and Fainsod, A. (2005). Ethanol exposure affects gene expression in the embryonic organizer and reduces retinoic acid levels. Dev. Biol. 279, 193-204.

47. Zalik, S. E., Lewandowski, E., Kam, Z. and Geiger, B. (1999). Cell adhesion and the actin cytoskeleton of the enveloping layer in the zebrafish embryo during epiboly. Biochem. Cell Biol. 77, 527-542.

48. Zhang, Y., Shao, M., Wang, L., Liu, Z., Gao, M., Liu, C. and Zhang, H. (2010). Ethanol exposure affects cell movement during gastrulation and induces split axes in zebrafish embryos. Int. J. Dev. Neurosci. 28, 283-288.

SECTION V. GENE EDITING IN VERTEBRATE EMBRYOS

15

CHAPTER

DEVELOPMENT OF A NEW APPROACH FOR TARGETED GENE EDITING IN PRIMORDIAL GERM CELLS USING TALENS IN XENOPUS

Keisuke Nakajima and Yoshio Yaoita

Division of Embryology and Genetics, Institute for Amphibian Biology, Graduate School of Science, Hiroshima University, Higashihiroshima 739-8526, Japan

ABSTRACT

A gene of interest can be efficiently modified using transcription activator-like effector nucleases (TALENs) (Christian et al., 2010; Li et al., 2011). However, if a target gene is essential for development, growth and fertility, use of TALENs with high mutagenic activity in F0 frogs could result in developmental disorders or sterility, which would reduce the number of F1 progeny and make F1 phenotypical analysis difficult. We used the 39 untranslated region of DEADSouth gene (DS-39) of Xenopus tropicalis to solve this problem, because the addition of the DS-39 to mRNA is known to induce primordial germ cell (PGC)-specific expression and reduce the stability in somatic cells of mRNA in Xenopus laevis. At first, we inserted the X. tropicalis DS-39 downstream of the EGFP termination codon and confirmed that the EGFP expression was specifically detected in PGCs for three weeks. Therefore, we inserted the DS-39 downstream of the termination codon of the TALEN coding sequence. The tyrosinase gene was selected as the target gene for TALEN because the bi-allelic mutation of this gene is easily discernible by the albino phenotype. When fertilized eggs were microinjected with TALEN mRNAs fused to the DS-39, their sperm and oocytes had a high rate (84–100%) of target-gene modification in contrast to the lower rate (0–45%) of nucleotide alteration observed in somatic cells.

Citation: Keisuke Nakajima and Yoshio Yaoita, Development of a new approach for targeted gene editing in primordial germ cells using TALENs in Xenopus, doi:10.1242/bio.201410926.

INTRODUCTION

Targeted gene disruption is becoming a common, facile and essential method for demonstrating the function of a specific gene and is currently performed using transcription activator-like effector nucleases (TALENs) (Cermak et al., 2011; Christian et al., 2010; Li et al., 2011) or the CRISPR/Cas (clustered regularly interspaced short palindromic repeats/CRISPRassociated proteins) system (Jinek et al., 2012). TALENs are fusion proteins, consisting of a nuclear localization signal, a target DNA-binding domain and the nuclease domain of FokI, that enter the nucleus and recognize 15–24 nucleotides of both the left and right halves of the target sequence to form a dimer in the nuclease domain and catalyze double-strand breaks between the two halves of the target sequence. In the CRISPR/Cas system, ,20 nucleotides of the target sequence are complementary and bind to the 59 end of a single synthetic guide RNA, which recruits Cas9 to generate double-strand breaks near the target site. The double strand cleavages formed by these tools are often repaired by non-homologous end-joining, which is an error-prone repair mechanism, resulting in nucleotide deletion and/or insertion. When gene-modified frogs are generated using the TALEN method or CRISPR/Cas system, different F0 frogs have different mosaic patterns of target gene modification (Blitz et al., 2013; Guo et al., 2014; Ishibashi et al., 2012; Nakajima et al., 2012; Nakayama et al., 2013; Suzuki et al., 2013), and different groups of cells in a single F0 frog possess different mutations at the target locus. Some of the cells in the F0 frog may maintain their functions because they contain an in-frame mutation or lack a mutation, whereas other cells may lose the expression of the target gene because of premature translational termination or nonsense-mediated mRNA decay caused by an out-of-frame mutation. Additionally, some cells may lose the function of the target-gene product because of changes in functionally essential amino acids caused by the insertion and deletion of nucleotides. It is impossible to determine the types of mutation that occur in every cell of the body of an F0 animal, and F0 animals are not suitable for phenotypical analysis to elucidate the function of a gene of interest. It is important to obtain an F1 generation with a bi-allelic null mutation of the target locus to analyze the precise function of a gene because an F1 frog is composed of cells with the common bi-allelic mutation of the target gene, and these mutations can be elucidated. However, when the function of the target gene is indispensable to viability and reproduction, F0 frogs injected with highly active TALEN mRNAs cannot grow to sexual maturity or become fertile adults because of developmental abnormalities and growth arrest. In contrast, if less active TALEN mRNAs are used, it is more difficult to obtain offspring with the bi-allelic mutation. One method of overcoming this dilemma is germ cell-specific disruption of a target gene without somatic cell mutation.

The germ plasm is a cytoplasmic region of the oocyte containing germ cell lineage determinants, including unique mRNAs, proteins, and granules and has been observed in Drosophila, Xenopus and Caenorhabditis elegans (Ikenishi et al., 1986; Illmensee and Mahowald, 1974; Okada et al., 1974). DEADSouth mRNA is localized in the germ plasm of Xenopus laevis oocytes (MacArthur et al., 2000; Mosquera et al., 1993) and encodes a putative RNA helicase, a member of the DEAD-box protein family. Primordial germ cells (PGCs) can be visualized in living X. laevis embryos by injecting mRNA encoding the coding region of a fluorescent protein and the 39 untranslated region of DEADSouth gene (DS-39) of X. laevis into the vegetal pole of fertilized eggs (Kataoka et al., 2006). The present study was undertaken to test the hypothesis that adding the DS-39 to TALEN mRNAs may direct the PGC-specific expression of TALENs. In this study, we succeeded in preferentially editing the genome of the germ cells by injecting TALEN mRNAs fused to the DS-39 into Xenopus embryos.

MATERIALS AND METHODS

Animals

The Ivory Coast line of X. tropicalis was provided by the Institute for Amphibian Biology (Graduate School of Science, Hiroshima University) through the National Bio-Resource Project of the MEXT, Japan. The frogs were maintained at 24°C. For the experiments, we used albino X. tropicalis frogs that we generated (Nakajima et al., 2012). The male and female frogs were injected with 200 U of human chorionic gonadotropin (ASKA, Tokyo, Japan) dissolved in 0.45% NaCl. The eggs were manually removed from the adult females by squeezing 4–5 hours after the injection. A testis was dissected from a male 2–3 hours after the injection and was suspended in 1 ml of 16MBS (88 mM NaCl/1 mM KCl/1 mM MgSO4/ 5 mM HEPES (pH 7.8)/2.5 mM NaHCO3) containing 0.1% BSA. The testis suspension (100–300 ml) was placed on the eggs, mixed, and allowed to settle in 0.16MMR [MMR; 100 mM NaCl/2 mM KCl/2 mM CaC12/ 1 mM MgC12/5 mM HEPES (pH 7.4)] at 22°C for 8 minutes; then, the fertilized eggs were dejellied using 3% L-cysteine (Sigma) in 0.16MBS. All of the animals were maintained and used in accordance with the guidelines established by Hiroshima University for the care and use of experimental animals

Construction of the Reporter and the TALENs

A DNA fragment containing the X. tropicalis DS-39 was amplified using TaKaRa LA Taq Hot Start Version (TaKaRa) and the primers XhoXtDS5 (59-GGCTCGAGTAGGTGTGGCAGCACAA-39) and XtDS3BamHI (59-GGGGATCCGAATTTCCCTAAATTGTCTTTACAAAG-39) in a threestep protocol (94°C for 60 seconds; 35 cycles of 98°C for 10 seconds, 55°C for 30 seconds, and 72°C for 30 seconds; and 72°C for 10 minutes). The product was digested using XhoI and BamHI and then ligated into pEGFPC3 (Clontech). The presence of the DS-39 was confirmed by DNA sequencing. This construct was designated pEGFP-DS. The DS-39 fragment was inserted into the XbaI site of pCS2-Flag-TALEN-ELD/ KKR (Lei et al., 2012) using the In-Fusion Advantage PCR cloning kit to obtain pTALEN-ELD/ KKR-DS. The DNA-binding domains were designed to target the first exon of the tyrosinase gene (Tyr-B) (Nakajima et al., 2013). TALEN repeats were assembled as previously described (Cermak et al., 2011), with minor modifications (Nakajima et al., 2013), and were inserted into pTALEN-ELD-DS and pTALEN-KKR-DS to generate the Tyr-TALEN-DS expression constructs. The target sequences of Tyr-TALEN-ELD and -KKR were 59-GGCCCTCAGTTTCCAT-39 and 59-GGCCAGTTCTCTCTAT-39, respectively.

RNA Microinjection

The EGFP-DS DNA fragment was amplified via PCR using the primers T3-pCMV (59-CGAAATTAACCCTCACTAAAGGGAGGTCTATATAAGCAGAG-39) and EGFPC3-polyA (59-TTTTTTTTTTTTTTTTTTTTTTTTTCCACAACTAGAATG CA GTG-39). mRNA was transcribed from the EGFP-DS DNA fragment and XbaI-digested Tyr-TALEN-DS expression constructs in vitro using the mMESSAGE mMACHINE kits (Ambion). Each Tyr-TALEN-DS mRNA (4 nl; 2 or 0.2 ng/ml) and EGFP-DS mRNA (4 nl; 25 ng/ml) dissolved in nuclease-free water (Ambion) was injected into the cortical region of the vegetal pole of fertilized X. tropicalis eggs suspended in 6% Ficoll PM 400 (Sigma)/ 0.16MMR/0.1% BSA (supplementary material Movie 1). Fluorescence was used to identify embryos that had been successfully injected and thus contained EGFP-positive PGCs (Kataoka et al., 2006). The embryos were raised at 22–24°C in 0.16MMR containing 0.1% BSA and 50 mg/ml gentamycin. Tadpoles were anesthetized, and the EGFP expression in PGC was photographed under a fluorescent dissecting microscope (MZ FLIII, Leica) equipped with a color CCD camera DP70 (Olympus).

DNA Extraction

The phenotypes of the F1 tadpoles were assessed seven to ten days after fertilization (at stages 47–48). The tip of the tail of each tadpole was homogenized in 90 ml of 50 mM NaOH and incubated for 10 minutes at 95°C. The homogenate was mixed with 10 ml of 1 M Tris-Cl (pH 8.0) and centrifuged at 15006g for 10 minutes at room temperature. The supernatant was extracted using phenol and chloroform. An organ or tissue was mixed with 80 ml of 50 mM Tris-HCl (pH 7.4) containing 1 mM CaCl2. After the addition of 5 ml of 10% SDS and 3 ml of 10 mg/ ml proteinase K, the mixture was incubated at 65°C for a few hours. Then, 2 ml of 0.1 M PMSF and 0.5 ml of 20 mg/ml RNase A were added, and the mixture was incubated at 37°C for 10 minutes. The genomic DNA was purified using Wizard PCR Preps DNA Purification Resin (Promega).

Mutation Analysis

DNA fragments containing the target site were amplified using the EmeraldAmp MAX PCR Master Mix (TaKaRa) and the primers Tyr-F1 (59-TCACAGAAAGGGTTAAGGGGAAG-39) and Tyr-R1 (59-GCACCCCTACAACAGCCTTC-39) for 25 cycles (95°C, 30 seconds; 58°C, 30 seconds; 72°C, 120 seconds). The second round of PCR was performed using Tyr-F2 (59-GTGAGGAGCAGCATGGAA-39) and TyrR2 (59-CTGCCATGAAGCGAAGAAGGATG-39) for 20 cycles (95°C, 30 seconds; 58°C, 30 seconds; 72°C, 60 seconds) (Fig. 3A). The PCR products were subcloned into the pGEM-T Easy vector (Promega), and the nucleotide sequences were subsequently determined. When we identified tyrosinase mutations derived only from albino mates (Fig. 3F), PCR was conducted using Tyr-F1, Tyr-R1 and TaKaRa Ex Taq Hot Start Version (TaKaRa) for 35 cycles (95°C, 30 seconds; 58°C, 30 seconds; 72°C, 120 seconds), and DNA fragments with 620-bp and 952-bp deletions were cloned. When we did not observe two different alleles in individual offspring of the m1 and f1 frogs via a first round of PCR using the Tyr-FI and -R1 primers and a second round of PCR using the Tyr-F2 and -R2 primers (Fig. 4B), the second round of PCR was exchanged for a protocol that used the Tyr-F1 and Tyr-R2 primers for 20 cycles (95°C, 30 seconds; 58°C, 30 seconds; 72°C, 120 seconds), and a fragment with a 339-bp deletion was cloned. Alternatively, the target DNA was amplified using genomic DNA and the Tyr-F3 (59-TGCGACGAAATAATACCCGCA-39) and Tyr-R1 primers for 35 cycles (95°C, 30 seconds; 58°C, 30 seconds; 72°C, 180 seconds) using TaKaRa Ex Taq Hot Start Version (TaKaRa). A DNA fragment with a 615-bp deletion was obtained. The PCR products were subcloned and their nucleotide sequences were determined.

RESULTS

We hypothesized that injecting TALEN mRNAs fused to the DS- 39 would result in PGC-specific TALEN expression. To test this hypothesis, we constructed an EGFP plasmid, pEGFP-DS, in which the X. tropicalis DS-39 was positioned downstream of the termination codon of the EGFP fluorescent protein. EGFP and EGFP-DS mRNAs were transcribed from pEGFP-C3 and pEGFP-DS, respectively, and injected vegetally into fertilized X. tropicalis eggs. EGFP fluorescence was frequently observed in the PGCs in the mesentery of eight-day-old tadpoles (19 EGFPpositive tadpoles on the eighth day out of 39 tadpoles injected with EGFP-DS mRNA) and in the genital ridges of 21-day-old tadpoles (13 EGFP-positive tadpoles on the 21st day out of 13 EGFP-positive tadpoles on the eighth day). In contrast, when EGFP mRNA was injected, EGFP expression was rarely detected in the PGCs of eight-day-old tadpoles (two EGFP-positive tadpoles on the eighth day out of 36 tadpoles injected with EGFP mRNA) and was detected in the PGCs of only one 21-dayold tadpole (one EGFP-positive tadpole on the 21st day out of two EGFP-positive tadpoles on the eighth day) (Fig. 1). These results showed that the EGFP protein that was translated from EGFP-DS mRNA was stable in a PGC-specific manner for three weeks, and this expression pattern could be ascribed to the presence of DS-39 in the injected mRNA. The mRNAs that were synthesized using the Tyr-TALEN-DS expression constructs were designed to target the X. tropicalis tyrosinase gene; these constructs contained the obligatory heterodimeric nuclease domains (Lei et al., 2012; Nakajima and Yaoita, 2013) and the DS-39 of X. tropicalis. Tyrosinase is essential for melanin synthesis, and the bi-allelic disruption of the tyrosinase gene leads to the albino phenotype, which is easily discernible (Blitz et al., 2013; Guo et al., 2014; Ishibashi et al., 2012; Nakajima et al., 2012; Nakayama et al., 2013). Although fertilized X. tropicalis eggs are generally injected with 400 pg of TALEN mRNAs, in this study, we injected 16 pg or 1.6 pg of Tyr-TALEN-DS mRNAs because the translated proteins were expected to be expressed in the PGCs for more than three weeks and because injecting a smaller amount of exogenous mRNA is more conducive to embryonic viability and development. The TALEN mRNAs were co-injected with EGFP-DS mRNA to confirm that the injected EGFP-DS mRNA was translated in the migratory PGCs. The tadpoles that contained EGFP-positive PGCs two days after injection developed into sexually mature adult frogs that had a few tiny patches of depigmentation in their skin (Fig. 2A).

The male F0 frogs were mated to albino females to estimate the frequency of tyrosinase gene disruption in the germ cells based on the ratio of albino to non-albino offspring. The allele from a male F0 frog was easily distinguished from the allele from albino female in offspring because the albino mates used in the crosses have two out of three types of mutations that are located far from the targeting site of Tyr-TALEN (Fig. 3A). Albinism was observed in 80%, 63% and 21% of the offspring of albino females crossed to three different males injected with 16 pg mRNA, which were designated as m1, m2 and m3, respectively (Fig. 3B– E); these results imply that the tyrosinase gene was mutated in ,21–80% of the spermatozoa. Some of the resultant tadpoles were lightly pigmented (Fig. 3C). To determine their individual genotypes, the tyrosinase gene was cloned and sequenced. In the lightly pigmented offspring, almost all of the tyrosinase alleles that were expected to have been derived from the m2 and m3 males had nucleotide deletions or substitutions, but maintained an in-frame coding region (Table 1), suggesting that these in-frame mutations reduced the activity of the enzyme but did not result in complete gene inactivation. Furthermore, genotypic analysis of F1 wild-type offspring revealed that 21%, 88% and 86% of the alleles derived from the m1, m2 and m3 males, respectively, were modified and were still in-frame, implying that these in-frame mutations had little effect on the activity of the enzyme. Based on these data, we estimated the mutation hit rates of the tyrosinase gene to be 84%, 96% and 89% in the spermatozoa of the m1, m2 and m3 males, respectively (Table 1).

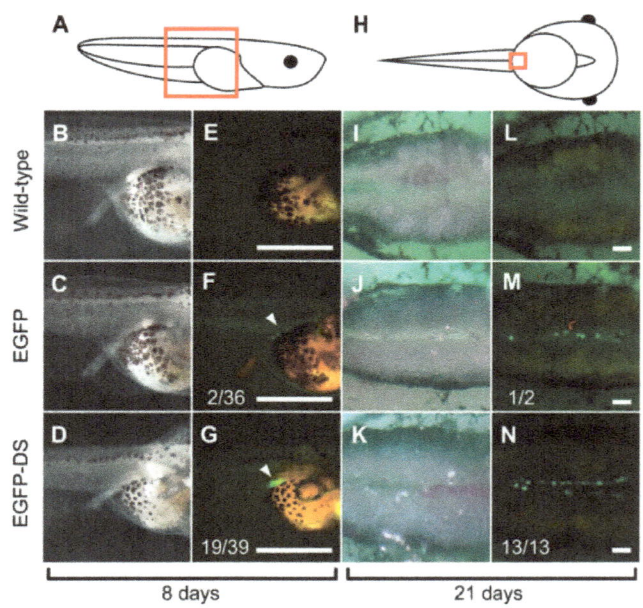

Figure. 1: Expression of EGFP in X. tropicalis tadpoles injected with EGFP-DS mRNA. (A,H) A schematic representation of a lateral view of an 8- day-old tadpole (A) and a ventral view of a 21-day-old tadpole (H). (B–G) Higher-magnification view of A. (I–N) Higher-magnification view of H. (B,E,I,L) Images of wild-type tadpoles. (C,F,J,M) Images of tadpoles injected with EGFP mRNA. (D,G,K,N) Images of tadpoles injected with EGFP-DS mRNA. (B–D) Brightfield

images of live 8-day-old tadpoles. (E–G) EGFP expression in B, C and D, respectively. (I–K) Brightfield images of the mesonephros of 21-day-old tadpoles. (L–N) EGFP expression in I,J and K, respectively. F and G show the ratio of the number of EGFP-positive tadpoles on the eighth day to the number of mRNA-injected tadpoles. M and N show the ratio of the number of EGFP-positive tadpoles on the 21st day to the number of EGFP-positive tadpoles on the eighth day. The white arrowheads indicate PGCs. Scale bars51 mm in (E–G) and 0.1 mm in (L–N).

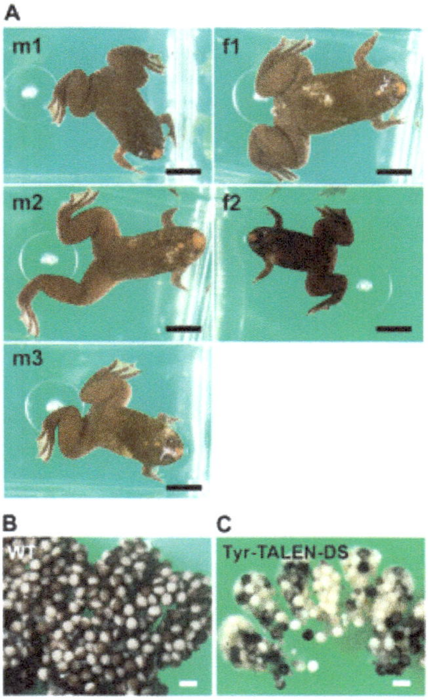

Figure. 2: Phenotype of F0 frogs. (A) Photographs of F0 frogs. Tyr-TALEN-DS mRNAs were co-injected with EGFP-DS mRNA at the vegetal pole of fertilized eggs. The frogs were allowed to reach sexual maturity. Scale bars51 cm. (B) Ovary of a wild-type one-year-old frog. (C) Ovary of a oneyear-old F0 frog that was injected with Tyr-TALEN-DS mRNAs. (B,C) Scale bars51 mm.

Albino tadpoles comprised 26% and 16% of the offspring derived from crosses between albino females and 1.6-pg-mRNAinjected m4 and m5 males, respectively, and the gene modification rate in the m5 spermatozoa was estimated to be 28% using genotypic analysis (Table 1). When the amount of injected TALEN mRNAs was reduced to one-tenth of the original amount, the frequency of the albino phenotype and the mutation hit rate both decreased to approximately one-third of their prior levels. Conversely, the albino phenotype appeared in 54% and 42% of offspring resulting from crosses between albino males and 16 pg-mRNA-injected f1 and f2 female frogs, respectively, and all of the examined tyrosinase alleles derived from f1 and f2 oocytes were found to be modified (Table 1). F0 female frogs had many albino oocytes in their ovary compared to the wild-type ovary (Fig. 2B,C).

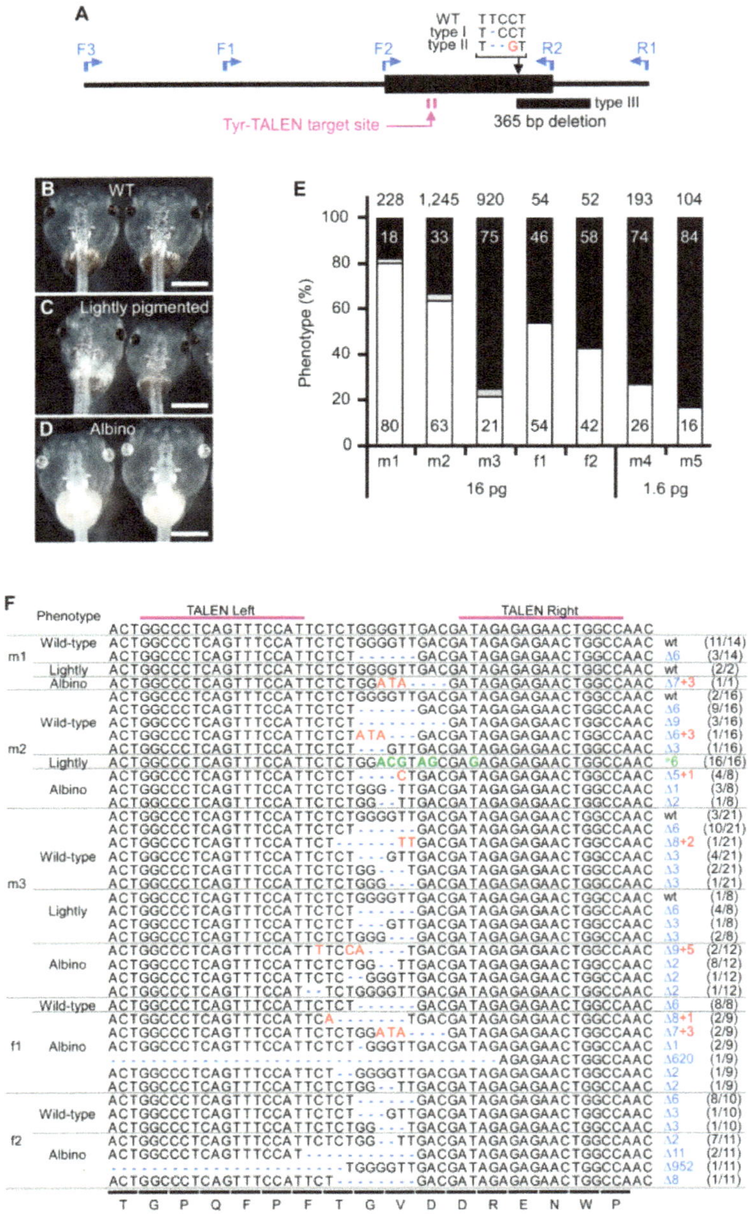

Figure. 3: Analysis of individual offspring obtained from crosses between F0 and albino frogs. (A) Schematic representation showing both the gene mutations of the albino mates used in the crosses and the locations of primers used in the mutation analysis. The bold horizontal line indicates the DNA region from the initiation codon to the end of the first exon. The mutated sequences of the albino mates are shown as type I, type II and type III with the wild-type sequence. The inserted and deleted nucleotides are indicated with a red character and blue dashes, respec-

tively. The locations of the mutations and the 365-bp deletion in the albino mates are indicated by a black arrow and rectangle, respectively. The Tyr-TALEN target site and the direction of the primers are indicated by a purple arrow and blue arrows, respectively. (B–D) Brightfield images of live wild-type (B), lightly pigmented (C) and albino (D) tadpoles. Note that the abdominal region is lightly pigmented in C. Scale bars51 mm. (E) Phenotypic analysis of the offspring of F0 and albino frogs. The percentages of wild-type, lightly pigmented and albino F1 tadpoles are indicated by black, gray and white bars, respectively (Table 1). The numbers in the bars are the percentages of tadpoles with the indicated phenotype. The total number of tadpoles is shown at the top of the bars. (F) Genotypic analysis of individual offspring of F0 and albino frogs. The target DNA fragment was amplified using genomic DNA purified from individual wild-type, lightly pigmented and albino F1 tadpoles and was recloned for sequence determination. All of the mating data are shown. Different sequences were bequeathed by F0 frogs to different F1 offspring, which also had the other mutated tyrosinase allele inherited from the albino parent. The target sequences derived from the albino mates are not shown. The wild-type target DNA and the amino acid sequences are indicated at the top and bottom of the panel, respectively. A pair of purple bars denotes the TALEN-binding sites. The gaps resulting from a deletion (D), the inserted nucleotides (+) and the exchanged nucleotides (*) are indicated with blue dashes, red characters and green characters, respectively. The mutation types are indicated on the right. The ratio of the number of tadpoles with the indicated sequence to the total number of tadpoles with the same phenotype is shown in parentheses on the right. Large deletions of 620 and 952 nucleotides were observed in the offspring derived from the f1 and f2 frogs, respectively, and both of these deletions contained the start codon.

Mating of two F0 frogs, m1 and f1, resulted in 44% albino offspring, which was consistent with the frequency estimated from the ratios of albino to non-albino offspring obtained by mating m1 or f1 with albino partners (Fig. 3E and Fig. 4A). This result shows that it is possible to efficiently generate an F1 bi-allelic gene-knockout using our method (Table 2; Fig. 4). To confirm that the targeted gene mutation was induced specifically in germ cells by the Tyr-TALEN-DS mRNAs, the genomic DNA was extracted from several organs and tissues of the m2 frog and subjected to simple direct sequencing of the PCR-amplified targeted genome region (DSP assay) (Nakayama et al., 2013). A sequence comparison showed that the tyrosinase gene was modified in the genomic DNA of the right and left testis but was not often modified in the genomic DNA of other organs (Fig. 5A). To examine the organ specificity of the mutation of this gene, the tyrosinase gene was cloned from the genomic DNA of various organs and tissues and then sequenced (Fig. 5B,C). The mutation hit rate was higher in the testes (67–89%) than in the other organs (0–45%). No mutation was detected in randomly selected eight clones containing the tyrosinase gene derived from genomic DNA of m2 frog skin. This may be because the prospective epidermis is located far away from the cortical region of the vegetal pole, which is the mRNA injection site. It is possible that the m2 sperm were derived mainly from the right testis during mating with albino females because the spermatozoa and the right testis had higher mutation rates (96% and 89%, respectively) than the left testis (67%). Alternatively, the testes may include many non-germ cells containing the wild-type tyrosinase gene even though nearly all of the germ cells have mutations in this gene.

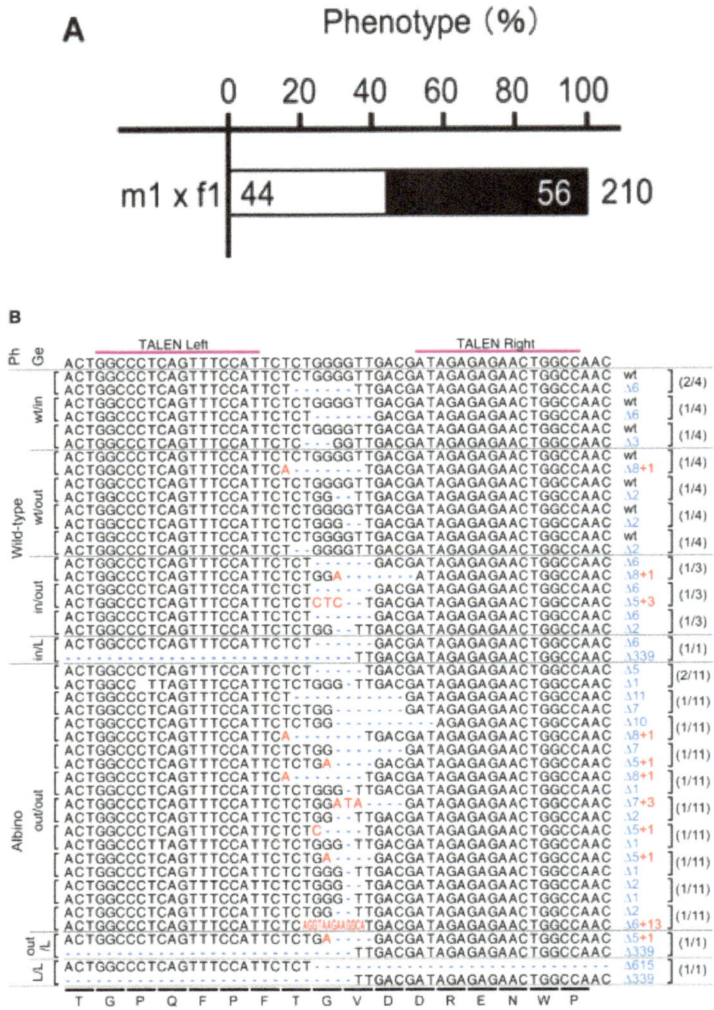

Figure. 4: Analysis of individual offspring of F0 frogs. (A) Phenotypic analysis of the offspring obtained by mating two F0 frogs, m1 and f1. The percentages of wild-type and albino F1 tadpoles are indicated by black and white bars, respectively, and the values are denoted in the bars. The total number of tadpoles is shown on the right. (B) Genotypic analysis of individual offspring obtained from mating the m1 and f1 frogs. The target DNA fragment was amplified using genomic DNA that was purified from individual F1 tadpoles and was recloned for sequence determination. The phenotype (Ph) and genotype (Ge) of F1 tadpoles are indicated on the left. Each F1 tadpole had a pair of sequence lines as shown in brackets. wt, wild-type target sequence; in, in-frame mutation; out, out-of-frame mutation; L, large deletion containing the start codon or the exon-intron boundary. The ratio of the number of tadpoles with a pair of the indicated sequences to the total number of tadpoles with the same phenotype and genotype is shown in parentheses on the right. The alignment is labeled as described in the legend of Fig. 3F.

DISCUSSION

We report a simple and efficient method of preferentially editing the genome of germ cells using TALEN, which may enable the generation of F1 frogs with a bi-allelic target-gene mutation through mating of healthy and fertile F0 frogs even when the gene of interest is necessary for viability, normal development, homeostasis or reproduction. In the m2 male, the hit rate for tyrosinase gene mutation was 96% in spermatozoa and ,40% in the pancreas, intestine, brain and stomach. Therefore, the bi-allelic gene modification rate may have been 16% in pancreatic cells and cells of other organs. Even if the 16% of the cells with the bi-allelic target-gene mutation did not survive within the pancreas or other organs, the remaining cells with one or two copies of the wild-type gene may compensate for the dead cells. However, when the target gene is necessary for the survival of germ cells or the fertilization itself, F1 offspring cannot be obtained by our method. We have not performed the side-by-side experiment to examine the ratios of germ line transmission in TALEN-DSmRNA-injected embryos to those in TALEN-mRNA-injected embryos. However, the result in embryos injected with EGFPDS and EGFP mRNAs strongly suggests that the efficiency of germ line transmission is enhanced by adding DS-39 to TALEN mRNAs.

Table 1: Genotypic analysis of individual offspring of F0 and albino frogs

Phenotype	m1	m2	m3	f1	f2	m5
Wild-type tadpoles	3/14 21%[a] (n=41)	14/16 88%[a] (n=415)	18/21 86%[a] (n=693)	8/8 100%[a] (n=25)	10/10 100%[a] (n=30)	2/15 13%[a] (n=87)
Lightly pigmented tadpoles	0/2 0% (n=5)	16/16 100%[a] (n=43)	7/8 88%[a] (n=35)	(n=0)	(n=0)	(n=0)
Albino tadpoles	1/1 100%[b] (n=182)	8/8 100%[b] (n=787)	12/12 100%[b] (n=192)	9/9 100%[b,c] (n=29)	11/11 100%[b,c] (n=22)	11/11 100%[b] (n=17)
Estimated mutation rate of germ cells	84%	96%	89%	100%	100%	28%

The mutation hit rate as determined by sequencing is indicated (Fig. 3F). The mutated target sequences derived from the albino mates are not shown. The number of tadpoles with the indicated phenotype is shown in parentheses. The estimated mutation rate of the germ cells is presented in the bottom row.
[a]in-frame mutation;
[b]out-of-frame mutation;
[c]large deletion containing the start codon.

Table 2: Results of the genotypic analysis of individual offspring of F0, m1 and f1 frogs

	Genotype						
Phenotype	wt/in	wt/out	in/out	in/L	out/out	out/L	L/L
Wild type	4/12 33%	4/12 33%	3/12 25%	1/12 8%	0/12 0%	0/12 0%	0/12 0%
Albino	0/13 0%	0/13 0%	0/13 0%	0/13 0%	11/13 85%	1/13 8%	1/13 8%

The types and frequencies of the mutations were determined by sequencing (Fig. 4B). The ratio of the number of tadpoles with the indicated phenotype and genotype to the total number of tadpoles with the indicated phenotype is shown. wt, wild-type target sequence; in, in-frame mutation; out, out-of-frame mutation; L, large deletion containing the start codon or the exon-intron boundary.

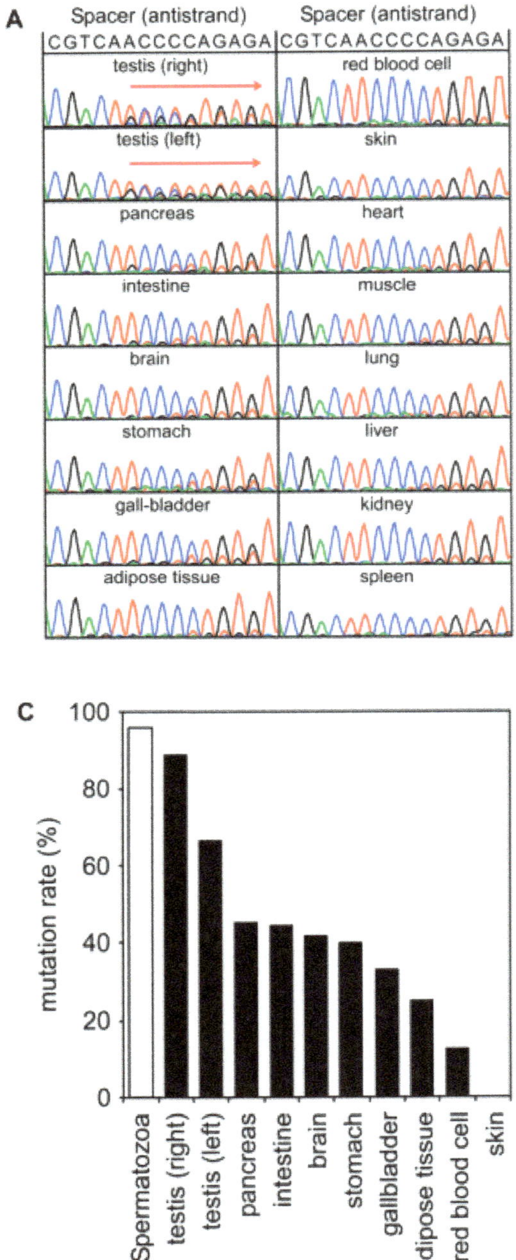
A
Spacer (antistrand)
CGTCAACCCCAGAGA
testis (right)
testis (left)
pancreas
intestine
brain
stomach
gall-bladder
adipose tissue
Spacer (antistrand)
CGTCAACCCCAGAGA
red blood cell
skin
heart
muscle
lung
liver
kidney
spleen
C
mutation rate (%)
0
20
40
60
80
100
Spermatozoa
testis (right)
testis (left)
pancreas
intestine
brain
stomach
gallbladder
adipose tissue
red blood cell
skin

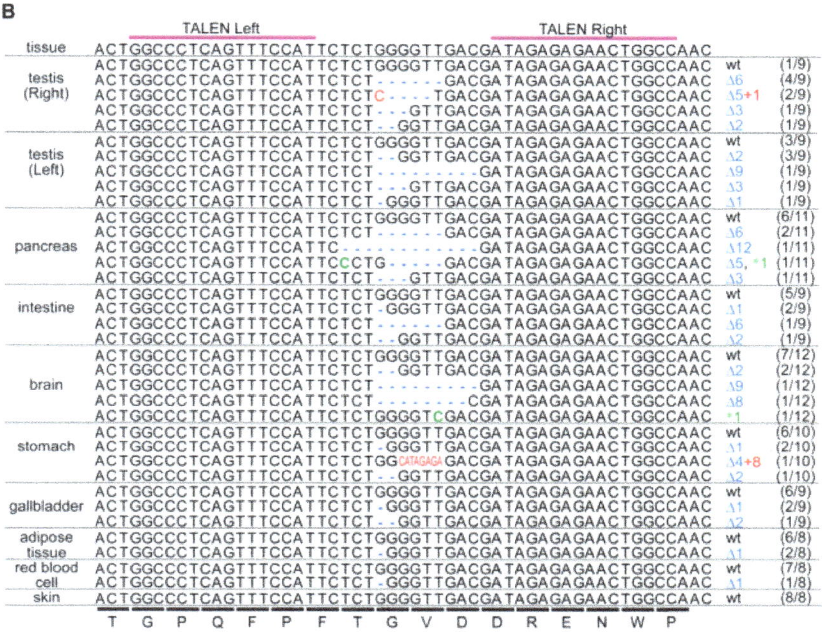

Figure. 5: Preferential mutagenesis in the testes of the m2 frog. (A) Simple direct sequencing of the PCRamplified targeted genome region (DSP assay) using genomic DNA obtained from several organs and tissues of the Tyr-TALEN-DS-mRNA-injected m2 frog. The spacer sequence located between the Tyr-TALEN binding sites is shown at the top of the panel. The ambiguous sequence in the right and left testes begins with "A" (a red wave) and extends to the right, as shown by the red arrows, suggesting that these amplicons are mixtures of heterogeneous fragments with different mutations. (B) Mutated target sequences in organs and tissues of the m2 frog. The target DNA fragment was amplified using genomic DNA that was purified from organs and tissues and subcloned. The sequences of eight to 12 clones per organ or tissue were determined. The ratio of the number of the indicated sequence to the total number of sequences in an organ or tissue is shown in parentheses on the right. The alignment is labeled as described in the legend of Fig. 3F. (C) Mutation hit rate in several organs and tissues of m2. The mutation hit rate in an organ or tissue is indicated by a black bar, whereas the estimated mutation rate in the m2 spermatozoa is indicated by a white bar.

The germ line of the F0 animals was highly mosaic, as demonstrated by the genotyping of F1 animals generated by mating two F0 frogs, m1 and f1. Twenty different types of mutation were observed in the F1 offspring. Furthermore, the genetic mosaicism of the F0 animals is consistent with the results of many studies in which zinc-finger nuclease, TALEN and CRISPR/Cas systems were used (Blitz et al., 2013; Guo et al., 2014; Ishibashi et al., 2012; Nakajima et al., 2012; Nakayama et al., 2013; Suzuki et al., 2013). Although genetically heterogeneous F0 animals show a similar phenotype, this result suggests that knocking out the gene of interest induced this phenotype. Furthermore, if the F0 frogs do not show any phenotype, no conclusions regarding the gene in question can be drawn. However, the necessity and functional redundancy of the gene can be evaluated, when no phenotype is evident in the F1 animals with a bi-allelic out-of-frame mutation. When fertilized eggs were injected vegetally with mRNA encoding a fluorescent protein fused to the DS-39 of X. laevis, a diffuse signal was observed in the vegetal blastomeres of the blastula embryos, and this signal appeared to be restricted to the PGCs from the mid-blastula transition (MBT) stage onward (Kataoka et al., 2006). The mRNA fused to the DS-39 was degraded in the somatic cells via miR-427-mediated mRNA clearance after the MBT stage, but not in PGCs (Yamaguchi et al., 2014). Because the TALEN mRNAs fused to the DS-39 would be translated before the MBT stage, the target gene should be modified in somatic cells using our method, which would impair the germ-cell specificity of the genomic editing. If translation of the TALEN-DS mRNAs could be repressed using anti-sense oligo-deoxynucleotides or RNA for several hours until the MBT stage is reached, the target gene may be protected from degradation in somatic cells. Two modes of germ-cell specification have been proposed: the inductive (regulative) mode and the predetermined (germ plasm) mode (Bachvarova et al., 2009; Extavour and Akam, 2003). In the inductive mode, cells of the epiblast or the animal cap would not normally form germ cells at the late blastula to the early gastrula stages, but they can be induced to form these cells in the presence of the appropriate tissue or growth factors. In the predetermined mode, germ cells are specified before gastrulation by the germ plasm, the maternally inherited determinants present in the egg. This mode is used by nematodes, flies, tunicates, teleosts, frogs and birds for PGC specification. Our method should be applicable for genomic editing in animals that employ the predetermined mode but not in those using the inductive mode because the germ plasm contains localized maternal mRNAs. For example, nanos1 mRNA is one component of zebrafish germ plasm, and its localization is restricted to the PGCs by its 39UTR, which is similar to the DS-39 (Ko¨prunner et al., 2001). If zebrafish embryos are injected with TALEN mRNAs fused to the nanos1 gene 39UTR, this mRNA should be localized and translated specifically in PGCs, leading to digestion of the target sequence within the genome. In Drosophila, a genomic editing method was recently developed for specifically targeting germline cells; in this system, a single guide RNA-expression construct is injected into the embryos of a transgenic line expressing Cas9 in the germline (Ren et al., 2013). Although the overall heritable mutation rates are high (12.4–74.2%), the germ-cell specificity of genomic editing was not examined, and this system is available only for Cas9-expressing transgenic lines, in contrast to our method, which uses wild-type animals.

Our TALEN system simply and efficiently generated heritable mutant alleles in germ cells with less frequent somatic mutations using only TALEN mRNAs fused to the DS-39. This method may enable the generation of F1 offspring possessing a bi-allelic mutation of a lethal gene through mating of fertile F0 individuals. Furthermore, this method is applicable to genomic editing in animals that use the predetermined mode of germ-cell specification if the 39UTR of a germ plasm-specific mRNA, similar to the DS- 39, determines the location of this mRNA in the PGCs.

ACKNOWLEDGEMENTS

We thank Dr H. Ogino for allowing K.N. to visit his laboratory to learn the microinjection technique using X. tropicalis; Dr D. Voytas for supplying the Golden Gate TALEN and the TAL effector kit (Addgene, #1000000016); Dr C.H.K. Cheng for providing the obligate heterodimeric TALEN scaffolds; and Ms. T. Nakajima for her technical assistance and frog husbandry.

REFERENCES

1. Bachvarova, R. F., Crother, B. I. and Johnson, A. D. (2009). Evolution of germ cell development in tetrapods: comparison of urodeles and amniotes. Evol. Dev. 11, 603-609.
2. Blitz, I. L., Biesinger, J., Xie, X. and Cho, K. W. (2013). Biallelic genome modification in F(0) Xenopus tropicalis embryos using the CRISPR/Cas system. Genesis 51, 827-834.
3. Cermak, T., Doyle, E. L., Christian, M., Wang, L., Zhang, Y., Schmidt, C., Baller, J. A., Somia, N. V., Bogdanove, A. J. and Voytas, D. F. (2011). Efficient design and assembly of custom TALEN and other TAL effector-based constructs for DNA targeting. Nucleic Acids Res. 39, e82.
4. Christian, M., Cermak, T., Doyle, E. L., Schmidt, C., Zhang, F., Hummel, A., Bogdanove, A. J. and Voytas, D. F. (2010). Targeting DNA double-strand breaks with TAL effector nucleases. Genetics 186, 757-761.
5. Extavour, C. G. and Akam, M. (2003). Mechanisms of germ cell specification across the metazoans: epigenesis and preformation. Development 130, 5869-5884.
6. Guo, X., Zhang, T., Hu, Z., Zhang, Y., Shi, Z., Wang, Q., Cui, Y., Wang, F., Zhao, H. and Chen, Y. (2014). Efficient RNA/Cas9-mediated genome editing in Xenopus tropicalis. Development 141, 707-714.
7. Ikenishi, K., Nakazato, S. and Okuda, T. (1986). Direct evidence for the presence of germ cell determinant in vegetal pole cytoplasm of Xenopus laevis and in a subcellular fraction of it. Dev. Growth Differ. 28, 563-568.
8. Illmensee, K. and Mahowald, A. P. (1974). Transplantation of posterior polar plasm in Drosophila. Induction of germ cells at the anterior pole of the egg. Proc. Natl. Acad. Sci. USA 71, 1016-1020.
9. Ishibashi, S., Cliffe, R. and Amaya, E. (2012). Highly efficient bi-allelic mutation rates using TALENs in Xenopus tropicalis. Biol. Open 1, 1273-1276.
10. Jinek, M., Chylinski, K., Fonfara, I., Hauer, M., Doudna, J. A. and Charpentier, E. (2012). A programmable dual-RNA-guided DNA endonuclease in adaptive bacterial immunity. Science 337, 816-821.
11. Kataoka, K., Yamaguchi, T., Orii, H., Tazaki, A., Watanabe, K. and Mochii, M. (2006). Visualization of the Xenopus primordial germ cells using a green fluorescent protein controlled by cis elements of the 39 untranslated region of the DEADSouth gene. Mech. Dev. 123, 746-760.

12. Ko¨prunner, M., Thisse, C., Thisse, B. and Raz, E. (2001). A zebrafish nanosrelated gene is essential for the development of primordial germ cells. Genes Dev. 15, 2877-2885.

13. Lei, Y., Guo, X., Liu, Y., Cao, Y., Deng, Y., Chen, X., Cheng, C. H., Dawid, I. B., Chen, Y. and Zhao, H. (2012). Efficient targeted gene disruption in Xenopus embryos using engineered transcription activator-like effector nucleases (TALENs). Proc. Natl. Acad. Sci. USA 109, 17484-17489.

14. Li, T., Huang, S., Jiang, W. Z., Wright, D., Spalding, M. H., Weeks, D. P. and Yang, B. (2011). TAL nucleases (TALNs): hybrid proteins composed of TAL effectors and FokI DNA-cleavage domain. Nucleic Acids Res. 39, 359- 372.

15. MacArthur, H., Houston, D. W., Bubunenko, M., Mosquera, L. and King, M. L. (2000). DEADSouth is a germ plasm specific DEAD-box RNA helicase in Xenopus related to eIF4A. Mech. Dev. 95, 291-295.

16. Mosquera, L., Forristall, C., Zhou, Y. and King, M. L. (1993). A mRNA localized to the vegetal cortex of Xenopus oocytes encodes a protein with a nanos-like zinc finger domain. Development 117, 377-386.

17. Nakajima, K. and Yaoita, Y. (2013). Comparison of TALEN scaffolds in Xenopus tropicalis. Biol. Open 2, 1364-1370.

18. Nakajima, K., Nakajima, T., Takase, M. and Yaoita, Y. (2012). Generation of albino Xenopus tropicalis using zinc-finger nucleases. Dev. Growth Differ. 54, 777-784.

19. Nakajima, K., Nakai, Y., Okada, M. and Yaoita, Y. (2013). Targeted gene disruption in the Xenopus tropicalis genome using designed TALE nucleases. Zoolog. Sci. 30, 455-460.

20. Nakayama, T., Fish, M. B., Fisher, M., Oomen-Hajagos, J., Thomsen, G. H. and Grainger, R. M. (2013). Simple and efficient CRISPR/Cas9-mediated targeted mutagenesis in Xenopus tropicalis. Genesis 51, 835-843.

21. Okada, M., Kleinman, I. A. and Schneiderman, H. A. (1974). Restoration of fertility in sterilized Drosophila eggs by transplantation of polar cytoplasm. Dev. Biol. 37, 43-54.

22. Ren, X., Sun, J., Housden, B. E., Hu, Y., Roesel, C., Lin, S., Liu, L. P., Yang, Z., Mao, D., Sun, L. et al. (2013). Optimized gene editing technology for Drosophila melanogaster using germ line-specific Cas9. Proc. Natl. Acad. Sci. USA 110, 19012-19017.

23. Suzuki, K. T., Isoyama, Y., Kashiwagi, K., Sakuma, T., Ochiai, H., Sakamoto, N., Furuno, N., Kashiwagi, A. and Yamamoto, T. (2013). High efficiency TALENs enable F0 functional analysis by targeted gene disruption in Xenopus laevis embryos. Biol. Open 2, 448-452.

24. Yamaguchi, T., Kataoka, K., Watanabe, K. and Orii, H. (2014). Restriction of the Xenopus DEADSouth mRNA to the primordial germ cells is ensured by multiple mechanisms. Mech. Dev. 131, 15-23.

16
CHAPTER

EFFICIENT GENERATION OF MYOSTATIN KNOCK-OUT SHEEP USING CRISPR/CAS9 TECHNOLOGY AND MICROINJECTION INTO ZYGOTES

M. Crispo[1], A. P. Mulet[1], L. Tesson[3], N. Barrera[2], F. Cuadro[2], P. C. dos Santos-Neto[2], T. H. Nguyen[3], A. Crénéguy[3], L. Brusselle[3] , I. Anegón3, A. Menchaca[2]

[1]Unidad de Animales Transgénicos y de Experimentación (UATE), Institut Pasteur de Montevideo, Montevideo, Uruguay

[2]Instituto de Reproducción Animal Uruguay, Fundación IRAUy, Montevideo, Uruguay

[3]INSERM UMR 1064, Center for Research in Transplantation and Immunology-ITUN, Nantes, France

ABSTRACT

While CRISPR/Cas9 technology has proven to be a valuable system to generate gene-targeted modified animals in several species, this tool has been scarcely reported in farm animals. Myostatin is encoded by *MSTN* gene involved in the inhibition of muscle differentiation and growth. We determined the efficiency of the CRISPR/Cas9 system to edit *MSTN* in sheep and generate knock-out (KO) animals with the aim to promote muscle development and body growth. We generated CRISPR/Cas9 mRNAs specific for ovine *MSTN* and microinjected them into the cytoplasm of ovine zygotes. When embryo development of CRISPR/Cas9 microinjected zygotes (n = 216) was compared with buffer injected embryos (n = 183) and non microinjected embryos (n = 173), cleavage rate was lower for both microinjected groups (P<0.05) and neither was affected by CRISPR/Cas9 content in the injected medium. Embryo development to blastocyst was not affected by microinjection and was similar among the experimental groups. From 20 embryos analyzed by Sanger sequencing, ten were mutant (heterozygous or mosaic; 50% efficiency). To

Citation: Crispo M, Mulet AP, Tesson L, Barrera N, Cuadro F, dos Santos-Neto PC, et al. (2015) Efficient Generation of Myostatin Knock-Out Sheep Using CRISPR/Cas9 Technology and Micro-injection into Zygotes. PLoS ONE 10(8): e0136690. doi:10.1371/journal.pone.0136690.

obtain live *MSTN* KO lambs, 53 blastocysts produced after zygote CRISPR/Cas9 microinjection were transferred to 29 recipient females resulting in 65.5% (19/29) of pregnant ewes and 41.5% (22/53) of newborns. From 22 born lambs analyzed by T7EI and Sanger sequencing, ten showed indel mutations at *MSTN* gene. Eight showed mutations in both alleles and five of them were homozygous for indels generating out-of frame mutations that resulted in premature stop codons. Western blot analysis of homozygous KO founders confirmed the absence of myostatin, showing heavier body weight than wild type counterparts. In conclusion, our results demonstrate that CRISPR/Cas9 system was a very efficient tool to generate gene KO sheep. This technology is quick and easy to perform and less expensive than previous techniques, and can be applied to obtain genetically modified animal models of interest for biomedicine and livestock.

INTRODUCTION

Modification of the genome in large animals has always been a difficult task, with few efficient technologies available until recently [1–4]. The generation of genetically modified large animals has several advantages for different applications not only in meat, dairy or wool production but also in the generation of bioreactors to produce recombinant proteins of biomedical interest or for the production of models of human diseases [5]. Nevertheless, the massive application of these technologies has been limited due in part to the low efficiency to obtain transgenic founders [6, 7]. While lentiviral transgenesis does show high efficiency [8], site-specific gene integration is still not possible, among other problems. Gene edition in general and in particular the generation of genetically-modified animals has experienced a major advance in the last few years due to the incorporation of novel technologies of direct microinjection into zygotes of gene-specific nucleases such as zinc finger nucleases (ZFN) [9], TALENs [10] and homing endonucleases [11, 12]. Although the use of these gene-specific nucleases made possible the generation of KO animals, they are difficult to engineer. Lately, a new gene-specific nuclease is being applied with high efficiency. Known as clustered regularly interspaced short palindromic repeats (CRISPR/Cas9), this system is taken from the adaptive immune system of bacteria and archae [13], and is based in small RNAs and Cas endonucleases proteins that can induce site-specific DNA double-stranded breaks. CRISPR/Cas9 has been rapidly tested in several species to date [14–19] and promises to revolutionize the field of genome editing in general and in particular transgenic vegetal and animal fields. Genetically-modified farm animals using zygote microinjection of CRISPR/Cas9 have been reported recently in pigs [20], goats [17] and sheep [21], but more information is needed to extend the generated information.

According to official data of FAO estimations, in order to feed a larger, more urban and richer human population, food production must be increased around 70 percent before the year 2050 [22]. According to meat production, the improvement of traditional animal breeds with modern technologies should be implemented. Myostatin is a member of the transforming growth factor beta (TGF-β) superfamily, involved in the inhibition of muscle differentiation and growth [23]. Previous studies have reported that the inhibition of this protein induces a significant increase in the muscle volume and mass producing more meat in animals known as double-muscle animals [24–27]. In farm animals, sheep are good models to be modified genetically due to small size and short gestation period compared to cows. However, until now their use was limited due to the lack of efficient gene editing tools in livestock. In addition, the silencing of this gene in breeds as Australian Merino, the most specialized breed in the production of superfine or ultrafine wool, could be an interesting model to produce more meat in a high-quality wool producing animal.

The objective of this study was to demonstrate the efficiency of the CRISPR/Cas9 gene editing technology coupled to zygote microinjection in the generation of *MSTN* KO sheep. We show that this approach induces mutations with a high efficiency (45.5%), resulting in live offspring carrying single-gene mono and bi-allelic mutations.

MATERIALS AND METHODS

Ethics Statement

All procedures that include animal handling were carried out in strict accordance with the recommendations in the Guide for the Care and Use of Laboratory Animals of the National Institutes of Health. Protocol # 001/2014 was approved by the Animal Care Committee of the Fundación IRAUy certified by the National Council of Animal Care of Uruguay and have therefore been performed in accordance with the ethical standards laid down in the 1964 Declaration of Helsinki and its later amendments. All surgeries were performed under strict aseptic conditions, and all efforts were made to minimize animal suffering.

Generation of the Plasmid Co-Expressing Cas9 and sgRNA

The pX330-U6-Chimeric_BB-CBh-hSpCas9 plasmid (a gift from Feng Zhang, Addgene plasmid # 42230) was digested with BbsI, dephosphoryated using Antarctic Phosphatase (NEB, UK), and the linearized vector was gel purified. To generate the bicistronic vector (pX330-cas9-MSTN) expressing Cas9 and sgRNA against *MSTN* (GGCTGTGTAATGCATGCTTG), a pair of oligos for targeting myostatin exon 1 (5'-CACCGGCTGTGTAATGCATGCTTG-3' and 5'-CACCGGCGAAGCTTACTGAGGATT-3') was annealed, phosphorylated and ligated to a linearized vector (details of the vector plasmid is described by Cong et al [28]).

Genome Editing Assay in Cells

The A15 astroglial sheep cell line (kindly provided by J. Chapuis, UR 892, INRA, Jouy-en-Josas, France) [29] was maintained in Dulbecco's modified Eagle medium (DMEM) in 10% Fetal Bovine Serum, 2mM glutamine, 1% sodium pyruvate and 1% penicillin/streptomycin. Cells were transfected in 24-well plates with 2μg of pX330-cas9-MSTN co-expressing Cas9 and sgRNA against myostatin using lipofectamine LTX reagent (Life technologies, NY, USA) according to the manufacturer's manual. Three days later, genomic DNA from transfected cells was extracted using a kit from Macherey

Nagel (NucleoSpin Tissue, Duren, Germany) and quantified using the NanoDrop2000 spectrophotometer (Thermo Fisher Scientific, Illkirch, France). A260/A280 and A260/A230 ratios accounted for sample purity.

Gene mutation activity of sgRNA sequence at the target locus of *MSTN* exon 1 was quantified using the T7EI mismatch detection assay. DNA sequence of interest was PCR-amplified with a high-fidelity polymerase (Herculase II fusion polymerase) using specific primers (forward primer 5'-tcactggtgtggcaagttgt-3' and reverse primer 5'-aaaagctctttgccctcctc-3'). The 634bp PCR product was then denatured and slowly re-annealed (95°C, 2min; 95°C to 85°C, -2°C/sec; 85°C to 25°C, -1°C/sec) to produce homoduplex/heteroduplex mix. This was then digested by 5U of T7EI restriction enzyme (NEB) at 37°C for 30 minutes. Digestion products were separated by 2% agarose gel electrophoresis. The ratio of cleaved (255-bp and 379-bp) to uncleaved (634-bp) products was used to calculate NHEJ frequency as previously described using Image J software [30]. NHEJ frequency was calculated as follow: % gene modification = 100 X (1-(1-fraction cleaved)^1/2.

Production of sgRNA and Cas9 mRNA

These procedures have been described in detail elsewhere [31]. Briefly, T7 promoter was added to sgRNA template by PCR amplification of pX330-cas9-*MSTN* plasmid using the following primers: 5'-ttaatacgactcactataggctgtgtaatgcatgcttg-3' and 5'-tttaaaagcaccgactcggtgcc-3'. The PCR product was purified using NucleoSpin Gel and PCR Clean-up (Macherey Nagel). It was used as the template for *in vitro* transcription using MEGAshortscript T7 kit (Life Technologies) according to the manufacturer's manual. Following completion of transcription, DNase I treatment was performed.

The Cas9 mRNA was transcribed using PmeI-digested Cas9 expression JDS246 plasmid (a gift from Keith Joung, Addgene plasmid # 43861) and the mMESSAGE mMACHINE T7 ULTRA Transcription Kit (Life Technologies) according to the manufacturer's manual. Following completion of transcription, the poly(A) tailing reaction and DNase I treatment were performed.

Both the Cas9 mRNA and the sgRNAs were purified using MEGAclear kit (Life Technologies) and eluted in elution buffer.

Prediction of Potential Off Target Sites

Potential targets of the CRISPR sgRNA *MSTN* were found using the rules outlined in [32] since this is the only web tool that includes the sheep genome for predicting potential off-targets. The sequences TAATGCATGCTTGTGG, TAATGCATGCTTGAGG, TAATGCATGCTTGCGG, TAATGCATGCTTGGGG (underlined is the PAM, preceded by the N13 sequence of the sgRNA MSTN) were blasted in the genome database of the sheep (http://blast.ncbi.nlm.nih.gov/Blast.cgi). Seven potential off-target sites were detected, that were amplified in the DNA samples from all the mutant lambs, and subjected to Sanger sequencing.

In vitro Production of Embryos

Unless otherwise indicated, chemicals were purchased from Sigma Chemical Company (St Louis, MO, USA). The embryos were produced by *in vitro* fertilization according to routine procedure performed in our lab as described earlier [8]. Briefly, ovaries from slaughterhouse were transported to the laboratory and cumulus oocyte complexes (COCs) were aspirated in recovery medium. The selected COCs were placed in maturation medium for 24 h in 5% CO_2in humidified air atmosphere at 39°C. Then, expanded COCs were inseminated in 100 µl drops with 1 x 10^6 dose of frozen-thawed semen selected by ascendant migration on a swim up method. Fertilization was carried out in 5% CO_2 with humidified atmosphere at 39°C for 22 h.

Microinjection into Zygotes

Soon after fertilization, 572 presumptive zygotes were randomly assigned to three experimental groups to be microinjected (CRISPR group, n = 216; and Buffer group, n = 183) or not (Control group, n = 173). Microinjection of CRISPR group was performed into the cytoplasm with 5 ng/µl of sgRNA and 20ng/µl of Cas9 mRNA diluted in injection buffer (10mM Tris pH 7.5, 0.1mM EDTA), while Buffer group was injected with the same procedure but with buffer alone. Lastly, injected and non-injected embryos were transferred to culture medium under mineral oil, in 5% CO_2, 5% O_2 and 90% N_2 in humidified atmosphere at 39°C. Cleavage rate on Day 2 (cleaved zygotes per total oocytes) and development rate on Day 6 (morulae and blastocysts per total oocytes) were recorded for all experimental groups. After Day 6, DNA from 20 CRISPR group embryos was analyzed by Sanger sequencing to detect the mutation at the *MSTN* gene level.

To determine the *in vivo* efficiency of the system, 53 blastocysts produced by CRISPR/Cas9 zygote microinjection were transferred to 29 recipient females. Only early blastocysts, blastocysts and expanded blastocysts classified as excellent or good (i.e. Grade 1) [33] were transferred on Day 6 after fertilization. Embryo transfer was performed by minimally invasive surgery assisted by laparoscopy to place the embryos into the cranial side of the ipsilateral uterine horn to the corpus luteum. Recipient ewes were previously synchronized to be on Day 6 of the estrous cycle using a standard protocol to control ovulation described previously [34].

Monitoring of Fetuses and Lambs

Pregnancy diagnosis and fetal development were performed on Day 30 and 105, respectively, by using B-mode ultrasonography equipped with a 5 and 3.5 MHz probe (Well-D, Shenzhen, China). Day 0 of the experiment was defined as the moment of embryo fertilization. Several parameters were measured to study the development of fetuses at Day 105 of gestation: thoracic diameter, biparietal diameter, occipitonasal length and heart rate. At delivery, length of gestation, gender, rectal temperature, heart and respiratory rates, body weight, thoracic perimeter, biparietal diameter, crown-rump and occipitonasal length, height at withers, height at hips, width at hips and width at chest were recorded. Body weight and morphometric variables were determined at birth, and 15, 30 and 60 days later.

Identification and Genotyping of Transgenic Animals

Samples from skin and limb muscle of the lambs were taken seven days after birth and T7EI assay, western blot test and histology examinations were performed in order to identify and characterize KO founders and off-target sites. Total DNA was isolated from skin biopsies for all animals and from muscle for some animals. PCR and T7EI assay were performed using the same protocol than from cells samples. Samples were analyzed using capillary electrophoresis (Caliper, PerkinElmer, Hopkinton, MA). Genotyping of *MSTN* exon 1 was performed by direct sequencing of PCR amplicons using the same primers described above and in muscle biopsies by additional sequencing of isolated bacterial clones with individual amplicon sequences, as described in detail previously [30, 35].

Analysis of Myostatin Expression

Western blotting was performed to determine the presence of myostatin in the muscle fiber. Equal amounts of total proteins were run on 12% (v/v) gel electrophoresis and electrophoretically transferred to a PVDF membrane. Monoclonal mouse anti-myostatin (sc-398333, Santacruz) and anti-GAPDH (G8795, Sigma) antibodies were used in the western blotting. The washed membranes were incubated with 1:50000 dilution of secondary antibody linked to horseradish peroxidase (HPR). HPR activity was detected using western blot chemiluminescence (Promega, MA, USA).

Muscle Fiber Histology

Samples from deltoid and biceps femoris muscles were fixed in 4% formaldehyde solution until analysis. Samples were included in paraffin, sectioned in 1 mm slides and stained with haematoxilin-eosin to study muscle morphology. Muscle fibers (min 250 per sample) were measured with respect to their minimum Feret (MinFeret) diameter. The mean area was also calculated.

Statistical Analysis

In vitro embryo development on Day 2 and Day 6 among experimental groups was compared by logistic regression. Continues variables measured during pregnancy and body growth after birth in mutant *vs.* wild type lambs were analyzed by one-way ANOVA or by non-parametric Kruskal Wallis test. Differences were considered significant when $P<0.05$.

RESULTS

Efficiency of Cas9/sgRNA Activity in Sheep Fetal Fibroblast

Before microinjection of the sgRNA and Cas9 coding sequences into the ovine zygotes, fetal ovine fibroblasts were transfected with the pX330-cas9-*MSTN* plasmid co-expressing Cas9 and sgRNA, showing mutations at the targeted *MSTN* exon 1 in 10% of the total DNA (Fig 1). Although it is difficult to compare with other cell lines, this efficacy *in vitro* is comparable to what we and others have previously observed as being sufficient to obtain mutated animals by microinjection into zygotes of not only CRISPRs/Cas9 but also ZFNs or TALENs [15, 35]. Thus, we proceeded to microinject ovine zygotes with both mRNAs encoding for Cas9 and sgRNA.

A

ATGCAAAAACTGCAAATCTTTGTTTATATTTACCTATTTATGCTGCTTGTTGCTGGCCCAGTG
GATCTGAATGAGAACAGCGAGCAGAAGGAAAATGTGGAAAAAAAGG***GGCTGTGTAATGCA***
TGCTTG<u>TGG</u>AGACAAAACAATAAATCCTCAAGACTAGAAGCCATAAAAATCCAAATCCTCAG
TAAGCTTCGCCTGGAAACAGCTCCTAACATCAGCAAAGATGCTATAAGACAACTTTTGCCC
AAGGCTCCTCCACTCCGGGAACTGATTGATCAGTACGATGTCCAGAGAGATGACAGCAGC
GACGGCTCCTTGGAAGACGATGACTACCACGTTACGACGGAAACGGTCATTACCATGCCC
ACGGAGT

B

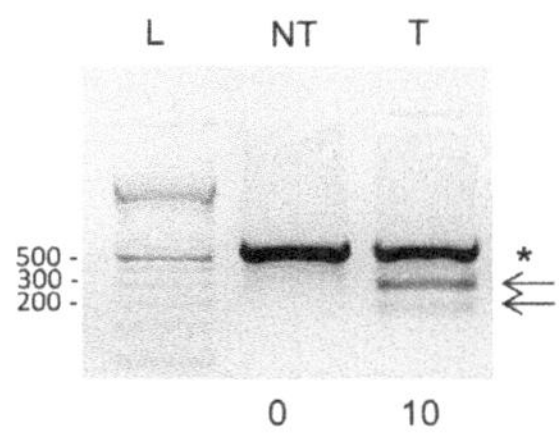

Figure 1: CRISPR-Cas9 genome editing activity in transfected ovine cells.

A) Exon 1 of the ovine *MSTN* gene. The sequence recognized by sgRNA is in italic and bold, from nt 110–130, the PAM sequence TGG is underlined and the rest of exonic sequences in smaller font. B) Ovine cells were transfected with pX330-cas9-*MSTN*plasmid co-expressing Cas9 and sgRNA and three days later cellular DNA was isolated and subjected to T7EI assay for detecting mutation at the targeted *MSTN* exon 1. The PCR amplifies a band of 634 bp and uncleaved (asterisk) and cleaved (arrows, 255 bp and 379 bp) products are indicated. Numbers below each lane indicate the percentage of targeted indels. NT: non-transduced cells, T: transfected cells, L: 1Kb DNA. The experiment is representative of two replicates performed with similar results.

http://dx.doi.org/10.1371/journal.pone.0136690.g001

Analysis of *MSTN* Mutations in Embryos Microinjected with Cas9/sgRNA mRNAs

We first determined whether zygotes microinjected with Cas9 and sgRNA mRNAs and further cultured *in vitro* showed normal development and *MSTN* mutations. From 216 injected zygotes 63.9% survived to micromanipulation and reached cleavage stage on Day 2, and 39.1% achieved morula or blastocyst stage on Day 6. Cleavage rate and development rate for control non-injected embryos were significantly higher than for CRISPR and buffer microinjected embryos (Table 1). However, no differences were found on embryo development rate among control and microinjected zygote groups after cleavage. The total number of embryos on Day 6 from cleaved embryos did not show statistical differences among groups (Table 1). Thus, microinjection resulted in some embryo lethality up to the cleavage state but Cas9 and sgRNA mRNAs microinjection had no adverse effects on subsequent *in vitro* embryo development.

Table 1: Cleavage rate on Day 2 and development rate on Day 6 of embryos that were microinjected into cytoplasm with CRISPR/Cas9 RNA system, buffer injection solution, or non-injected embryos (Control group)

http://dx.doi.org/10.1371/journal.pone.0136690.t001

	No. zigotes	Cleavage rate on Day 2	Morulae and Blastocysts on Day 6	No. of embryos on Day 6/ clived	Mutant embryos at blastocyst stage
CRISPR injection	216	63,9% (138/216)[a]	25,0% (54/216)[a]	39,1% (54/138)[a]	50.0% (10/20)
Buffer injection	183	60,7% (111/183)[a]	20,2% (37/183)[a]	33,3% (37/111)[a]	—
Non-injected	173	86,1% (149/173)[b]	35,8% (62/173)[b]	41,6% (62/149)[a]	—

For different superscripts, $P<0.05$.

doi:10.1371/journal.pone.0136690.t001

DNA from developed CRISPR injected embryos were extracted and analyzed by Sanger sequencing. From 20 embryos analyzed, 10 showed mutations at the *MSTN* gene (50%). Two of those mutants were heterozygous (wt + mutated allele) and eight were mosaic presenting more than two sequences. The deletions observed ranged from 6 to 350 bp, besides just 1 bp insertions were observed.

PCR and T7EI Genotyping of Lambs following Microinjection of Cas9/ sgRNA mRNAs

From 53 Grade 1 *in vitro* produced sheep blastocyst transferred into 29 recipient females, 41.5% of them were detected at 30 days of gestation, 65.5% of the recipients were pregnant and 22 lambs were delivered (Table 2). From these 22 lambs, three (#43 49 and 53) died at delivery or within the first day after birth. Skin and muscle biopsies from deltoid and biceps femoris were taken from 22 lambs within 1 week after birth. DNA from the 22 skin biopsies was extracted and analyzed by PCR and T7EI assay followed by capillary electrophoresis (Fig 2). PCR analysis showed the presence of main band of the expected size in the absence of mutations (634 bp) and bands of apparent higher molecular weight, due to open angles formed by heteroduplexes of DNA strands with mismatches [36] in nine animals, suggesting the presence of mutations (Fig 2A). In an additional animal (#43) a single band of slightly smaller size as compared to the other animals was observed (Fig 2A).

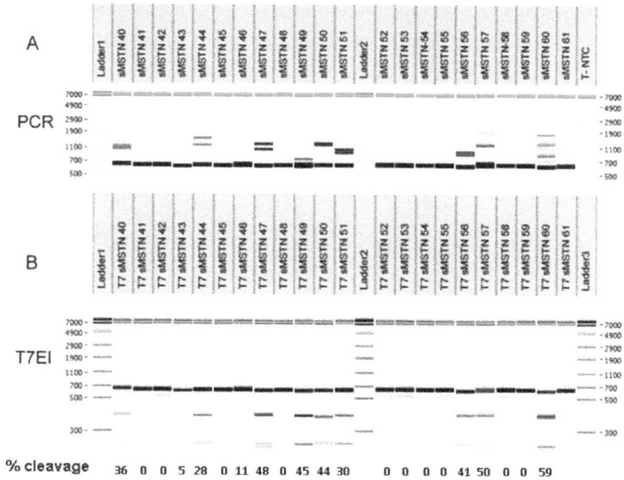

Figure 2: Genotype of lambs born following microinjection of sgRNA for *MSTN* **exon 1 and Cas9.**

DNA was extracted from lambs' skin biopsies one week after birth, with the exception of lambs #43 and #49 that were stillborn and were biopsied 1–2 hours after delivery. A) Results from PCR amplifying 634 bp. Additional larger bands represent formation of heteroduplex DNA double strands with mismatches between the two strands resulting in slower migration during electrophoresis. B) T7EI assay of lambs produced by zygote injection of CRISPR/Cas9 mRNA, indicating numbers of mutant animals with bands of 255, 379 and 634 bp (#40, 43, 44, 47, 49, 50, 51, 56, 57 and 60) and WT animals with only one band of 634 bp (#41, 42, 45, 46, 48, 52, 53, 54, 55, 58, 59 and 61).

http://dx.doi.org/10.1371/journal.pone.0136690.g002

Table 2: Efficiency obtained with CRISPR/Cas9 system injected into cytoplasm of ovine zygotes to produce *MSTN* mutant lambs

Data obtained from 53 embryos transferred into 29 recipient ewes.

http://dx.doi.org/10.1371/journal.pone.0136690.t002

	Embryos on Day 30	Pregnant ewes	Fetal loss	Mutant/ born lambs	Biallelic/ mutant lambs	Homozygous/ mutant lambs
CRISPR/Cas9 efficiency	41.5% (22/53)	65.5% (19/29)	0.0% (0/22)	45.5% (10/22)	80.0% (8/10)	50.0% (5/10)

doi:10.1371/journal.pone.0136690.t002

Since T7EI digestion of amplicon allows detection of mutations when both DNA strands show mismatches, animals with WT sequences or with the same mutation in both alleles should not generate new bands and animals which have different alleles, either one WT and the other mutated or both with different mutations, should generate new bands. The T7EI assay revealed that the same nine animals with heteroduplexes in the PCR assay generated bands of lower molecular weight upon T7EI digestion (Fig

2B), confirming the presence of mutations in these animals. Animal #43 did not show new smaller bands but if both alleles contained the same mutation the T7EI assay was not expected to be positive. The analysis of DNA cleavage showed that several animals (#47, 49, 50, 56, 57 and 60) had around 50% cleavage corresponding to half of the alleles being mutated and are thus not mosaic. Some animals (#40, 44, 46 and 51) had cleavage of less than 50% and are likely mosaic.

DNA sequence Genotyping of Lambs following Microinjection of Cas9/sgRNA mRNAs

Sequencing of PCR amplicons compassing the targeted *MSTN* exon 1 sequence showed that ten animals contained mutations; nine identified using capillary electrophoresis and animal #43 which showed an identical deletion of 20 nt in both chromosomes (Fig 3A and 3B), explaining the smaller size of the amplicons in capillary electrophoresis and the negative T7EI assay. The other 12 animals showed only WT alleles. Among the ten mutated animals, eight showed both alleles mutated (#40, 43, 47, 49, 50, 56, 57 and 60) and two only one allele (#44 and 51). These introduced mutations disrupted the coding frame of exon 1 (number of indels different of 3nt or multiples of 3nt) with the generation of premature stop codons in nine mutated animals in at least one of the two alleles. The exception was animal #44 which showed an in-frame shift mutation (Fig 3A). Newly generated stop codons (Fig 3A) are recognized by cells as premature stop codons due to the absence of other normal *Cis* sequences targeting these mRNAs for degradation through nonsense-mediate decay pathway resulting in the complete absence of the protein [36]. Some animals showed biallelic frameshift mutations introducing premature stop codons (#40, 43, 47, 57 and 60), and are thus homozygous KO animals. Other animals showed one frameshift mutation and another with inframe mutations (#49, 50 and 56) and are thus heterozygous KO animals. Analogously, one animal (#51) showed one allele with a frameshift mutation and a WT allele and thus it is also a heterozygous KO animal. Finally, one animal (#44) had one allele with an inframe mutation and one wild-type allele and thus it is not a KO animal.

The same DNAs analyzed in Fig 2 were PCR amplified and the amplicons directly sequenced. In some animals (#40, 47, 49, 50, 56, 57 and 60) DNA from muscle biopsies were PCR amplified and amplicons were cloned into plasmids by TA cloning and electroporated into bacteria, followed by sequencing of 8–10 bacterial clones. A)Depicts for each of the 22 delivered lambs the flanking DNA sequences (in blue) close to the targeted sgRNA sequence (in red) and the PAM sequence (violet); missing nucleotides are represented by spaces, added ones in green and small characters and stop codons are labeled in black. The column Genotype recapitulates the genotype found for each lamb. The column TA indicates the number of bacterial colonies that were sequenced for each allele of the muscle biopsies. The column Translation depicts the aminoacids translated; spaces for the missing ones, in green the ones that are new due to the shift in the coding reading frame and the * represents the stop of the aminoacid sequence due to the premature stop codons. Results are representative of two different PCR amplicons sequencing for all animals. B) A representative sequence electrophoresis, in this case the one of animal #43 which has a biallelic identical deletion of 20 nt.

A

Animal	sgRNA PAM	Indels	TA	Translation
WT	AGGGGCTGTGTAATGCATGCTTGTGGAGACAAAACAATAAATCCTCAAGAC	WT		KGLCNACLWRQNN
40	AGGGGCTGTGTAATGCATGCatgtaatgcatgttTTGTGGAGACAAAACAA	ins 14 bp	x2	KGLCNACM*
	AGGGGCTGTGTAATGCATG.ttTTGTGGAGACAAAACAATAAATCCTCAAG	del 1 bp + ins 2 bp	x3	KGLCNACFVETKQ*
41	AGGGGCTGTGTAATGCATGCTTGTGGAGACAAAACAATAAATCCTCAAGAC	WT		KGLCNACLWRQNN
42	AGGGGCTGTGTAATGCATGCTTGTGGAGACAAAACAATAAATCCTCAAGAC	WT		KGLCNACLWRQNN
43	AGGGGCTGTGTAA....................ACAATAAATCCTCAAGAC	del 20 bp biallelic		KGLCKQ*
44	AGGGGCTGTGTAATGCATGCTTGTGGAGACAAAACAATAAATCCTCAAGAC	WT		KGLCNACLWRQNN
	AGGGGCTGTGTAATGCATG......GAGACAAAACAATAAATCCTCAAGAC	del 6 bp		KGLCNA..WRQNN
45	AGGGGCTGTGTAATGCATGCTTGTGGAGACAAAACAATAAATCCTCAAGAC	WT		KGLCNACLWRQNN
46	AGGGGCTGTGTAATGCATGCTTGTGGAGACAAAACAATAAATCCTCAAGAC	WT		KGLCNACLWRQNN
47	AGGGGCTGTGTAATGCAT..TTGTGGAGACAAAACAATAAATCCTCAAGAC	del 2 bp	x6	KGLCNAFVETKQ*
	AGGGGCTGTGT........CTTGTGGAGACAAAACAATAAATCCTCAAGAC	del 8 bp	x4	KGLCLVETKQ*
48	AGGGGCTGTGTAATGCATGCTTGTGGAGACAAAACAATAAATCCTCAAGAC	WT		KGLCNACLWRQNN
49	AGGGGCTGTGTAATG.....TTGTGGAGACAAAACAATAAATCCTCAAGAC	del 5 bp	x5	KGLCNACLVETKQ*
	AGGGGCTGTGTAATGCA......TGGAGACAAAACAATAAATCCTCAAGAC	del 6 bp	x3	KGLCNA..WRQNN
50	AGGGGCTGTGTAATGCATGCatTTGTGGAGACAAAACAATAAATCCTCAAG	ins 2 bp	x5	KGLCNAC[illegible]
	AGGGGCTGTGTAATGCATGC......AGACAAAACAATAAATCCTCAAGAC	del 6 bp	x5	KGLCNAC..RQNNKSSR
51	AGGGGCTGTGTAATGCATGCTTGTGGAGACAAAACAATAAATCCTCAAGAC	WT		KGLCNACLWRQNN
	AGGGGCTGTGTAATGCATG....TGGAGACAAAACAATAAATCCTCAAGAC	del 4 bp		KGLCNACGDKTINPQD*
52	AGGGGCTGTGTAATGCATGCTTGTGGAGACAAAACAATAAATCCTCAAGAC	WT		KGLCNACLWRQNN
53	AGGGGCTGTGTAATGCATGCTTGTGGAGACAAAACAATAAATCCTCAAGAC	WT		KGLCNACLWRQNN
54	AGGGGCTGTGTAATGCATGCTTGTGGAGACAAAACAATAAATCCTCAAGAC	WT		KGLCNACLWRQNN
55	AGGGGCTGTGTAATGCATGCTTGTGGAGACAAAACAATAAATCCTCAAGAC	WT		KGLCNACLWRQNN
56	AGGGGCTGTGTAATGCATGCT.......gaaACAAAACAATAAATCCTCAA	del 7 bp + ins 3 bp	x4	KGLCNAC*
	AGGGGCTG............gggcctTTGTGGAGACAAAACAATAAATCCT	del 12 bp + ins 6bp	x6	KGL...GPLWRQNN
57	AGGGGCTGTGTAATGCATGCacatgcacattggggctcctatgcaTTGTGG	ins 25 bp	x2	KGLCNAC[illegible]
	AGGGGC.............atgcatTTGTGGAGACAAAACAATAAATCCT	del 14 bp + ins 6 bp	x10	KG[illegible]
58	AGGGGCTGTGTAATGCATGCTTGTGGAGACAAAACAATAAATCCTCAAGAC	WT		KGLCNACLWRQNN
59	AGGGGCTGTGTAATGCATGCTTGTGGAGACAAAACAATAAATCCTCAAGAC	WT		KGLCNACLWRQNN
60	AGGGGCTGTGT.........ctcctTTGTGGAGACAAAACAATAAATCCTC	del 9 bp + ins 5 bp	x5	KGLCLLGGDKTINPQD*
	AGGG...........................ttatacagcccAAACAATAA	del 27 + ins 11 bp	x3	KGYTAQTINPQD*
61	AGGGGCTGTGTAATGCATGCTTGTGGAGACAAAACAATAAATCCTCAAGAC	WT		KGLCNACLWRQNN

B

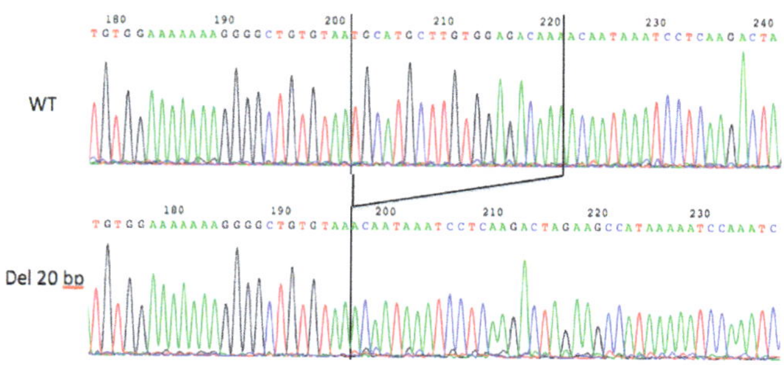

Figure 3: Sequence analysis of lambs' *MSTN* **exon 1.**

http://dx.doi.org/10.1371/journal.pone.0136690.g003

The DNA sequences performed in skin and muscle from the same lambs showed the same mutations, indicating the absence of mosaicism in ectoderm and mesoderm-derived tissues and thus indicating the action of Cas9 at the zygote stage. Thus, sgRNA and Cas9 mRNA microinjection into ovine zygotes achieved high efficient introduction of mutations resulting in inactivation of the *MSTN* gene.

Primers pairs were designed and the possible off target sites were amplified and sequenced in order to assess the presence of possible off target effects. In the animals #40 and #57 one 27 pb deletion and one 1 bp insertion were found respectively for the off target region number 3. In both cases this mutation was heterozygous. None other off target effects were identified in the seven regions assessed.

Analysis of Myostatin Expression

Homozygous KO lambs with two frameshift mutations should show absence of myostatin whereas heterozygous KO lambs with one coding allele should show myostatin. Western blot for myostatin analysis showed undetectable levels in homozygous KO lamb #43 and the presence of myostatin in heterozygous (1 WT and 1 in-frame shift allele) lamb #44 (Fig 4) and these levels were comparable to those of lambs with 2 WT alleles.

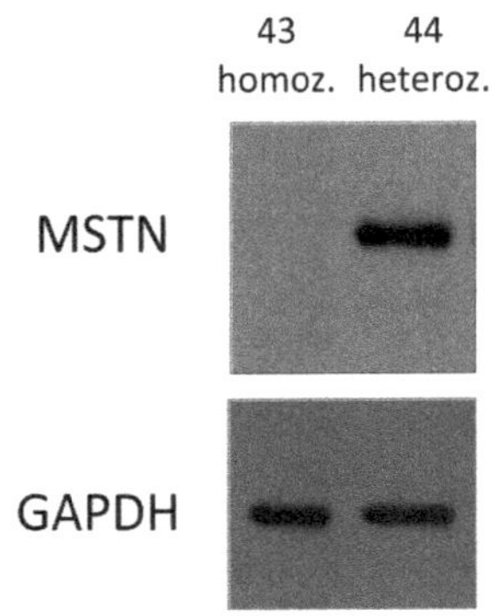

Figure 4: Myostatin western blot.

Muscle biopsies from two representative animals, one homozygous (homoz.) for a frameshift mutation and another heterozygous (heteroz.) for a frameshift mutation and a WT copy in the second allele (# 43 and 44, respectively, the numbers of animals correspond to those of Fig 3) analyzed by western blot using an anti-MSTN monoclonal antibody. After stripping of the anti-MSTN antibody from the membrane, an anti-GAPDH was used as loading control. One representative western blot experiment out of three performed with the same results.

http://dx.doi.org/10.1371/journal.pone.0136690.g004

Fetal Development and Delivery

Fetal determinations by ultrasonography on Day 105 of gestation showed no statistical differences for thoracic and biparietal diameter, occipito-nasal length or heart rate among WT, homozygous KO or heterozygous KO mutant animals, indicating that the mutation did not affect *in utero* measurements although the *MSTN* gene was already mutated.

For mutant lambs, # 43 and 49 died at delivery and lamb # 57 died within first month after birth due to a wound infection. For WT lambs, #53 and 58 died at delivery and within first month after birth, respectively. The rest of the lambs developed normaly until at least 60 days of life. Length of gestation was 147.0 ± 0.9, 149.8 ± 0.6 and 149.4 ± 1.0 days for WT, homozygous KO and heterozygous KO mutant lambs respectively (P = NS). Rectal temperature at birth was 40.2 ± 0.2, 39.9 ± 0.3 and 39.9 0078 0.4°C for WT, homozygous KO and heterozygous KO mutant lambs, respectively (P = NS). Heart rate and respiratory rate did not differ among groups and was 189.8 ± 3.1 and 93.3 ± 1.8 for WT lambs, 184.0 ± 5.7 and 86.0 ± 13.1 for homozygous KO lambs, and 203.0 ± 11.8 and 75.0 ± 5.0 for heterozygous KO lambs, respectively.

Postnatal Phenotype

Morphometric variables registered from birth to 60 days later are shown in Fig 5. Body weight from homozygous KO lambs was not different than WT animals at birth (P = NS) and was greater at 15 and 30 days ($P<0.05$) with a statistical tendency at 60 days after birth (P = 0.08). This faster growth showed by homozygous KO lambs within two months of age averaged an increase between 20 and 30%. Body weight for heterozygous KO mutant lambs was intermediate (P = NS). The rest of the variables did not show statistical differences among the three genotypes, indicating that the mutation at the *MSTN* gene affects body weight mainly by increasing muscle mass. However, a statistical tendency ($P<0.1$) was found at 60 days for height and width at withers, and width at chest among homozygous KO and WT lambs, which gave a very clear phenotype in some *MSTN* KO lambs as shown in Fig 6.

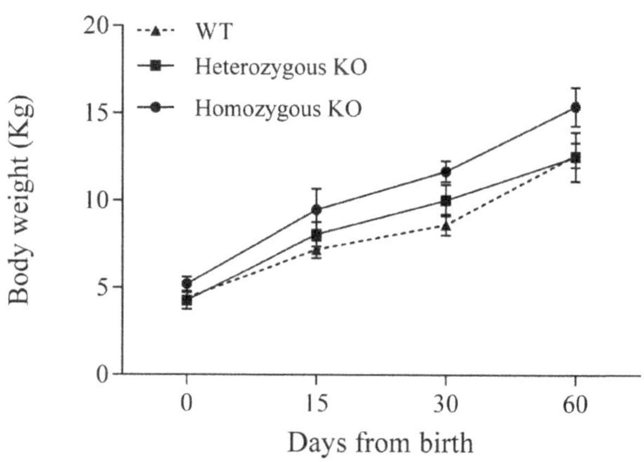

Figure 5: Body weight in lambs produced by CRISPR/Cas9 system.

Significant diference (P<0.05) from 15 to 60 days after birth was found among homozygous KO and wild-type lambs.

http://dx.doi.org/10.1371/journal.pone.0136690.g005

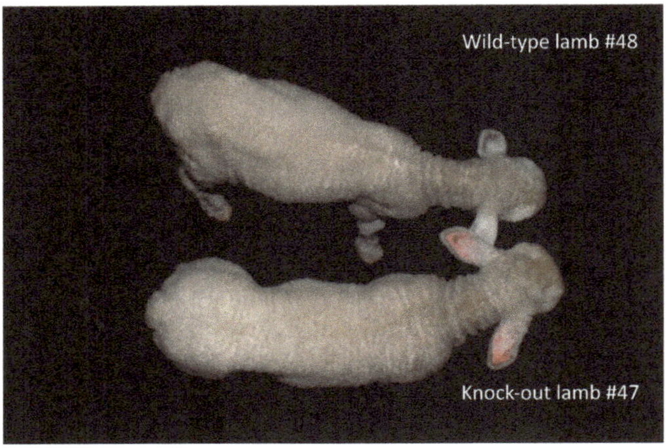

Figure 6: Phenotype of lambs 30 days after birth.

The lamb at the top is wild type (#48), the lamb at the bottom is a homozygous KO (#47). Note the differences in muscle mass from the posterior limbs and loin from the homozygous KO lamb when compared to wild-type. At the time of this picture, with 30 days old the body weight was 8.750 (#48) and 11.150 kg (#47).

http://dx.doi.org/10.1371/journal.pone.0136690.g006

Muscle Fiber Histology

As expected from previous results of lambs with spontaneous *MSTN* mutations [37], histological analyses of muscle biopsies showed muscle fibers of lower diameter although due to the relatively low number of homozygous KO animals analyzed (n = 4) it did not reach statistical significance when compared to WT animals (data not shown).

DISCUSSION

The current study demonstrates that microinjection into zygotes of sgRNA and Cas9 mRNAs was a very efficient approach to obtain myostatin KO sheep. Embryo survival and embryo development were not affected by the microinjection of sgRNA and Cas9 mRNAs into the cytoplasm. Also, no differences in fetal loss, birth rate and postnatal survival rates were detected for both groups, suggesting that mutant animals are as healthy as WT animals. These results are in agreement with our recent report where the microinjection of a GFP lentiviral construct into ovine zygotes did not affect embryo development, fetal growth or postnatal variables when compared to WT lambs [8]. Ten lambs with mutations in the *MSTN*gene were obtained (45.5%), eight had biallelic mutations and five of them were biallelic out of frame shift mutations. These lambs resulted in myostatin deficiency and a double-muscle phenotype with a significant increase in body weight when compared to control wild type animals.

During recent years gene editing nucleases have revolutionized the field of transgenesis, decreasing the time, cost and effort to obtain genetically-modified animals [38]. However, the use of CRISPR/Cas9 system has not been reported in many ruminant species to date [17, 38,21]. We demonstrated that this technology does not affect embryo survival and *in vitro*development rate when injected into the cytoplasm of zygotes. However, the cleavage rate was affected by the embryo manipulation required for the microinjection *per se*, which has been reported previously in zygotes submitted to zone pellucidae break and microinjection [39]. No effect of microinjection was found in the ability of those surviving embryos 48h after fertilization to continue its development and to reach blastocyst stage on Day 6 after fertilization. In addition, the overall procedure allowed a high fertility rate since 41.5% of the microinjected embryos transferred to recipients were pregnant and developed to term. The efficiency of the technique is also seen in the fact that no gestation loss occurs during pregnancy avoiding a well-known issue reported for somatic cell nuclear transfer and other techniques [40, 41]. In addition, the production of mutant animals was remarkably high. We found that 50% of *in vitro* produced blastocysts showed some type of mutations. This result was confirmed after birth, since from 22 born lambs analyzed by T7EI assay and sequencing, ten were mutant (45.5%). Five of them (i.e. 50% of mutated ones) were biallelic KO inducing a myostatin deficient phenotype. The performance showed in this study with CRISPR/Cas9 system is clearly higher than the reported for other previous techniques in farm animals. In general, very low outcomes of pronuclear microinjection, several difficulties with nuclear transfer, and the technical challenge to produce lentiviral vectors of good quality, have limited the spread of transgenic technologies to new laboratories around the world [42]. Previous work has shown that mice or rats mutated in the *mdx* gene generated using CRISPR/Cas9 [43] or TALENs [44] respectively, showed mosaicism that resulted in some cases in absence of transmission of the mutation to the offspring. For our mutated lambs, their age and time of gestation after mating preclude this analysis. Nevertheless, the fact that both skin and muscle, developing from ectoderm and mesoderm, showed mutations and that the percentage of DNA mutated analyzed using the T7EI assay showed a high degree of mutation, it is likely that the mutations will be transmitted to the offspring by at least some of the founders. Our results show that CRISPR/Cas9 is an easier, faster and more effective technique to produce a high proportion of mutant animals, including biallelic founders, when compared to somatic cell nuclear transfer, the only technique available up to date in sheep to perform targeted gene KO.

Off target effects using CRISPR/Cas9 system as well as ZFNs or TALENs could be a problem when trying to obtain KO animals for a specific gene. However it should be pointed out that most of these off effects have been shown in cells transfected *in vitro* [45] and not in genetically-modified animals [15, 46], possibly due to a larger amount of transfected nucleic acids as compared to microinjection. Furthermore, breeding of mutated founders with WT partners results in dilution of the potential off target mutations in the offspring. In our case, we only found one non coding region that was mutated in two animals. Thus, the potential off effects seem low in this report. If more specificity is desired, strategies have been designed to overcome this difficulty, such as the generation of Cas9 mutants able to cleave only one DNA strand [47] or shorter sgRNAs [48] that show lower off effects.

Our study describes the use of CRISPR/Cas9 gene editing system in sheep, targeting the*MSTN* gene to modify the phenotype related to body weight and growth rate. Myostatin mutant lambs produced in this study were significantly heavier than control lambs, and moreover mutant animals developed normally until today. Myostatin have been previously modified by genetic engineering in ruminant species using technologies such as ZFN and TALEN [49, 50]. However, the use of the CRISPR/Cas9 system is easier and is now available for livestock production. Interestingly, while no differences were found at birth in body size and weight among homozygous KO lambs and wild type counterparts, KO lambs were 20–30% heavier 60 days later with no differences in body size. The production of newborns with low body weight at birth and faster growth rate, obtaining heavy lambs showing similar body size, represent an interesting productive feature in terms of lambing ease, meat production, and dressing percentage for livestock industry. Moreover, double muscle animals had shown a lower content of intramuscular fat, with more unsaturated fatty acid [51]. These characteristics make this meat healthier for the consumers, reaffirming the relevance of applying this technology to improve these kind of traits [52]. In addition, this mutation applied in Merino superfine animals could exemplify a new system for genetic improvement programs as never described before, with specifically designed dual or multi-purpose animals to produce more meat maintaining very high quality wool production. It is interesting to point out that the use of gene editing technologies, with the generation of mutations without introducing exogenous DNA, overcome the issue of introducing an exogenous gene into the animal genome. In this sense, a spontaneous mutation of *MSTN* gene that modify myostatin expression is present in some cattle breeds (e.g. Belgian Blue and Piedmont breeds) [53], as well as in sheep (e.g. Texel breed) [54] and goats [55]. Since these animals are already present in livestock production, their meat has been consumed for a long time. The easy access to CRISPR/Cas9 technology by more laboratories, together with more reasonable regulatory standards for its approval since no exogenous genes are inserted, suggests a promising future for gene editing systems in animal industry.

The CRISPR/Cas9 technology will also be likely used in the future to generate lambs harboring exogenous sequences to produce recombinant proteins with much higher efficiency through the introduction of these sequences by homologous recombination into *loci* that are permissive for frequent and high transgene expression, such as it has been done in other species using the *Rosa26* locus [35].

In conclusion, our study reports the production of healthy myostatin KO lambs using the CRISPR/Cas9 system in an efficient way to increase muscle growth and body weight. CRISPR/Cas9 system could be easily applied to obtain mutant farm animals for several genes of productive or biomedical interest.

ACKNOWLEDGMENTS

The authors wish to thank Magdalena Cárdenas for technical support in Western Blot design, Richard Núñez for ultrasound scanning of recipients and Marcela Alonso and Patricia Armand Ugon for lamb assistance after birth. Sequencing experiments were performed by the integrative genomic facility of Nantes.

AUTHOR CONTRIBUTIONS

Conceived and designed the experiments: MC IA AM. Performed the experiments: MC APM NB FC PCdSN TN IA AM. Analyzed the data: MC APM LT TN AC LB IA AM. Contributed reagents/materials/analysis tools: MC IA AM. Wrote the paper: MC APM TN IA AM.

REFERENCES

1. Hammer RE, Pursel VG, Rexroad CE Jr. , Wall RJ, Bolt DJ, Ebert KM et al. Production of transgenic rabbits, sheep and pigs by microinjection. Nature. 1985;315(6021):680–3. pmid:3892305 doi: 10. 1038/315680a0
2. Whitelaw CB, Lillico SG, King T. Production of transgenic farm animals by viral vector-mediated gene transfer. Reprod Dom Anim. 2008;43(2):355–8. doi: 10. 1111/j. 1439-0531. 2008. 01184. x
3. Hyun S, Lee G, Kim D, Kim H, Lee S, Nam D et al. Production of nuclear transfer-derived piglets using porcine fetal fibroblasts transfected with the enhanced green fluorescent protein. Biol Reprod. 2003;69(3):1060–8. pmid:12773429 doi: 10. 1095/biolreprod. 102. 014886
4. Lois C, Hong EJ, Pease S, Brown EJ, Baltimore D. Germline transmission and tissue-specific expression of transgenes delivered by lentiviral vectors. Science. 2002;295(5556):868–72. pmid:11786607 doi: 10. 1126/science. 1067081
5. Kues WA, Niemann H. Advances in farm animal transgenesis. Prev Vet Med. 2011;102(2):146–56. doi: 10. 1016/j. prevetmed. 2011. 04. 009. pmid:21601297
6. Vajta G. Somatic cell nuclear transfer in its first and second decades: successes, setbacks, paradoxes and perspectives. Reprod Biomed Online. 2007;15(5):582–90. pmid:18028751 doi: 10. 1016/s1472-6483(10)60391-4
7. Chan A. Transgenic animals: current and alternative strategies. Cloning. 1999;1:25–46. pmid:16218828 doi: 10. 1089/15204559950020076
8. Crispo M, Vilarino M, Dos Santos-Neto PC, Nunez-Olivera R, Cuadro F, Barrera N et al. Embryo development, fetal growth and postnatal phenotype of eGFP lambs generated by lentiviral transgenesis. Transgenic Res. 2014;24(1):31–41. doi: 10. 1007/s11248-014-9816-x. pmid:25048992
9. Remy S, Tesson L, Menoret S, Usal C, Scharenberg AM, Anegon I. Zinc-finger nucleases: a powerful tool for genetic engineering of animals. Transgenic Res. 2010;19(3):363–71. doi: 10. 1007/s11248-009-9323-7. pmid:19821047
10. Boch J. TALEs of genome targeting. Nat Biotech. 2011;29(2):135–6. doi: 10.

1038/nbt. 1767

11. Delacote F, Perez C, Guyot V, Duhamel M, Rochon C, Ollivier N et al. High Frequency Targeted Mutagenesis Using Engineered Endonucleases and DNA-End Processing Enzymes. PloS one. 2013;8(1):e53217. doi: 10. 1371/journal. pone. 0053217
12. Menoret S, Fontaniere S, Jantz D, Tesson L, Thinard R, Remy S et al. Generation of Rag1-knockout immunodeficient rats and mice using engineered meganucleases. Faseb J. 2013;27(2):703–11. doi: 10. 1096/fj. 12-219907. pmid:23150522
13. Horvath P, Barrangou R. CRISPR/Cas, the immune system of bacteria and archaea. Science. 2010;327(5962):167–70. doi: 10. 1126/science. 1179555. pmid:20056882
14. Whitworth K, Lee K, Benne J, Beaton B, Spate L, Murphy S et al. Use of the CRISPR/Cas9 System to Produce Genetically Engineered Pigs from In Vitro-Derived Oocytes and Embryos. Biol Reprod. 2014;91(3):78, 1–13. doi: 10. 1095/ biolreprod. 114. 121723. pmid:25100712
15. Zhou X, Xin J, Fan N, Zou Q, Huang J, Ouyang Z et al. Generation of CRISPR/ Cas9-mediated gene-targeted pigs via somatic cell nuclear transfer. Cel Mol Life Sci. 2014: 72(6):1175–84. doi: 10. 1007/s00018-014-1744-7
16. Honda A, Hirose M, Sankai T, Yasmin L, Yuzawa K, Honsho K et al. Single-Step Generation of Rabbits Carrying a Targeted Allele of the Tyrosinase Gene Using CRISPR/Cas9. Exp Anim. 2014; 64(1):31–7. doi: 10. 1538/expanim. 14-0034. pmid:25195632
17. Ni W, Qiao J, Hu S, Zhao X, Regouski M, Yang M et al. Efficient gene knockout in goats using CRISPR/Cas9 system. PloS one. 2014;9(9):e106718. doi: 10. 1371/journal. pone. 0106718. pmid:25188313
18. He Z, Proudfoot C, Mileham AJ, McLaren DG, Whitelaw CBA, Lillico SG. Highly efficient targeted chromosome deletions using CRISPR/Cas9. Biotechnol Bioeng. 2015 May;112(5):1060–4 doi: 10. 1002/bit. 25490. pmid:25362885
19. Tan W, Carlson DF, Lancto CA, Garbe JR, Webster DA, Hackett PB et al. Efficient nonmeiotic allele introgression in livestock using custom endonucleases. PNAS. 2013;110(41):16526–31. doi: 10. 1073/pnas. 1310478110. pmid:24014591
20. Hai T, Teng F, Guo R, Li W, Zhou Q. One-step generation of knockout pigs by zygote injection of CRISPR/Cas system. Cell research. 2014;24(3):372–5. doi: 10. 1038/cr. 2014. 11. pmid:24481528
21. Han H, Ma Y, Wang T, Lian L, Tian X, Hu R et al. One-step generation of myostatin gene knockout sheep via the CRISPR/Cas9 system. Front Agr Sci Eng. 2014;1(1):2–5. doi: 10. 15302/j-fase-2014007.
22. FAO. How to Feed the World in 2050. FAO. 2009. http://www. fao. org/fileadmin/ templates/wsfs/docs/expert_paper/How_to_Feed_the_World_in_2050. pdf.
23. Lee S-J. Regulation of muscle mass by myostatin. Annu Rev Cell Dev Biol. 2004;20:61–86. pmid:15473835 doi: 10. 1146/annurev. cellbio. 20. 012103. 135836

24. McPherron AC, Lawler AM, Lee S-J. Regulation of skeletal muscle mass in mice by a new TGF-p superfamily member. Nature. 1997;387(6628):83–90. pmid:9139826 doi: 10. 1038/387083a0

25. Grobet L, Royo Martin LJ, Poncelet D, Pirottin D, Brouwers B, Riquet J et al. A deletion in the bovine myostatin gene causes the double-muscled phenotype in cattle. Nat Genet. 1997;17(1):71–4. pmid:9288100 doi: 10. 1038/ng0997-71

26. Johnson PL, McEwan JC, Dodds KG, Purchas RW, Blair HT. Meat quality traits were unaffected by a quantitative trait locus affecting leg composition traits in Texel sheep. J Anim Sci. 2005;83(12):2729–35. pmid:16282610

27. LeBrasseur NK, Schelhorn TM, Bernardo BL, Cosgrove PG, Loria PM, Brown TA. Myostatin inhibition enhances the effects of exercise on performance and metabolic outcomes in aged mice. J Gerontol A Biol Sci Med Sci. 2009;64(9):940–8. doi: 10. 1093/gerona/glp068. pmid:19483181

28. Cong L, Ran FA, Cox D, Lin S, Barretto R, Habib N et al. Multiplex Genome Engineering Using CRISPR/Cas Systems. Science. 2013;339(6121):819–23. doi: 10. 1126/science. 1231143. pmid:23287718

29. Vilette D, Madelaine MF, Laude H. Establishment of astrocyte cell lines from sheep genetically susceptible to scrapie. In Vitro Cell Dev Biol Anim. 2000;36(1):45–9. pmid:10691040 doi: 10. 1290/1071-2690(2000)036<0045:eoaclf>2. 0. co;2

30. Ménoret S, Remy S, Tesson L, Usal C, Iscache A, Anegon I. Generation of Transgenic Rats Using Microinjection of Plasmid DNA or Lentiviral Vectors. In: Pease S, Saunders T, editors. Advanced Protocols for Animal Transgenesis. An ISTT Manual. Heidelberg: Springer; 2011. p. 117–36.

31. Bellec J, Bacchetta M, Losa D, Anegon M, Chanson T, Nguyen H. CFTR Inactivation by lentiviral vector-mediated RNA Interference and CRISPR-Cas9 genome editing in human airway epithelial cells. Current gene therapy. 2015:In press.

32. Mali P, Yang L, Esvelt KM, Aach J, Guell M, DiCarlo JE et al. RNA-Guided Human Genome Engineering via Cas9. Science. 2013;339(6121):823–6. doi: 10. 1126/science. 1232033. pmid:23287722

33. Stringfellow D, Givens M. Manual of the International Embryo Transfer Society. 4 ed. IETS. Savoy, IL: 2010.

34. Menchaca A, Rubianes E. New treatments associated with timed artificial insemination in small ruminants. Reprod Fertil Dev. 2004;16(4):403–13. pmid:15315739 doi: 10. 1071/rd04037

35. Remy S, Tesson L, Menoret S, Usal C, De Cian A, Thepenier V et al. Efficient gene targeting by homology-directed repair in rat zygotes using TALE nucleases. Genome Res. 2014;24(8):1371–83. doi: 10. 1101/gr. 171538. 113. pmid:24989021

36. Wagner E, Lykke-Andersen J. mRNA surveillance: the perfect persist. J Cell Sci. 2002;115(15):3033–8.

37. Haynes FE, Greenwood Pl, McDonagh M, McMahon C, Nicholas G, Berry C et al. Lack of association between allelic status and myostatin content in lambs with

the myostatin g+6723G>A allele. J Anim Sci. 2013;91(1):78–89. doi: 10. 2527/jas. 2012-5482. pmid:23048142

38. Petersen B, Niemann H. Molecular scissors and their application in genetically modified farm animals. Trans Research. 2015; 24(3):381–96. doi: 10. 1007/s11248-015-9862-z

39. Makarevich AV, Chrenek P, Zilka N, Pivko J, Bulla J. Preimplantation development and viability of in vitro cultured rabbit embryos derived from in vivo fertilized gene-microinjected eggs: apoptosis and ultrastructure analyses. Zygote. 2005;13(2):125–37. pmid:16128408 doi: 10. 1017/s0967199405003138

40. Liu JH, Yin S, Xiong B, Hou Y, Chen DY, Sun QY. Aberrant DNA methylation imprints in aborted bovine clones. Mol Reprod Dev. 2008;75(4):598–607. pmid:17886268 doi: 10. 1002/mrd. 20803

41. Niemann H, Tian XC, King WA, Lee RSF. Epigenetic reprogramming in embryonic and foetal development upon somatic cell nuclear transfer cloning. Reprod. 2008;135(2):151–63. doi: 10. 1530/rep-07-0397.

42. Niemann H, Kues WA. Transgenic farm animals: an update. Reprod Fertil Dev. 2007;19(6):762–70. http://dx. doi. org/10. 1071/RD07040. pmid:17714630

43. Long C, McAnally JR, Shelton JM, Mireault AA, Bassel-Duby R, Olson EN. Prevention of muscular dystrophy in mice by CRISPR/Cas9-mediated editing of germline DNA. Science. 2014;345(6201):1184–8. doi: 10. 1126/science. 1254445. pmid:25123483

44. Larcher T, Lafoux A, Tesson L, Remy S, Thepenier V, Francois Vet al. Characterization of dystrophin deficient rats: a new model for Duchenne muscular dystrophy. PloS one. 2014;9(10):e110371. doi: 10. 1371/journal. pone. 0110371. pmid:25310701

45. Fu Y, Foden JA, Khayter C, Maeder ML, Reyon D, Joung JKet al. High-frequency off-target mutagenesis induced by CRISPR-Cas nucleases in human cells. Nature biotechnology. 2013;31(9):822–6. doi: 10. 1038/nbt. 2623. pmid:23792628

46. Niu Y, Shen B, Cui Y, Chen Y, Wang J, Wang Let al. Generation of gene-modified cynomolgus monkey via Cas9/RNA-mediated gene targeting in one-cell embryos. Cell. 2014;156(4):836–43. doi: 10. 1016/j. cell. 2014. 01. 027. pmid:24486104

47. Ran FA, Hsu PD, Lin CY, Gootenberg JS, Konermann S, Trevino AEet al. Double nicking by RNA-guided CRISPR Cas9 for enhanced genome editing specificity. Cell. 2013;154(6):1380–9. doi: 10. 1016/j. cell. 2013. 08. 021. pmid:23992846

48. Fu Y, Sander JD, Reyon D, Cascio VM, Joung JK. Improving CRISPR-Cas nuclease specificity using truncated guide RNAs. Nat Biotech. 2014;32(3):279–84. doi: 10. 1038/nbt. 2808

49. Proudfoot C, Carlson D, Huddart R, Long C, Pryor J, King T et al. Genome edited sheep and cattle. Trans Research. 2014: 24(1):147–53. doi: 10. 1007/s11248-014-9832-x

50. Zhang C, Wang L, Ren G, Li Z, Ren C, Zhang T et al. Targeted disruption of the sheep MSTN gene by engineered zinc-finger nucleases. Mol Biol Rep.

2014;41(1):209–15. doi: 10. 1007/s11033-013-2853-3. pmid:24197697

51. Wiener P, Woolliams JA, Frank-Lawale A, Ryan M, Richardson RI, Nute GR et al. The effects of a mutation in the myostatin gene on meat and carcass quality. Meat Science. 2009;83(1):127–34. doi: 10. 1016/j. meatsci. 2009. 04. 010. pmid:20416780
52. Fiems LO. Double Muscling in Cattle: Genes, Husbandry, Carcasses and Meat. Animals. 2012;2(3):472–506. doi: 10. 3390/ani2030472
53. Grobet L, Martin L, Poncelet D, Pirottin D, Brouwers B, Riquet J et al. A deletion in the bovine myostatin gene causes the double-muscled phenotype in cattle. Nat Genet. 1997;17(1):71–4. pmid:9288100 doi: 10. 1038/ng0997-71
54. Clop A, Marcq F, Takeda H, Pirottin D, Tordoir X, Bibe B et al. A mutation creating a potential illegitimate microRNA target site in the myostatin gene affects muscularity in sheep. Nat Genet. 2006;38(7):813–8. pmid:16751773 doi: 10. 1038/ng1810
55. Zhang C, Liu Y, Xu D, Wen Q, Li X, Zhang W et al. Polymorphisms of myostatin gene (MSTN) in four goat breeds and their effects on Boer goat growth performance. Mol Biol Rep. 2012;39(3):3081–7. doi: 10. 1007/s11033-011-1071-0. pmid:21710248

INDEX